Lecture Notes in Mathematics

A collection of informal reports and seminars
Edited by A. Dold, Heidelberg and B. Eckmann, Zürich

330

Proceedings of the Second Japan-USSR Symposium on Probability Theory

Edited by G. Maruyama, Tokyo University of Education, Tokyo/Japan
Yu. V. Prokhorov, Academy of Sciences of the USSR, Moscow/USSR

Springer-Verlag
Berlin · Heidelberg · New York 1973

AMS Subject Classifications (1970): 60-02, 60Bxx, 60Fxx, 60Gxx, 60Hxx, 60Jxx, 62Exx, 62F10, 62L12, 93E20

ISBN 3-540-06358-7 Springer-Verlag Berlin · Heidelberg · New York
ISBN 0-387-06358-7 Springer-Verlag New York · Heidelberg · Berlin

Offsetdruck: Julius Beltz, Hemsbach/Bergstr.

PREFACE

The Second Japan-USSR Symposium on Probability Theory was held at Kyoto, Japan, August 2-9, 1972. Eleven probabilists attended from USSR and 113 from Japan. The present volume contains most of the papers presented to the symposium. Record of meetings and lists of the organizing committee and participants are attached at the end of the volume. We had the First Symposium at Khabarovsk, USSR, August 16-25, 1969, with 32 attendants from Japan and about 80 from USSR. In both symposia many young probabilists participated and vigorous informal discussions took place.

The Second Symposium was planned by Kakuritsuron Seminar, an organization of Japanese probabilists, with the support of the Mathematical Society and Science Council of Japan. It is a great pleasure for us to express our gratitude to the organizing committee and all those who have contributed to the success of the symposium and the preparation of this volume.

Academician Professor Yu. V. Linnik died shortly before the Second Symposium. He was a member of the organizing committee and was expected to attend the symposium. His death is a great loss to our science.

G. Maruyama　　　　Yu. V. Prokhorov

CONTENTS

ENTROPY AND HAUSDORFF DIMENSION OF A SEQUENCE OF COORDINATE FUNCTIONS IN BASE-r EXPANSION

Yoshikazu Baba

1. Definitions and notations. Let A be the set $\{0,1, \ldots, r-1\}$ and $\{P_k\}_{k\geq 1}$ be a sequence of probability measures on A. We consider the infinite direct product probability measure space $(\Omega, \mathcal{B}, P) = \Pi_{k=1}^{\infty}(A, P_k)$, where $\mathcal{B}$ is the σ-field generated by all cylinder sets. Denote the n-th coordinate of $\omega = (\omega_1, \omega_2, \ldots, \omega_n, \ldots)$ in Ω by $x_n(\omega)$. Then, the sequence of coordinate functions $x = \{x_n(\omega)\}_{n\geq 1}$ is a stochastic process which is not stationary except for the case of $P_k = P_1$, for any $k \geq 2$. We introduce a probability measure $\overline{P}$ on the Borel field $\mathcal{B}_{[0,1]}$ of the unit interval $[0,1]$ by the mapping ϕ of Ω into $[0,1]$:

$$\omega = (x_1(\omega), x_2(\omega), \ldots) \longrightarrow \phi(\omega) \equiv \overline{\omega} = \Sigma_{n=1}^{\infty} \frac{x_n(\omega)}{r^n}.$$

This is the base-r expansion of $\overline{\omega} \in [0,1]$ and ϕ is one-to-one except for countable $\overline{\omega}$'s. $\overline{P}$ is defined by $\overline{P}(E) = P(\phi^{-1}(E))$ for all $E \in \mathcal{B}_{[0,1]}$. We shall call a set of the form $[\frac{j}{r^n}, \frac{j+1}{r^n}) \equiv I_{nj}$, $0 \leq j \leq r^n-1$, $n \geq 1$, an r-cylinder. $\overline{P}(I_{nj})$ can be expressed, if P is nonatomic, as

$$\overline{P}(I_{nj}) = P\{\omega |\ x_1(\omega) = i_1, \ldots, x_n(\omega) = i_n\} = p_{1i_1} \cdots p_{ni_n}$$

where $p_{ki} = P_k(\{i\})$ and $\frac{j}{r^n} = \frac{i_1}{r} + \frac{i_2}{r^2} + \ldots + \frac{i_n}{r^n}$. Clearly, $\overline{P}$ is Lebesgue measure Λ if and only if $p_{ki} = \frac{1}{r}$, $k \geq 1$, $0 \leq i \leq r-1$ and in this case we shall write $P = P_\Lambda$.

Let $H_n(P)$ be the entropy of the space of all messages of the first n letters:

$$H_n(P) = -\Sigma_{i_k \varepsilon A} P\{x_1(\omega)=i_1,\ldots,x_n(\omega)=i_n\}\log P\{x_1(\omega)=i_1,\ldots,x_n(\omega)=i_n\}$$

$$= \Sigma_{k=1}^{n}(-\Sigma_{i=0}^{r-1} p_{ki}\log p_{ki}).$$

If the process $x = \{x_n(\omega)\}_{n\geq 1}$ is stationary, then there exists the limit:

$$\lim_n \frac{1}{n} H_n(P) = H(P).$$

Although in our case $x = \{x_n(\omega)\}_{n\geq 1}$ is not stationary in general, we may define $H(P)$ as above if the limit exists. $H(P)$ is the entropy per symbol of the information source x. Obviously, $0 \leq H_n(P) \leq n\log r$ and hence $0 \leq H(P) \leq \log r$.

Finally, we define, following [2], the Hausdorff dimension of a subset M of $[0,1]$ with respect to a nonatomic probability measure μ on $[0,1]$:

$$\dim_\mu M = \sup\{\alpha \mid \mu_\alpha(M) = +\infty\}$$

where $\mu_\alpha(M) = \lim_{\rho\to 0} \mu_\alpha(M,\rho)$ and $\mu_\alpha(M,\rho) = \inf \Sigma_{n\geq 1} \mu(M_n)^\alpha$ where the infimum extends over all μ-ρ-coverings of M, a μ-ρ-covering being an atmost countable covering by r-cylinders with $\mu(M_n) < \rho$. We notice here the two properties of $\dim_\mu M$: (i) if μ is Lebesgue measure Λ then $\dim_\Lambda M = \dim M$ where $\dim M$ is the usual Hausdorff dimension of a set M, (ii) $0 \leq \dim_\mu M \leq 1$ and if $\mu(M) > 0$, then $\dim_\mu M = 1$.

2. Results. Let $P = \Pi_{k=1}^{\infty} P_k$ and $Q = \Pi_{k=1}^{\infty} Q_k$ be two infinite direct product probability measures on Ω where $\{P_k\}_{k\geq 1}$ and $\{Q_k\}_{k\geq 1}$ are given sequences of probabilities on A. Then, according to Kakutani's dichotomy theorem [3], either P and Q are absolutely continuous with each other (in this case, we shall write $P \sim Q$) or P and Q are mutually singular (in this case, we shall write $P \perp Q$), under the condition that $P_k \sim Q_k$ for any $k \geq 1$. The following theorem gives necessary conditions for P and Q to be absolutely

continuous with each other in terms of entropy and Hausdorff dimension.

Theorem 1. Suppose $P \sim Q$. Then, we have (i) either $H(P) = H(Q)$ or both of them do not exist and (ii) if further P and Q are nonatomic, $\dim_{\overline{P}}(\mathrm{supp}\ \overline{Q}) = \dim_{\overline{Q}}(\mathrm{supp}\ \overline{P})$ where $\mathrm{supp}\ \mu$ denotes the support of a measure μ.

In the case of $P \perp Q$, we can say nothing in general. Next, we give a version of the Shannon-McMillan-Breiman theorem ([2], Theorem 13.1). If the process $x = \{x_n(\omega)\}_{n\geq 1}$ is stationary, the next theorem reduces to a special case of the Shannon-McMillan-Breiman theorem.

Theorem 2. Let $u_n(\omega) = \{\omega' \mid \omega' \in \Omega,\ x_k(\omega') = x_k(\omega),\ 1 \leq k \leq n\}$. Then, we have

$$P\{\omega \mid \lim_n \frac{-\log P(u_n(\omega)) - H_n(P)}{n} = 0\} = 1,$$

in particular, if P has the entropy $H(P)$, then

$$P\{\omega \mid \lim_n [-\frac{1}{n}\log P(u_n(\omega))] = H(P)\} = 1.$$

As a corollary of Theorem 2, we have a generalization of Eggleston's theorem [1].

Corollary. Suppose that P is nonatomic and has the entropy $H(P)$ and that M is a set in $\mathcal{B}_{[0,1]}$ such that

$$M \subset \{\overline{\omega} \mid \lim_n [-\frac{1}{n}\log \overline{P}(u_n(\overline{\omega}))] = H(P)\}$$

with $\overline{P}(M) > 0$ where $u_n(\overline{\omega}) = \phi(u_n(\omega))$. Then, $\dim M = \frac{H(P)}{\log r}$, especially $\dim(\mathrm{supp}\ \overline{P}) = \frac{H(P)}{\log r}$.

If $x = \{x_n(\omega)\}_{n\geq 1}$ is stationary, that is, $P_n = P_1$ for any $n \geq 2$, then this reduces to the result of Eggleston.

3. Proofs.

Proof of Theorem 1. (i) If $P \sim Q$, then by Kakutani's theorem [3] we have $\Sigma_{k=1}^{\infty}(\Sigma_{i=0}^{r-1}[\sqrt{p_{ki}} - \sqrt{q_{ki}}]^2) < \infty$ where p_{ki} and q_{ki} denote $P_k(\{i\})$ and $Q_k(\{i\})$, $k \geq 1$, $0 \leq i \leq r-1$, respectively. Then, $\lim_k |p_{ki} - q_{ki}| = 0$, $0 \leq i \leq r-1$. Hence, for any $\varepsilon > 0$ we can take N such that if $k > N$, then $|-p_{ki}\log p_{ki} + q_{ki}\log q_{ki}| < \frac{\varepsilon}{r}$, $0 \leq i \leq r-1$. From this we have easily $\left|\frac{H_n(P)}{n} - \frac{H_n(Q)}{n}\right| < 2\varepsilon$ for sufficiently large n, which proves (i). (ii) By the condition that P and Q are nonatomic, follows at once $\overline{P} \sim \overline{Q}$, hence $\overline{P}(\text{supp } \overline{Q}) = \overline{Q}(\text{supp } \overline{P}) = 1$, from which follows (ii).

Proof of Theorem 2. Put, for any $k \geq 1$, $y_k(\omega) = -\log p_{ki}$ if $x_k(\omega) = i$. Then, $\{y_k(\omega)\}_{k\geq 1}$ is a sequence of independent random variables. Here, if $p_{ki} = 0$ for some $k \geq 1$ and $0 \leq i \leq r-1$, then $y_k(\omega) = +\infty$ for such ω as $x_k(\omega) = i$. However, since the set of all such ω's has P-measure zero, the proof is not affected. We have $-\log P(u_n(\omega)) = \Sigma_{k=1}^{n} y_k(\omega)$, and we see that

$$E[y_k] = -\Sigma_{i=0}^{r-1} p_{ki}\log p_{ki}$$

$$\operatorname{Var}[y_k] = \Sigma_{i=0}^{r-1} p_{ki}(\log p_{ki})^2 - (\Sigma_{i=0}^{r-1} p_{ki}\log p_{ki})^2.$$

By using the inequality $0 \leq x(\log x)^2 \leq 4e^{-2}$ for $0 \leq x \leq 1$, we have

$$\Sigma_{k=1}^{\infty}\frac{\operatorname{Var}[y_k]}{k^2} < \infty$$

and so we can apply the strong law of large numbers to the sequence of independent random variables $\{y_k(\omega)\}_{k\geq 1}$:

$$P\{\omega \mid \lim_n \frac{\Sigma_{k=1}^{n} y_k(\omega) - \Sigma_{k=1}^{n} E[y_k]}{n} = 0\}$$

$$= P\{\omega \mid \lim_n \frac{-\log P(u_n(\omega)) - H_n(P)}{n} = 0\} = 1.$$

<u>Proof of Corollary</u>. Since P is nonatomic, we have

$$P\{\omega|\ \lim_n [-\frac{1}{n}\log P(u_n(\omega))] = H(P)\}$$

$$= \overline{P}\{\overline{\omega}|\ \lim_n [-\frac{1}{n}\log \overline{P}(u_n(\overline{\omega}))] = H(P)\} = 1,$$

therefore we can take M with $\overline{P}(M) > 0$. Billingsley's theorem ([2], Theorem 14.1) states that if ν and μ are two probability measures on $[0,1]$ and $0 \leq \delta$ is a constant and if $M \subset \{\overline{\omega}|\ \lim_n \frac{\log \nu(u_n(\overline{\omega}))}{\log \mu(u_n(\overline{\omega}))} = \delta\}$, then $\dim_\mu M = \delta \dim_\nu M$. Since $\{\overline{\omega}|\ \lim_n [-\frac{1}{n}\log \overline{P}(u_n(\overline{\omega}))] = H(P)\}$ $= \{\overline{\omega}|\ \lim_n \frac{\log \overline{P}(u_n(\overline{\omega}))}{\log \Lambda(u_n(\overline{\omega}))} = \frac{H(P)}{\log r}\}$, if we take $\nu = \overline{P}$ and $\mu = \Lambda$, then we have $\dim_\Lambda M = \dim M = \frac{H(P)}{\log r}\dim_{\overline{P}} M = \frac{H(P)}{\log r}$.

4. <u>Example and remarks</u>.

<u>Example</u>. Let $r = 2$, $A = \{0,1\}$ and $\{p_{k0}\}_{k\geq 1}$ is any sequence of p and q with $p,q > 0$ and $p + q = 1$ (for example, $P_{2k}(\{0\}) = P_{2k-1}(\{1\}) = p$ and $P_{2k}(\{1\}) = P_{2k-1}(\{0\}) = q$, $k \geq 1$). Then, $H(P) = -p\log p - q\log q$ and the support of $\overline{P}$ has Hausdorff dimension $\frac{-p\log p - q\log q}{\log 2}$.

<u>Remark 1</u>. G.Marsaglia [4] showed that P has an atom if and only if $\Sigma_{n=1}^{\infty}(1 - \max(p_{n0},p_{n1}, \dots ,p_{n,r-1})) < \infty$, and that under the condition P is purely atomic.

<u>Remark 2</u>. In Theorem 1-(i) and Theorem 2, the set $A = \{0,1, \dots ,r-1\}$ may be replaced by any finite set $\{a_1,a_2, \dots ,a_r\}$.

<u>Remark 3</u>. Even in the case of infinite $A = \{a_1,a_2, \dots \}$, Theorem 2 remains valid under additional conditions: (i)$\Sigma_{i=1}^{\infty} p_{ki}(\log p_{ki})^2 < \infty$, $k \geq 1$ and (ii) $\Sigma_{k=1}^{\infty}\frac{\mathrm{Var}[y_k]}{k^2} < \infty$.

Remark 4. If supp P is a countable set, then from Remark 1, we have easily $H(P) = 0$. If $P \sim P_\Lambda$, then by Kakutani's theorem [3] we have $H(P) = H(P_\Lambda) = \log r$. Both are extreme cases. In an intermediate case, where $\bar{P}$ is nonatomic and $\bar{P} \perp \Lambda$, the situation is not so simple.

References

[1] H.G. Eggleston: The fractional dimension of a set defined by decimal properties, Quart. J. Math. Oxford Ser. 20 (1949), 31-36.

[2] P. Billingsley: Ergodic theory and information, John Wiley, New York, 1965.

[3] S. Kakutani: On equivalence of infinite product measures, Ann. of Math. 49 (1948), 214-224.

[4] G. Marsaglia: Random variables with independent binary digits, Ann. Math. Statist. 42 (1971), 1922-1929.

Department of Mathematics
Shizuoka University
Shizuoka, Japan.

GENERAL THEOREMS OF CONVERGENCE FOR RANDOM PROCESSES

A.A. Borovkov

This paper consists of two parts. The first part is based on results due to A.A. Borovkov and E.A. Pechersky about general conditions of weak convergence of measures in σ-topological spaces. (see definition 2). Here, while studying the convergence of random processes, one should take as a topology the weakest σ-topology, induced by a functional or a class of functionals the distribution of which we study. The theorems that have been obtained in this part generalize, in particular, the well-known results of Yu.V. Prokhorov [1].

In the second part, we deal with an approximation method which is essentially more elementary and transparent. In concrete functional spaces it leads to conditions of convergence that are very close to those appearing in the first part. This circumstance is quite natural because the conditions of convergence in both the cases are, or close to, necessary ones.

1. Let $\langle X, \mathfrak{A} \rangle$ be a measurable sampling space on which probability measures P and P_n, $n=1,2,\ldots$, are given. Along with the distributions P and P_n we shall consider corresponding to them random elements ξ and ξ_n on X and call the latter random processes.

Let $\mathcal{F}$ be a certain class of measurable functionals f. ($\mathcal{F}$ may consist of only one functional.) The main problem we consider in the paper is the following: under what conditions the distributions of $f(\xi_n)$ weakly converge to the distribution of $f(\xi)$:

(1) $$P_n\{f(x)<t\} \Rightarrow P\{f(x)<t\}$$

for every $f \in \mathcal{F}$.

Evidently, for (1) to be satisfied it is necessary and sufficient that

(2) $$\int f^N(x)\,dP_n(x) \to \int f^N(x)\,dP(x)$$

for every $f \in \mathcal{F}$ and every $N>0$, where $f^N = \max\{-N, \min\{N, f\}\}$.

In the problem stated, the case when X is a metric space and $\mathcal{F} = C(X)$ is the set of all continuous functions on X has been studied most thoroughly. In this case relations (1),(2) mean usual weak convergence $P_n \Rightarrow P$ in space X. An exposition and a review of the results for this case can be found, for instance, in Gikhman and Skorokhod [1](1971) and in Borovkov [2](1972).

Let us return to the problem (1),(2) where P, P_n denote arbitrary bounded measures not necessarily probability ones. To formulate the main result we need some difinitions.

<u>Definition 1</u>. The class of functionals $\mathcal{G}$ defined on X, is called <u>complete</u> (in the sense of Hausdorff) if

(1) $g(x) = const \in \mathcal{G}$;

2) $g_1, g_2 \in \mathcal{G}$ implies

$$\max(g_1, g_2) \in \mathcal{G},\ g_1 g_2 \in \mathcal{G},\ g_1+g_2 \in \mathcal{G},\ g_1/g_2 \in \mathcal{G}$$

(the latter must hold true if $g_2(x) \neq 0$ on X);

3) $\lim_{n\to\infty} g_n(x) \in \mathcal{G}$ if $g_n \in \mathcal{G}$ and $g_n \to g$ uniformly.

<u>Definition 2</u>. Let $\mathcal{U}$ be a class of sets in X. The pair $(X, \mathcal{U})$ is called a <u>σ-topological space</u> or a <u>space with σ-topology</u> if the class $\mathcal{U}$ has the properties:

1) any countable union of sets of $\mathcal{U}$ belongs to $\mathcal{U}$;

2) any finite intersection of sets of $\mathcal{U}$ belongs to $\mathcal{U}$;

3) X and the empty set belong to $\mathcal{U}$.

This is A.D. Alexandrov's space ([3]), which we denote $(X, \mathcal{U})_\sigma$. The sets from $\mathcal{U}$ are called open.

Now let us consider an arbitrary class of functionals $\mathcal{G}$ and the sets $U = \{g(x) > 0\}$, $g \in \mathcal{G}$. The union of all such sets we denote by $\mathfrak{g}$.

<u>Theorem A</u>. (Hausdorff, [3]). <u>If the class $\mathcal{G}$ of functions on X is complete, then the pair $(X, \mathfrak{g})$ is a σ-topological space $(X, \mathfrak{g})_\sigma$. The set $C(X)$ of all continuous functions on $(X, \mathfrak{g})_\sigma$ coincides with $\mathcal{G}$</u>.

Let us return to the initial class of functionals $\mathcal{F}$ given on X. It is clear, that we can always extend $\mathcal{F}$ to a complete class. To avoid new notation, we assume $\mathcal{F}$ to be a complete class of functions. Then, if we denote the union of U-sets (i.e. the sets of the type $\{f(x) > 0\}$ by $\mathfrak{f}$ we get a σ-topological space $(X, \mathfrak{f})_\sigma$. It is clear that we have thus introduced the weakest σ-topology in which $f \in \mathcal{F}$ are continuous.

Let us denote by $\mathcal{B} = \sigma(\mathfrak{f})$ the σ-algebra generated by sets from $\mathfrak{f}$. Since $f \in \mathcal{F}$ are measurable with respect to $\mathfrak{A}$, U-sets belong to $\mathfrak{A}$ and $\mathcal{B} \subset \mathfrak{A}$.

Now the space $(X, \mathfrak{f})_\sigma$, by its construction, is completely normal (any open set is an U-set). That means that any measure on $\langle X, \mathcal{B} \rangle$ is regular: for any $B \in \mathcal{B}$, $P(B) = \inf_{U \supseteq B} P(U)$.

We shall denote the measures induced by P and P_n on $\langle X, \mathcal{B} \rangle$ by the same letters P and P_n. Since $\mathcal{F} = C(X)$, the relations (1),(2) mean the usual weak convergence $P_n \Rightarrow P$ in $(X, \mathfrak{f})_\sigma$.

<u>Definition 3</u>. We shall say that a set $B \in \mathcal{B}$ <u>belongs to the class</u> $\mathcal{D}_P$ of all P-continuous sets, if there exists a U-set and set Z of the form $Z = \{f(x) = 0\}$ with the properties:

$$U \subset B \subset Z, \qquad P(Z - U) = 0.$$

Theorem B. (Alexandrov, [3]). $P_n \Rightarrow P$ iff $P_n(A) \to P(A)$ for all $A \in \mathcal{D}_P$.

However this criterion of convergence uses the too broad class of sets $\mathcal{D}_P$ and is therefore not convenient for testing the convergence of random processes. If the limiting measure P is concentrated on a compact, the class of sets for which we require the convergence can be considerably narrowed.

Definition 4. The set $K \subset X$ is called compact in $(X, \mathfrak{f})_\sigma$ if, from any cover of K by open sets (i.e. U-sets), we can choose a finite subcover.

Definition 5. The sequence of measures $\{P_n\}$ is called weakly tight in $(X, \mathfrak{f})_\sigma$ if, for any $\varepsilon > 0$, there exists a compact $K = K_\varepsilon$ in $(X, \mathfrak{f})_\sigma$ such that, for any neighbourhood $U(K)$ (that is, for any open set containing K),

(3) $$\liminf_{n \to \infty} P_n(U(K)) \geq 1 - \varepsilon .$$

If the measure P satisfies inequality (3), it is called tight in $(X, \mathfrak{f})_\sigma$.

Now we can state the main result.

Theorem 1. $P_n \Rightarrow P$ (that is, (1),(2) hold true) and the measure P is tight in $(X, \mathfrak{f})_\sigma$ iff the following two conditions are satisfied:

1) the sequence $\{P_n\}$ is weakly tight in $(X, \mathfrak{f})_\sigma$;

2) there exists a class of sets $\mathcal{A} \subset \mathcal{L}$ such that

a) for any $A \in \mathcal{A}$,
$$\lim_n P_n(A) = P(A),$$

b) the σ-algebra $\sigma(\mathcal{A}\mathcal{D}_Q)$ generated by $\mathcal{A}\mathcal{D}_Q$, for any measure Q, coincides with $\mathcal{L}$.

This theorem generalizes the well-known results due to Prokhorov on weak convergence of measures in metric spaces [4] . Note that various sufficient convergence conditions obtained earlier contained, instead of condition 1, the stronger requirement of tightness of the measures. Usual tightness means that the probabilities of the compact are uniformly large, that is $\sup_n P_n(K) > 1-\varepsilon$. (For simplicity we assume K to be measurable.)

The weakening of this requirement allowed us to receive simple necessary and sufficient conditions of convergence. A further weakening of this requirement allows to get rid of the tightness condition for the measure P in Theorem 1.

From theorem 1 one can get many results on convergence of processes in special functional spaces, in particular, the results established in Borovkov's paper [2].

The criteria of weak tightness for different concrete functional spaces are similar to those of ordinary tightness (but are wider), and naturally they essentially coincide with the latter when weak tightness implies ordinary one.

Conditions of convergence in ordinary topological spaces follow from Theorem 1. This problem was studied before by Prokhorov, Le Cam, Varadarajan, Topsol and others. The methods of their works are very delicate because the notion of measure in an ordinary topological (not σ-toplological) space is not very natural and not convenient for studying the convergence of processes.

Now let $(X, \mathcal{G})$ be an ordinary topological space, where $\mathcal{G}$ is the class of open sets, and let $\mathcal{F}$ coincide with the set of all the functionals continuous in the given topology ($\mathcal{F}$ is evidently a complete class). We shall call a sequence $\{P_n\}$ weakly tight in $(X, \mathcal{G})$ if it satisfies the condition of Definition 5 with the space $(X, \mathfrak{f})_\sigma$ replaced by $(X, \mathcal{G})$.

Then, from Theorem 1, it follows:

Corollary 1. $P_n \Rightarrow P$ in $(X, \mathcal{G})$, if

1) the sequence $\{P_n\}$ is weakly tight;

2) the condition 2 of Theorem 1 is satisfied.

This assertion is true since any compact in $(X, \mathcal{g})$ is a compact in $(X, \mathfrak{f})_\sigma$ and any U-set in $(X, \mathfrak{f})_\sigma$ is an open set in $(X, \mathcal{g})$; therefore condition 1 of Corollary 1 implies condition 1 of Theorem 1.

In the theory of random processes conditions of convergence connected with cylindrical sets are of particular interest. Let $X = X(T) = \{x(t), t \in T\}$ be a functional space on parametric set T and let S be a subset of T . Let us denote by $\mathcal{C}_S$ the algebra generated by the events $\{x(t) < u\}, t \in S, -\infty < u < \infty$. (By the definition of a random process, σ-algebra $\sigma(\mathcal{C}_T) \subset \mathcal{O}\!\mathcal{L}$).

Corollary 2. Let ξ and ξ_n be random processes in $X(T)$ and P and P_n be their distributions. Let us assume that the set $S \subset T$ and space $(X, \mathfrak{f})_\sigma$ satisfy the condition $\mathcal{L} \subset \sigma(\mathcal{C}_S \mathcal{D}_Q)$ for any measure Q . Then $P_n \Rightarrow P$ in $(X, \mathfrak{f})_\sigma$ if the sequence $\{P_n\}$ is weakly tight in $(X, \mathfrak{f})_\sigma$ and

$$\lim_n P_n(A) = P(A)$$

for any $A \in \mathcal{C}_S$

All the above results remain valid for the nets of measures $\{P_\theta\}_{\theta \in \Theta}$, where Θ is a filtering set with respect to a certain order.

2. A very interesting fact is that, along with the results described, there is a very simple criterion of convergence which I call an "approximative" criterion.

The proof of this criterion is short and simple. At the same time the verification of its conditions for concrete functional spaces is not more difficult than the verification of conditions of Theorem 1.

One may illustrate this by the results on convergence in spaces **C**,**D**,**E**,**F**, obtained in my work [2]. (**C**,**D** are known spaces; **E**,**F** are spaces of functions with discontinuounces of the second kind and with the metric defined by the distance between functions on the plane).

Let, as before, ξ and ξ_n, $n=1,2,\dots$, be arbitrary random elements on $\langle X, \mathfrak{A} \rangle$ and f be a measurable function on X.

<u>Theorem 2.</u> $P_n(f(x)<t) \Rightarrow P(f(x)<t)$ (that is, (2) is satisfied) <u>if and only if there exists the system of sets</u> $K_\theta(\delta)$ <u>from</u> $\mathfrak{A}$ <u>and sequence of functions</u> $f_N(x)$, $N=1,2,\dots$, <u>such that:</u>

1) <u>distribution of</u> $f_N(\xi_n)$ <u>for every</u> N <u>weakly converges as</u> $n \to \infty$ <u>to the distribution of</u> $f_N(\xi)$;

2) $\sup_{x \in K_\theta(\delta)} |f(x)-f_N(x)| \to 0$ <u>as</u> $\delta \to 0$, $N \to \infty$;

3) <u>for any</u> $\delta>0$, $\theta>0$,

$$\liminf_{n\to\infty} P_n(K_\theta(\delta)) > 1-\theta, \quad P(K_\theta(\delta)) > 1-\theta.$$

<u>An example</u>. Let $X=C(0,1)$ be the space of continuous functions on segment $[0,1]$ and $\mathfrak{A}$ be the σ-algebra, generated by cylindrical sets. Let f be a function on X, continuous in uniform metric $\rho(x,y) = \sup|x(t)-y(t)|$.

Along with $f_N(x)$, it is natural to consider $f(x_N)$ where $x_N = x_N(x)$ is a broken line with the step $1/N$ generated by x. If finite-dimensional distributions of ξ_n converge to the distributions of ξ, then the condition 1 of Theorem 2 for f_N is satisfied.

The condition 2 can be also checked, if we consider the sets

$$K_\theta = \left\{ x : \omega_\Delta(x) = \sup_{|t-t'|<\Delta} |x(t)-x(t')| \le \varepsilon_\theta(\Delta) \quad \forall \Delta > 0 \right\}$$

where $\varepsilon_\theta(\Delta) \to 0$ as $\Delta \to 0$ (they are compacts in $C(0,1)$).

Finally, the condition 3 is satisfied if, for any $\varepsilon>0$,

$$\inf_n P_n(\omega_\Delta(x)<\varepsilon) \to 1, \quad P(\omega_\Delta(x)<\varepsilon) \to 1$$

as $\Delta \to 0$.

Thus, we obtain from Theorem 2 well-known conditions of convergence of random processes in $C(0,1)$.

In the same way, we may obtain the conditions of convergence of distributions in the space $D(0,1)$ etc. In a somewhat different form, this has been done in [2].

I have already mentioned that the proof of Theorem 2 is short, and I can reproduce it here.

Let t be a point of continuity of the distribution of $f(\xi)$. For a given $\varepsilon > 0$, we choose δ and N such that $\sup_{K_\theta(\delta)} |f(x) - f_N(x)| < \varepsilon$. Then

$$P_n(f(x) < t) \leq P_n(\overline{K}_\theta(\delta)) + P_n(f(x) < t, K_\theta(\delta)).$$

Thus, if $t+\varepsilon$ is a point of continuity of the distribution of $f_N(\xi)$, then

$$\limsup_{n\to\infty} P_n(f(x) < t) \leq \theta + \limsup_{n\to\infty} P_n(f_N(x) < t+\varepsilon) =$$

$$= \theta + P(f_N(x) < t+\varepsilon) \leq 2\theta + P(f_N(x) < t+\varepsilon, K_\theta(\delta)) \leq$$

$$\leq 2\theta + P(f(x) < t + 2\varepsilon).$$

Similarly we can establish the converse inequality for $\liminf_{n\to\infty} P_n(f(x) < t)$. θ and ε being arbitrary, we obtain:

$$\lim_n P_n(f(x) < t) = P(f(x) < t).$$

The necessity of the condition of Theorem 2 is trivial since $K_\theta(\delta)$ are not obliged to satisfy any structural conditions: It suffices to take $f_N(x) = f(x)$, $K_\theta(\delta) = X$.

The significance of the approximative method essentially increases, when investigating the convergence of random processes, defined on the whole axis, for example, of processes in $C(0,\infty)$. For this space with the metric $\rho(x,y)=\sup_t \left|\frac{x(t)-y(t)}{\psi(t)}\right|, \psi(t)\leq 1 \vee \sqrt{t\ln\ln t}$, Prokhorov's theorems on convergence in metric spaces are not applicable. These theorems would give us the convergence of the most interesting class of functionals, continuous in this metric.

Impossibility to use Prokhorov's theorems is explained by the following fact: in such nonseparable metric spaces, it is impossible to continue the measure of random processes toaBorel measure. For instance, in these spaces, there does not exist a Wiener Borel measure. The last remark was made together with Sachanenko, Pechersky and Lev.

References

[1] I.I. Gikhman, A.V. Skorokhod: Theory of random processes, Nauka, Moscow, 1971 (Russian).

[2] A.A. Borovkov: Convergence of distributions of functionals of random processes, Uspehi Mat. Nauk, 27, 1(1972), 3-41. (Russian)

[3] A.D. Alexandrov: Additive set functions in abstract spaces, Mat. zb. 8(1940), 307-348; 9(1941), 563-628; 13(1943),169-238. (Russian)

[4] P. Billingsley: Convergence of probability measures, J. Wiley &Sons, New York, 1968.

Institute of Mathematics of the Academy of Sciences of the USSR, Novosibirsk

ASYMPTOTIC EXPANSIONS FOR NEYMAN'S $C(\alpha)$ TESTS

D.M.Chibisov

1. Introduction. Let $X_1, \ldots, X_n$ be a sample from a distribution with a density function $p(x, \theta, \vartheta)$, $\theta \in R^s$, $\vartheta \in R^1$, and the hypothesis $H_0: \vartheta = 0$ be tested against $H_1: \vartheta \neq 0$. We consider the class $C(\alpha)$ of the tests proposed by J.Neyman [8]. These tests are based on statistics of the form

$$Z_n(\hat{\theta}_n) = n^{-1/2} \sum_{i=1}^{n} g(X_i, \hat{\theta}_n)$$

where $\hat{\theta}_n$ is an estimator for θ and $g(x, \theta)$ a function satisfying certain conditions (see Section 2) which ensure that under H_0 and under local alternatives $\vartheta = \eta n^{-1/2}$, $\eta = const$, $Z_n(\hat{\theta}_n)$ has the same (normal) asymptotic distribution as $Z_n(\theta_0)$, θ_0 being the true value of θ. Moreover, a particular form of $g(x, \theta)$ provides an asymptotically optimal test.

In the present paper, asymptotic expansions of the distributions of $Z_n(\hat{\theta}_n)$ for the null hypothesis and local alternatives are obtained. The term of order $n^{-1/2}$ is given explicitly and some properties of $C(\alpha)$ tests in this order of approximation are studied. It is shown that, up to this order, the power of an optimal $C(\alpha)$ test does not depend on the choice of the estimator $\hat{\theta}_n$ (in a wide class of estimators). Moreover, an upper bound for the power of any test for H_0 against $(\theta_0, \eta n^{-1/2})$ is obtained which turns out to coincide with the power of the optimal $C(\alpha)$ test up to order $n^{-1/2}$. Thus for power comparisons between $C(\alpha)$ tests with different estimators $\hat{\theta}_n$ or between $C(\alpha)$ and other asymptotically optimal tests (see, e.g., [6]) a more accurate approximation for the power (up to $O(n^{-1})$) is necessary. Such approximation can be obtained by the methods of the present paper but very cumbersome calculations are required. However the present results suggest, first, that no other test can have a substantially greater power than the optimal $C(\alpha)$ test and, secondly, that the choice of $\hat{\theta}_n$ is not essential for the power of the optimal $C(\alpha)$ test so that more easily computable estimators may be used.

The term of order $n^{-1/2}$ was obtained by the author in [1]. Now we use another method which enables us to obtain expansions of any order and requires weaker assumptions for the order $n^{-1/2}$. A part of the results of the present paper were published without proofs in [3].

In what follows, let R^s be the space of s-dimensional column-vectors $x=(x_1,\dots,x_s)'$, a prime denoting the transposition, $\|x\|=(x'x)^{\frac{1}{2}}$ for $x\in R^s$. For the simplicity of notation, let $\tau=n^{-1/2}$. The subscript i will always run from 1 to n, the limits of summation from $i=1$ to $i=n$ will be omitted.

2. $C(\alpha)$ tests. Let $X_1,\dots,X_n$ be independent observations of a real-valued random variable X with the distribution function (d.f.) $F(x,\theta,\vartheta)$ which depends on parameters $\theta=(\theta_1,\dots,\theta_s)'$ and ϑ taking their values in open sets $\Theta\subset R^s$ and $\Omega\subset R^1$ respectively. The hypothesis $H_0:\ \vartheta=0$ is to be tested against $\vartheta\neq 0$. $F(x,\theta,\vartheta)$ will be assumed to have the density $p(x,\theta,\vartheta)$ with respect to the Lebesgue measure. The expectation and probability corresponding to parameter values θ,ϑ will be denoted by $E_{\theta,\vartheta}$ and $P_{\theta,\vartheta}$. We shall write E_θ, P_θ, instead of $E_{\theta,0}$, $P_{\theta,0}$. Sometimes a true value $\theta=\theta_0$ will be fixed; then we shall write E and P instead of E_{θ_0} and P_{θ_0}. Similarly, we write $p(x,\theta)$ for $p(x,\theta,0)$ and $p(x)$ for $p(x,\theta_0,0)$. Let

$$\ell(x,\theta,\vartheta)=\log p(x,\theta,\vartheta) \tag{2.1}$$

with a similar convention about omitting $\vartheta=0$ and $\theta=\theta_0$, and

$$\ell_{\theta_j}=\frac{\partial \ell}{\partial \theta_j},\ j=1,\dots,s,\qquad \ell_\vartheta=\frac{\partial \ell}{\partial \vartheta},\qquad \ell_\theta=(\ell_{\theta_1},\dots,\ell_{\theta_s})'. \tag{2.2}$$

The $C(\alpha)$ tests are constructed as follows. Let a function $g^*(x,\theta)$ be such that

(2.3) $$E_\theta g^*(X,\theta)=0,\quad E_\theta g^*(X,\theta)^2=\sigma^2(\theta)<\infty,\quad \theta\in\Theta.$$

Put

(2.4) $$g(x,\theta)=g^*(x,\theta)/\sigma(\theta),$$

(2.5) $$Z_n(\theta)=\tau\sum g(X_i,\theta).$$

Let $\hat\theta_n$ be a locally root n consistent estimator of θ (which means that $\sqrt{n}(\hat\theta_n-\theta_0)$ is bounded in probability when θ_0 is the true value of θ ; for a precise definition see [8]). It was shown in [8] that $Z_n(\hat\theta_n)$ is asymptotically normally distributed if and only if

(2.6) $$E_\theta g(X,\theta)\,\ell_{\theta_j}(X,\theta)=0,\quad \theta\in\Theta,\ j=1,\dots,s.$$

A further result of [8] gives a rule for constructing an asymptotically optimal $C(\alpha)$ test. Namely, let

(2.7) $$g^*(x,\theta)=\ell_\vartheta(x,\theta)-c'(\theta)\,\ell_\theta(x,\theta)$$

where the vector $c(\theta)$ is determined by the "orthogonality" condition (2.6). Then the $C(\alpha)$ test based on this $g^*(x,\theta)$ is an asymptotically optimal one.

In view of (2.3), (2.4) we shall always assume that

(2.8) $$E_\theta g(X,\theta)=\int g(x,\theta)p(x,\theta)\,dx=0,\quad \theta\in\Theta,$$

(2.9) $$E_\theta g^2(X,\theta)=\int g^2(x,\theta)p(x,\theta)\,dx=1,\quad \theta\in\Theta.$$

Let

(2.10) $$g_{\theta_j}=\frac{\partial g}{\partial\theta_j},\ j=1,\dots,s,\quad g^{(1)}=(g_{\theta_1},\dots,g_{\theta_s})'.$$

If (2.8) may be differentiated under the integral sign, we have

(2.11) $$\int g_{\theta_j}(x,\theta)\,p(x,\theta)\,dx + \int g(x,\theta)\,l_{\theta_j}(x,\theta)\,p(x,\theta)\,dx = 0,\quad j=1,\dots,s.$$

Then (2.6) is equivalent to

(2.12) $$E_\theta\, g_{\theta_j}(X,\theta)=0,\quad j=1,\dots,s.$$

<u>3. A theorem on asymptotic expansions.</u> In this section, we state a theorem the construction of asymptotic expansions for the distributions of $C(\alpha)$ test statistics will be based on. First, we introduce some convenient notation.

<u>Definition 3.1. Let</u> $\{\zeta_n\}$, $n=1,2,\dots$, <u>be a sequence of random variables (r.v.'s). We shall write</u> $\zeta_n=\omega(k)$ <u>if, for any</u> $\delta>0$,

(3.1) $$P\{|\zeta_n|>n^\delta\}=o(\tau^k).$$

<u>We say that</u> $\zeta_n=\omega(k)$ <u>uniformly in</u> $P\in\mathcal{P}$ <u>where</u> $\mathcal{P}$ <u>is a family of distributions if, for any</u> $\delta>0$, <u>(3.1) holds uniformly in</u> $P\in\mathcal{P}$.

<u>For a sequence of random vectors</u> $\zeta_n=(\zeta_{1n},\dots,\zeta_{pn})'$ <u>we say that</u> $\zeta_n=\omega(k)$ <u>if</u> $\zeta_{jn}=\omega(k)$, $j=1,\dots,p$.

Note that in [1] a similar symbol was introduced with a different meaning of its argument.

A typical example of a sequence $\omega(k)$ is given by the following

<u>Lemma 3.1. Let</u> $Y_1,\dots,Y_n$ <u>be independent identically distributed r.v.'s</u>. (i) <u>Let</u> $EY_1=0$, $E|Y_1|^r<\infty$, $r\geq 2$, <u>and</u> $\xi_n=\tau\sum Y_i$. <u>Then</u> $\xi_n=\omega(r-2)$ (ii) <u>Let</u> $E|Y_1|^{r/2}<\infty$, $r\geq 2$, <u>and</u> $\zeta_n=n^{-1}\sum Y_i$. <u>Then</u> $\zeta_n=\omega(r-2)$. <u>If, for a family</u> $\mathcal{P}=\{P\}$ <u>of distributions</u>, $E|Y_1|^r$ <u>in (i)</u> ($E|Y_1|^{r/2}$ <u>in (ii)</u>) <u>are uniformly bounded then</u> $\xi_n=\omega(r-2)$ ($\zeta_n=\omega(r-2)$) <u>uniformly in</u> $P\in\mathcal{P}$.

This lemma is a direct consequence of the following

<u>Lemma 3.2. Let</u> $Y_1,\dots,Y_n$ <u>be independent identically distributed r.v.'s and</u> $\Sigma_n=\sum Y_i$. <u>Suppose that</u> $\vartheta_r=E|Y_1|^r<\infty$ <u>for</u>

some $r>0$. Then there exists a constant $C(r,\vartheta_r)$ which depends only on r and ϑ_r such that

(3.2) $$P\{|\Sigma_n|>x\}\leq C(r,\vartheta_r)\,n\,x^{-r}$$

provided one of the following conditions is satisfied: (i) $0<r<1$, (ii) $1\leq r\leq 2$, $EY_1=0$, (iii) $r>2$, $EY_1=0$, $x>(K_r\vartheta_r)^{1/r}\sqrt{n}\log n$ where K_r depends only on r.

Proof. Part (iii) follows from [7] , Corollary 1, where an explicit form of $C(r,\vartheta_r)$ and K_r is given. The case $r=2$ follows from the Chebyshev inequality. In the case $0<r<2$ consider the characteristic function (ch.f.), $\varphi(t)$, of Y_1 . The results of [11] (for $0<r<1$) and [10] (for $1\leq r<2$) imply that there exists $C_1(r,\vartheta_r)$ such that

$$|1-\varphi(t)|\leq C_1(r,\vartheta_r)\,t^r.$$

Now for the ch.f. of Σ_n , $\varphi^n(t)$, we have

$$1-\operatorname{Re}\varphi^n(t)\leq|1-\varphi^n(t)|\leq n\,|1-\varphi(t)|\leq C_1(r,\vartheta_r)\,n\,t^r,$$

and (3.2) follows from the Truncation Inequality ([12] , 12.4B').

It is easily seen that $\xi_n=\omega(k)$, $\zeta_n=\omega(k)$ imply $\xi_n+\zeta_n=\omega(k)$ and $\xi_n\zeta_n=\omega(k)$. More generally, we have

Lemma 3.3. Let ξ_n, $n=1,2,\dots$, be a sequence of random vectors in R^p and $\xi_n=\omega(k)$ uniformly in $P\in\mathcal{P}$. Let $Q_n(y)$, $y\in R^p$, be a sequence of polynomials of a fixed power with uniformly bounded coefficients. Then $Q_n(\xi_n)=\omega(k)$ uniformly in $P\in\mathcal{P}$.

Definition 3.2. A statistic Z_n will be said to belong to the class $SE(r,k)$, $k\geq 0$, integer, $r\geq 2$, if it is expressible in the form

(3.3) $$Z_n=S_{on}+\sum_{\ell=1}^{k}\tau^\ell H_\ell(S_n)+\tau^{k+1}\zeta_n$$

where

(3.4) $$S_n = (S_{on}, S_{1n}, \dots, S_{pn})' = \tau \sum Y_i,$$

$$Y_i = (Y_{oi}, Y_{1i}, \dots, Y_{pi})', \quad i=1,2,\dots,$$

being a sequence of independent identically distributed random vectors such that $EY_1 = 0$, $EY_{01}^2 > 0$ and $E|Y_{j1}|^r < \infty$, $j=0,1,\dots,p$, and $H_\ell(x)$, $\ell=1,\dots,k$, $x \in R^{p+1}$, are some polynomials. When $k=0$ the sum in (3.3) is understood to be zero.

An expression of the form (3.3) will be called a stochastic expansion for Z_n. Suppose now that the distribution of Y_i's and ζ_n, P_θ, say, and the coefficients of H_ℓ's may depend on some parameter θ (which is of arbitrary nature and need not be the same as in the preceding and subsequent sections).

Definition 3.2'. We say that $Z_n \in SE(r,k)$ uniformly in $\theta \in K$ if $Z_n \in SE(r,k)$ in the sense of Definition 3.2 for each $\theta \in K$, the coefficients of H_ℓ's are bounded uniformly in $\theta \in K$, $\inf_{\theta \in K} E_\theta Y_{01}^2 > 0$, $|Y_{j1}|^r$, $j=0,1,\dots,p$, are uniformly integrable with respect to P_θ, $\theta \in K$, and $\zeta_n = \omega(r-2)$ uniformly in $\theta \in K$.

Lemmas 3.1 and 3.3 immediately imply

Lemma 3.4. If $Z_n \in SE(r,k)$ uniformly in $\theta \in K$ then $Z_n = \omega(r-2)$ uniformly in $\theta \in K$.

Definition 3.3. P_θ will be said to satisfy the condition (AC) uniformly in $\theta \in K$ if there exist m, α and b, $m \geq 1$, integer, $0 < \alpha \leq 1$, $0 < b < \infty$, such that, for any Borel set $A \in R^p$,

$$P_\theta\{S_m \in A\} = a \int_A p_{m,\theta}(y)\,dy + (1-a)\,Q_\theta(A)$$

where Q_θ is a probability measure on R^p and $p_{m,\theta}$ a probability density on R^p, $0 \leq p_{m,\theta}(y) \leq b$, $y \in R^p$, $\theta \in K$.

This is a uniform version of the condition that, for some m, the m-fold convolution of the distribution of Y_1 with itself contains an absolutely continuous component. Some results useful for verifying the conditions of definition 3.3 are contained in [9].

Theorem 3.1. Let $Z_n \in SE(k+2,k)$ uniformly in $\theta \in K$ and P_θ satisfy the condition (AC) uniformly in $\theta \in K$. Then

$$P_\theta\{Z_n < z\} = \Phi\left(\frac{z}{\sigma_{0,\theta}}\right) + \sum_{\ell=1}^{k} \tau^\ell Q_{\ell,\theta}(z)\varphi\left(\frac{z}{\sigma_{0,\theta}}\right) + o(\tau^k) \tag{3.5}$$

with the remainder $o(\tau^k)$ uniform in $z \in R^1$, $\theta \in K$ where $\sigma_{0,\theta}^2 = E_\theta Y_{01}^2$, $\Phi(\cdot)$ is the (0,1) normal d.f. and $\varphi(\cdot)$ its density, and $Q_{\ell,\theta}(\cdot)$ are polynomials with coefficients dependent on the moments of Y_{j1}'s and the coefficients of the polynomials H_j.

This theorem (without uniformity in θ) was proved in [4]. By looking through this proof the uniformity assertion can easily be seen to hold true. A general rule for constructing the polynomials Q_ℓ is also given in [4] (now we suppress the subscript θ). In particular, if $\sigma_0 = 1$ (as will be the case when applying the theorem to $Z_n(\hat{\theta}_n)$) then

$$Q_1(z) = -\frac{\mu_3}{6}(z^2-1) - E[H_1(S) \mid S_0 = z] \tag{3.6}$$

where $\mu_3 = EY_{01}^3$ and $S = (S_0, S_1, \ldots, S_p)'$ is a normally distributed random vector with zero mean and covariance matrix $\Sigma = EY_1Y_1'$.

Sometimes we shall write a stochastic expansion (3.3) as

$$Z_n \simeq S_{0n} + \sum_{\ell=1}^{k} \tau^\ell H_\ell(S_n). \tag{3.7}$$

4. Asymptotic expansion for $Z_n(\hat{\theta}_n)$ under H_0. By Theorem 3.1 the problem of obtaining an asymptotic expansion for the distribution of $Z_n(\hat{\theta}_n)$ is reduced to obtaining a stochastic expansion for $Z_n(\hat{\theta}_n)$. For a vector $\alpha = (\alpha_1, \ldots, \alpha_s)$ with nonnegative integer components, let $|\alpha| = \alpha_1 + \ldots + \alpha_s$, $x^\alpha = x_1^{\alpha_1} \ldots x_s^{\alpha_s}$, $x \in R^s$, and $f^{(\alpha)}(\theta) = (\partial^{|\alpha|}/\partial\theta^\alpha) f(\theta)$, $\theta \in R^s$.

Definition 4.1. (i) Denote by $\mathcal{D}_{r,k}(\theta_0)$ the class of functions

$f(x,\theta)$, $\theta\in\Theta$, satisfying the following conditions: in a neighbourhood $U_{\theta_0}\subset\Theta$ of θ_0 for almost all $x\in R^1$ (with respect to $dF(x,\theta_0)$) the derivatives $f^{(\alpha)}(x,\theta)$ exist and are continuous (in θ) for all α with $|\alpha|=1,\dots,k$ and

(4.1) $$E_{\theta_0}|f^{(\alpha)}(X,\theta_0)|^r<\infty,\quad |\alpha|=0,1,\dots,k,$$

(4.2) $$|f^{(\alpha)}(x,\theta)-f^{(\alpha)}(x,\theta_0)|\le\|\theta-\theta_0\|R(x,\theta_0),\ \theta\in U_{\theta_0},\ |\alpha|=k,$$

(4.3) $$E_{\theta_0}[R(X,\theta_0)]^{r/2}<\infty.$$

(ii) We say that $f(x,\theta)\in\mathcal{D}_{r,k}(K)$, $K\subset\Theta$, if the conditions of (i) are satisfied for each $\theta_0\in K$, the neighbourhoods U_{θ_0} may be chosen so that $\{\theta:\|\theta-\theta_0\|<\delta_K\}\subset U_{\theta_0}$ for some $\delta_K>0$ independent of $\theta_0\in K$, the expectations in (4.1) converge uniformly in $\theta_0\in K$ and the left hand side of (4.3) is bounded uniformly in $\theta_0\in K$.

Theorem 4.1. Let the common d.f. of $X_1,\dots,X_n$ be $F(x,\theta_0,0)$. Suppose that $g(x,\theta)$ satisfies the conditions (2.8), (2.9) and (2.12) for all $\theta\in K\subset\Theta$ and $g(x,\theta)\in\mathcal{D}_{k+2,k+1}(K)$, $k\ge1$, integer. Let $T_n^*=\sqrt{n}(\hat\theta_n-\theta_0)\in SE(k+2,k-1)$ uniformly in $\theta_0\in K$. Then $Z_n(\hat\theta_n)\in SE(k+2,k)$ uniformly in $\theta_0\in K$.

Remark 4.1. If $s>1$ then T_n^* is vector-valued and a modification of Definition 3.2 is required. We suppose that

(4.4) $$T_n^*=h_0(\xi_n)+\sum_{\ell=1}^{k-1}\tau^\ell h_\ell(\xi_n)+\tau^k\rho_n$$

where ξ_n is a vector of the same type as S_n in Definition 3.2, h_0 is a linear function from R^p to R^s, $h_\ell=(h_{\ell1},\dots,h_{\ell s})'$ with $h_{\ell j}$'s polynomials and $\rho_n=\omega(k)$. The distribution of ξ_n and ρ_n and the coefficients of h_ℓ, $\ell=0,1,\dots,k-1$, depend on θ_0 and satisfy the conditions of Definition 3.2', the condition $\inf_{\theta\in K}E_\theta Y_{01}^2>0$ being replaced by $\inf_{\theta_0\in K}\lambda_{min}[E(h_0(\xi_n))(h_0(\xi_n))']>0$ where

$\lambda_{min}(A)$ denotes the minimal characteristic root of the matrix A.

The proof of Theorem 4.1 will be given for $s=1$ and θ_0 fixed. The modifications for the general case are straightforward. For simplicity of notation, let $\theta_0 = 0$.

Lemma 4.1. For a function $f(x,\theta) \in \mathcal{D}_{r,k+1}(0)$, $k \geq 1$, integer, $r \geq 2$, let

(4.5) $$a^{(\ell)} = E f^{(\ell)}(X,0), \quad \xi_n^{(\ell)} = \tau \sum [f^{(\ell)}(X_i,0) - a^{(\ell)}]$$

(the superscript (0) corresponds to f itself). Suppose that $a^{(0)}=0$ and there is a sequence of r.v.'s $T_n = \omega(r-2)$. Then

(4.6) $$\tau \sum f(X_i, \tau T_n) = \sum_{j=0}^{k} \tau^j \left(\frac{T_n^j \xi_n^{(j)}}{j!} + \frac{T_n^{j+1}}{(j+1)!} a^{(j+1)} \right) + \tau^{k+1} \zeta_n$$

where $\zeta_n = \omega(r-2)$. If $\tau T_n \in U_{\theta_0}$ then

(4.7) $$\zeta_n = \frac{T_n^{k+1}}{(k+1)!} [\xi_n^{(k+1)} + \tau \sum (f^{(k+1)}(X_i, \tilde{\theta}_i) - f^{(k+1)}(X_i,0))], \quad |\tilde{\theta}_i| \leq \tau |T_n|.$$

Proof. If $\tau T_n \notin U_{\theta_0}$ define ζ_n to satisfy (4.6). It follows from $T_n = \omega(r-2)$ that

(4.8) $$P\{\tau T_n \in U_{\theta_0}\} = o(\tau^{r-2}).$$

If $\tau T_n \in U_{\theta_0}$ then (4.6), (4.7) are obtained by the Taylor formula. By (4.2),

(4.9) $$\tau \sum [f^{(k+1)}(X_i, \tilde{\theta}_i) - f^{(k+1)}(X_i, 0)] \leq |T_n| \, n^{-1} \sum R(X_i, 0).$$

By Lemmas 3.1 and 3.3, (4.1) and (4.3) imply $\zeta_n = \omega(r-2)$.

Proof of Theorem 4.1. By Lemma 3.4, $T_n^* = \omega(k)$. Apply Lemma 4.1 to $Z_n(\hat{\theta}_n)$ taking $g(x,\theta)$ and T_n^* for $f(x,\theta)$ and T_n and letting $r = k+2$. Put

(4.10) $$b^{(j)} = E g^{(j)}(X,0), \quad S_n^{(j)} = \tau \sum [g^{(j)}(X_i,0) - b^{(j)}].$$

By (2.8) and (2.9), $b^{(0)} = b^{(1)} = 0$, and we have by Lemma 4.1

(4.11) $$Z_n(\hat{\theta}_n) = S_n^{(0)} + \sum_{j=1}^{k} \tau^j \left(\frac{(T_n^*)^j S_n^{(j)}}{j!} + \frac{(T_n^*)^{j+1}}{(j+1)!} b^{(j+1)} \right) + \tau^{k+1} \omega(k).$$

By (4.4) we have

(4.12) $$(T_n^*)^j = h_{0j}(\xi_n) + \sum_{l=1}^{k-1} \tau^l h_{lj}(\xi_n) + \tau^k \omega(k), \quad j = 1, \dots, k+1,$$

where h_{lj}, $l = 0, 1, \dots, k-1$, are polynomials. Substituting (4.12) into (4.11) we get a sum of terms

(4.13) $$\tau^{j+l} \frac{h_{lj}(\xi_n) S_n^{(j)}}{j!}, \quad \tau^{j+l} \frac{h_{l,j+1}(\xi_n)}{(j+1)!} b^{(j+1)},$$

(4.14) $$\tau^{j+k} \frac{\omega(k) S_n^{(j)}}{j!}, \quad \tau^{j+k} \frac{\omega(k)}{(j+1)!} b^{(j+1)},$$

$j = 1, \dots, k$; $l = 0, 1, \dots, k-1$. All the terms (4.14) and the terms (4.13) with $j + l \geq k+1$ are $\tau^{k+1} \omega(k)$. Rearranging the remaining terms according the powers of τ we obtain the representation (3.3) for $Z_n(\hat{\theta}_n)$. The conditions of the Definition 3.2 follow immediately from the assumptions of the theorem. Thus the theorem is proved.

5. Asymptotic expansion for $Z_n(\hat{\theta}_n)$ under local alternatives.

Consider an alternative $(\theta, \vartheta) = (\theta_0, \tau\eta)$. We shall write $F(x, \theta, \vartheta)$ as $F_\vartheta(x, \theta)$. Let $F_\vartheta^{-1}(u, \theta_0) = \inf\{x : F_\vartheta(x, \theta_0) \geq u\}$ and

(5.1) $$G_\vartheta(x, \theta_0) = F_\vartheta^{-1}(F_0(x, \theta_0), \theta_0).$$

As a rule, the argument θ_0 will be suppressed. The r.v. $G_\vartheta(X)$ has the d.f. $F_\vartheta(x)$ if X has the d.f. $F_0(x)$. Therefore, the distribution of

$$Z_n(\hat\theta_n) = Z_n(X_1,\dots,X_n;\hat\theta_n(X_1,\dots,X_n))$$

under $F_\vartheta(x)$ is the same as that of

$$Z_n(\hat\theta_n,\vartheta) = Z_n(G_\vartheta(X_1),\dots,G_\vartheta(X_n);\hat\theta_n(G_\vartheta(X_1),\dots,G_\vartheta(X_n))) \tag{5.2}$$

under $F_0(x)$. In what follows we assume as before that $X_1,\dots,X_n$ have the d.f. $F_0(x)$ but consider the statistic $Z_n(\hat\theta_n,\vartheta)$.

Usually for a further development of the expansion (4.4) an extension of the vector ξ_n is required, so that $h_\ell(\xi_n)$ with a smaller ℓ depends on a smaller number of components of ξ_n. Let the components of ξ_n be indexed as follows:

$$\xi_n = \{\xi_{jn}^{(\ell)}\},\quad \ell=0,1,\dots,k-1;\ j=1,\dots,j_\ell,$$

so that $h_0,\dots,h_\ell$ in (4.4) do not depend on $\xi_{jn}^{(m)}$ with $m\geq \ell+1$. Let $\xi_{jn}^{(\ell)}$ have the form

$$\xi_{jn}^{(\ell)} = \tau\sum f_j^{(\ell)}(X_i),\quad E f_j^{(\ell)}(X_i)=0. \tag{5.3}$$

Suppose that

$$g^{(\alpha)}(G_\vartheta(x),\theta_0)\in \mathcal{D}_{k+2,k+1-|\alpha|}(0),\quad |\alpha|=0,1,\dots,k+1, \tag{5.4}$$

$$f_j^{(\ell)}(G_\vartheta(x),\theta_0)\in\mathcal{D}_{k+2,k+1-\ell}(0),\quad \ell=0,1,\dots,k-1;\ j=1,\dots,j_\ell. \tag{5.5}$$

Definition 4.1 is applied here to functions of (x,ϑ), $\vartheta\in\Omega$ rather than (x,θ), $\theta\in\Theta$. For $\rho_n=\rho_n(X_1,\dots,X_n)$ in (4.4) put

$$\rho_{n,\vartheta} = \rho_n(G_\vartheta(X_1),\dots,G_\vartheta(X_n)).$$

Theorem 5.1. Let the assumptions of Theorem 4.1 be fulfilled. Let $R(x,\theta_0)$ from the definition of $\mathcal{D}_{k+2,k+1}(K)$ for $g(x,\theta)$ satisfy the condition: there exist a $C<\infty$ and a neighbourhood of zero, V, such that

(5.6) $$E_{\theta_0}\left[R(G_\vartheta(X),\theta_0)\right]^{\frac{k+2}{2}} \le C, \quad (\theta_0,\vartheta)\in K\times V.$$

Let (5.4) and (5.5) hold uniformly in $\theta_0 \in K$ and $\rho_{n,\tau\eta} = \omega(k)$ uniformly in $(\theta_0,\eta)\in K\times B$ for any bounded $B\in R^1$. Then $Z_n(\hat\theta_n,\tau\eta) - \eta b^{(0,1)} \in SE(k+2,k)$ uniformly in $(\theta_0,\eta)\in K\times B$ for any bounded $B\subset R^1$; $b^{(0,1)}$ to be defined later, see (5.10).

The proof will also be given for $s=1$ and θ_0 fixed. Let $S_{n,\vartheta}^{(j)}$ and $T_{n,\vartheta}^{*}$ be defined as $S_n^{(j)}$ and T_n^* with X_i replaced by $G_\vartheta(X_i)$, $i=1,\dots,n$. Consider $S_{n,\tau\eta}^{(j)}$. Applying Lemma 4.1 with $k-j$ instead of k, $T_n=\eta$ and $r=k+2$ to $g^{(j)}(G_{\tau\eta}(x))$ we get

(5.7) $$S_{n,\tau\eta}^{(j)} = \sum_{l=0}^{k-j}\tau^l\left(\frac{\eta^l S_n^{(j,l)}}{l!} + \frac{\eta^{l+1} b^{(j,l+1)}}{(l+1)!}\right) + \tau^{k-j+1}\omega(k)$$

where

(5.8) $$b^{(j,l)} = Eg^{(j,l)}(X), \quad S_n^{(j,l)} = \tau\sum\left[g^{(j,l)}(X_i) - b^{(j,l)}\right],$$

(5.9) $$g^{(j,l)}(x) = \frac{\partial^l}{\partial\vartheta^l} g^{(j)}(G_\vartheta(x))\Big|_{\vartheta=0}.$$

By Lemma 3.4, it follows from (5.7) that

(5.10) $$S_{n,\tau\eta}^{(j)} = \omega(k) \qquad \text{uniformly in } \eta\in B.$$

In a similar way expansions for

$$\xi_{jn,\tau\eta}^{(l)} = \tau\sum f_j^{(l)}(G_{\tau\eta}(X_i))$$

are obtained which imply that

$$(5.11) \qquad T^{*}_{n,\tau\eta} = \omega(k) \qquad \text{uniformly in } \eta \in B.$$

An application of Lemma 4.1 gives

$$(5.12) \qquad Z_n(\hat{\theta}_n, \tau\eta) = S^{(0)}_{n,\tau\eta} + \sum_{j=1}^{k} \tau^j \left(\frac{(T^{*}_{n,\tau\eta})^j S^{(j)}_{n,\tau\eta}}{j!} + \frac{(T^{*}_{n,\tau\eta})^{j+1} b^{(j+1)}}{(j+1)!} \right) + \tau^{k+1}\omega(k)$$

with $\omega(k)$ uniform in $\eta \in B$ by (5.6). Substituting (5.7) and the expansions for $\xi^{(l)}_{jn,\tau\eta}$ into (5.12) and proceeding as in the proof of Theorem 4.1 we obtain the theorem.

Remark 5.1. A theorem giving sufficient conditions for $T^{*}_{n} \in$ $\in SE(k+2, k-1)$ required in Theorem 4.1 is stated without proof in [2] (Theorem 1). (Note that in the second line of Theorem 1 in [2] $k \geq 3$ should be replaced by $k \geq 2$.) The theorem is stated for a one-dimensional parameter but it can be extended in an obvious way to the vector-valued case. The condition of Theorem 5.1 that $\rho_{n,\tau\eta} =$ $= \omega(k)$ is also fulfilled under the conditions of Theorem 1 of [2] with the exception that instead of $E_0 K^{m_1}(X) < \infty$ (in the notation of [2]) one should assume that $K^{m_1}(X)$ is uniformly integrable with respect to $P_{0,\vartheta}$ for $\vartheta \in V$. The proofs of this assertions will be given in a separate paper.

Remark 5.2. We imposed the condition (2.8) in Theorems 4.1 and 5.1 having in mind the application of the theorems to $C(\alpha)$ test statistics. However, the theorems remain true without this assumption. In that case the main term in (4.11) would be $S^{(0)}_n + T^{*}_n b^{(1)}$ and we should require that $T^{*}_n \in SE(k+2, k)$ with corresponding modifications in the proofs.

Remark 5.3. In Theorems 4.1 and 5.1 $\hat{\theta}_n$ may be any statistic (not necessary an estimator) and $G_\vartheta(x)$ any function (not necessary of the form (5.1)) satisfying the conditions of the theorems.

6. The term of order $n^{-1/2}$ for the distribution of a $C(\alpha)$ test statistic. Consider the case $k=1$ in Theorems 4.1 and 5.1. Introduce the following notation. Let $g(x)$ and $g^{(1)}(x)$ stand for $g(x,\theta_0)$ and $g^{(1)}(x,\theta_0)$ (see (2.10)). For a function $f(x), x\in R^1$, write Ef instead of $Ef(X)$. Put

$$(6.1)\qquad g_{\theta_\ell\theta_j}(x)=\frac{\partial^2 g(x,\theta)}{\partial\theta_\ell\,\partial\theta_j}\Big|_{\theta=\theta_0},\quad b_{\ell j}^{(2)}=Eg_{\theta_\ell\theta_j},\quad B^{(2)}=\|b_{\ell,j}^{(2)}\|_{\ell,j=1}^{s};$$

$$(6.2)\qquad g^{(0,j)}(x)=\frac{\partial}{\partial\vartheta}g(G_\vartheta(x),\theta_0)\Big|_{\vartheta=0},\quad b^{(0,j)}=Eg^{(0,j)};$$

$$(6.3)\qquad g_{\theta_j,\vartheta}(x)=\frac{\partial}{\partial\vartheta}g_{\theta_j}(G_\vartheta(x))\Big|_{\vartheta=0},\quad g^{(1,1)}(x)=(g_{\theta_1,\vartheta}(x),\dots,g_{\theta_s,\vartheta}(x))',$$

$$(6.4)\qquad b^{(1,1)}=(Eg_{\theta_1,\vartheta},\dots,Eg_{\theta_s,\vartheta})'.$$

(A single or a first superscript refer to the total order of derivatives with respect to θ_j's and a second one to the order of a derivative with respect to ϑ.) Put

$$(6.5)\qquad S_n^{(0)}=\tau\sum g(X_i),\quad S_n^{(0,1)}=\tau\sum[g^{(0,1)}(X_i)-b^{(0,1)}],$$

$$(6.6)\qquad S_{n,j}^{(1)}=\tau\sum g_{\theta_j}(X_i),\quad S_n^{(1)}=(S_{n1}^{(1)},\dots,S_{ns}^{(1)})'.$$

When $k=1$, (4.4) may be written as

$$(6.7)\qquad T_n^*=T_n+\tau\rho_n,\qquad T_n=\tau\sum f(X_i)$$

where $f(x)=(f_1(x),\dots,f_s(x))'$, $\rho_n=(\rho_{n1},\dots,\rho_{ns})'$. Put

$$(6.8)\qquad f_{j,\vartheta}(x)=\frac{\partial}{\partial\vartheta}f_j(G_\vartheta(x))\Big|_{\vartheta=0},\quad a_j^{(0,1)}=Ef_{j,\vartheta},\quad a^{(0,1)}=(a_1^{(0,1)},\dots,a_s^{(0,1)})'.$$

We shall use the symbol $\simeq$ in the sense of (3.7). For real numbers $a_n\simeq b_n$ will mean that $a_n=b_n+o(\tau)$.

When $k=1$, an analogue of (5.12) for the case of vector-valued parameter is

$$(6.9)\qquad Z_n(\hat\theta_n,\tau\eta)\simeq S_{n,\tau\eta}^{(0)}+\tau[T_{n,\tau\eta}'S_{n,\tau\eta}^{(1)}+\tfrac{1}{2}T_{n,\tau\eta}'B^{(2)}T_{n,\tau\eta}].$$

Now, similarly to (5.7), we have

$$S^{(0)}_{n,\tau\eta} \simeq S^{(0)}_n + \eta b^{(0,1)} + \tau(\eta S^{(0,1)}_n + \frac{1}{2}\eta^2 b^{(0,2)}), \tag{6.10}$$

$$T_{n,\tau\eta} \simeq T_n + \eta a^{(0,1)}, \quad S^{(1)}_{n,\tau\eta} \simeq S^{(1)}_n + \eta b^{(1,1)}. \tag{6.11}$$

Substituting (6.10) and (6.11) into (6.9), we get

$$Z_n(\hat{\theta}_n, \tau\eta) \simeq S^{(0)}_n + \eta b^{(0,1)} + \tau H_1(S_n) \tag{6.12}$$

where $S_n = (S^{(0)}_n, S^{(1)\prime}_n, T^{\prime}_n, S^{(0,1)}_n)^{\prime}$ and

$$H_1(S_n) = T^{\prime}_n S^{(1)}_n + \frac{1}{2} T^{\prime}_n B^{(2)} T_n + \eta(S^{(0,1)}_n + a^{(0,1)\prime} S^{(1)}_n + T^{\prime}_n b^{(1,1)} + T^{\prime}_n B^{(2)} a^{(0,1)}) + \eta^2(\frac{1}{2} b^{(0,2)} + a^{(0,1)\prime} b^{(1,1)} + \frac{1}{2} a^{(0,1)\prime} B^{(2)} a^{(0,1)}). \tag{6.13}$$

Put $Y = (Y_0, Y_1, \ldots, Y_{2s+1})^{\prime}$ where

$$Y_0 = g(X), \quad (Y_1, \ldots, Y_s)^{\prime} = g^{(1)}(X), \tag{6.14}$$
$$(Y_{s+1}, \ldots, Y_{2s})^{\prime} = h(X), \quad Y_{2s+1} = g^{(0,1)}(X) - b^{(0,1)}.$$

Then $EY = 0$. Put

$$\Sigma = EYY^{\prime} = \begin{Vmatrix} 1 & \Sigma_{01} & \Sigma_{02} & \sigma_{0,2s+1} \\ \Sigma_{10} & \Sigma_{11} & \Sigma_{12} & \Sigma_{1,2s+1} \\ \Sigma_{20} & \Sigma_{21} & \Sigma_{22} & \Sigma_{2,2s+1} \\ \sigma_{0,2s+1} & \Sigma_{2s+1,1} & \Sigma_{2s+1,2} & \sigma^2_{2s+1} \end{Vmatrix} \tag{6.15}$$

where the partition of the matrix Σ corresponds to the partition (6.14) of Y . Obviously, the same covariance matrix has the vector S_n . Let

$$Q^*_1(\eta, y) = E[H_1(S) \mid S^{(0)} = y], \tag{6.16}$$

$S = (S^{(0)}, S^{(1)\prime}, T^{\prime}, S^{(0,1)})^{\prime}$ being a random vector with a normal $(0,\Sigma)$ distribution. By the well-known formulas for the conditional expectations and covariances of a multivariate normal distribution, we obtain

(6.17) $$Q_1^*(\eta,y) = tr(\Sigma_{12} + \frac{1}{2}B^{(2)}\Sigma_{22}) + (y^2-1)(\Sigma_{01}\Sigma_{20} + \frac{1}{2}\Sigma_{02}B^{(2)}\Sigma_{20}) +$$
$$+ \eta y(\sigma_{0,2s+1} + \Sigma_{01}a^{(0,1)} + \Sigma_{02}b^{(1,1)} + \Sigma_{02}B^{(2)}a^{(0,1)}) +$$
$$+ \eta^2(\frac{1}{2}b^{(0,2)} + a^{(0,1)\prime}b^{(1,1)} + \frac{1}{2}a^{(0,1)\prime}B^{(2)}a^{(0,1)}).$$

Put

(6.18) $$\mu_3 = Eg^3(X), \quad Q_1(\eta,y) = \frac{\mu_3}{6}(y^2-1) + Q_1^*(\eta,y).$$

Theorem 6.1. Let the conditions of Theorem 4.1 with $k=1$ be fulfilled and the distribution of the vector $(Y_0,\dots,Y_{2s})'$ (see (6.14)) satisfy the condition (AC) uniformly in $\theta_0 \in K$. Then uniformly in $(z,\theta_0) \in R^1 \times K$

(6.19) $$P_{\theta_0,0}\{Z_n(\hat{\theta}_n) < z\} \simeq \Phi(z) - \tau Q_1(0,z)\varphi(z).$$

Theorem 6.2. Let the conditions of theorem 5.1 with $k=1$ be fulfilled and the distribution of the vector Y (see (6.14)) satisfy the condition (AC) uniformly in $\theta_0 \in K$. Then uniformly in $(z,\theta_0,\eta) \in R^1 \times K \times B$

(6.20) $$P_{\theta_0,\tau\eta}\{Z_n(\hat{\theta}_n) < z\} \simeq \Phi(z-\eta b^{(0,1)}) - \tau Q_1(\eta, z-\eta b^{(0,1)})\varphi(z-\eta b^{(0,1)}).$$

Theorems 6.1 and 6.2 follow directly from Theorems 4.1, 5.1 and 3.1 and formula (3.6).

One can rewrite (6.20) as

(6.21) $$P_{\theta_0,\tau\eta}\{Z(\hat{\theta}_n) < z\} \simeq \Phi(z-\eta b^{(0,1)} - \tau Q_1(\eta, z-\eta b^{(0,1)})).$$

Let the hypothesis H_0 be tested against the one-sided alternative $\vartheta > 0$. Suppose for a moment that we use a test of size α based on $Z_n(\hat{\theta}_n)$ for testing a simple hypothesis $(\theta,\vartheta) = (\theta_0,0)$ against

$(\theta,\vartheta)=(\theta_0,\tau\eta)$. Let t_α be determined by $\Phi(t_\alpha)=1-\alpha$. Put

$$z_{n,\alpha}=t_\alpha+\tau Q_1(0,t_\alpha). \tag{6.22}$$

Then we have from (6.19) (or (6.21) with $\eta=0$)

$$P\{Z_n(\hat\theta_n)<z_{n,\alpha}\}\simeq 1-\alpha, \tag{6.23}$$

i.e. $z_{n,\alpha}$ may serve as a critical value for our order of approximation. Denote the power of the test by $\beta_n(\eta)$. Then $\beta_n(\eta)\simeq$ $\simeq P_{\theta_0,\tau\eta}\{Z_n(\hat\theta_n)\geq z_{n,\alpha}\}$ and we obtain from (6.21)

$$\beta_n(\eta)\simeq 1-\Phi(t_\alpha-\eta b^{(0,1)}+\tau(Q_1(0,t_\alpha)-Q_1(\eta,t_\alpha-\eta b^{(0,1)}))). \tag{6.24}$$

When testing the composite hypothesis H_0 this test is generally not valid because $Q_1(0,y)$ depends on the unknown value θ_0 In order to indicate this dependence explicitly we shall write in the remainder of this section $Q_1(y,\theta_0)$ instead of $Q_1(0,y)$. Consider the test with the critical region

$$Z_n(\hat\theta_n)\geq\hat z_{n,\alpha}, \quad \hat z_{n,\alpha}=t_\alpha+\tau Q_1(t_\alpha,\hat\theta_n). \tag{6.25}$$

<u>Theorem 6.3. Let the conditions of Theorem 6.2 be satisfied. Suppose that M_3 and the elements of the matrices $\Sigma_{01},\Sigma_{02},\Sigma_{12},\Sigma_{22},$ $B^{(2)}$ satisfy the Lipschitz condition uniformly in $\theta_0\in K$. Then the size of the test (6.25) is $\alpha+o(\tau)$ uniformly in $\theta_0\in K$ and its power differs from the right hand side of (6.24) by $o(\tau)$ uniformly in $(\theta_0,\eta)\in K\times B$.</u>

<u>Proof</u>. We have, for any $\rho>0$,

$$P_{\theta_0,\tau\eta}\{Z_n(\hat\theta_n)<\hat z_{n,\alpha}\}\leq P_{\theta_0,\tau\eta}\{Z_n(\hat\theta_n)<z_{n,\alpha}+\tau\rho\}+ \tag{6.26}$$
$$+P_{\theta_0,\tau\eta}\{|Q_1(t_\alpha,\hat\theta_n)-Q_1(t_\alpha,\theta_0)|>\rho\}=p_1+p_2 \quad \text{, say.}$$

Take $\rho=n^{-\gamma},\ 0<\gamma<1/2$. Then using (6.20), we obtain

(6.27) $$p_1 = P_{\theta_0,\tau\eta}\{Z_n(\hat{\theta}_n) < z_{n,\alpha}\} + o(\tau)$$

uniformly in $(\theta_0,\eta) \in K \times B$. It follows from (6.17), (6.18) and the conditions of the theorem that $Q_1(t_\alpha,\theta)$ satisfies the Lipschitz condition uniformly in $\theta \in K$. Therefore there exist a $\delta > 0$ and a $C < \infty$ such that

(6.28) $$|Q_1(t_\alpha,\theta) - Q_1(t_\alpha,\theta_0)| \le C\|\theta-\theta_0\| \quad \text{for } \|\theta-\theta_0\| \le \delta.$$

Hence

$$p_2 \le P_{\theta_0,\tau\eta}\{\|\hat{\theta}_n - \theta_0\| > C^{-1}n^{-\gamma}\}$$

and by (5.11) $p_2 = o(\tau)$. Now (6.26), (6.27) and an estimate from below similar to (6.26) imply the assertion of the theorem.

Remark 6.1. Let $\beta_n(\theta_0,\eta)$ be the power of a test for H_0 when the alternative $(\theta,\vartheta) = (\theta_0,\tau\eta)$ holds, and $\beta_n(\theta_0,0)$ be the size of the test. When testing a composite hypothesis $H_0: \theta \in \Theta, \vartheta = 0$ with a prescribed significance level $\alpha > 0$ it is required that $\beta_n(\theta,0) \le \alpha$ for all $\theta \in \Theta$. Asymptotic considerations lead usually to the assertion that $\beta_n(\theta,0) \to \alpha$ for all $\theta \in \Theta$ though a more desirable property would be $\sup_{\theta\in\Theta} \beta_n(\theta,0) \to \alpha$. The latter is not usually proved and probably does not always hold true. Similarly, in our case Theorem 6.3 gives for the size of a $C(\alpha)$ test

(6.29) $$\sup_{\theta \in K} \beta_n(\theta,0) = \alpha_n = \alpha + o(\tau).$$

The conditions of Theorem 6.3 are typically fulfilled for a compact $K \subset \Theta$. These are simple sufficient conditions and a question when the supremum in (6.29) can be taken over $\theta \in \Theta$ requires a special investigation. However, there is an important class of problems where the distribution of $Z_n(\hat{\theta}_n)$ under H_0 does not depend on θ_0 and the critical value (6.22) may be used. Namely, this is the case when $\theta = (\theta_1, \theta_2)$ with θ_1 and θ_2 location and scale parameters, $g(x,\theta)=g((x-\theta_1)/\theta_2)$ and $\hat{\theta}_n=(\hat{\theta}_{n1},\hat{\theta}_{n2})$ is such that

$$\hat{\theta}_{n1}(aX_1+b,\dots,aX_n+b)=a\hat{\theta}_{n1}(X_1,\dots,X_n)+b,$$

$$\hat{\theta}_{n2}(aX_1+b,\dots,aX_n+b)=a\hat{\theta}_{n2}.$$

Now we shall show that $b^{(0,1)}, b^{(0,2)}, a^{(0,1)}, b^{(1,1)}$ can be evaluated without differentiating functions like $g(G_\vartheta(x))$ with respect to ϑ. Let for a function $h(x)$

$$h^{(0,\ell)}(x)=\frac{\partial^\ell}{\partial\vartheta^\ell}h(G_\vartheta(x))\Big|_{\vartheta=0}. \tag{6.30}$$

Since $h(G_\vartheta(X))$ has under $(\theta_0,0)$ the same distribution as $h(X)$ under (θ_0,ϑ), we have the identity

$$\int h(G_\vartheta(x))p(x)\,dx=\int h(x)p(x,\theta_0,\vartheta)\,dx \;\left(=E_{\theta_0,\vartheta}h(X)\right). \tag{6.31}$$

<u>Definition 6.1. We say that</u> $h\in\mathcal{E}_\ell$ <u>if (6.31) may be differentiated with respect to</u> ϑ <u>under the integral sign</u> ℓ <u>times, i.e. if for</u> $j=1,\dots,\ell$

$$\int h^{(0,j)}(x)p(x)\,dx=\int h(x)\left(\frac{\partial^j}{\partial\vartheta^j}p(x,\theta_0,\vartheta)\Big|_{\vartheta=0}\right)dx. \tag{6.32}$$

<u>If</u> $h(x)$ <u>is vector-valued then</u> $h\in\mathcal{E}_\ell$ <u>means that this condition is fulfilled for each component</u>.

In particular, if $g\in\mathcal{E}_1$, $f\in\mathcal{E}_1$, or $g^{(1)}\in\mathcal{E}_1$ then, respectively,

$$b^{(0,1)}=\int g(x)\left(\frac{\partial p(x,\theta_0,\vartheta)}{\partial\vartheta}\Big|_{\vartheta=0}\right)dx=Eg\,l_\vartheta, \tag{6.33}$$

$$a^{(0,1)}=Ef\,l_\vartheta,\qquad b^{(1,1)}=Eg^{(1)}l_\vartheta; \tag{6.34}$$

moreover, $\partial^2p/\partial\vartheta^2=(l_{\vartheta\vartheta}+l_\vartheta^2)p$ where $l_{\vartheta\vartheta}=\partial^2 l/\partial\vartheta^2$ and if $g\in\mathcal{E}_2$ then

$$b^{(0,2)}=Eg\,(l_{\vartheta\vartheta}+l_\vartheta^2). \tag{6.35}$$

We shall not give sufficient conditions for the possibility of differentiations under the integral sign. Rather these and other relations themselves obtained by the differentiation under the integral sign will be used as conditions of theorems.

7. Maximum likelihood estimator and optimal $C(\alpha)$ test. Differentiating (2.6) and (2.12) under the integral sign with respect to θ_l, $l=1,\ldots,s$, and letting $\theta=\theta_0$, we obtain

(7.1) $$\int g_{\theta_l}(x)\, l_{\theta_j}(x)\, p(x)\, dx + \int g(x)\left(\frac{\partial^2 p(x,\theta)}{\partial\theta_l\,\partial\theta_j}\bigg|_{\theta=\theta_0}\right) dx = 0,$$

(7.2) $$\int g_{\theta_l\theta_j}(x)\, p(x)\, dx + \int g_{\theta_j}(x)\, l_{\theta_l}(x)\, p(x)\, dx = 0,$$

$j, l = 1,\ldots,s$. (7.2) may be rewritten as

(7.3) $$B^{(2)} = -E g^{(1)} l'_{\theta}.$$

On differentiating (2.9) we obtain

(7.4) $$2\int g(x)\, g_{\theta_j}(x)\, p(x)\, dx + \int g^2(x)\, l_{\theta_j}(x)\, p(x)\, dx = 0.$$

Let $c = (c_1,\ldots,c_s)'$ with $c_j = c_j(\theta_0)$, $j=1,\ldots,s$ (see(2.7)). Then

(7.5) $$c = A^{-1} E\, l_{\theta}\, l_{\theta}$$

where

(7.6) $$A = E\, l_{\theta}\, l'_{\theta}.$$

Consider the case when $\hat\theta_n$ is the maximum likelihood estimator or an estimator equivalent to it, namely, assume that in (6.7)

(7.7) $$f(x) = A^{-1} l_{\theta}(x).$$

Then in (6.15)

(7.8) $$\Sigma_{22} = A^{-1}, \quad \Sigma_{20} = \Sigma'_{02} = A^{-1} E g\, l_{\theta} = 0$$

and, by (7.3),

(7.9) $$\Sigma_{12} = E g^{(1)} f' = -B^{(2)} A^{-1}.$$

These relations imply

Proposition 7.1. Let $\hat{\theta}_n$ satisfy (6.7) and (7.7), the conditions of Theorem 6.1 be fulfilled and (7.3) hold. Then $P_{\theta_0,0}\{Z_n(\hat{\theta}_n) < z\}$ is given by (6.19), (6.18) with $Q_1^*(0,y) = -\frac{1}{2} B^{(2)} A^{-1}$.

If $l_\theta \in \mathcal{E}_1$ then $f \in \mathcal{E}_1$ and (6.33), (7.5) and (7.7) imply

(7.10) $$a^{(0,1)} = c.$$

The relations (7.8), (7.9), (7.10) can be used also for an alternative distribution (see (6.17)).

All the results so far concern a general $C(\alpha)$ test. Consider now an optimal $C(\alpha)$ test where $g(x,\theta)$ is given by (2.4), (2.7). Then

(7.11) $$g(x) = \frac{1}{\sigma} \left(l_\vartheta(x) - c' l_\theta(x) \right)$$

with $\sigma = \sigma(\theta_0)$ (see (2.3)).

Proposition 7.2. Let $g(x,\theta)$ be given by (2.4), (2.7).

(i) If $l_\vartheta, l_\theta \in \mathcal{E}_1$ then

(7.12) $$b^{(0,1)} = \sigma.$$

(ii) If (7.3) holds and $g^{(1)} \in \mathcal{E}_1$ then

(7.13) $$b^{(1,1)} + B^{(2)} c = \sigma \Sigma_{10}.$$

(iii) If (7.4) holds and $g^2(x) \in \mathcal{E}_1$ then

(7.14) $$\sigma_{0,2s+1} + c' \Sigma_{10} = \frac{1}{2} \sigma \mu_3.$$

Proof. (i) follows from (6.32), (2.6) and (7.11); (6.34), (7.3) and (7.11) imply (ii); (6.31) with $l = 1$ applied to $g^2(x)$ provides

(7.15) $$\int g(x) g^{(0,1)}(x) p(x)\, dx = \frac{1}{2} \int \left(\frac{\partial}{\partial \vartheta} g^2(G_\vartheta(x)) \Big|_{\vartheta=0} \right) p(x)\, dx = \frac{1}{2} E g^2 l_\vartheta$$

which together with (7.4) and (7.11) gives (7.14).

These relations applied to (6.17),(6.18) and (6.20) imply

Proposition 7.3. Let $g(x,\theta)$ be given by (2.4), (2.7), $\hat{\theta}_n$ satisfy (6.7) and (7.7) and the conditions of Theorem 6.2 be fulfilled. Let $\ell_\vartheta, \ell_\theta \in \mathcal{E}_1$ and (7.3) hold. Then $P_{\theta_0,\tau\eta}\{Z_n(\hat{\theta}_n)<z\}$ is given by (6.20), (6.18) with $b^{(0,1)}=\sigma$ and

(7.16) $$Q_1^*(\eta,y) = -\frac{1}{2}BA^{-1} + \eta y(\sigma_{0,2s+1} + c'\Sigma_{01}) + \\ +\eta^2\left(\frac{1}{2}b^{(0,2)} + c'b^{(1,1)} + \frac{1}{2}c'B^{(2)}c\right).$$

Under the conditions of this proposition the power, $\beta_n(\eta)$, of the optimal $C(\alpha)$ test (see Thorem 6.3 and Remark 6.1) satisfies (6.24) with $b^{(0,1)}=\sigma$,

(7.17) $$Q_1(0,t_\alpha) - Q_1(\eta,t_\alpha-\eta\sigma) = \frac{\mu_3}{6}(2t_\alpha\eta\sigma - \eta^2\sigma^2) + \\ +Q_1^*(0,t_\alpha) - Q_1^*(\eta,t_\alpha-\eta\sigma),$$

Q_1^* being given by (7.16). Using (7.13), (7.14) one can get several expressions for $Q_1^*(\eta,y)$ and $\beta_n(\eta)$. They can be used, in particular, in numerical applications for checking the computations. The following expression to be used in the sequel is obtained by an application of Proposition 7.2(ii) to (7.16).

Proposition 7.4. Let the conditions of Proposition 7.3 be satisfied and $g^{(1)}\in\mathcal{E}_1$. Then

(7.18) $$Q_1^*(0,t_\alpha) - Q_1^*(\eta,t_\alpha-\eta\sigma) = -(\sigma_{0,2s+1} + c'\Sigma_{10})\eta t_\alpha + \\ +\eta^2\left(\sigma\sigma_{0,2s+1} - \frac{1}{2}b^{(0,2)} + \frac{1}{2}c'B^{(2)}c\right).$$

8. An optimal $C(\alpha)$ test with a general estimator. Suppose that $\hat{\theta}_n$ satisfies (6.7) with

(8.1) $$f(x) = A^{-1}\ell_\theta(x) + d(x)$$

where $d(X)$ is uncorrelated with $l_\theta(X)$:

(8.2) $$E\,d(X)\,l_\theta'(X) = 0.$$

This condition is satisfied for a very broad class of estimators, e. g., for minimum contrast estimators (see [5], [2]). Now we have

(8.3) $$\Sigma_{22} = A^{-1} + \Sigma_{dd}\,, \quad \Sigma_{dd} = E\,dd',$$

(8.4) $$\Sigma_{20} = \Sigma_{do} = \Sigma_{od}'\,, \quad \Sigma_{do} = E\,g\,d,$$

(8.5) $$\Sigma_{12} = -BA^{-1} + \Sigma_{1d}\,, \quad \Sigma_{1d} = E\,g^{(1)} d',$$

(8.6) $$a^{(0,1)} = c + \delta, \qquad \delta = E\,l_\theta d.$$

It follws from (8.6), (8.4), (8.2) and (7.11) that

(8.7) $$\delta = \sigma\,\Sigma_{do}.$$

<u>Theorem 8.1. Let $\hat\theta_n$ satisfy (6.7), (8.1), (8.2) and otherwise the conditions of Proposition 7.3 be fulfilled. Then for the power of the optimal $C(\alpha)$ test, $\beta_n(\eta)$, the following relation holds</u>

(8.8) $$\beta_n(\eta) \simeq 1 - \Phi\big(t_\alpha - \eta\sigma + \tau(Q_1(0,t_\alpha) - Q_1(\eta, t_\alpha - \eta\sigma))\big)$$

<u>where $Q_1(0,t_\alpha) - Q_1(\eta, t_\alpha - \eta\sigma)$ is given by (7.17), (7.16).</u>

<u>Proof</u>. Denote now by $Q_{10}^*(\eta,y)$ the function $Q_1^*(\eta,y)$ corresponding to the maximum likelihood estimator as given by (7.16). Then we have from (6.17): $Q_1^*(\eta,y) = Q_{10}^*(\eta,y) + Q_{11}^*(\eta,y)$ where

(8.9) $$\begin{aligned} Q_{11}^*(\eta,y) = {} & tr\big(\Sigma_{1d} + \tfrac{1}{2}B^{(2)}\Sigma_{dd}\big) + (y^2-1)\big(\Sigma_{01}\Sigma_{do} + \\ & + \tfrac{1}{2}\Sigma_{od}B^{(2)}\Sigma_{do}\big) + \eta y\big(\Sigma_{01}\delta + \Sigma_{od}b^{(1,1)} + \Sigma_{od}B^{(2)}(c+\delta)\big) + \\ & + \eta^2\big(\delta' b^{(1,1)} + \delta' B^{(2)} c + \tfrac{1}{2}\delta' B^{(2)}\delta\big). \end{aligned}$$

We shall show that

$$Q^*_{11}(0,t_\alpha) - Q^*_{11}(\eta, t_\alpha - \eta\sigma) = 0. \tag{8.10}$$

Then the theorem will follow from Theorem 6.3 and Proposition 7.3. On substituting (8.7) and (7.13) into (8.9) we obtain

$$Q^*_{11}(\eta,y) = tr(\Sigma_{1d} + \tfrac{1}{2}B^{(2)}\Sigma_{dd}) - (\Sigma_{01}\Sigma_{d0} + \tag{8.11}$$
$$+\tfrac{1}{2}\Sigma_{0d}B^{(2)}\Sigma_{d0}) + (y+\eta\sigma)^2(\Sigma_{01}\Sigma_{d0} + \tfrac{1}{2}\Sigma_{0d}B^{(2)}\Sigma_{d0}).$$

Since $y+\eta\sigma = t_\alpha$ for both $(\eta,y)=(0,t_\alpha)$ and $(\eta,y)=(\eta,t_\alpha-\eta\sigma)$, (8.11) implies (8.10).

9. Upper bound for the power function. Let $\beta_n(\theta,\eta)$ be the power of a test for H_0 (see Remark 6.1) satisfying (6.29). Let θ_0 be an interior point of K. Denote by $\beta^*_n(\theta,\eta)$ the power of the most powerful (Neyman-Pearson) test of size α_n (see (6.29)) for the simple hypothesis $(\theta,0)$ against the simple alternative $(\theta_0,\tau\eta)$ (the dependence on θ_0 will not be indicated explicitly). Then

$$\beta_n(\theta_0,\eta) \le \beta^*_n(\theta,\eta) \quad \text{for any } \theta\in K. \tag{9.1}$$

In particular, under the conditions of Theorem 6.3 the power of an optimal $C(\alpha)$ test, $\beta_n(\eta)$, satisfies (9.1). We shall show now that this power satisfies the relation

$$\beta_n(\eta) \simeq \beta^*_n(\theta_n,\eta) \tag{9.2}$$

for a certain sequence $\theta_n \to \theta_0$ (so that $\theta_n \in K$ for large enough n). Namely, under certain conditions, $\beta^*_n(\theta_n,\eta)$ with $\theta_n = \theta_0 + \tau\eta c$ will be shown to satisfy (8.8). Then (9.2) holds, provided the conditions of Theorem 8.1 are fulfilled. Note that the size of the Neyman-Pearson test in the subsequent theorem is not exactly α_n but $\alpha_n + o(\tau)$. However, this differehce affects the power by $o(\tau)$, so that the above assertions remain true.

For the sake of simplicity consider the case $s=1$. The modifications for the general case are obvious.

Denote the derivatives of $\ell(G_\gamma(x),\theta,\vartheta)$ with respect to θ and ϑ by the corresponding subscripts; a superscript will indicate the order of a derivative with respect to γ , e.g.,

$$\ell^{(1)}_{\vartheta\vartheta}(x,\theta,\vartheta)=\frac{\partial^3}{\partial\vartheta^2\partial\gamma}\ell(G_\gamma(x),\theta,\vartheta)\Big|_{\gamma=0}. \tag{9.3}$$

When $(\theta,\vartheta)=(\theta_0,0)$ these arguments will be omitted. Put $\theta_0=0$.

For (9.2) we need the distribution of $Z_n(\hat\theta_n)$ corresponding to $(\theta,\vartheta)=(\tau\eta c,0)$. We shall treat this as a local alternative to $(\theta,\vartheta)=(0,0)$ and obtain the distribution by the same method as for $(0,\tau\eta)$ (Section 5). Let $F_\theta(\cdot)=F(\cdot,\theta,0)$ and

$$\tilde F^{-1}_\theta(u)=\inf\{x: F_\theta(x)\geq u\}. \tag{9.4}$$

Put

$$\tilde G_\theta(x)=\tilde F^{-1}_\theta(F_0(x)); \tag{9.5}$$

for a variable defined through $G_\gamma(x)$ (like $\ell^{(1)}_{\vartheta\vartheta}$ in (9.3)) we shall denote the variable with G_γ replaced by $\tilde G_\gamma$ in its definition by the same symbol with a wave (e.g. $\tilde\ell^{(1)}_{\vartheta\vartheta}$). We have similarly to (6.31)

$$\int h(\tilde G_\theta(x))p(x)\,dx=\int h(x)p(x,\theta)\,dx. \tag{9.6}$$

We say that $h\in\tilde{\mathcal E}_\ell$ if for $j=1,\dots,\ell$

$$\int\Big(\frac{\partial^j}{\partial\theta^j}h(\tilde G_\theta(x))\Big|_{\theta=0}\Big)p(x)\,dx=\int h(x)\Big(\frac{\partial^j}{\partial\theta^j}p(x,\theta)\Big|_{\theta=0}\Big)dx. \tag{9.7}$$

<u>Theorem 9.1. Let</u> $\ell_\theta(G_\gamma(x),\theta,0)$, $\ell_\theta(\tilde G_\gamma(x),\theta,0)$ <u>as functions of</u> x <u>and</u> (θ,γ) <u>and</u> $\ell_\vartheta(G_\gamma(x),0,\vartheta)$, $\ell_\vartheta(\tilde G_\gamma(x),0,\vartheta)$ <u>as functions of</u> x <u>and</u> (ϑ,γ) <u>belong to</u> $\mathcal D_{3,2}(0,0)$ (<u>see Definition 4.1); let</u> $\ell_\vartheta,\ell_\theta,\ell_{\theta\theta},\ell_{\vartheta\vartheta},g^2\in\mathcal E_1$, $\ell_{\vartheta\vartheta},\ell_{\theta\theta}\in\tilde{\mathcal E}_1$, $\ell_\vartheta,\ell_\theta\in\tilde{\mathcal E}_2$ <u>and (7.1), (7.3) and (7.4) hold. Then</u> $\beta^*_n(\tau\eta c,\eta)$ <u>satisfies</u>

(8.8),(7.17), (7.18).

Proof. Consider the statistics

$$\Lambda(\lambda,\eta) = \sum [\ell(G_{\tau\lambda}(X_i),0,\tau\eta) - \ell(G_{\tau\lambda}(X_i),\tau\eta c,0)], \tag{9.8}$$

$$\tilde{\Lambda}(\lambda,\eta) = \sum [\ell(\tilde{G}_{\tau\lambda}(X_i),0,\tau\eta) - \ell(\tilde{G}_{\tau\lambda}(X_i),\tau\eta c,0)]. \tag{9.9}$$

Then $\Lambda(0,\eta)$ ($=\tilde{\Lambda}(0,\eta)$)is the statistic of the Neyman-Pearson test for the hypothesis $(\tau\eta c,0)$ against $(0,\tau\eta)$ and $\Lambda(\eta,\eta)$, $\tilde{\Lambda}(c\eta,\eta)$ have under $(\theta,\vartheta)=(0,0)$ the same distributions as $\Lambda(0,\eta)$ under $(0,\tau\eta)$ and $(c\tau\eta,0)$ respectively. In the sequel we shall use Taylor expansions formally discarding the remainders in $\Lambda(\lambda,\eta)$ and $\tilde{\Lambda}(\lambda,\eta)$ which would contribute $o(\tau)$ into the corresponding distributions and denoting this by $\simeq$. These operations may be justified by Lemma 4.1.

Using the notation (7.11), (5.8) and (5.9) we have

$$\begin{aligned} \ell(G_{\tau\lambda},0,\tau\eta) - \ell(G_{\tau\lambda},\tau\eta c,0) &\simeq \tau\eta\sigma g + \tau^2\eta\lambda\sigma g^{(0,1)} + \\ &+ \tfrac{1}{2}\tau^3\eta\lambda^2\sigma g^{(0,2)} + \tfrac{1}{2}\tau^2\eta^2(\ell_{\vartheta\vartheta} - c^2\ell_{\theta\theta}) + \\ &+ \tfrac{1}{2}\tau^3\eta^2\lambda(\ell^{(1)}_{\vartheta\vartheta} - c^2\ell^{(1)}_{\theta\theta}) + \tfrac{1}{6}\tau^3\eta^3(\ell_{\vartheta\vartheta\vartheta} - c^3\ell_{\theta\theta\theta}). \end{aligned} \tag{9.10}$$

Put

$$d_{\vartheta\vartheta} = E\ell_{\vartheta\vartheta}, \quad U_{n,\vartheta\vartheta} = \tau\sum(\ell_{\vartheta\vartheta}(X_i) - d_{\vartheta\vartheta}) \tag{9.11}$$

and similarly for the ℓ's with other sub- and superscripts. Using the notation $S_n^{(0)}$ from (4.10), $S_n^{(0,1)}$, $f^{(0,1)}$, $f^{(0,2)}$ from (5.8) and the equality (7.12) we obtain

$$\begin{aligned} \Lambda(\lambda,\eta) = \eta\sigma[& S_n^{(0)} + \tau\lambda S_n^{(0,1)} + \lambda\sigma + \tfrac{1}{2}\tau\lambda^2 f^{(0,2)} + \\ &+ \frac{\tau\eta}{2\sigma}(U_{n,\vartheta\vartheta} - c^2U_{n,\theta\theta}) + \frac{\eta}{2\sigma}(d_{\vartheta\vartheta} - c^2d_{\theta\theta}) + \\ &+ \frac{\tau\eta\lambda}{2\sigma}(d^{(1)}_{\vartheta\vartheta} - c^2d^{(1)}_{\theta\theta}) + \frac{\tau\eta^2}{6\sigma}(d_{\vartheta\vartheta\vartheta} - c^3d_{\theta\theta\theta})]. \end{aligned} \tag{9.12}$$

The power of the test will not be affected if we divide $\Lambda(\lambda,\eta)$ by $\eta\sigma$ and drop the terms which do not depend on the observations and on λ. Then we obtain from $\Lambda(\lambda,\eta)$ the statistic

(9.13) $$\Lambda^*(\lambda,\eta) = S_n^{(0)} + \lambda\sigma + \tau\left[\lambda S_n^{(0,1)} + \frac{1}{2}\lambda^2 b^{(0,2)} + \right.$$

$$\left. + \frac{\eta}{2\sigma}\left(U_{n,\vartheta\vartheta} - c^2 U_{n,\theta\theta}\right) + \frac{\eta\lambda}{2\sigma}\left(d_{\vartheta\vartheta}^{(1)} - c^2 d_{\theta\theta}^{(1)}\right)\right].$$

Denote the expression in the brackets in (9.13) by $H_1^*(S_n,\lambda,\eta)$ where $S_n = (S_n^{(0)}, S_n^{(0,1)}, U_{n,\vartheta\vartheta}, U_{n,\theta\theta})'$ and let

(9.14) $$Q_1^*(\lambda,\eta,y) = E[H_1^*(S,\lambda,\eta) \mid S^{(0)} = y],$$

$S = (S^{(0)}, S^{(0,1)}, U_{\vartheta\vartheta}, U_{\theta\theta})'$ being a normally distributed random vector with zero mean and the covariance matrix equal to that of S_n. Then letting $\sigma(g,h) = Egh$ we have

(9.15) $$Q_1^*(\lambda,\eta,y) = \lambda y\,\sigma(g,g^{(0,1)}) + \frac{1}{2}\lambda^2 b^{(0,2)} +$$

$$+ \frac{\eta y}{2\sigma}\left(\sigma(g,l_{\vartheta\vartheta}) - c^2\sigma(g,l_{\theta\theta})\right) + \frac{\eta\lambda}{2\sigma}\left(d_{\vartheta\vartheta}^{(1)} - c^2 d_{\theta\theta}^{(1)}\right).$$

Consider now $\tilde\Lambda(\lambda,\eta)$. Proceeding in the same way we obtain as an analogue of (9.13)

(9.16) $$\tilde\Lambda^*(\lambda,\eta) = S_n^{(0)} + \lambda\tilde b^{(0,1)} + \tau\left[\lambda\tilde S_n^{(0,1)} + \frac{1}{2}\lambda^2\tilde b^{(0,2)} + \right.$$

$$\left. + \frac{\eta}{2\sigma}\left(U_{n,\vartheta\vartheta} - c^2 U_{n,\theta\theta}\right) + \frac{\eta\lambda}{2\sigma}\left(\tilde d_{\vartheta\vartheta}^{(1)} - c^2\tilde d_{\theta\theta}^{(1)}\right)\right].$$

The assumptions $l_\vartheta, l_\theta \in \tilde{\mathcal{E}}_2$ imply $g \in \tilde{\mathcal{E}}_2$ and we have from (9.7)

(9.17) $$\tilde b^{(0,1)} = \int g(x)\, l_\theta(x)\, p(x)\, dx = 0,$$

(9.18) $$\tilde b^{(0,2)} = \int g(x)\left(\frac{\partial^2 p(x,\theta)}{\partial\theta^2}\bigg|_{\theta=0}\right) dx = b^{(2)}.$$

The last equality follows from the comparison of (7.1) and (7.2). For $\tilde{Q}_1^*$ defined similarly to (9.14) we have

$$(9.19)\quad \tilde{Q}_1^*(\lambda,\eta,y)=\lambda y\,\sigma(g,\tilde{g}^{(0,1)})+\frac{1}{2}\lambda^2 b^{(2)}+ + \frac{\eta y}{2\sigma}\left(\sigma(g,l_{\vartheta\vartheta})-c^2\sigma(g,l_{\theta\theta})\right)+\frac{\eta\lambda}{2\sigma}\left(\tilde{d}^{(1)}_{\vartheta\vartheta}-c^2\tilde{d}^{(1)}_{\theta\theta}\right).$$

Similarly to (6.24), we have for the power

$$(9.20)\quad \beta_n^*(\tau\eta c,\eta)\simeq 1-\Phi\Big(t_\alpha-\eta\sigma+\tau\Big(\frac{\mu_3}{6}(2t_\alpha\eta\sigma-\eta^2\sigma^2)+ +\tilde{Q}_1^*(\eta c,\eta,t_\alpha)-Q_1^*(\eta,\eta,t_\alpha-\eta\sigma)\Big)\Big).$$

The assumption $g^2\in\mathcal{E}_1$ and (7.4) imply (cf. (7.15))

$$(9.21)\quad \sigma(g,\tilde{g}^{(0,1)})=\frac{1}{2}Eg^2 l_\theta=-\sigma(g,g^{(1)}).$$

Now using (9.15) and (9.19) we obtain

$$(9.22)\quad \tilde{Q}_1^*(\eta c,\eta,t_\alpha)-Q_1^*(\eta,\eta,t_\alpha-\eta\sigma)=-\eta t_\alpha[\sigma(g,g^{(0,1)})+ +c\sigma(g,g^{(1)})]+\eta^2[\sigma\sigma(g,g^{(0,1)})-\frac{1}{2}b^{(0,2)}+\frac{1}{2}c^2 b^{(2)}+R],$$

$$(9.23)\quad R=\frac{c}{2\sigma}\left(\tilde{d}^{(1)}_{\vartheta\vartheta}-c^2\tilde{d}^{(1)}_{\theta\theta}\right)-\frac{1}{2\sigma}\left(d^{(1)}_{\vartheta\vartheta}-c^2 d^{(1)}_{\theta\theta}\right)+ +\frac{1}{2}\left(\sigma(g,l_{\vartheta\vartheta})-c^2\sigma(g,l_{\theta\theta})\right).$$

The assumptions $l_{\vartheta\vartheta},\ l_{\theta\theta}\in\mathcal{E}_1,\ l_{\vartheta\vartheta},l_{\theta\theta}\in\tilde{\mathcal{E}}_1$ imply

$$\tilde{d}^{(1)}_{\vartheta\vartheta}=El_{\vartheta\vartheta}l_\theta,\ \tilde{d}^{(1)}_{\theta\theta}=El_{\theta\theta}l_\theta,\ d^{(1)}_{\vartheta\vartheta}=El_{\vartheta\vartheta}l_\vartheta,\ d^{(1)}_{\theta\theta}=El_{\theta\theta}l_\vartheta.$$

Using these relations and (7.11) in (9.23) we obtain $R=0$. Since $\Sigma_{01}=\sigma(g,g^{(1)})$ for $s=1$ and $\sigma_{0,2s+1}=\sigma(g,g^{(0,1)})$, (9.22) is equal to (7.18). Thus the theorem is proved.

References

[1] D.M.Chibisov: On the normal approximation for a certain class of statistics, Proc. 6-th Berkeley Symp. Math. Statist. and Prob., 1972, vol.1, 153-174.

[2] D.M.Chibisov: Asymptotic expansions for the maximum likelihood estimate, Teor. Verojatnost. i Primenen., 17, 2 (1972), 387-388.

[3] D.M.Chibisov: Asymptotic expansions for some test statistics for testing composite hypotheses, Teor. Verojatnost. i Primenen., 17, 3 (1972), 600-602.

[4] D.M.Chibisov: An asymptotic expansion for the distribution of a statistic admitting an asymptotic expansion, Teor. Verojatnost. i Primenen., 17, 4 (1972), 658-668.

[5] K.Michel, J. Pfanzagl: The accuracy of the normal approximation for minimum contrast estimates, Z. Wahrscheinlichkeitstheorie verw. Geb., 18 (1970), 73-84.

[6] P.A.P.Moran: On asymptotically optimal tests of composite hypotheses, Biometrika, 57, 1 (1970), 47-55.

[7] S.V.Nagaev: Some limit theorems for large deviations, Teor. Verojatnost. i Primenen., 10, 2 (1965), 231-254.

[8] J.Neyman: Optimal asymptotic tests of composite statistical hypotheses, Probability and statistics, Uppsala, Almquist and

Wiksells, (The Harald Cramér Volume), 1959, 213-234.

[9] V.V.Yurinsky: Bounds for characteristic functions of certain degenerate multidimensional distributions, Teor. Verojatnost. i Primenen., 17, 1 (1972), 99-110.

[10] K.G.Binmore, H.H.Stratton: A note on characteristic functions, Ann. Math. Statist., 40 (1969), 303-307.

[11] R.P.Boas: Lipschitz behaviour and integrability of characteristic functions, Ann. Math. Statist., 38 (1967), 32-36.

[12] M.Loève: Probability Theory, Van Nostrand, 1960.

Steklov Mathematical Institute
of the Academy of Sciences of the USSR
Moscow

ON THE GENERATION OF MARKOV PROCESSES BY SYMMETRIC FORMS

Masatoshi Fukushima

§1. Introduction

The advance of the theory of Markov processes during the past 15 years was principally led by the notion of the strong Markov property. Among others, Hunt built up a probabilistic potential theory based on a certain strong Markov process now called a Hunt process (cf. R.M. Blumenthal and R.K. Getoor [2]) and Dynkin established the theory of transformations of a slightly more general strong Markov process named a standard process (cf. E.B. Dynkin [5]).

As compared with such an intensive development of the theory, the existence theorems of strong Markov processes are not rich enough. In this article we will give a method of producing a wide class of symmetric Hunt processes from some concrete analytic data.

In order to illustrate our method, take a look at a formally self-adjoint second order elliptic partial differential operator

$$(1.1) \qquad Au(x) = \sum_{i,j=1}^{N} \frac{\partial}{\partial x_i} \left(a_{ij}(x)\frac{\partial u(x)}{\partial x_j}\right) - c(x)u(x)$$

acting on functions u defined on an Euclidean domain $D \subset R^N$. Kolmogorov first discovered that the transition function of a given Markov process satisfies under certain regularity conditions a parabolic differential equation, which is of the form $\frac{\partial u(t,x)}{\partial t} = Au(t,x)$ in the symmetric case.

Given conversely coefficients $\{a_{ij}, c\}$, there have been two ways of constructing a strong Markov process whose transition function satisfies the given parabolic equation:

1°. an analytical method relying essentially upon the <u>theory of partial differential equations</u>,

2°. a probabilistic method of solving Ito's <u>stochastic differential equations</u>.

In the following, we will present another method:

3°. an analytical method relying essentially upon an <u>analytic potential theory</u>.

Instead of thinking about the elliptic operator (1.1) directly, we consider the bilinear form

$$(1.2) \qquad \varepsilon(u, v) = (-Au, v)$$

$$= \int_D \sum_{i,j=1}^{N} \frac{\partial u}{\partial x_i} \frac{\partial v}{\partial x_j} a_{ij}(x)dx + \int_D u(x)v(x)c(x)dx$$

defined for $u, v \in C_0^\infty(D)$, $C_0^\infty(D)$ being the space of all infinitely differentiable functions with compact supports in D. Much more generally, we start with the integro-differential expression of the type

$$(1.3) \qquad \varepsilon(u, v) = \sum_{i,j=1}^{N} \int_D \frac{\partial u}{\partial x_i} \frac{\partial v}{\partial x_j} \nu_{ij}(dx)$$

$$+ \int_{D \times D} (u(x)-u(y))(v(x)-v(y))\Phi(dx,dy) + \int_D u(x)v(x)n(dx),$$

$u, v \in C_0^\infty(D)$, and regard this as a symmetric bilinear form on the space $L^2(D\, ;\, m)$ based on another measure m.

We formulate our problem as follows; given a set of measures

$\{\nu_{ij}, \Phi, n, m\}$, how can we produce a strong Markov process on D governed by the form (1.3) in a certain sense? Our procedure consists of two steps. The first step is, given $\{\nu_{ij}, \Phi, n, m\}$, to construct a regular Dirichlet form (§2, 3, 6). The second step is, given a regular Dirichlet form, to construct a Hunt process (§4, 5). In 1959, A. Beurling and J. Deny [1] introduced for the first time the axioms of a Dirichlet space and developed an associated potential theory. It is this axiomatic potential theory that we are going to use in the second step.

The important notions in the above theory are, among others, the capacity of sets and the quasi-continuity of functions. Given a regular Dirichlet form, there corresponds uniquely a Markov semigroup $\{T_t\}$ on L^2-space (§3). By virtue of the potential theory, the functions $T_t u$ for sufficiently many u have the following properties : each $T_t u$ has a quasi-continuous modification and $\lim_{t_n \to 0} T_{t_n} u = u$ except on a set of zero capacity.

Accordingly we encounter in the second step a semigroup which is not very much worse than a strongly continuous Feller semigroup. Going along a similar line as in Blumenthal-Getoor [2 ; pp 46-50] but ignoring successively the sets of capacity zero on which things might go wrong, we can finally get a Hunt process outside some Borel set of capacity zero (§5). We regard two such Hunt processes as equivalent if they obey the same law outside some Borel set of capacity zero. Our Hunt process to be constructed should be considered as a representative of an equivalence class.

By allowing the state space to narrow this way, we are able to obtain a considerably wider class of symmetric Hunt processes than we ever know. We can observe this in §6 by several examples.

The idea for the first step appeared already in [9 ; Appendix], but we will exploit the method further by introducing the notion of Markov symmetric forms (§2).

The second step has been carried out in [10] for the first time but by taking a rather indirect course. Quite recently M. Silverstein [16] pointed out that a much more direct course is available. Our present purely potential theoretic approach is based on Silverstein's idea, although ours is different in many technical points.

It is conjectured that, if the measure Φ vanishes identically in the expression (1.3), then the constructed process is a diffusion : almost all sample paths are continuous. This condition for the form ε is closely related to its local property already defined by Beurling-Deny [1]. In this connection, we refer the reader to the article by N. Ikeda and S. Watanabe [11].

The third case (1°.c) in Example 1 of §6 is due to K. Sato who permits me to mention it in this paper.

§2. Markov symmetric forms

Let X be a locally compact separable Hausdorff space and m be a positive Radon measure on X. A symmetric form ε on the real L^2-space $L^2(X ; m)$ is, by definition, a non-negative definite symmetric bilinear form defined on $\mathcal{D}[\varepsilon] \times \mathcal{D}[\varepsilon]$, $\mathcal{D}[\varepsilon]$ being a dense linear subspace of $L^2(X ; m)$.

Given a symmetric form ε, we say that every unit contraction operates on ε if, for any $u \in \mathcal{D}[\varepsilon]$, the function $v = (0 \vee u) \wedge 1$ is again in $\mathcal{D}[\varepsilon]$ and $\varepsilon(v, v) \leqq \varepsilon(u, u)$. We now define a notion of Markovity of ε which is more useful and general than the above one. They are equivalent however when ε is a closed form in the sense that will be specified later.

A symmetric form ε is called Markov if, for any $\delta > 0$, there exists a non-decreasing function $\phi_\delta(t)$, $-\infty < t < \infty$, satisfying the following conditions.

(2.1) $\phi_\delta(t) = t$ for $0 \leqq t \leqq 1$. Further $|\phi_\delta(t)| \leqq t$ and $-\delta \leqq \phi_\delta(t) \leqq 1 + \delta$ for all t.

(2.2) If a function u belongs to $\mathcal{D}[\varepsilon]$, then so is the composite function $\phi_\delta(u)$. Moreover $\varepsilon(\phi_\delta(u), \phi_\delta(u)) \leqq \varepsilon(u, u)$.

We are particularily interested in the following examples. Let D be a domain of the N-space R^N.

Example 1. The form

(2.3) $$\varepsilon(u, v) = \int_D \sum_{i,j=1}^N \frac{\partial u(x)}{\partial x_i} \frac{\partial v(x)}{\partial x_j} a_{ij}(x)dx + \int_D \int_D (u(x) - u(y))(v(x) - v(y))\Phi(dx, dy) + \int_D u(x)v(x)n(dx)$$

(2.4) $\mathcal{D}[\varepsilon] = C_0^\infty(D)$

is a Markov symmetric form on $L^2(D ; m)$. Here $a_{ij}(x)$, $1 \leqq i,j \leqq N$, $x \in D$, is a symmetric non-negative definite matrix in i,j, and a locally integrable function in x. Φ is a positive symmetric measure on $D \times D$ such that $\int_K \int_K |x - y|^2 \Phi(dx, dy) < \infty$ for any compact set $K \subset D$. m and

n are positive Radon measures on D, m being everywhere dense. In order to verify the Markovity of ε, it suffices to take a C^∞-function $\phi_\delta(t)$ satisfying not only (2.1) but also the property $0 \leq \phi_\delta'(t) \leq 1$ for all t.

The next example is a case that the measure ν_{ij} in the expression (1.3) is not necessarily absolutely continuous with respect to the Lebesgue measure.

Example 2. On R^2, we define

$$(2.5) \quad \varepsilon(u, v) = \int_{-\infty}^{\infty}\int_{-\infty}^{\infty} \frac{\partial u(x)}{\partial x_1} \frac{\partial v(x)}{\partial x_1} dx_1 \mu(dx_2) + \int_{-\infty}^{\infty}\int_{-\infty}^{\infty} \frac{\partial u(x)}{\partial x_2} \frac{\partial v(x)}{\partial x_2} \nu(dx_1)dx_2$$

$$(2.6) \quad \mathcal{D}[\varepsilon] = C_0^\infty(R^2),$$

where μ and ν are positive Radon measures on R^1. This is a Markov symmetric form on $L^2(R^2) = L^2(R^2\ ;\ dx)$. The proof is the same as in Example 1.

Example 3. The form

$$(2.7) \quad \varepsilon(u, v) = \frac{1}{2}\int_D \sum_{i=1}^N \frac{\partial u(x)}{\partial x_i} \frac{\partial v(x)}{\partial x_i} dx$$

with $\mathcal{D}[\varepsilon]$ given by (2.4) is a Markov symmetric form on $L^2(D) = L^2(D\ ;\ dx)$. This is a very special case of Example 1. But we can get Markov symmetric forms on $L^2(D)$ of quite different characters if other domains $\mathcal{D}[\varepsilon]$ rather than (2.4) are adopted. For instance

(2.8) $\mathcal{D}[\varepsilon] = \hat{C}^{\infty}(D)$

(2.9) $\mathcal{D}[\varepsilon] = H^1(D)$,

where $\hat{C}^{\infty}(D)$ is the restrictions to D of functions in $C_0^{\infty}(R^N)$ and $H^1(D)$ is the space of those functions of $L^2(D)$ whose distribution derivatives are also in $L^2(D)$. The Markovity of the form (2.7) with the domain (2.8) or (2.9) is verified in the same way as before.

§ 3. Generation of Dirichlet forms by a Markov one

A symmetric form ε on $L^2(X ; m)$ is called closed if $\mathcal{D}[\varepsilon]$ is complete with metric $\sqrt{\varepsilon(u, u) + (u, u)}$, where (,) denotes the L^2-inner product. The space $\mathcal{D}[\varepsilon]$ is then a real Hilbert space with the inner product

(3.1) $\varepsilon_{\alpha}(u, v) = \varepsilon(u, v) + \alpha(u, v)$

for each $\alpha > 0$.

Let B be a non-negative definite self-adjoint linear operator on $L^2(X ; m)$ and $\{E_{\lambda}\}$ be the associated spectral family. For any non-negative continuous function $\phi(t)$ on $[0, \infty)$, the operator $\phi(B)$ defined by $\mathcal{D}(\phi(B)) = \{u \in L^2 ; \int_0^{\infty} \phi(\lambda)^2 d(E_{\lambda}u, u) < +\infty\}$, $(\phi(B)u, v) = \int_0^{\infty} \phi(\lambda) d(E_{\lambda}u, v)$, $u \in \mathcal{D}(\phi(B))$, $v \in L^2$, is again a non-negative definite self-adjoint operator. Now we proceed to

Theorem 3.1. All closed symmetric forms ε on $L^2(X ; m)$ and all-non-negative definite self-adjoint operators $-A$ on

$L^2(X ; m)$ stand in one to one correspondence. The correspondence is given by

(3.2) $\quad \varepsilon(u, v) = (\sqrt{-A}\, u, \sqrt{-A}\, v)$

(3.3) $\quad \mathcal{D}[\varepsilon] = \mathcal{D}(\sqrt{-A})$.

Let $-A$ be a non-negative definite self-adjoint operator, then so is $\sqrt{-A}$. Consequently $\sqrt{-A}$ is a closed linear operator, which means that the symmetric form ε defined by (3.2) and (3.3) is closed, proving the half of Theorem 3.1.

Before completing the proof of Theorem 3.1, we will give two lemmas. A family $\{G_\alpha, \alpha > 0\}$ of symmetric operators on $L^2(X ; m)$ is called a symmetric strong continuous contraction resolvent on L^2 if $G_\alpha - G_\beta + (\alpha - \beta)G_\alpha G_\beta = 0$, $(\alpha G_\alpha u, \alpha G_\alpha u) \leqq 1$, $(\alpha G_\alpha u - u, \alpha G_\alpha u - u) \longrightarrow 0$, $\alpha \longrightarrow \infty$, $u \in L^2$. Its generator A is defined by $A = \alpha I - G_\alpha^{-1}$, $\mathcal{D}(A) = \mathcal{R}(G_\alpha)$. A family $\{T_t, t > 0\}$ of symmetric operators on $L^2(X ; m)$ is called a symmetric strongly continuous contraction semigroup on L^2 if $T_t T_s = T_{t+s}$, $(T_t u, T_t u) \leqq 1$, $(T_t u - u, T_t u - u) \longrightarrow 0$, $t \longrightarrow 0$, $u \in L^2$. Its generator A is defined by $Au = \lim_{t \downarrow 0} \frac{T_t u - u}{t}$, $\mathcal{D}(A) = \{u \in L^2 ; Au \in L^2\}$. In this case, A coincides with the generator of the resolvent $G_\alpha = \int_0^\infty e^{-\alpha t}\, T_t\, dt$.

Lemma 3.1. (i) For a given symmetric strongly continuous contraction semigroup or resolvent on L^2, let us denote its generator by A. Then $-A$ is a non-negative definite self-adjoint operator. (ii) Conversely, given a non-negative definite self-adjoint operator $-A$ on L^2, then $T_t = \exp(tA)$ (resp. $G_\alpha = (\alpha - A)^{-1}$) becomes a symmetric strongly continuous contraction semigroup (resp. resolvent) on L^2 whose generator coincides with the given A.

We only note the following : given any symmetric strongly continuous contraction resolvent $\{G_\alpha, \alpha > 0\}$, then $\frac{d}{d\alpha}(G_\alpha u, u) \leqq 0$ and $\lim_{\alpha\to\infty}(G_\alpha u, u) = 0$, $u \in L^2$. Hence G_α is non-negative definite. Let A be the generator of G_α. A is then self-adjoint and $-A$ is non-negative definite because $(-Au, u) + (\alpha u, u) = (G_\alpha^{-1}u, u) \geqq 0$, $u \in \mathcal{D}(A)$, for any $\alpha > 0$.

Lemma 3.2. Let $-A$ be a non-negative definite self-adjoint operator on L^2, ε be the closed symmetric form generated by $-A$ according to (3.2) and (3.3) and finally T_t (resp. G_α) be the semigroup (resp. resolvent) generated by A according to Lemma 3.1 (ii). Then

(i) $T_t(L^2) \subset \mathcal{D}[\varepsilon]$,

$$\varepsilon(T_t u, T_t u) \leqq \frac{1}{2t}((u, u) - (T_t u, T_t u)), \quad u \in L^2,$$

(ii) $G_\alpha(L^2) \subset \mathcal{D}[\varepsilon]$. For any fixed $u \in L^2$,

$$\varepsilon_\alpha(G_\alpha u, v) = (u, v), \quad v \in \mathcal{D}[\varepsilon].$$

(iii) For any $u \in \mathcal{D}[\varepsilon]$, $T_t u \longrightarrow u$ and

$$\frac{1}{t}(G_1 u - e^{-t}G_1 T_t u) \longrightarrow u \text{ as } t \longrightarrow 0 \text{ in } \varepsilon_1\text{-norm}.$$

Proof. All statements can be obtained simply by using the spectral family $\{E_\lambda\}$ associated with $-A$. For instance, integrating the inequality $\lambda e^{-2t\lambda} \leqq \frac{1}{2t}(1 - e^{-2t\lambda})$ with the measure $d(E_\lambda u, u)$, we arrive at (i).

We now return to the proof of the remaining half of Theorem 3.1. Given a symmetric form ε, there exists, for each $\alpha > 0$ and $u \in L^2$, a unique element $G_\alpha u \in \mathcal{D}[\varepsilon]$ such as $\varepsilon_\alpha(G_\alpha u, v) = (u, v)$, $v \in \mathcal{D}[\varepsilon]$, in view of Riesz representation theorem. It is quite easy to see that this $\{G_\alpha, \alpha > 0\}$ is a symmetric contraction resolvent on L^2. For $u \in \mathcal{D}[\varepsilon]$, $\beta(\beta G_\beta u - u, \beta G_\beta u - u) \leqq \varepsilon_\beta(\beta G_\beta u - u, \beta G_\beta u - u) = \beta^2(G_\beta u, u) - \beta(u, u) + \varepsilon(u, u) \leqq \varepsilon(u, u)$, and hence $(\beta G_\beta u - u, \beta G_\beta u - u) \leqq \frac{1}{\beta}\varepsilon(u, u) \longrightarrow 0$, $\beta \longrightarrow \infty$. Since $\mathcal{D}[\varepsilon]$ is dense in L^2 and $\{G_\alpha\}$ is contraction, we can see that $\{G_\alpha\}$ is also strongly continuous.

Let A be the generator of $\{G_\alpha\}$. $-A$ is then non-negative definite and self-adjoint by Lemma 3.1 (i). Let ε' be the closed symmetric form generated by $-A$ according to (3.2) and (3.3). By virtue of Lemma 3.2 (ii), $G_\alpha(L^2) \subset \mathcal{D}[\varepsilon']$ and $\varepsilon'_\alpha(G_\alpha u, G_\alpha u) = (G_\alpha u, u)$. Therefore $\varepsilon'_\alpha = \varepsilon_\alpha$ on the space $G_\alpha(L^2)$, which is however dense in $\mathcal{D}[\varepsilon]$ (resp. $\mathcal{D}[\varepsilon']$) with respect to the metric ε_α (resp. ε'_α), getting $\varepsilon' = \varepsilon$. Any closed symmetric form is thus generated by a non-negative definite self-adjoint operator, completing the proof of Theorem 3.1.

A bounded linear operator S on $L^2(X ; m)$ is called <u>Markov</u> if $0 \leqq Su \leqq 1$ m - a.e. whenever $u \in L^2$ and $0 \leqq u \leqq 1$ m - a.e.

Suppose that a symmetric form ε has the following property :

if $u \in \mathcal{D}[\varepsilon]$ and $v \in L^2$ are such that there exist their Borel modifications $\tilde{u}$ and $\tilde{v}$ satisfying the inequalities $|\tilde{v}(x)| \leqq |\tilde{u}(x)|$, $|\tilde{v}(x) - \tilde{v}(y)| \leqq |\tilde{u}(x) - \tilde{u}(y)|$ for every $x, y \in X$, then $v \in \mathcal{D}[\varepsilon]$ and $\varepsilon(v, v) \leqq \varepsilon(u, u)$. In this case we say that <u>every normal contraction operates on</u> ε.

<u>Theorem 3.2. Let</u> ε <u>be a closed symmetric form and</u> $\{T_t, t > 0\}$, $\{G_\alpha, \alpha > 0\}$ <u>be the associated semigroup and resolvent according to Theorem</u> 3.1 <u>and Lemma</u> 3.1. <u>Then the following five conditions are equivalent to each other</u>:

(a) ε <u>is Markov</u>,

(b) T_t <u>is Markov for each</u> $t > 0$,

(c) αG_α <u>is Markov for each</u> $\alpha > 0$,

(d) <u>every unit contraction operates on</u> ε,

(e) <u>every normal contraction operates on</u> ε.

<u>Proof</u>. (a) $\Longrightarrow$ (c). Take any function $u \in L^2$ such as $0 \leqq u \leqq 1$ m - a.e. For any $\delta > 0$ and $\alpha > 0$, let $\phi_\delta(t)$ be the function in the definition of the Markovity of ε (§2) and put $\phi_{\frac{1}{\alpha},\delta}(t) = \frac{1}{\alpha}\phi_{\alpha\delta}(\alpha t)$. Since $G_\alpha u \in \mathcal{D}[\varepsilon]$, the composite function $w = \phi_{\frac{1}{\alpha},\delta}(G_\alpha u)$ also belongs to $\mathcal{D}[\varepsilon]$, and $\varepsilon(w, w) \leqq \varepsilon(G_\alpha u, G_\alpha u)$.

Define a quadratic form $\Psi = \Psi_{\alpha,u}$ on $\mathcal{D}[\varepsilon]$ by

$$(3.4) \quad \Psi(v) = \varepsilon(v, v) + \alpha\left(v - \frac{u}{\alpha}, v - \frac{u}{\alpha}\right), \quad v \in \mathcal{D}[\varepsilon],$$

then

(3.5) $\Psi(G_\alpha u) + \varepsilon_\alpha(G_\alpha u - v, G_\alpha u - v) = \Psi(v)$,

namely, $G_\alpha u$ is a unique element in $\mathcal{D}[\varepsilon]$ minimizing the quadratic form Ψ. However it is easy to see that $(w - \frac{u}{\alpha}, w - \frac{u}{\alpha}) \leqq (G_\alpha u - \frac{u}{\alpha}, G_\alpha u - \frac{u}{\alpha})$. Hence $\Psi(w) \leqq \Psi(G_\alpha u)$, from which follows $G_\alpha u = w$. Therefore $-\delta \leqq G_\alpha u \leqq \frac{1}{\alpha} + \delta$ for any $\delta > 0$, proving the Markovity of αG_α.

(c) $\Longrightarrow$ (b). This follows from

$$T_t u = \lim_{\beta\to\infty} e^{-t\beta} \sum_{n\geqq 0} \frac{(t\beta)^n}{n!} (\beta G_\beta)^n u.$$

The implication (b) $\Longrightarrow$ (c) and (e) $\Longrightarrow$ (d) $\Longrightarrow$ (a) are trivial. As for the proof of the implication (c) $\Longrightarrow$ (e), we refer to J.Deny [3 ; pp 155].

Turning to the main task of this section, let us introduce an important notion.

We call a symmetric form ε <u>closable</u> if $\varepsilon(u_n, u_n) \longrightarrow 0$ whenever $u_n \in \mathcal{D}[\varepsilon]$ satisfies $\varepsilon(u_n - u_m, u_n - u_m) \longrightarrow 0$ and $(u_n, u_n) \longrightarrow 0$.

Given a symmetric form ε, any closed symmetric form $\tilde{\varepsilon}$ is called a <u>closed extension</u> of ε if $\mathcal{D}[\tilde{\varepsilon}] \supset \mathcal{D}[\varepsilon]$ and $\tilde{\varepsilon} = \varepsilon$ on $\mathcal{D}[\varepsilon] \times \mathcal{D}[\varepsilon]$.

The closability of a symmetric form is a necessary and sufficient condition for it to admit at least one closed extension. For a closable symmetric form ε, its smallest closed extension $\bar{\varepsilon}$ can be defined as follows: the domain $\mathcal{D}[\bar{\varepsilon}]$ of $\bar{\varepsilon}$ is just the abstract completion of $\mathcal{D}[\varepsilon]$ by means of the metric $\sqrt{\varepsilon(u, u) + (u, u)}$.

Theorem 3.3. If a symmetric form ε is Markov and closable, then its smallest closed extension $\overline{\varepsilon}$ is also Markov.

Proof. Let $\{G_\alpha, \alpha > 0\}$ be the resolvent associated with the closed symmetric form $\overline{\varepsilon}$ according to Theorem 3.1 and Lemma 3.1. On account of Theorem 3.2, it suffices to show that αG_α is Markov. Take any function $u \in L^2$ such as $0 \leqq u \leqq 1$ m - a.e. Then by making use of the identity (3.5) and following essentially the same line as in [9 ; Appendix], we can get $-\delta \leqq G_\alpha u \leqq \frac{1}{\alpha} + \delta$ m - a.e. for any $\delta > 0$.

A closed Markov symmetric form is called a Dirichlet form. Theorem 3.3 provides us with a method of generating a Dirichlet form starting with a form of the type in the preceding examples. Once we get a Dirichlet form, then we have a symmetric Markov semigroup on L^2 by virtue of Theorem 3.2. We will assert in § 5 that we can even get a Hunt process provided that the Dirichlet form is regular. In the final section, the examples of § 2 will be examined to see whether they generate regular Dirichlet forms.

Incidentally we mention some more about closed extensions. Suppose that a symmetric form ε satisfies the following:

(3.6) if $u_n \in \mathcal{D}[\varepsilon]$ converges to zero in L^2, then $\varepsilon(u_n, v) \longrightarrow 0$ for any $v \in \mathcal{D}[\varepsilon]$.

Then ε is readily seen to be closable. In particular this criterion applies to the case when a symmetric form is expressible by some symmetric operator.

Assume that S is a symmetric linear operator densely defined on $L^2(X ; m)$ such as $(-Su, u) \geqq 0$ for all $u \in \mathcal{D}(S)$. Then

(3.7) $\varepsilon_S(u, v) = (-Su, v)$, $\mathcal{D}[\varepsilon_S] = \mathcal{D}(S)$, is a closable symmetric form. Let $-A_F$ be the non-negative definite self-adjoint operator associated with the smallest closed extension of ε_S. A_F turns out to be a self-adjoint extension of S. A_F is called <u>Friedrichs extension</u> of S.

To any non-negative definite self-adjoint extension $-A$ of $-S$, there corresponds a closed symmetric extension ε_A of ε_S. Among them, there is one, say A_K, which is maximum in the sense that $\mathcal{D}[\varepsilon_{A_K}] \supset \mathcal{D}[\varepsilon_A]$, $\varepsilon_{A_K}(u, u) \leq \varepsilon_A(u, u)$, $u \in \mathcal{D}[\varepsilon_A]$, for every A's. We call A_K <u>Krein extension</u> of S ([13]).

Suppose that the given ε_S is Markov. Theorem 3.2 and 3.3 tell us that A_F then generates a Markov semigroup. However the same statement does not hold for A_K in general. In a sense ε_{A_K} is too big to be Markov. As for a description of all possible closed Markov extensions ε_A of ε_S, see the papers by the auther [8] and by J. Elliott [6]. Among those extensions, there is the maximum one, which is related to the <u>reflecting barrier Markov process</u> (c.f. § 6, Example 3).

§ 4. Potential theoretic preparations

From now on, we assume that $m(A) > 0$ for any non-empty open set $A \subset X$. Let us consider a Dirichlet form ε on $L^2(X ; m)$ which is <u>regular</u> in the following sense : the space $\mathcal{D}[\varepsilon] \cap C(X)$ is dense both in $\mathcal{D}[\varepsilon]$ with metric ε_1 and in $C(X)$ with the uniform norm. Here $C(X)$ is the space of all continuous functions on X <u>vanishing at infinity</u>.

Denote by $\mathcal{O}$ the class of all open subsets of X. (1-)<u>capacity</u> of a set $A \in \mathcal{O}$ is defined by

$$(4.1)\qquad \mathrm{Cap}(A) = \begin{cases} \inf_{u\in\mathcal{L}_A} \varepsilon_1(u, u) & \mathcal{L}_A \neq \phi \\ \infty & \mathcal{L}_A = \phi, \end{cases}$$

where $\mathcal{L}_A = \{u \in \mathcal{D}[\varepsilon]\ ;\ u \geq 1\ \ m\text{ - a.e. on } A\}$. The capacity of any subset $B \subset X$ is defined by $\mathrm{Cap}(B) = \inf_{B\subset A,\ A\in\mathcal{L}_A} \mathrm{Cap}(A)$. This gives rise to a strongly subadditive Choquet capacity [10 ; Theorem 1.1].

Let us put $\Theta_0 = \{A \in \Theta\ ;\ \mathcal{L}_A \neq \phi\}$. For $A \in \Theta_0$, there is a unique element $e_A \in \mathcal{D}[\varepsilon]$ called the (1-)<u>equilibrium potential</u> which minimizes $\varepsilon_1(u, u)$ on $\mathcal{L}_A[\varepsilon]$. It has the following properties.

(4.2) $0 \leq e_A \leq 1$ m - a.e. on X

$e_A = 1$ m - a.e. on A.

(4.3) $\varepsilon_1(e_A, v) \geq 0$ for any $v \in \mathcal{D}[\varepsilon]$ such as $v \geq 0$ m - a.e. on A.

(4.4) $e^{-t}T_t e_A \leq e_A$ m - a.e. on X.

Here T_t is the Markov semigroup associated with ε. In fact, for any $v \in L^2$ such as $v \geq 0$ m - a.e., $(e_A - e^{-t}\, T_t e_A,\ v) = \varepsilon_1(e_A - e^{-t}\, T_t e_A,\ G_1 v) = \varepsilon_1(e_A,\ G_1 v - e^{-t}\, T_t G_1 v)$ which is non-negative in view of (4.3) and the inequality $G_1 v \geq e^{-t}\, T_t G_1 v$.

(4.5) If $A, B \in \Theta_0$ and $A \subset B$, then $e_A \leq e_B$ m - a.e.

This follows from $\varepsilon_1(e_A - e_A \wedge e_B,\ e_A - e_A \wedge e_B)$
$= \varepsilon_1(-e_A \wedge e_B,\ (e_A - e_B)^+)$

$= \varepsilon_1((e_A - e_B)^-, (e_A - e_B)^+) - \varepsilon_1(e_B, (e_A - e_B)^+) \leq 0.$

Now let us introduce several notions. A set A is called <u>almost polar</u> if $Cap(A) = 0$. "<u>Quasi-everywhere</u>" or "<u>q.e.</u>" means "except on an almost polar set". Denote by $X \cup \partial$ the one-point compactification of X. ∂ is adjoined as an isolated point if X is compact already. A function u defined q.e. on X is called <u>quasi-continuous</u> (<u>in a restricted sense</u>) if, for any $\delta > 0$, there is an open set G with $Cap(G) < \delta$ such that the restriction of u to $X - G$ is continuous and continuously extendable to $X \cup \partial - G$ by setting $u(\partial) = 0$.

An increasing family $\{F_k\}$ of closed sets such as $Cap(X - F_k) \longrightarrow 0$ is called a <u>nest</u>. A closed set F is said to be <u>m-regular</u> if $m(U(x) \cap F) \neq 0$ for any $x \in F$ and any its neighbourhood $U(x)$. A <u>regular nest</u> $\{F_k\}$ is a nest such that each F_k is m-regular.

The notion of m-regularity has been introduced in [10]. See [10 ; pp 198-199] for the proof of the next theorem.

<u>Theorem 4.1</u>. (i) <u>Let</u> Q <u>be a countable family of quasi-continuous functions</u>. <u>Then there exists a nest</u> $\{F_k\}$ <u>such that</u> $Q \subset C(\{F_k\})$, <u>where</u>

(4.6) $C(\{F_k\}) = \{u$; <u>the restriction of</u> u <u>to each</u> F_k <u>is continuous and continuously extendable to</u> $F_k \cup \partial$ <u>by setting</u> $u(\partial) = 0\}$.

(ii) <u>Let</u> $\{F_k\}$ <u>be a nest</u>. <u>Then</u> $F'_k = \{x \in F_k$; $m(F_k \cap U(x)) \neq 0$ <u>for any neighbourhood</u> $U(x)$ <u>of</u> $x\}$ <u>defines a regular nest</u> $\{F'_k\}$.

(iii) <u>Let</u> $\{F_k\}$ <u>be a regular nest</u>. <u>If</u> $u \in C(\{F_k\})$ <u>and</u> $u \geq 0$

m - a.e., then $u(x) \geqq 0$ for every $x \in \bigcup_{k=1}^{\infty} F_k$.

This theorem particularily implies the following : if a quasi-continuous function u is non-negative m - a.e., then $u \geqq 0$ q.e. We note that this statement can be localized to any open set (see [10 ; Theorem 1.2]).

The next theorem due to Deny is fundamental for regular Dirichlet forms (see [4], [10]).

Theorem 4.2. (i) Any function $u \in \mathcal{D}[\varepsilon]$ admits a quasi-continuous version $\tilde{u}$: $\tilde{u}$ is quasi-continuous and $\tilde{u} = u$ m - a.e.
(ii) If quasi-continuous functions $\tilde{u}_n \in \mathcal{D}[\varepsilon]$ form a Cauchy sequence with respect to the norm ε_1, then there is a subsequence n_k such that $\tilde{u}_{n_k}$ converges q.e. to a quasi-continuous function $\tilde{u}$. Furthermore u_n converges to $\tilde{u}$ in ε_1-norm.
(iii) If functions $u_n \in \mathcal{D}[\varepsilon]$ form a Cauchy sequence in norm ε_1 and if their suitable quasi-continuous versions $\tilde{u}_n$ converges to a function $\tilde{u}$ q.e., then $\tilde{u}$ is quasi-continuous and u_n converges to $\tilde{u}$ in ε_1-norm.

§ 5. Generation of a Hunt process by a regular Dirichlet form

We follow Blumenthal-Getoor [2 ; Chap. I] about the definitions of a Markov process and a Hunt process except that we allow the state space of a Hunt process to be an arbitrary Borel subset of X. See P.A. Meyer [15 ; Chap. XIV] where the definition of a Hunt process is relaxed in this respect.

For a Borel set $A \subset X$, we denote by $\mathcal{B}(A)$ the topological σ-field of subsets of A and by $\underset{\sim}{B}(A)$ the space of all bounded $\mathcal{B}(A)$-measurable functions on A. $\mathcal{B}(X)$ and $\underset{\sim}{B}(X)$ are simply denoted by $\mathcal{B}$ and $\underset{\sim}{B}$ respectively.

Suppose that we are given a regular Dirichlet form ε on $L^2(X\ ;\ m)$ and a Markov process $\underset{\sim}{M} = (\Omega, \mathcal{M}, \mathcal{M}_t, X_t, \theta_t, P_x)$ with state space $(Y, \mathcal{B}(Y))$, Y being some Borel subset of X. We adjoin a "death" point ∂ to Y regarding $Y \cup \partial$ as the topological subspace of the one-point compactification $X \cup \partial$ of X. The Markov semigroup on $L^2(X\ ;\ m)$ generated by the form ε will be denoted by $\{T_t, t > 0\}$, while the transition semigroup of the process $\underset{\sim}{M}$ will be denoted by $\{P_t, t > 0\}$: $P_t f(x) = E_x(f(X_t))$, $x \in Y$, $f \in \underset{\sim}{B}(Y)$. Let us agree to say that <u>the Markov process $\underset{\sim}{M}$ is properly associated with the Dirichlet form</u> ε if

(5.1) $\quad \mathrm{Cap}(X - Y) = 0,$

(5.2) $\quad P_t f$ is a quasi-continuous version of $T_t f$ for each $f \in L^2 \cap \underset{\sim}{B}$ and $t > 0$.

Our main theorem is the following.

<u>Theorem 5.1</u>. <u>For any regular Dirichlet form</u> ε <u>on</u> $L^2(X\ ;\ m)$, <u>there exists a Hunt process properly associated with</u> ε.

This theorem can be reduced to the next proposition. For a Borel set $Y \subset X$ and a nest $\{F_k\}$ such as $\bigcup_{k=1}^{\infty} F_k \supset Y$, let $C(\{F_k\}, Y)$ be the restrictions to Y of those functions in $C(\{F_k\})$. Q^+ will stand for the set of all positive rational numbers.

<u>Proposition 5.1</u>. <u>For any regular Dirichlet form</u> ε <u>on</u> $L^2(X\ ;\ m)$, <u>there exists a normal Markov process</u> $\underset{\sim}{M} = (\Omega, \mathcal{M}, \mathcal{M}_t,$

$X_t, \theta_t, P_x)$ with state space $(Y, \mathcal{B}(Y))$ satisfying the following conditions.

(i) $\mathrm{Cap}(X - Y) = 0$.

(ii) For each $w \in \Omega$, the sample path $X_t(w)$ is right continuous $(t \geqq 0)$ and has the left limits $(t > 0)$ on $Y \cup \partial$.

(iii) $P_t f$ is a version of $T_t f$ for each $f \in L^2 \cap C(X)$ and $t \in Q^+$.

(iv) There is a regular nest $\{F_k\}$ such that

(a) $\bigcup_{k=1}^{\infty} F_k \supset Y$,

(b) $P_t(C(X)) \subset C(\{F_k\} ; Y)$ for each $t \in Q^+$,

(c) $\lim_{n\to\infty} \sigma_k(w) = \infty$ for each $w \in \Omega$, where $\sigma_k(w) = \inf \{t > 0; X_t(w) \in Y - F_k\}$.

It is easy to see that the Markov process $\underset{\sim}{M}$ described in Proposition 5.1 is properly associated with ε. To prove this, take any $t > 0$ and $f \in L^2 \cap C(X)$. Then $P_{t_n} f(x) \longrightarrow P_t f(x)$, $x \in Y$, as $t_n \in Q^+$ decreases to t. On the other hand $T_{t_n} f = T_{t_n - t} T_t f$ converges to $T_t f$ in ε_1-norm by virtue of Lemma 3.2. Hence $P_t f$ is a quasi-continuous version of $T_t f$ on account of Theorem 4.2. The same statement can be proved for any $f \in L^2 \cap \underset{\sim}{B}$ in the same way as in Lemma 5.1 below.

Proposition 5.1 implies Theorem 5.1. To see this, it now suffices to show that the Markov process $\underset{\sim}{M}$ of Proposition 5.1 gives rise to a Hunt process. Take the canonical modification of $\underset{\sim}{M}$ exactly in the same manner as in Blumenthal-Getoor

[2 ; pp 49-50]. It is then a Hunt process satisfying Proposition 5.1 with (iv)(c) replaced by a.e. statement. Its strong Markovity and quasi-left continuity follow easily from the following observation for $\underset{\sim}{M}$: $P_s f(X_t(w))$ is right continuous in $t \geq 0$ and $\lim_{t' \uparrow t} P_s f(X_{t'}(w)) = P_s f(X_{t-}(w))$, $t > 0$, for each fixed $w \in \Omega$, $s \in Q^+$ and $f \in C(X)$.

From now on, we will concentrate our attention on the proof of Proposition 5.1. Suppose that we are given a regular Dirichlet form ε on $L^2(X ; m)$ with the associated semigroup $\{T_t, t > 0\}$, resolvent $\{G_\alpha, \alpha > 0\}$ and equilibrium potentials $\{e_A ; A \in \mathcal{O}_0\}$. We will produce a Markov process $\underset{\sim}{M}$ of Proposition 5.1 by six steps (I) $\sim$ (VI).

(I) <u>Integral operators</u> $\widetilde{P_t}$, $t \in Q^+$, <u>and</u> $\widetilde{G_1}$.

Since the form ε is regular, we can find a countable sub-collection $\underset{\sim}{B}_0$ of $\mathcal{D}[\varepsilon] \cap C(X)$ such that $\underset{\sim}{B}_0$ is linear over rationals, uniformly dense in $C(X)$ and closed under the operation of taking the absolute value. We then put $H_0 = (\bigcup_{t \in Q^+} T_t(\underset{\sim}{B}_0)) \cup G_1(\underset{\sim}{B}_0)$. H_0 is a countable subset of $\mathcal{D}[\varepsilon]$ by Lemma 3.2, and consequently each element $u \in H_0$ admits a quasi-continuous version $\tilde{u}$ according to Theorem 4.2. Applying Theorem 4.1 to $\tilde{H}_0 = \{\tilde{u} ; u \in H_0\}$, find a regular nest $\{F_k{}^0\}$ such that $\tilde{H}_0 \subset C(\{F_k{}^0\})$. Let us put $Y_0 = \bigcup_{k=1}^{\infty} F_k{}^0$.

By virtue of Theorem 4.1, $\widetilde{T_t(u + v)}(x) = \widetilde{T_t u}(x) + \widetilde{T_t v}(x)$, $\widetilde{T_t(au)}(x) = a\, \widetilde{T_t u}(x)$ for every $u, v \in \underset{\sim}{B}_0$, rational a and $x \in Y_0$. Further $0 \leq u \leq 1$, $u \in \underset{\sim}{B}_0$, implies $0 \leq \widetilde{T_t u}(x) \leq 1$,

$x \in Y_0$. Therefore there exist unique stochastic measures $\{\widetilde{P_t}(x, \cdot), t \in Q^+, x \in Y_0\}$ on $\mathcal{B}(X)$ such that $\widetilde{T_t u}(x) = \int_X \widetilde{P_t}(x, dy) u(y)$, $u \in \underset{\sim}{B}_0$, $t \in Q^+$, $x \in Y_0$. In the same way, we can introduce unique substochastic measures $\{\widetilde{G_1}(x, \cdot), x \in Y_0\}$ such that $\widetilde{G_1 u}(x) = \int_X \widetilde{G_1}(x, dy) u(y)$, $u \in \underset{\sim}{B}_0$, $x \in Y_0$.

Let us define

$$(5.3) \qquad \widetilde{P_t} u(x) = \begin{cases} \int \widetilde{P_t}(x, dy) u(y), & x \in Y_0 \\ 0, & x \in X - Y_0. \end{cases}$$

$\widetilde{G_1} u$ is similarily defined. It is easy to see that $\widetilde{P_t}$ and $\widetilde{G_1}$ are then linear operators from $\underset{\sim}{B}$ into $\underset{\sim}{B}$. Furthermore

<u>Lemma 5.1</u>. <u>For each</u> $u \in L^2 \cap \underset{\sim}{B}$, <u>the functions</u> $\widetilde{P_t} u$ <u>and</u> $\widetilde{G_1} u$ <u>are quasi-continuous versions of</u> $T_t u$ <u>and</u> $G_1 u$ <u>respectively</u> $(t \in Q^+)$.

Since $\widetilde{P_t}(C^+(X))$ and $\widetilde{G_1}(C^+(X))$ are subsets of $C(\{F_k^0\})$, this lemma can be shown just in the same manner as in [10; § 3] by making use of Lemma 3.2.

(II) <u>A regular nest</u> $\{F_k\}$.

Let $\{A_n\}$ be a countable open base of X such that each A_n is relatively compact. Put $\Theta_1 = \{A ; A$ is a finite union of A_n's$\}$. Obviously $\Theta_1 \subset \Theta_0$. For each $A \in \Theta_1$, choose a quasi-continuous version $\widetilde{e_A}$ of e_A. Let $\widetilde{\mathcal{D}[\varepsilon]}$ be the collection of all quasi-continuous versions of elements in $\mathcal{D}[\varepsilon]$. Define $\widetilde{H}$ as the smallest subfamily of $\widetilde{\mathcal{D}[\varepsilon]}$ satisfying the following. $\widetilde{H}$ is then countable.

($\widetilde{H}$.1) $\widetilde{H} \supset \underset{\sim}{B}_0$, $\{\widetilde{e_A} ;\ A \in \Theta_1\}$.

($\widetilde{H}$.2) $\widetilde{P_t}(\widetilde{H}) \subset \widetilde{H}$, $t \in Q^+$, and $\widetilde{G_1}(\widetilde{H}) \subset \widetilde{H}$.

($\widetilde{H}$.3) $\widetilde{H}$ is an algebra over rationals.

Lemma 5.2. There exists a regular nest $\{F_k\}$ satisfying the following. Put $Y_1 = \bigcup_{k=1}^{\infty} F_k$.

(5.4) $\widetilde{H} \subsetneq C(\{F_k\})$, $F_k \subset F_k{}^0$, $k = 1, 2, \cdots$.

(5.5) $\widetilde{e_A}(x) = 1$, $x \in A \cap Y_1$, $A \in \Theta_1$.

(5.6) There exists a sequence of rationals $t_k \downarrow 0$ such that, as $t_k \downarrow 0$, $\widetilde{P_{t_k}}u(x) \longrightarrow u(x)$ and $\frac{1}{t_k}(\widetilde{G_1}u(x) - e^{-t_k}\widetilde{G_1}\widetilde{P_{t_k}}u(x)) \longrightarrow u(x)$ for every $x \in Y_1$ and $u \in \widetilde{H}$.

(5.7) $\widetilde{P_t}\widetilde{P_s}u(x) = \widetilde{P_{t+s}}u(x)$, $x \in Y_1$, $s,t \in Q^+$, $u \in \widetilde{H}$.

(5.8) $e^{-t}\widetilde{P_t}\widetilde{G_1}u(x) \leq \widetilde{G_1}u(x)$, $x \in Y_1$, $t \in Q^+$, $u \in \widetilde{H}$.

(5.9) $e^{-t}\widetilde{P_t}\widetilde{e_A}(x) \leq \widetilde{e_A}(x)$, $x \in Y_1$, $t \in Q^+$, $A \in \Theta_1$.

(5.10) $0 \leq \widetilde{e_A}(x) \leq 1$, $x \in Y_1$, $A \in \Theta_1$.

(5.11) $\widetilde{e_A}(x) \leq \widetilde{e_B}(x)$, $x \in Y_1$, $A,B \in \Theta_1$, $A \subset B$.

Proof. Use Theorem 4.1 (i), the equality of (4.2), the remark preceding to Theorem 4.2, Lemma 3.2 and Theorem 4.2 (ii) to find a nest $\{F_k\}$ satisfying (5.4) $\sim$ (5.6) and then pass to its m-regularization (Theorem 4.1 (ii)). Theorem 4.1 (iii) implies the remaining properties of this lemma. We can use (4.4), (4.2) and (4.5) for (5.9), (5.10) and (5.11).

(III) A Markov process with time parameter Q^+

Lemma 5.3. There exists a Borel set $Y_2 \subset Y_1$ such that

$\mathrm{Cap}(X - Y_2) = 0$ _and_ $\tilde{P}_t(x, X - Y_2) = 0$ _for every_ $x \in Y_2$ _and_ $t \in Q^+$.

Proof. Since $m(X - Y_1) = 0$, $\tilde{P}_t(x, X - Y_1) = 0$ for m - a.e. $x \in X$. Lemma 5.1 implies that there is a Borel set $Y_1^{(1)} \subset Y_1$ such that $\mathrm{Cap}(X - Y_1^{(1)}) = 0$ and $\tilde{P}_t(x, X - Y_1) = 0$ for all $x \in Y_1^{(1)}$ and $t \in Q^+$. Apply the same argument to $X - Y_1^{(1)}$ to get a Borel set $Y_1^{(2)} \subset Y_1^{(1)}$. Finally $Y_2 = \bigcap_{k=1}^{\infty} Y_1^{(k)}$ works.

Let us define

$$(5.12) \qquad P_t(x, B) = \begin{cases} \tilde{P}_t(x, B) & x \in Y_2, \quad B \in \mathcal{B} \\ 0 & x \in X - Y_2, \quad B \in \mathcal{B}, \end{cases}$$

and put $P_t u(x) = \int_X P_t(x, dy) u(y)$. Then

$$(5.13) \qquad P_t u(x) = \tilde{P}_t u(x), \quad x \in Y_2, \quad u \in \underset{\sim}{B},$$

$$(5.14) \qquad P_t P_s u(x) = P_{t+s} u(x), \quad x \in X, \quad u \in \underset{\sim}{B}.$$

This follows from (5.7) and the above lemma. Extending P_t to $(X \cup \partial, \mathcal{B}(X \cup \partial))$ by $P_t(x, \{\partial\}) = 1 - P_t(x, X)$ and $P_t(\partial, \{\partial\}) = 1$, we now have a transition probability over $X \cup \partial$.

We put $\Omega_0 = (X \cup \partial)^{Q^+}$, $X_t^0(w) = w(t)$, $w \in \Omega_0$, $t \in Q^+$, $\mathcal{M}_t^0 = \sigma\{X_s^0 ; 0 < s \leq t, \ s \in Q^+\}$, $t \in Q^+$, and $\mathcal{M} = \bigvee_{t \in Q^+} \mathcal{M}_t^0$. Then there are uniquely probability measures P_x, $x \in X \cup \partial$, over $(\Omega_0, \mathcal{M})$ such that

$$(5.15) \qquad E_x(f(X_t^0)) = P_t f(x)$$

$$(5.16) \qquad E_x(f(X_{t+s}^0) / \mathcal{M}_t^0) = E_{X_t^0}(f(X_s^0))$$

$$x \in X \cup \partial, \quad t, s \in Q^+, \quad f \in \underset{\sim}{B}(X \cup \partial).$$

Y_2 is obviously an invariant set for the Markov process $\underline{M}^0 = (\Omega_0, \mathcal{M}, \mathcal{M}_t^0, X_t^0, P_x)$, $t \in Q^+$:

(5.17) $P_x(X_t^0 \in Y_2 \cup \partial$ for every $t \in Q^+) = 1$, $x \in Y_2$.

(IV) Supermartingale $Y_t^0 = e^{-t}\widetilde{e}_A(X_t^0)$, $t \in Q^+$.

Take $A \in \Theta_1$ and fix a point $x \in Y_2$. Then $(Y_t^0, \mathcal{M}_t^0, P_x)$, $t \in Q^+$, is a positive bounded supermartingale because of (5.9), (5.16) and (5.17). Therefore, for almost all $w \in \Omega_0$, the right limits $Y_t = \lim_{s \in Q^+,\, s \downarrow t} Y_s^0$ exist for all $t \geq 0$ and $(Y_t, \mathcal{M}_t', P_x)$, $t \geq 0$, is again a positive bounded supermartin gale.

Here

(5.18) $\mathcal{M}_t = \bigcap_{s \in Q^+,\, s>t} \mathcal{M}_s^0$, $\mathcal{M}_t' = \mathcal{M}_t \vee \mathcal{N}$,

$\mathcal{N}$ being the collection of those sets $\Gamma \in \mathcal{M}$ such as $P_x(\Gamma) = 0$ for every $x \in Y_2$ (c.f. Meyer [14 ; VI]).

Lemma 5.4. $P_x(Y_t = Y_t^0$ for all $t \in Q^+) = 1$.

Proof. For any $f \in C[0, 1]$ and any polynomial g with rational coefficients, we have $E_x(f(Y_t^0)g(Y_{t+t_k}^0)) = E_x(f(Y_t^0) P_{t_k}g(Y_t^0))$, $t \in Q^+$. Since $g \circ \widetilde{e}_A$ is an element of $\widetilde{H}$, $P_{t_k}g(Y_t^0) = \widetilde{P}_{t_k}(g \circ \widetilde{e}_A)(X_t^0)$ converges to $g \circ \widetilde{e}_A(X_t^0) = g(Y_t^0)$ in view of (5.6). Hence $E_x(f(Y_t^0)g(Y_t)) = E_x(f(Y_t^0)g(Y_t^0))$, completing the proof.

Let us define, for an open set $G \subset X$,

(5.19) $\sigma_G{}^0 = \inf\{t \in Q^+ ; X_t{}^0 \in G\}$.

$\sigma_G{}^0$ is an $\mathcal{M}_t$-stopping time. We are ready to prove the following.

Lemma 5.5. (i) For each $A \in \mathcal{O}_1$ and $x \in Y_2$, $E_x(e^{-\sigma_A{}^0}) \leq \widetilde{e}_A(x)$.

(ii) For any $G \in \mathcal{O}_0$, e_G admits a quasi-continuous version $\widetilde{e}_G$ such that $E_x(e^{-\sigma_G{}^0}) \leq \widetilde{e}_G(x)$, $x \in Y_2$.

(iii) For any decreasing sequence $G_n \in \mathcal{O}_0$ such as $\mathrm{Cap}(G_n) \longrightarrow 0$ $n \longrightarrow \infty$, we have $P_x(\lim_{n\to\infty} \sigma_{G_n}^0 = \infty) = 1$ for q.e. $x \in Y_2$.

Proof. (i) When either $\sigma_A{}^0$ is irrational or $\sigma_A{}^0$ is rational but $X^0_{\sigma_A{}^0} \notin A$, then (5.5) implies $Y_{\sigma_A{}^0} = e^{-\sigma_A{}^0}$. When $\sigma_A{}^0$ is rational and $X^0_{\sigma_A{}^0} \in A$, then Lemma 5.4 means that $Y_{\sigma_A{}^0} = Y^0_{\sigma_A{}^0} = e^{-\sigma_A{}^0}$. Hence, by the supermartingale inequality ([14 ; VI]), $E_x(e^{-\sigma_A{}^0}) = E_x(Y_{\sigma_A{}^0}) \leq E_x(Y_0)$ $= \lim_{s\downarrow 0} e^{-s}P_s\widetilde{e}_A(x) \leq \widetilde{e}_A(x)$.

(ii) It suffices to choose $A_n \in \mathcal{O}_1$ such as $A_n \uparrow G$ and put $\widetilde{e}_G(x) = \lim_{n\to\infty} \widetilde{e}_{A_n}(x)$, $x \in Y_2$. The limit exists in view of (5.11). Moreover $\varepsilon_1(e_G - \widetilde{e}_{A_n}, e_G - \widetilde{e}_{A_n}) = \mathrm{Cap}(G) - \mathrm{Cap}(A_n) \longrightarrow 0$.

(iii) Since $\varepsilon_1(\widetilde{e}_{G_n}, \widetilde{e}_{G_n}) = \mathrm{Cap}(G_n) \longrightarrow 0$, a subsequence of $\widetilde{e}_{G_n}$ converges to zero q.e. on X. Hence (ii) implies (iii).

(V) Regularity of sample paths

Lemma 5.6. There exists a Borel set $Y_3 \subset Y_2$ with $\mathrm{Cap}(X - Y_3) = 0$ and the following statements hold.

(i) Put $\Omega_{01} = \{w \in \Omega_0 \ ; \ \lim_{k\to\infty} \sigma^0_{X-F_k} = \infty\}$, $\Omega_{02} = \{w \in \Omega_0 \ ; \ X^0_s(w)$ has the right and left limits in $Y_1 \cup \partial$ at every $t \geqq 0$ through $Q^+\}$, $\Omega_{03} = \{w \in \Omega_0 \ ; \ w([0,t] \cap Q^+)$ is bounded in X if $X^0_t(w) \in X$, $t \in Q^+\}$ and $\Omega_1 = \Omega_{01} \cap \Omega_{02} \cap \Omega_{03}$. Then $P_x(\Omega_1) = 1$ for every $x \in Y_3$.

(ii) For $w \in \Omega_1$ and $t \geqq 0$, we put

(5.20) $$X_t(w) = \lim_{s \in Q^+,\ s \downarrow t} X^0_s(w).$$

Then $P_x(X_t = X^0_t$ for every $t \in Q^+) = 1$, $x \in Y_3$.

(iii) $P_x(X_0 = x) = 1$, $x \in Y_3$.

Proof. (i) Lemma 5.5 (iii) implies that there is a Borel set $Y_3 \subset Y_2$ such that $\mathrm{Cap}(X - Y_3) = 0$ and $P_x(\Omega_{01}) = 1$ for $x \in Y_3$. Next (5.4) and (5.6) imply that $\widetilde{G}_1(H^+)$ is contained in $C(\{F_k\})$ and separates points of Y_1. Therefore we have $\Omega_{01} - \Omega_{02} \subset \bigcup_{u \in \widetilde{H}^+} \{w \in \Omega_0 \ ; \ \widetilde{G}_1 u(X^0_s),\ s \in Q^+,$ has an oscillatory discontinuity at some $t \geqq 0\}$. However $\{e^{-s}\widetilde{G}_1 u(X^0_s), \mathcal{M}^0_s, P_x\}$ $s \in Q^+$, is a bounded positive supermartingale for each $u \in \widetilde{H}^+$ and $x \in Y_2$ on account of (5.8). Hence $P_x(\Omega_{01} - \Omega_{02}) = 0$, $x \in Y_2$. Finally there is, by Lemma 5.2, a function $v \in C^+$ such that $\widetilde{G}_1 v \in C(\{F_k\})$ is strictly positive on Y_1 and satisfies (5.8). We have then $\Omega_{01} - \Omega_{03} = \bigcup_{t \in Q^+} \{w \in \Omega_{01} \ ; \ \widetilde{G}_1 v(X^0_t) > 0,$ $\inf_{s \in Q^+, s \leqq t} \widetilde{G}_1 v(X^0_s) = 0\}$, which has zero P_x-measure $(x \in Y_2)$ in view of [2; Chap. 0, (1.6)]. (ii) For $u, v \in B_0$, $t \in Q^+$ and $x \in Y_2$, $E_x(u(X^0_t)v(X_t)) = \lim_{t_k \downarrow 0} E_x(u(X^0_t)P_{t_k}v(X^0_t)) = E_x(u(X^0_t)v(X^0_t))$ by virtue of (5.6). (iii) $E_x(u(X_0)) = \lim_{t_k \downarrow 0} P_{t_k}u(x) = u(x)$,

$u \in B_0$, $x \in Y_2$.

Lemma 5.7. *There exists a Borel set* $Y \subset Y_3$ *with* $\mathrm{Cap}(X - Y) = 0$ *satisfing the following condition: the set* $\Gamma = \{w \in \Omega_1\ ;\ X_t(w)$ *or* $X_{t-}(w)\quad X - Y$ *for some* $t \geqq 0\}$ *is contained in a set* $\Gamma_0 \in \mathcal{M}$ *such that* $P_x(\Gamma_0) = 0$ *for every* $x \in Y$.

Proof. Choose a decreasing sequence of open sets $G_n \supset X - Y_3$ such as $\mathrm{Cap}(G_n) \longrightarrow 0$. Put $\Gamma_3 = \{w \in \Omega_1\ ;\ X_t$ or $X_{t-} \in X - Y_3$ for some $t \geqq 0\}$, $\Gamma_3^0 = \{w \in \Omega_0\ ;\ \lim_{n\to\infty} \sigma^0_{G_n} < +\infty\}$. Then $\Gamma_3 \subset \Gamma_3^0$ and Lemma 5.5 (iii) further implies $P_x(\Gamma_3^0) = 0$ if x is in some Borel set $Y_4 \subset Y_3$ with $\mathrm{Cap}(X - Y_4) = 0$. Apply the same argument to Y_4. We thus get sequences $Y_3 \supset Y_4 \supset \cdots$, $\Gamma_3 \subset \Gamma_4 \cdots$ and $\Gamma_3^0 \subset \Gamma_4^0 \subset \cdots$. Put $Y = \bigcap_{k=3}^{\infty} Y_k$, then obviously $\Gamma = \bigcup_{k=3}^{\infty} \Gamma_k$. Now $\Gamma_0 = \bigcup_{k=3}^{\infty} \Gamma_k^0$ works.

(VI) Extended Markov property

Let us put $\Omega = \Omega_1 - \Gamma_0$. The restrictions to Ω of $X_t^0, \mathcal{M}_t^0$ $(t \in Q^+)$, $X_t, \mathcal{M}_t$ $(t \geqq 0)$, $\mathcal{M}$ and P_x $(x \in Y \cup \partial)$ are again denoted by the same notations.

Lemma 5.8. $\underset{\sim}{M} = (\Omega, \mathcal{M}, \mathcal{M}_t, X_t, P_x)$ *is a normal Markov process with state space* $(Y, \mathcal{B}(Y))$ *satisfying*

$$(5.21)\qquad P_x(X_t \in B) = P_t(x, B), \quad t \in Q^+, \quad x \in Y, \quad B \in \mathcal{B}(Y).$$

Moreover $\underset{\sim}{M}$ *possesses all the properties of Proposition 5.1.*

Proof. (5.21) and the normality of $\underset{\sim}{M}$ follow from Lemma 5.6 (ii) and (iii) respectively. Other properties of $\underset{\sim}{M}$ are now evident except for its Markov property.

Take $t, s \geq 0$, $f \in C(X)$ and $\Lambda \in \mathcal{M}_t$. We have, by (5.16) $E_x(f(X^0_{t'+s'}) ; \Lambda) = E_x(E_{X^0_{t'}}(f(X^0_{s'})) ; \Lambda)$ for any $t' > t$ and $s' > s$, $t', s' \in Q^+$, because $\Lambda \in \mathcal{M}^0_{t'}$. Note that the function $v(x) = E_x(f(X^0_{s'})) = P_{s'}f(x)$ is an element of $C(\{F_k\} ; Y)$. Hence, by letting t' decrease to t and then s' to s, we arrive at $E_x(f(X_{t+s}) ; \Lambda) = E_x(E_{X_t}(f(X_s)) ; \Lambda)$ the Markovity of $\underset{\sim}{M}$.

§ 6. Examples

We now return to the examples of Markov symmetric forms in § 2 to see if they give rise to regular Dirichlet forms. According to Theorem 3.3, it suffices to check the closability of a given Markov symmetric form and then the regularity of its smallest closed extension.

1°. Consider Example 1. Let us confine ourselves to the case when $\Phi \equiv 0$, $n \equiv 0$ and m is the Lebesgue measure on D. If either

(1°. a) the first order (Schwartz) distribution derivatives of $a_{ij}(x)$ are locally integrable functions

or

(1°. b) $a_{ij}(x)$ is uniformly elliptic : there is a positive constant δ such that for any N-vector ξ,

$$\sum_{i,j=1}^{N} a_{ij}(x)\xi_i\xi_j \geq \delta|\xi|^2 , \quad x \in D,$$

then ε is closable and the smallest closed extension $\bar{\varepsilon}$ is clearly a regular Dirichlet form. An associated Hunt process (according to Theorem 5.1) is the well-known absorbing barrier diffusion process on D if $A_{ij}(x)$ are sufficiently smooth.

In case of (1°. a), ε can be expressed as (3.7) with the symmetric operator $Su = \sum_{i,j=1}^{N} \frac{\partial}{\partial x_j}(a_{ij}(x)\frac{u(x)}{x_j})$, $\mathcal{D}(S) = C_0^\infty(D)$. Hence ε is closable. Next assume that (1°. b) is satisfied. Consider $u_n \in C_0^\infty(D)$ such that $\varepsilon(u_n - u_m, u_n - u_m) \to 0$ and $u_n \to 0$ in $L^2(D)$. Then $\{u_n\}$ forms a Cauchy sequence with respect to the usual Dirichlet integral $\underset{\sim}{D}$. Since $\underset{\sim}{D}$ is a special form satisfying (1°.a), $\underset{\sim}{D}(u_n, u_n) \to 0$, which in turn implies that a subsequence n_k exists and $\frac{\partial u_{n_k}}{\partial x_i} \to 0$ a.e. on D, $i = 1, 2, \cdots, N$. By Fatou's lemma,

$$\varepsilon(u_m, u_m) = \int_D \underline{\lim}_{n_k\to\infty} \left(\sum_{i,j=1}^{N} \frac{\partial(u_{nk}-u_m)}{\partial x_i}(x)\, \frac{\partial(u_{n_k}-u_m)}{\partial x_j}(x) a_{ij}(x)\right)dx$$

$\leqq \underline{\lim}_{n_k\to\infty} \varepsilon(u_{n_k} - u_m, u_{n_k} - u_m)$, which is small if m is sufficiently large. Hence ε is closable.

Here is an example of K.Sato. Assume that $D = R^N$ and

(1°.c) $\frac{\partial}{\partial x_k} a_{ij}(x)$ is locally integrable for $1 \leqq k \leqq N-1$ but $\frac{\partial}{\partial x_N} a_{ij}(x)$ is defined and continuous only on $\{x \in R^N; x_N \neq 0\}$, then ε is closable (and $\bar{\varepsilon}$ is a regular Dirichlet form).

To prove this, let us put $\mathcal{F}^0 = \{u \in C_0^\infty(R^N)\ ;\ u(x) = u(x_1,x_2,\cdots,x_{N-1})$ if $x_N \in (-\delta,\delta)$ for some $\delta > 0\}$. We can see that our form ε with domain restricted to $\mathcal{F}^0$ satisfies (3.6). Hence this is closable. Moreover we can find, for any $u \in C_0^\infty(R^N)$, a sequence $u_n \in \mathcal{F}^0$ such that $\varepsilon(u_n - u, u_n - u) \to 0$ and $(u_n - u, u_n - u) \to 0$, proving that ε is closable. For instance, it suffices to take $u_n \in \mathcal{F}^0$ such that $u_n(x) \to u(x)$,

$\frac{\partial u_n(x)}{\partial x_i} \longrightarrow \frac{\partial u(x)}{\partial x_i}$ boundedly on R^N-F and all u_n have a common compact support.

2°. Consider Example 2. We assume that there is a positive constant δ such that $\mu(E) \geqq \delta|E|$, $\nu(E) \geqq \delta|E|$ for any linear Borel set E, $|E|$ being the Lebesgue measure. ε is then closable and the smallest closed extension $\bar{\varepsilon}$ is a regular Dirichlet form. An associated Hunt process on R^2 is one of the diffusions that are constructed more concretely by N. Ikeda and S. Watanabe [11].

Suppose $u_n \in C_0^\infty(R^2)$ satisfies $\varepsilon(u_n - u_m, u_n - u_m) \longrightarrow 0$ and $u_n \longrightarrow 0$ in $L^2(R^2)$, then $\frac{1}{2}\underset{\sim}{D}(u_n, u_n) + (u_n, u_n) \longrightarrow 0$, which means, by virtue of Theorem 4.2 and [10 ; Theorem 4.5], that a subsequence $u_{n_k}(x)$ converges to zero on R^2 except on a Borel polar set $\mathcal{N}$ of the two-dimensional standard Brownian motion. In particular the linear Lebesgue measure of $\mathcal{N} \cap \ell$ vanishes for any straight line ℓ on R^2.

There is a Borel function $\phi(x_1, x_2)$ on R^2 such that $\frac{\partial u_n}{\partial x_1}$ converges to ϕ in $L^2(R^2 ; dx_1 \times d\mu)$. This implies $\underline{\lim}_{n_k\to\infty} \int_{-\infty}^{\infty} (\frac{\partial u_n(x_1,x_2)}{\partial x_1} - \phi(x_1, x_2))^2 dx_1 = 0$ for μ-almost all $x_2 \in R^1$. Hence, by making use of the above observation and the equality

$$\int_a^x \phi(x_1, x_2)dx_1 = u_{n_k}(x_1, x_2) - u_{n_k}(a, x_2) - \int_a^x (\frac{\partial u_{n_k}}{\partial x_1} - \phi)dx_1,$$

we can see that $\phi(x_1, x_2) = 0$ for a.e. $x_1 \in R^1$.

Thus $\frac{\partial u_n}{\partial x_1}$ converges to zero in $L^2(R^2 ; d\nu \times dx_2)$. In this way we get $\varepsilon(u_n, u_n) \longrightarrow 0$ the closability of ε.

3°. Consider Example 3. The form (2.7) with the domain $H^1(D)$ is closed already. Hence this is a Dirichlet form, which is not regular however unless R^N-D has zero Newtonian outer capacity (in case $N \geq 3$).

More generally consider a linear space $\mathscr{D}$ such that $C_0^\infty(D) \subset \mathscr{D} \subset H^1(D)$ and denote by $\varepsilon_{\mathscr{D}}$ the form (2.7) with the domain $\mathscr{D}$. We suppose that $\varepsilon_{\mathscr{D}}$ is Markov. Then $\varepsilon_{\mathscr{D}}$ is closable and its smallest closed extension $\bar{\varepsilon}_{\mathscr{D}}$ is a Dirichlet form. $\bar{\varepsilon}_{\mathscr{D}}$ is not regular in general, nevertheless we can regard this as a regular Dirichlet form by a suitable enlargement of the underlying space D.

A locally compact separable Hausdorff space D^* is called an admissible enlargement of D relative to $\mathscr{D}$ if D is continuously embedded onto a dense subset of D^* and if the intersection $\mathscr{D} \cap C(D^*)$ is dense in the domain of $\bar{\varepsilon}_{\mathscr{D}}$ and is uniformly dense in $C(D^*)$.

Given $\mathscr{D}$ and D^* as above, let m be the measure on D^* induced by the Lebesgue measure on D : $m(E) = |E \cap D|$. Identify the space $L^2(D)$ with $L^2(D^*) = L^2(D^* ; m)$. Then the form $\bar{\varepsilon}_{\mathscr{D}}$ turns out to be a regular Dirichlet form on $L^2(D^*)$ (in the terminology of [9], $(D^*, m, \bar{\varepsilon}_{\mathscr{D}})$ is a regular representation of $\bar{\varepsilon}_{\mathscr{D}}$). In this way, we can get a strong Markov process on D^* which may be considered as an extension of the absorbing barrier Brownian motion on D.

Let us examine three cases.

(3°. a) $\mathcal{D} = C_0^\infty(D)$. D itself is an admissible enlargement of D relative to $C_0^\infty(D)$. A process associated with $\bar{\varepsilon}_{\mathcal{D}}$ is the absorbing barrier Brownian motion on D.

(3°. b) $\mathcal{D} = \hat{C}^\infty(D)$. The closure $\bar{D}$ of D in R^N is an admissible enlargement of D relative to $\hat{C}^\infty(D)$ in view of Tietze extension theorem of a continuous function. The associated process on $\bar{D}$ (minus an exceptional set) may be considered as a reflecting barrier Brownian motion if the domain of $\bar{\varepsilon}_{\mathcal{D}}$ coincides with $H^1(D)$. This is the case if the boundary $\partial D = \bar{D} - D$ is sufficiently smooth.

(3°. c) $\mathcal{D} = H^1(D)$. As we have just mentioned, $\bar{D}$ is an admissible enlargement of D relative to $H^1(D)$ if ∂D is sufficiently smooth. In the general case, we can take as an admissible enlargement D^* the space constructed in [9 ; §6]. The associated process on D^* is, by definition, a reflecting barrier Brownian motion. If D_1^* and D_2^* are admissible enlargements of D relative to the same space $H^1(D)$, then D_1^* and D_2^* are related to each other by a capacity preserving quasi-homeomorphism [10 ; § 2] which is the identity on D. This transformation interrelates the reflecting barrier processes on the respective spaces.

References

[1] A. Beurling and J. Deny: Dirichlet spaces, Proc. Nat. Acad. Sci. U.S.A. 45 (1959), 208-215.

[2] R.M. Blumenthal and R.K. Getoor: Markov processes and potential theory, Academic Press, New York and London, 1968.

[3] J. Deny: Méthodes Hilbertiennes en théorie du potentiel, Potential Theory, Centro Internazionale Matematico Estivo, Edizioni Cremonese, Roma, 1970, pp. 121-201.

[4] J. Deny - J.L. Lions: Les espaces du type de Beppo Levi, Ann. Inst. Fourier (Grenoble) 5 (1953/54), 305-370.

[5] E.B. Dynkin: Markov processes, Springer-Verlag, Berlin, 1965.

[6] J. Elliott: Dirichlet spaces and boundary conditions for submarkovian resolvents, J. Math. Anay. Appl. 36 (1971), 251-282.

[7] J. Elliott and M.L. Silverstein: On boundary conditions for symmetric submarkovian resolvents, Bull. Amer. Math. Soc. 76 (1970), 752-757.

[8] M. Fukushima: On boundary conditions for multi-dimensional Brownian motions with symmetric resolvent densities, J. Math. Soc. Japan 21 (1969), 58-93.

[9] M. Fukushima: Regular representations of Dirichlet spaces, Trans. Amer. Math. Soc. 155 (1971), 455-473.

[10] M. Fukushima: Dirichlet spaces and strong Markov processes, Trans. Amer. Math. Soc. 162 (1971), 185-224.

[11] N. Ikeda and S. Watanabe: The local structure of a class of diffusions and related problems, Proceedings of the Second Japan - USSR Symposium on Probability Theory.

[12] T. Kato: Perturbation theory for linear operators, Springer-Verlag, Berlin, 1966.

[13] M.G. Krein: The theory of self-adjoint extensions of semi-bounded Hermitian transformations and its applications, Mat. Sbornik 20 (1947), 431-495 (Russian).

[14] P.A. Meyer: Probability and Potentials, Ginn, Blaisdell, Waltham, Mass., 1966.

[15] P.A. Meyer: Processus de Markov, Lecture Notes in Math., no. 26, Springer-Verlag, Berlin, 1967.

[16] M.L. Silverstein: Dirichlet spaces and random time change, to appear.

Department of Mathematics
Osaka University
Toyonaka, Japan.

ON NONLINEAR FILTERING THEORY AND ABSOLUTE CONTINUITY OF MEASURES, CORRESPONDING TO STOCHASTIC PROCESSES

B. Grigelionis

Introduction

New important results obtained during the last ten years in the theory of martingales and stochastic integrals with respect to martingales ([1] - [3]) were a foundation for a powerful apparatus in statistics of stochastic processes. This apparatus is natural in investigation of such problems as nonlinear filtering of stochastic processes, absolute continuity of measures, corresponding to stochastic processes, calculation of their Radon-Nikodym densities, stochastic control, etc. It is important that we can consider a rather wide class of stochastic processes which are locally infinetely divisible, i.e. multidimensional stochastic processes without discontinuities of the second kind defined on some probability space and adapted to a given increasing family of σ-subalgebras; for such processes we can naturally define local coefficients of drift and diffusion and Levy's measure (for the fefinition see § 1).

The main purpose of our paper is a short survey of some results obtained by the author on non-linear filtering of stochastic processes and on absolute continuity of probability measures using the technique of stochastic integrals with respect to martingales (see [4] - [7] and the references there).

When deriving non-linear filtering equations, the question plays an important role, how the local characteristics (drift and diffusion coefficients and Levy's measure) of stochastic processes alter when we substitute a given increasing family of σ-subalgebras with another family of more "narrow" σ-subalgebras. The so-called innovation approach is essentially connected with it (see e.g. [8] - [10]). On the other hand, when we derive criteria of absolute continuity for probability measures, the transformation formulas for the local characteristics, when we change the measure on the basic probability space, are very important. We obtain explicit formulas for Radon-Nikodym densities by combining the two transformations: substitution of the family of σ-subalgebras and change of the probability measure by absolute continuity.

In § 1, a locally infinitely divisible stochastic process is defined, in § 2 the non-linear filtering problem for stochastic processes is considered, and in § 3 the problems of absolute continuity of probability measures are investigated.

§1. Locally infinitely divisible stochastic processes

Let $\{\mathcal{F}_t, t\geq 0\}$ be an increasing right continuous family of σ-algebras on the probability space $(\Omega, \mathcal{F}, P)$, $\mathcal{F}_t \subseteq \mathcal{F}$, where $\mathcal{F}_t$ and $\mathcal{F}$ are complete with respect to P. Denote by $\mathcal{M}^{(m)}$ the class of all m-dimensional right continuous square integrable martingales with respect to the family of σ-algebras $\{\mathcal{F}_t, t\geq 0\}$ and by $\mathcal{M}_c^{(m)}$ the subclass of continuous martingales.

Consider a stochastic process $X=\{X(t), t\geq 0\}$ with values in the m-dimensional Euclidean space $(R_m, \mathcal{B}_m)$ with right continuous and having left limits paths.

(I) Assume that X is adapted to the family $\{\mathcal{F}_t, t\geq 0\}$, and there exists a function $\Pi(t,\Gamma)=\Pi(t,\omega,\Gamma)$, $(t,\omega,\Gamma)\in[0,\infty)\times\Omega\times\mathcal{B}_m$, such that it is a measure on $\mathcal{B}_m$ for fixed (t,ω), it is $\mathcal{B}[0,\infty)\times\mathcal{F}$-measurable)^x for every $\Gamma\in\mathcal{B}_m$, $\Pi(\tau,\Gamma)$ is $\mathcal{F}_\tau$-measurable for

---

)^x $\mathcal{B}[0,\infty)$ is the σ-algebra of Borel subsets of the interval $[0,\infty)$.

every stopping time τ with respect to the family $\{\mathcal{F}_t, t \geq 0\}$,

$$E\Big[\int_0^t \Pi(s, \mathcal{U}_\varepsilon)\,ds\Big] < \infty, \quad E\Big[\int_0^t \int_{|x|\leq 1} |x|^2 \Pi(s, dx)\,ds\Big] < \infty$$

for all $t > 0$, $\varepsilon > 0$ and

$$q(t, \Gamma) = p(t, \Gamma) - \int_0^t \Pi(s, \Gamma)\,ds \in \mathcal{M}^{(1)}$$

for all $\Gamma \in \mathcal{B}_m \cap \mathcal{U}_\varepsilon$, $\varepsilon > 0$, where $\mathcal{U}_\varepsilon = \{x : |x| \geq \varepsilon\}$
and

$$p(t, \Gamma) = \sum_{0 \leq s \leq t} \chi_\Gamma\big(X(s) - X(s-0)\big),$$

$\chi_\Gamma(x)$ being the indicator of the set Γ.

Denote by $F^{(m)}$ the class of m-dimensional $\mathcal{B}[0,\infty) \times \mathcal{B}_m \times \mathcal{F}$-measurable functions $\varphi(t, x) = \varphi(t, x, \omega)$ such that $\varphi(\tau, x)$ is $\mathcal{F}_\tau$-measurable for each fixed x and stopping time τ. Let $F_P^{(m)}$ be a subclass of functions $\varphi \in F^{(m)}$ such that the sums

$$P_\varphi(t) = \sum_{0 \leq s \leq t} \varphi(s, X(s) - X(s-0))$$

converge almost everywhere with respect to the measure P (a.e.), and let $F_Q^{(m)}$ be the subclass of functions $\varphi \in F^{(m)}$ such that

$$\|\varphi\|_t^{(m)} = \Big[E\Big(\int_0^t \int_{R_m} |\varphi(s, x)|^2 \,\Pi(s, dx)\,ds\Big]^{1/2} < \infty, \quad t \geq 0.$$

The sums $P_\varphi(t)$ are called stochastic integrals of the functions $\varphi \in F_p^{(m)}$ with respect to the measure p and are denoted by

$$P_\varphi(t) = \int_0^t \int_{R_m} \varphi(s, x)\,p(ds, dx).$$

In the usual way, for each $\varphi \in F_Q^{(m)}$, the stochastic integral $Q_\varphi(t) \in \mathcal{M}^{(m)}$ with respect to the measure q is defined and denoted by

$$Q_\varphi(t) = \int_0^t \int_{R_m} \varphi(s, x)\,q(ds, dx)$$

Under assumption (I), it is easy to verify that the stochastic process

$$X_0(t) = X(t) - \int_0^t \int_{|x|\le 1} x q(ds, dx) - \int_0^t \int_{|x|>1} x p(ds, dx)$$

is well defined and continuous.

(II) Assume that there exist an m-dimensional function $a(t) = (a_1(t), \dots, a_m(t))$ and a matrix $\| a_{ij}(t) \|_1^m = A(t)$ such that the functions $a_i(t)$, $a_{ij}(t)$, $i, j = 1, \dots, m$, are $\mathcal{B}[0,\infty) \times \mathcal{F}$ -measurable, adapted to the family $\{\mathcal{F}_t, t \ge 0\}$, for all $t \ge 0$

$$E\Big[\int_0^t |a(s)|^2 ds\Big] < \infty, \quad E\Big[\int_0^t |a_{ij}(s)| ds\Big] < \infty,$$

$$\tilde{X}(t) = X_0(t) - \int_0^t a(s)\, ds \in \mathcal{M}_c^{(m)}$$

and a.e.

$$\langle \tilde{X}_i, \tilde{X}_j \rangle_t = \int_0^t a_{ij}(s)\, ds. \qquad)^x$$

We shall call a stochastic process $X = \{X(t), t \ge 0\}$ for which assumptions (I) and (II) are satisfied, in the sequel, locally infinitely divisible with the local characteristics (a, A, Π) with respect to the measure P and the family of σ-algebras $\{\mathcal{F}_t, t \ge 0\}$.

Example 1. If X is a process with independent increments and the characteristic function

$$E \exp\{i(X(t) - X(s), z)\} = \exp\Big\{ i \int_s^t (\hat{a}(u), z)\, du - \frac{1}{2} \int_s^t (z, z\hat{A}(u))\, du + \int_s^t \int_{|x|\le 1} (e^{i(x,z)} - 1 - i(x,z)) \hat{\Pi}(u, dx)\, du + \int_s^t \int_{|x|>1} (e^{i(x,z)} - 1) \hat{\Pi}(u, dx)\, du \Big\}$$

$)^x$ We use notation of [2] - [3].

then it is obvious that assumptions (I) and (II) are satisfied with $a(t)=\hat{a}(t)$, $A(t)=\hat{A}(t)$ and $\Pi(t,\Gamma)=\hat{\Pi}(t,\Gamma)$ (see [11]). It is interesting to note that the converse proposition is also true: a locally infinitely divisible stochastic process with independent of ω local characteristics has independent increments (see [4]).

Example 2. Let stochastic processes $\theta(t)=(\theta_1(t),\dots,\theta_n(t))$, $t\geq 0$ and $X(t)=(X_1(t),\dots,X_m(t))$, $t\geq 0$ be solutions of K. Ito's stochastic equations

$$\theta(t)=\theta(0)+\int_0^t a^{(1)}(s,\theta(s),X(s))ds+\sum_{r=1}^{m+n}\int_0^t b_r^{(1)}(s,\theta(s),X(s))dw_r(s)+$$
$$+\int_0^t\int_{|y|\leq 1}F^{(1)}(s,\theta(s),X(s),y)\,\hat{q}(ds,dy)+$$
$$+\int_0^t\int_{|y|>1}F^{(1)}(s,\theta(s),X(s),y)\hat{p}(ds,dy),$$

$$X(t)=X(0)+\int_0^t a^{(2)}(s,\theta(s),X(s))ds+\sum_{r=1}^{m+n}\int_0^t b_r^{(2)}(s,\theta(s),X(s))dw_r(s)+$$
$$+\int_0^t\int_{|y|\leq 1}F^{(2)}(s,\theta(s),X(s),y)\hat{q}(ds,dy)+$$
$$+\int_0^t\int_{|y|>1}F^{(2)}(s,\theta(s),X(s),y)\hat{p}(ds,dy),$$

where $W(t)=(w_1(t),\dots,w_{m+n}(t))$ and $\hat{p}(t,\Gamma)$ are mutually independent a standard $m+n$ -dimensional Wiener process and a standard Poisson measure on $\mathcal{B}[0,\infty)\times\mathcal{B}_{m+n}$ adapted to the family of σ -algebras $\{\mathcal{F}_t,\ t\geq 0\}$ and independent of $\theta(0)$ and $X(0)$. We assume that the coefficients of equations (2)-(3) satisfy the usual conditions of existence and uniqueness of solutions (see [12]).

Put

$$a^{(i)}(t)=a^{(i)}(t,\theta(t),X(t))+\int_{\substack{|y|>1\\ |F^{(i)}(t,\theta(t),X(t),y)|\leq 1}}F^{(i)}(t,\theta(t),X(t),y)\frac{dy}{|y|^{m+n+1}}-$$

$$-\int\limits_{\substack{|y|\leq 1 \\ |F^{(i)}(t,\theta(t),X(t),y)|>1}} F^{(i)}(t,\theta(t),X(t),y)\frac{dy}{|y|^{m+n+1}},$$

$$A^{(i)}(t)=b^{(i)}(t)\,b^{(i)}(t)', \quad b^{(i)}(t)=\|b^{(i)}_{jk}(t,\theta(t),X(t))\|,$$

$$\Pi^{(i)}(t,\Gamma)=\int\limits_{R_{m+n}}\chi_\Gamma(F^{(i)}(t,\theta(t),X(t),y))\frac{dy}{|y|^{m+n+1}}, \quad i=1,2,$$

where $'$ means transposing.

It is easy to prove that the stochastic process $X=\{X(t),\ t\geq 0\}$ is locally infinitely divisible with the local characteristics $(a^{(2)}, A^{(2)}, \Pi^{(2)})$ with respect to the measure P and the family of σ-algebras $\{\mathcal{F}_t,\ t\geq 0\}$.

Let $\mathcal{F}^X_t=\sigma(X(s),\ 0\leq s\leq t),\ t\geq 0$. Put

$$\bar{\Pi}(t,\Gamma)=E(\Pi(t,\Gamma)/\mathcal{F}^X_t)=E^t\Pi(t,\Gamma), \quad \bar{a}(t)=E^t a(t).$$

(III) Assume that the matrices $A(t)$ are adapted to the family of σ-algebras $\{\mathcal{F}^X_t,\ t\geq 0\}$, and there exist two functions: a function $\tilde{\Pi}(t,\Gamma)$ which is measure on $\mathcal{B}_m$ for fixed (t,ω), adapted to the family $\{\mathcal{F}^X_t,\ t\geq 0\}$ for every Γ and such that

$$\Pi(t,\Gamma)=\int_\Gamma \rho(t,x)\,\tilde{\Pi}(t,dx), \quad \Gamma\in\mathcal{B}_m, \quad t\geq 0$$

where the function $\rho(t,x)$ is (t,x,ω)-measurable and adapted to the family $\{\mathcal{F}_t,\ t\geq 0\}$; and a function $\Psi(t)$ which is (t,ω)-measurable, adapted to the family $\{\mathcal{F}_t,\ t\geq 0\}$ and such that

$$a(t)=\Psi(t)A(t)+\int\limits_{|x|\leq 1} x(\rho(t,x)-1)\,\tilde{\Pi}(t,dx)$$

and

$$E\rho(t,x)<\infty,\quad E\Big[\int_0^t\int_{R_m}\Big(\frac{\bar\rho(t,x)}{\rho(t,x)}-1\Big)^2\Pi(s,dx)ds\Big]<\infty,$$

$$E|\Psi(t)|<\infty,\quad E\Big[\int_0^t|(\Psi(s)-\bar\Psi(s))A(s)|^2ds\Big]<\infty$$

for all $x\in R_m$ and $t\geq 0$; $\bar\rho(t,x)=E^t\rho(t,x)$, $\bar\Psi(t)=E^t\Psi(t)$.

The following proposition is true.

Theorem 1 [6] . Under assumptions (I)-(III) the stochastic process X is locally infinitely divisible with the local characteristics $(\bar a, A, \bar\Pi)$ with respect to the measure P and the family of σ-algebras $\{\mathcal{F}_t^X, t\geq 0\}$.

Denote by $\Phi^{(m)}$ the class of m-dimensional stochastic functions $\varphi(t), t\geq 0$, which are $\mathcal{B}[0,\infty)\times\mathcal{F}$ -measurable and such that $\varphi(t)$ are $\mathcal{F}_\tau$ -measurable for all stopping times τ , by $\Phi_{rc}^{(m)}$ the class of m-dimensional right continuous processes and put $L^{(m)}=\Phi^{(m)}\cap\bar\Phi_{rc}^{(m)}$ where $\bar\Phi_{rc}^{(m)}$ is the closure with respect to the seminorms

$$\|\varphi\|_t=\Big[E\Big(\int_0^t(\varphi(s),\varphi(s)A(s)ds)\Big]^{\frac{1}{2}},\quad t\geq 0.$$

For each $\varphi\in L^{(m)}$ the stochastic integral $X_\varphi\in\mathcal{M}_c^{(1)}$ with respect to martingale $\tilde X$ is defined and denoted by

$$X_\varphi(t)=\int_0^t\varphi(s)\,d\tilde X(s)=\sum_{i=1}^m\int_0^t\varphi_i(s)\,d\tilde X_i(s).$$

Let

$$\alpha_t(\varphi,\psi)=\exp\Big\{X_\varphi(t)-\frac{1}{2}\langle X_\varphi\rangle_t+Q_{\chi\{|\psi|<1\}\psi}(t)+$$

$$+P_{\chi\{|\psi|\geq 1\}\psi}(t)-\int_0^t\int_{R_m}\chi_{\{|\psi|<1\}}(s,x)\big(e^{\psi(s,x)}-1-\psi(s,x)\big)\Pi(s,dx)ds-$$

$$-\int_0^t\int_{R_m}\chi_{\{|\psi|\geq 1\}}(s,x)\big(e^{\psi(s,x)}-1\big)\Pi(s,dx)ds\Big\},$$

where $\varphi \in L^{(m)}$, $\chi_{\{|\psi|<1\}}\psi \in F_a^{(1)}$ and $\chi_{\{|\psi|\geq 1\}}\psi \in F_p^{(1)}$.

Under assumption that $E\alpha_t(\varphi,\psi)=1$ for every $t\geq 0$, we define a probability measure $\tilde{P}$ on σ-algebra $\mathcal{F}_\infty = \sigma(\bigcup_{t\geq 0}\mathcal{F}_t)$:

$$\tilde{P}(A)=\int_A \alpha_t(\varphi,\psi)\,dP\,,\quad A\in\mathcal{F}_t\,,\quad t\geq 0.$$

Let

$$\tilde{a}(t)=a(t)+\varphi(t)A(t)+\int_{|x|\leq 1} x(e^{\psi(t,x)}-1)\,\Pi(t,dx)$$

and

$$\tilde{\Pi}(t,\Gamma)=\int_\Gamma e^{\psi(t,x)}\,\Pi(t,dx).$$

We shall use the following proposition.

<u>Theorem 2</u> [5]. <u>Under assumptions</u> (I) <u>and</u> (II) <u>the stochastic process</u> X <u>is locally infinitely divisible with the local characteristics</u> $(\tilde{a}, A, \tilde{\Pi})$ <u>with respect to the measure</u> $\tilde{P}$ <u>and the family of</u> σ<u>-algebras</u> $\{\mathcal{F}_t, t\geq 0\}$.

§2. On non-linear filtering stochastic equation

Let a stochastic process $\{X(t), t\geq 0\}$ satisfying assumptions (I)-(III) be observed, and let $\{\theta(t), t\geq 0\}$ be a non-observable stochastic process with values in some measurable space $(\mathfrak{X},\mathcal{O}\!\mathcal{L})$ adapted to the family of σ-algebras $\{\mathcal{F}_t, t\geq 0\}$. Following [13] we denote by $\mathcal{D}(\tilde{A})$ the space of the $\mathcal{O}\!\mathcal{L}$-measurable real valued functions $f(\theta)$ for which $E|f(\theta(t))|^2<\infty,\ t\geq 0$, and there exists a $\mathcal{B}[0,\infty)\times\mathcal{F}$-measurable function $\tilde{A}_t f$ adapted to the family $\{\mathcal{F}_t, t\geq 0\}$ such that, for each $t\geq 0$,

$$E\Big(\int_0^t |\tilde{A}_s f|^2 ds\Big)<\infty$$

and

$$M_f(t)=f(\theta(t))-\int_0^t \tilde{A}_s f\,ds \in \mathcal{M}^{(1)}$$

Our purpose is to derive a stochastic equation for the stochastic process $E^t f(\theta(t)),\ t \geq 0$.

We say that the functions $\varphi_1, \varphi_2 \in L^{(m)}$ $(\psi_1, \psi_2 \in F_Q^{(1)})$ are equivalent if $\|\varphi_1 - \varphi_2\|_t = 0$ $(\|\psi_1 - \psi_2\|_t^{(1)} = 0)$ for all $t \geq 0$.

The following assertion will be necessary.

Lemma 1 [6]. For every $M \in \mathcal{M}^{(1)}$, there exist unique, up to equivalence, functions $\varphi_M \in L_t^{(m)}$ and $\psi_M \in F_Q^{(1)}$ such that

$$E\left[M(t) X_\varphi(t)\right] = E\left[\int_0^t (\varphi_M(s), \varphi(s) A(s))\, ds\right]$$

and

$$E\left[M(t) Q_\psi(t)\right] = E\left[\int_0^t \int_{R^m} \psi_M(s,x)\, \psi(s,x)\, \Pi(s,dx)\, ds\right]$$

for all $\varphi \in L^{(m)}$ and $\psi \in F_Q^{(1)},\ t \geq 0$.

Under assumptions (I)-(III), the classes of functions $\bar{F}_P^{(m)}$, $\bar{F}_Q^{(m)}$ and $\bar{L}^{(m)}$ are defined similarly to the classes $F_P^{(m)}$, $F_Q^{(m)}$ and $L^{(m)}$, only with the family $\{\mathcal{F}_t,\ t \geq 0\}$ and the function $\Pi(t,\Gamma)$ substituted with the family $\{\mathcal{F}_t^X,\ t \geq 0\}$ and the function $\bar{\Pi}(t,\Gamma)$. Stochastic integrals with respect to the measure

$$\bar{q}(t,\Gamma) = p(t,\Gamma) - \int_0^t \bar{\Pi}(s,\Gamma)\, ds$$

and martingale

$$\bar{X}(t) = X(t) - \int_0^t \bar{a}(s)\, ds - \int_0^t \int_{|x| \leq 1} x\, \bar{q}(ds,dx) - \int_0^t \int_{|x| > 1} x\, p(ds,dx)$$

are denoted by

$$\bar{Q}_\psi(t) = \int_0^t \int_{R^m} \psi(s,x)\, \bar{q}(ds,dx), \quad \psi \in \bar{F}_Q^{(m)},$$

and

$$\bar{X}_\varphi(t) = \int_0^t \varphi(s)\, d\bar{X}(s) = \sum_{i=1}^m \int_0^t \varphi_i(s)\, d\bar{X}_i(s), \quad \varphi \in \bar{L}^{(m)}$$

Let $\mathcal{M}_X^{(m)}$ be the class of all m-dimensional right continuous square integrable martingales with respect to the family of σ-algebras $\{\mathcal{F}_t^X, t \geq 0\}$. We shall use the following assertion.

(A) If $\bar{M} \in \mathcal{M}_X^{(1)}$ and, for all $\varphi \in \bar{F}_Q^{(1)}$, $\phi \in \bar{L}^{(m)}$, $t \geq 0$,

$$E[\bar{M}(t)(\bar{Q}_\varphi(t) + \bar{X}_\phi(t))] = 0,$$

then $\bar{M}(t) \equiv \bar{M}(0)$ a.e.

This hypothesis is equivalent to the assumption that every martingale $\bar{M} \in \mathcal{M}_X^{(1)}$ can be expressed in the form

$$\bar{M}(t) = \bar{M}(0) + \bar{Q}_\varphi(t) + \bar{X}_\phi(t)$$

for some $\varphi \in \bar{F}_Q^{(1)}$ and $\phi \in \bar{L}^{(m)}$.

(IV) Assume that for every $t > 0$

$$E\Big[\int_0^t |f(\theta(s))|^2 (\Psi(s) - \bar{\Psi}(s), (\Psi(s) - \bar{\Psi}(s))A(s))\, ds\Big] < \infty$$

and

$$E\Big[\int_0^t \int_{R^m} \Big(|f(\theta(s))|^2 \frac{|\rho(s,x) - \bar{\rho}(s,x)|^2}{\rho(s,x)\,\bar{\rho}(s,x)} + |\tilde{F}_{s,x} f|^2 \frac{\rho(s,x)}{\bar{\rho}(s,x)}\Pi(s,dx)\, ds\Big] < \infty,$$

where $f \in D(\tilde{A})$, $\tilde{F}_{t,x} f \in F_Q^{(1)}$ and $\tilde{D}_t f \in L^{(m)}$ are defined from Lemma 1 uniquely, up to equivalence, by the equalities:

$$E[M_f(t) Q_\varphi(t)] = E\Big[\int_0^t \int_{R^m} \tilde{F}_{s,x} f\, \varphi(s,x)\Pi(s,dx)\, ds\Big]$$

and

$$E[M_f(t) X_\phi(t)] = E\Big[\int_0^t (\tilde{D}_s f, \phi(s) A(s))\, ds\Big],$$

$\varphi \in F_Q^{(1)}$, $\phi \in L^{(m)}$, $t \geq 0$. Using Theorem 1, Lemma 1 and some ideas of [13], we can prove the following theorem.

<u>Theorem 3</u> [6]. <u>Under hypothesis</u> (A) <u>and assumptions</u> (I)-(IV), <u>for all</u> $f \in D(\tilde{A})$, <u>we have the following non-linear filtering equation:</u>

$$E^t f(\theta(t)) = E^0 f(\theta(0)) + \int_0^t E^s(\tilde{A}_s f)\, ds + \int_0^t \phi_f(s)\, d\bar{X}(s) +$$

$$+ \int_0^t \int_{R^m} \varphi_f(s,x)\, \bar{q}(ds, dx),$$

where

$$\varphi_f(t) = E^t\left[f(\theta(t))(\Psi(t) - \widetilde{\Psi}(t)) + \widetilde{D}_t f\right]$$

and

$$\psi_f(t,x) = E^t\left[f(\theta(t))\left(\frac{\rho(t,x)}{\widetilde{\rho}(t,x)} - 1\right) + \widetilde{F}_{t,x} f \frac{\rho(t,x)}{\widetilde{\rho}(t,x)}\right].$$

<u>Example 3</u>. Let $\{\theta(t), t \geq 0\}$ and $\{X(t), t \geq 0\}$ be solutions of equations (2)-(3) and the functions $b_n^{(2)}(t,\theta,x)$ are independent of θ ! If $f(\theta) = f(\theta_1, \ldots, \theta_n)$ is a C^2 -class function, then $f \in D(\widetilde{A})$,

$$\widetilde{A}_t f = (\alpha^{(1)}(t), \nabla f(\theta(t)) + \tfrac{1}{2}(\nabla A^{(1)}(t), \nabla f(\theta(t))) +$$

$$+ \int_{|z| \leq 1} [f(\theta(t)+z) - f(\theta(t)) - (z, \nabla f(\theta(t)))]\, \Pi^{(1)}(t, dz) +$$

$$+ \int_{|z|>1} [f(\theta(t)+z) - f(\theta(t))]\, \Pi^{(1)}(t, dz),$$

where $\nabla = \left(\frac{\partial}{\partial \theta_1}, \ldots, \frac{\partial}{\partial \theta_n}\right)$, the vector $\widetilde{D}_t f$ is defined by the equality

$$\widetilde{D}_t f\, b^{(2)}(t)\, b^{(2)}(t)' = \nabla f(\theta(t))\, b^{(1)}(t)\, b^{(2)}(t)',$$

and

$$\widetilde{F}_{t,x} = \widetilde{\psi}_f(t, x, \theta(t), X(t)),$$

where $\widetilde{\psi}_f\,(t, x, \theta, \widetilde{x})$ is the Radon-Nikodym density of the generalized measure

$$\mu^f_{\theta,t,\widetilde{x}}(\Gamma) = \int_{R_{m+n}} \left(f(\theta + F^{(1)}(t,\theta,\widetilde{x},y)) - f(\theta)\right) \chi_\Gamma(F^{(2)}(t,\theta,\widetilde{x},y)) \frac{dy}{|y|^{m+n+1}}$$

with respect to the measure

$$\mu_{\theta,t,\widetilde{x}}(\Gamma) = \int_{R_{m+n}} \chi_\Gamma(F^{(2)}(t,\theta,\widetilde{x},y)) \frac{dy}{|y|^{m+n+1}}, \quad \Gamma \in \mathcal{B}_m.$$

§3. On absolute continuity of measures corresponding to the stochastic processes

Denote by $\mathcal{D}$ the space of all the right continuous and having left limits functions $\omega(t)$, defined on the interval $[0,\infty)$ with values in the m-dimensional Euclidean space $(R_m, \mathcal{B}_m)$, by $\mathcal{D}_t$ the σ-algebra generated by the cylindrical sets $C_1(\Gamma) = \{\omega : \omega \in \mathcal{D},\ \ \omega(s) \in \Gamma\},\quad 0 \le s \le t,\quad \Gamma \in \mathcal{B}_m,$

$\mathcal{D} = \sigma(\bigcup_{t\ge 0} \mathcal{D}_t),\quad Z(t,\omega) = \omega(t),\quad t \ge 0,\quad \omega \in \mathcal{D}.$

Let $\tilde{a}(t) = (\tilde{a}_1(t), \ldots, \tilde{a}_m(t)),\ \tilde{A}(t) = \|\tilde{a}_{ij}(t)\|,\ \tilde{\Pi}(t,\Gamma)$ be such that for every $\Gamma \in \mathcal{B}_m$ the functions $\tilde{\Pi}(t,\omega,\Gamma),\ \tilde{a}_i(t,\omega),$ $\tilde{a}_{ij}(t,\omega),\ i,j = 1,\ldots,m$ are $\mathcal{B}[0,\infty)\times\mathcal{D}$ -measurable and adapted to the family of σ-algebras $\{\mathcal{D}_t,\ t \ge 0\}$, and, for fixed (t,ω), $\tilde{\Pi}(t,\omega,\Gamma)$ is a measure on $\mathcal{B}_m$.

A stochastic m-dimensional process defined on some probability space with right continuous and having the left limits trajectories and the measure μ corresponding to it on the space $(\mathcal{D}, \mathfrak{D})$ will be called the Markov type process with the local characteristics $(\tilde{a}, \tilde{A}, \tilde{\Pi})$ if the process $\{Z(t),\ t \ge 0\}$ is locally infinitely divisible with the local characteristics $(\tilde{a}, \tilde{A}, \tilde{\Pi})$ with respect to the measure μ and the family of σ-algebras $\{\mathfrak{D}_t, t \ge 0\}$.

Let ν be the measure on $(\mathcal{D}, \mathfrak{D})$ corresponding to a locally infinitely divisible process X having the local characteristics (a, A, Π) with respect to the measure P and the family of σ-algebras $\{\mathcal{F}_t,\ t \ge 0\}$.

(III) Assume that

$$A(t) \equiv \tilde{A}(t, X(\cdot)),$$

$$\Pi(t,\Gamma) = \int_\Gamma \rho(t,x)\tilde{\Pi}(t, X(\cdot), dx),\ \Gamma \in \mathcal{B}_m,\ t \ge 0,$$

and

$$a(t) \equiv \tilde{a}(t, X(\cdot)) + \Psi(t)\tilde{A}(t, X(\cdot)) + \int_{|x|\le 1} x(\rho(t,x)-1)\,\tilde{\Pi}(t, X(\cdot), dx)$$

where $\rho(t,x)$ and $\Psi(t)$ satisfy conditions (6).

Denote by $P_T^{(1)}$ the restriction of the measure P to the σ -algebra $\mathcal{F}_T^X$ and define

$$P_T^{(2)}(A) = \mu(B), \quad A \in \mathcal{F}_T^X,$$

for $A = \{X(\cdot) \in B\}$, $B \in \mathcal{D}_T$.

We note that $P_T^{(2)}$ is a measure on $\mathcal{F}_T^X$ if $\mu(B) = 0$ for all $B \in \mathcal{D}_T$ such that $\{X(\cdot) \in B\} = \emptyset$. This condition is satisfied if, for instance, $\mu_T \ll \nu_T$, where the measures μ_T and ν_T are the restrictions of the measures μ and ν to the σ -algebra $\mathcal{D}_T$.

Let $\mathcal{L}_t(\varphi, \psi)$ be defined by formula (7) with the martingale $\tilde{X}$, the measure q and the function Π replaced by the martingale $\bar{X}$, the measure $\bar{q}$ and the function $\bar{\Pi}$.

(V) Assume that the initial distribution $\mu(C_0(\Gamma))$, $\Gamma \in \mathcal{B}_m$, has the density $\rho_0(x)$ with respect to the initial distribution $\nu(C_0(\Gamma))$, $\Gamma \in \mathcal{B}_m$ and, for every $t \ge 0$,

$$E\mathcal{L}_t(-\bar{\Psi}, -\ln\bar{\rho}) = 1.$$

Using Theorems 1 and2 it is easy to prove the following assertion.

<u>Theorem 4</u>. [7]. <u>If assumptions</u> (III) <u>and</u> (V) <u>are satisfied, the initial distribution</u> $\mu(C_0(\Gamma))$, $\Gamma \in \mathcal{B}_m$, <u>and the local characteristics</u> $(\tilde{a}, \tilde{A}, \tilde{\Pi})$ <u>uniquely determine the measure</u> μ, <u>then</u>

$$\mu_T \ll \nu_T, \quad P_T^{(2)} \ll P_T^{(1)}$$

<u>and</u>

$$\frac{dP_T^{(2)}}{dP_T^{(1)}}(X(\cdot)) = \rho_0(X(0))\,\mathcal{L}_T(-\bar{\Psi}, -\ln\bar{\rho}).$$

<u>If, in addition</u>,

$$P\{\rho_0(X(0))\,\bar{\mathcal{L}}_T(-\bar{\Psi}, -\ln\bar{\rho})>0\}=1,$$

then

$$\mu_T \sim \nu_T\ , \quad P_T^{(1)} \sim P_T^{(2)}$$

and

$$\frac{dP_T^{(1)}}{dP_T^{(2)}}(X(\cdot)) = \left[\rho_0(X(0))\right]^{-1} \exp\Big\{\sum_{K=1}^{m}\int_0^T \bar{\Psi}_K(s)\,d\bar{X}_K(s) +$$

$$+\frac{1}{2}\int_0^T(\bar{\Psi}(s),\bar{\Psi}(s)A(s))ds + \int_0^T\int_{R^m}\chi_{\{|\ln\bar{\rho}|<1\}}(s,x)\ln\bar{\rho}(s,x)\bar{q}(ds,dx) +$$

$$+\int_0^T\int_{R^m}\chi_{\{|\ln\bar{\rho}|\geq 1\}}(s,x)\ln\bar{\rho}(s,x)p(ds,dx) + \int_0^T\int_{R^m}\chi_{\{|\ln\bar{\rho}|<1\}}(s,x)\Big(\frac{1}{\bar{\rho}(s,x)} -$$

$$-1+\ln\bar{\rho}(s,x)\Big)\bar{\Pi}(s,dx)ds + \int_0^T\int_{R^m}\chi_{\{|\ln\bar{\rho}|\geq 1\}}(s,x)\Big(\frac{1}{\bar{\rho}(s,x)}-1\Big)\bar{\Pi}(s,dx)ds\Big\}.$$

The conditions for (8) to be satisfied are considered in [5]. The problem when the initial distribution and the local characteristics uniquely determine the measure on $\mathcal{D}$ corresponding to the Markov type process is connected with the conditions of uniqueness of a solution for the so-called martingale problem (for the details see [7]).

References

1. P.A. Meyer, Probability and Potentials, Blaisdell, 1966.

2. H. Kunita, S. Watanabe, On square integrable martingales, Nagoya Math. J., 30(1967), 209-245.

3. Meyer P.A. Intégrales stochastiques, Séminaire de Probabilites I, Lecture Notes in Math., 39(1967), Springer.

4. Б. Григелионис, О представлении целочисленных случайных мер как стохастических интегралов по пуассоновской мере, Литовский матет. сб., IX, I (I97I), 93-I08.

5. Б.Григелионис, Об абсолютной непрерывности мер, соответствующих случайным процессам, Литовский матем.сб., XI, 4 (I97I), 783-794.

6. Б.Григелионис, О стохастических уравнениях нелинейной фильтрации случайных процессов, Литовский матем.сб.,XII, 4 (I972).

7. Б.Григелионис, О структуре плотностей мер, соответствующих случайным процессам, Литовский матем.сб., XIII, I (I973).

8. А.Н.Ширяев, Стохастические уравнения нелинейной фильтрации скачкообразных марковских процессов, Проблемы передачи информации, II, 3 (I966), 3-22.

9. Р.Ш.Липцер, А.Н.Ширяев, Нелинейная фильтрация диффузионных марковских процессов, Труды МИАН им.В.А.Стеклова, CIУ, (I968), I35-I80.

IO. T.Kailath. An innovation approach to least squares estimation, Part I: Linear filtering with additive white noise IEEE Transactions on Automatic Control, AC-13, 6 (1969), 646-655

II. А.В.Скороход, Случайные процессы с независимыми приращениями, "Наука", М., I964.

I2. И.И.Гихман, А.В.Скороход, Стохастические дифференциальные уравнения, "Наукова думка", Киев, I968.

I3. M.Fujisaki, G.Kallianpur, H.Kunita, Stochastic differential equations for nonlinear filtering problem, Osaka J. of Math., 9 (1972), 19-40

Institute of Physics and Mathematics
of the Academy of Sciences of the Lithuanian SSR
Vilnius

ON THE CONTINUOUS PASSAGE THROUGH A FIXED LEVEL OF A HOMOGENEOUS PROCESS WITH INDEPENDENT INCREMENTS ON A MARKOV CHAIN

D.V. Gusak

The class of processes considered in present paper was described in [1 - 2]. We restrict ourselves by considering the processes which pass through a fixed positive level in continuous way only. Our aim is to study the distributions of the extrema of homogeneous processes with independent increments defined on a Markov chain. The case of continuous passage through a positive level has been studied in [3 - 4] for ordinary processes with independent increments.

Let $\{x_t\,,\ t>0\}$ be a homogeneous Markov chain with finite number of states $\kappa=\overline{1,n_0}$ and transition probability matrix $P(t)=\exp\{tQ\}$, $\xi_\kappa(t)$ $(\kappa=\overline{1,n_0})$ be homogeneous processes with independent increments and characteristic functions (ch.f.)

$$Me^{i\alpha\xi_\kappa(t)}=\exp\{t\Psi_\kappa(\alpha)\}\qquad(\kappa=\overline{1,n_0}),$$

$$\Psi_\kappa(\alpha)=i\alpha a_\kappa-\frac{1}{2}\alpha^2 b_\kappa^2+\int_{-\infty}^{0}[e^{i\alpha x}-1-i\alpha x\delta(x>-1)]\,d\Pi_\kappa(x)$$

($a_\kappa>0$ if $b_\kappa=0$; $\delta(\cdot)$ is the indicator function).

Denote by $A, B^2, \Pi(x)$ diagonal matrices with elements a_κ b_κ^2, $\Pi_\kappa(x)$ $(\kappa=\overline{1,n_0})$ respectively, and put

$$\Psi_0(\alpha)=i\alpha A-\frac{1}{2}\alpha^2B^2+\int_{-\infty}^{0}[e^{i\alpha x}-1-i\alpha x\delta(x>-1)]\,d\Pi(x).$$

A homogeneous process with independent increments $\xi(t)$ $(t>0$, $\xi(0)=0)$ on the chain x_t (one can call it process controlled by the chain x_t) is defined by the ch.f.

(1) $$\Phi_t(\alpha)=\| M(e^{i\alpha\xi(t)}, x_t=r/x_0=k)\| = exp\{t\Psi(\alpha)\},$$

$$\Psi(\alpha)=\Psi_0(\alpha)+N[\Phi(\alpha)-I], \quad N[\Phi(0)-I]=Q,$$

$$\Phi(\alpha)=\| M(e^{i\alpha\chi_{kr}})\mathcal{P}\{x_{\sigma_1+0}=r/x_0=k\}\|,$$

σ_1 is the first time when the chain x_t changes the state, $\{\chi_{kr};\ kr=\overline{1,n_0}\}$ are random non-positive jumps, their distribution on the transitions of the chain x_t is determined by the matrix

$$V(x)=\|\mathcal{P}\{\chi_{kr}\le x,\ x_{\sigma_1+0}=r/x_0=k\}\|.$$

The process $\xi(t)$ $(t>0,\ \xi(0)=0)$ reaches the level $z>0$ in continuous way. Denote by θ_s an exponentially distributed positive random variable with parameter $s>0$.

$$\tau_z=inf\{t:\ \xi(t)\ge z\}, \quad \xi^+(t)=\sup_{0\le u\le t}\xi(u)$$

are respectively the time of the first reaching the level $z>0$ and the maximum of the process $\xi(\cdot)$ on the time interval $[0,t]$ (sample functions of the process $\xi(t)$ are supposed to be continuous from the right).

In this paper, the result of [5] concerning the distribution of τ_z is developed and the fact that the ch.f. $M(e^{-u\xi^+(\theta_s)+i\alpha\xi(\theta_s)})$ can be determined by the product of $Me^{i\alpha\xi(\theta_s)}$ and a certain transformation of $Me^{-u\xi^+(\theta_s)}$ is proved. This result enables us to reveal the dependence between $\xi^+(\theta_s)$ and $\xi(\theta_s)$; it can be established without use of the factorization identity for $sI-\Psi(\alpha)$ which is usually used for determining the ch.f. of the joint distribution of $\{\xi(\theta_s),\ \xi^+(\theta_s)\}$ for processes $\xi(t)$ with arbitrary (i.e. positive and negative) jumps.

Put

$$P_+(s,x;z) = \| \mathcal{P}\{\xi(\theta_s) < x+z,\ \xi^+(\theta_s) < z,\ x_{\theta_s} = \imath / x_0 = \kappa\} \|,$$

$$F(s,x) = \| \mathcal{P}\{\xi(\theta_s) > x,\ x_{\theta_s} = \imath / x_0 = \kappa\} \|,\ F'(s,x) = \frac{\partial}{\partial x} F(s,x).$$

The equation and conditions determining $P_+(s,x,z)$ can be obtained from the integro-differential equation and corresponding conditions for the distribution of $\{\xi(t),\ \xi^+(t)\}$)(see [6]):

(2)
$$sP_+(s,x;z) = sI\delta(-z<x) - \frac{\partial P_+}{\partial x} A + \frac{1}{2}\frac{\partial^2 P_+}{\partial x^2} B^2 +$$

$$+\int_{-\infty}^{0} [P_+(s,x-y;z) - P_+(s,x;z) - \delta(x>-1)\frac{\partial}{\partial x} P_+(s,x;z)]\, d\Pi(y) +$$

$$+\int_{-\infty}^{0} P_+(s,dy;z) N V(x-y) - P_+(s,x;z) N, \qquad (x<0);$$

(3) $$P_+(s,x;z) = P_+(s,0;z)\ (x>0),\ \frac{\partial}{\partial x} P_+(s,x;z) = 0,\ (x>0).$$

Note that $F(s,x)$ and $P_+(s,x;z)$ with conditions under consideration ($a_\kappa > 0$ if $b_\kappa = 0$) possess bounded derivatives.

The values $\xi(\theta_s)$, $\xi^+(\theta_s)$, $\xi(\tau_z)$ are considered together with the corresponding values of the chain x_{θ_s} (x_{τ_z}) , but, for the sake of brevity, the terms $x_{\theta_s}(x_{\tau_z}) = \imath / x_0 = \kappa$ will be often omitted. For example, we write

$$\| M(e^{-s\tau_z},\ x_{\tau_z} = \imath / x_0 = \kappa) \| = Me^{-s\tau_z},$$

$$\| M(e^{i\alpha\xi(\theta_s)},\ x_{\theta_s} = \imath / x_0 = \kappa) \| = Me^{i\alpha\xi(\theta_s)},$$

$$\| \mathcal{P}\{\xi^+(\theta_s) > z,\ x_{\theta_s} = \imath / x_0 = \kappa\} \| = \mathcal{P}\{\xi^+(\theta_s) > z\}.$$

Note that

$$M[e^{i\alpha\xi(\theta_s)}] = P(\theta_s) = \| \mathcal{P}\{x_{\theta_s} = \imath / x_0 = \kappa\} \| \quad (\neq I).$$

<u>Theorem</u>. <u>For the process</u> $\xi(t)$ <u>with ch.f.</u> (1) <u>on the chain</u> x_t <u>the following relations hold</u>:

$$(4)\quad Me^{i\alpha\xi(\theta_s)-u\xi^+(\theta_s)}=\Big[I-u\int_0^\infty e^{(i\alpha-u)z}\,Me^{-s\tau_z}dz\Big]Me^{i\alpha\xi(\theta_s)}=$$

$$=\Big[I-u\int_0^\infty \mathcal{P}\{\xi^+(\theta_s)>z\}\,P^{-1}(\theta_s)e^{(i\alpha-u)z}dz\Big]\,Me^{i\alpha\xi(\theta_s)},$$

$$(5)\quad Me^{-s\tau_z}=e^{-zR(s)},\quad R(s)=-F'(s,0)\,F^{-1}(s,0)=P(\theta_s)[M\xi^+(\theta_s)]^{-1},$$

$$(6)\quad \frac{\partial}{\partial z}Me^{-s\tau_z}=\frac{\partial}{\partial z}\mathcal{P}\{\xi^+(\theta_s)>z\}P^{-1}(\theta_s)=F'(s,z)F^{-1}(s,0)\quad (z>0).$$

We shall begin our proof with establishing (4).

The equation (2) with conditions (3) being prolonged on the whole line $-\infty<x<\infty$ implies:

$$(7)\quad sP_+(s,x;z)=sI\delta(-z<x)-\frac{\partial P_+}{\partial x}A+\frac{1}{2}\frac{\partial^2P_+}{\partial x^2}+$$

$$+\int_{-\infty}^0[P_+(s,x-y;z)-P_+(s,x;z)-\delta(x>-1)\frac{\partial}{\partial x}P_+(s,x;z)]\,d\Pi(y)+$$

$$+\int_{-\infty}^0 P_+(s,dy;z)\mathcal{N}V(x-y)-P_+(s,x;z)\mathcal{N}+R_+(s,x;z),$$

$$R_+(s,x;z)=[P_+(s,0;z)(sI-Q)-sI]\,\delta(x>0).$$

Let $\quad \widetilde{P}_+(s,\alpha;z)=M\big(e^{i\alpha(\xi(\theta_s)-z)},\ \xi^+(\theta_s<z)\big).$

Consider the Fourier-Stielties transform in x of (6):

$$(8)\quad \widetilde{P}_+(s,\alpha;z)(sI-\Psi(\alpha))=s(e^{-i\alpha z}-1)I+P_+(s,0;z)(sI-Q).$$

Obviously (8) is equivalent to the equation

$$M\big(e^{i\alpha\xi(\theta_s)},\ \xi^+(\theta_s)<z\big)(sI-\Psi(\alpha))=$$

$$=s(1-e^{i\alpha z})I+e^{i\alpha z}\mathcal{P}\{\xi^+(\theta_s)<z\}(sI-Q),$$

which, in terms of the Laplace-Stielties transform, takes the form

(9) $$Me^{i\alpha\xi(\theta_s)-u\xi^+(\theta_s)} =$$

$$=\left\{I-\frac{u}{u-i\alpha}\left[I-Me^{(i\alpha-u)\xi^+(\theta_s)}P^{-1}(\theta_s)\right]\right\}Me^{i\alpha\xi(\theta_s)}.$$

Now we prove that

(10) $$\mathcal{P}\{\xi^+(\theta_s)>z\}=Me^{-s\tau_z}P(\theta_s).$$

In fact,

$$\mathcal{P}\{\xi^+(\theta_s)>z,\ x_{\theta_s}=r/x_0=\kappa\}=s\int_0^\infty e^{-st}\mathcal{P}\{\xi^+(t)>z,\ x_t=r/x_0=\kappa\}dt=$$

$$=-\int_0^\infty \mathcal{P}\{\tau_z<t,\ x_t=r/x_0=\kappa\}de^{-st}=\int_0^\infty e^{-st}d_t\,\mathcal{P}\{\tau_z<t,\ x_t=r/x_0=\kappa\}=$$

$$=\int_0^\infty e^{-st}d_t\int_0^t\sum_{j=1}^{n_0}\mathcal{P}\{\tau_z\in du,\ x_u=j/x_0=\kappa\}\ \mathcal{P}\{x_t=r/x_u=j\}=$$

$$=s\int_0^\infty e^{-st}\int_0^t\sum_{j=1}^{n_0}\mathcal{P}\{\tau_z\in du,\ x_u=j/x_0=\kappa\}\ \mathcal{P}\{x_{t-u}=r/x_0=j\}=$$

$$=\sum_{j=1}^{n_0}M(e^{-s\tau_z},\ x_{\tau_z}=j/x_0=\kappa)\ \mathcal{P}\{x_{\theta_s}=r/x_0=j\}.$$

Keeping in mind (10) and the equation

$$Me^{(i\alpha-u)\xi^+(\theta_s)}=P(\theta_s)+(i\alpha-u)\int_0^\infty e^{(i\alpha-u)z}\mathcal{P}\{\xi^+(\theta_s)>z\}dz,$$

the required formula (4) can be easily obtained from (9).

Note that, from (10), we have

$$P_+(s,0;z)=\mathcal{P}\{\xi^+(\theta_s)<z\}=P(\theta_s)-\mathcal{P}\{\xi^+(\theta_s)>z\}=$$

$$=P(\theta_s)-Me^{-s\tau_z}P(\theta_s),$$

and the equation (8) can be rewritten in the form

(11) $$\widetilde{P}_+(s,\alpha;z)(sI-\Psi(\alpha))=s(Ie^{-i\alpha z}-Me^{-s\tau_z}).$$

Now we write the equation connecting the generating function for τ_z with the distribution function $F(s,z)$ $(z>0)$. It can be derived using the fact that the process $\xi(t)$ reaches a positive level in continuous way. For $h>0,\ z>0$ we have:

$$\mathcal{P}\{z \le \xi(t) < z+h,\ x_t = r/x_0 = k\} =$$

$$= \sum_{j=1}^{n_0} \int_0^t \mathcal{P}\{\tau_z \in du,\ x_u = j/x_0 = k\}\, \mathcal{P}\{0 \le \xi(t-u) < h,\ x_t = r/x_u = j\}.$$

This is equivalent to

$$\mathcal{P}\{z \le \xi(\theta_s) < z+h,\ x_{\theta_s} = r/x_0 = k\} =$$

$$= \sum_{j=1}^{n_0} M(e^{-s\tau_z},\ x_{\tau_z} = j/x_0 = k)\, \mathcal{P}\{0 \le \xi(\theta_s) < h,\ x_{\theta_s} = r/x_0 = j\}$$

or to

$$F'(s,z) = Me^{-s\tau_z} F'(s,0).$$

It is established in [5] that $\tau_z\ (z>0)$ is a homogeneous process with independent increments controlled by the chain $x_{\tau_z} = \eta_z\ (z>0)$, its cumulant $R(s) = z^{-1} \ln Me^{-s\tau_z}$ satisfying the matrix equation

$$sI - \Psi(R(s)) = 0. \tag{13}$$

Generally speaking, this equation does not have a unique solution. By use of equation (12) one can easily give a probabilistic interpretation of $R(s)$, without the equation (13) being used. In fact, the equation

$$F'(s,z) = e^{-zR(s)} F'(s,0) \qquad (0 < z < \infty) \tag{14}$$

being integrated, one can establish

$$R(s) = -F'(s,0)\, F^{-1}(s,0).$$

On the other hand, the equations (12) and (10) imply

$$(15)\quad M(e^{i\alpha\xi(\theta_s)},\ \xi(\theta_s)>0)=\frac{1}{i\alpha}\left[Me^{i\alpha\xi^+(\theta_s)}-P(\theta_s)\right]P^{-1}(\theta_s)F'(s,0).$$

Letting in (15) $\alpha\to 0$, we establish

$$\mathcal{P}\{\xi(\theta_s)>0\}=M\xi^+(\theta_s)P^{-1}(\theta_s)F'(s,0),$$

that means: $R^{-1}(s)=M\xi^+(\theta_s)P^{-1}(\theta_s)$. Now (14) implies

$$(16)\quad F'(s,z)=\frac{\partial}{\partial z}e^{-zR(s)}R^{-1}(s)F'(s,0)=\frac{\partial}{\partial z}Me^{-s\tau_z}F^{-1}(s,0),$$

i.e. we came to (6). It enables to represent the distribution of $\xi^+(\theta_s)$ in terms of the normalized positive part of the distribution of $\xi(\theta_s)$.

The comparison of (4) with (23) in [6] which determines $M[exp\{i\alpha\xi(\theta_s)-u\xi^+(\theta_s)\}]$ with the help of the factorization method, gives us:

$$I-u\int_0^\infty e^{(i\alpha-u)z}Me^{-s\tau_z}dz=\Phi_+(s,\alpha+iu)\Phi_+^{-1}(s,\alpha),$$

where

$$\Phi_+(s,\alpha)=Me^{i\alpha\xi^+(\theta_s)}.$$

It should be pointed out that all the above proved results will be true in the case of countable number of states if to require in addition that x_t is stable: $0<\nu_K<\infty$ for all $K=1,2,\ldots$.

To define $\widetilde{P}_+(s,\alpha,z)$ from (8) when the jumps of the process $\xi(t)$ are of arbitrary sign, we use the representation

$$sI-\Psi(\alpha)=(sI+N)\left[Me^{i\alpha\zeta_s^-}\right]^{-1}\left[I-H(s,\alpha)\right]\left[Me^{i\alpha\zeta_s^+}\right]^{-1},$$

where

$$\zeta_s^{\pm}=\left\|\sup_{0\le u\le\theta_s}(\inf)\ \xi_K^{\pm}(\theta_s)\delta_{iK}\right\|$$

and $I-H(s,\alpha)$ admits the single canonical factorization. In this case, the ergodicity of the chain x_t implies that x_t is invertible and for

$$Me^{i\alpha\xi(\theta_s)} = s(sI-\Psi(\alpha))^{-1}$$

the relation

$$Me^{i\alpha\xi(\theta_s)} = Me^{i\alpha\xi^+(\theta_s)} P(\theta_s) Me^{i\alpha\hat{\xi}^-(\theta_s)},$$

where $\hat{\xi}(t)$ is a homogeneous process with independent increments defined on the inversed Markov chain $\hat{x}_t$. (For the definition of an inverted Markov chain see [7, p. 136-137]).

References

1. M. Fukushima, M. Hitsuda, On a class of Markov processes taking values on lines and the central limit theorem, Nagoya Math. J., 30(1967), 47-56.
2. И.И.Ежов, А.В.Скороход, Марковские процессы, однородные по второй компоненте, Теория вероят. и ее примен., I4 (I969), 3-I4, 679-692.
3. В.М.Золотарев, Момент первого достижения уровня и поведение в бесконечности для одного класса процессов с независимыми приращениями, 9 (I964), 653-662.
4. А.А.Боровков, О времени первого прохождения для одного класса процессов с независимыми приращениями, Теория вероят. и ее примен., I0 (I965), 360-364.
5. I.I.Єжов, В.С.Королюк, Е.С.Штатланд, Про розподIл максимуму процесIв з незалежними приростами, керованими ланцюгом Маркова, ДоповIдI АН УРСР, № 2 (I969), II5-II8.

6. Д.В.Гусак, Об одном классе процессов с независимыми приращениями на конечной цепи Маркова, Укр.матем.ж., № 2 (1973).

7. Дж.Кемени, Дж.Снелл, Конечные цепи Маркова, Москва, Изд."Наука", 1970.

Institute of Mathematics
of the Academy of Sciences of the Ukrain SSR
Kiev

STATISTICAL PROBLEMS IN QUANTUM PHYSICS

A.S.Holevo

§ 1. Introduction

In this paper, we give a general and unified mathematical treatment for a number of statistical problems concerning optimal quantum measurements. It has at least one field of application, namely, the theory of quantum communication channels and optimal receivers of optical signals [1] . Of course, if we are concerned with the data already obtained by a given measurement, then the classical theory of statistics is fully applicable; but the question of an optimal choice of such a measurement is out of the scope of classical statistics and needs special quantum-mechanical consideration. It is such a kind of problems that will be dealt with in this paper. The central role will play a generalisation of a notion of quantum measurement, introduced in § 2.

It is accepted in quantum theory [2]-[4] that the set of events (or propositions) related to a given quantum system may be described as a family $\mathcal{L}$ of all orthogonal projections in a separable Hilbert space H (corresponding to the quantum system under consideration). Thus, in contrast to the classical probability theory where events form a boolean σ-algebra, here the family of events is a complete lattice with orthocomplementation [4] .

A state is a function ρ on $\mathcal{L}$, possessing the usual properties of probability

1) $\rho(P) \geq 0 , \quad P \in \mathcal{L}$;

2) $\sum_i \rho(P_i) = 1$ for each countable orthogonal decomposition $\{P_i\}$ of the identity in H (i.e. $P_i P_j = \delta_{ij} P_i$ and $\sum_i P_i = I$ in the sense of strong operator topology; here I is the identity

operator in H). Gleason's theorem [5] asserts that every state on $\mathcal{L}$ is of the form

$$\rho(P) = Tr\,\rho P,$$

where ρ is a nonnegative trace class operator in H such that $Tr\,\rho = 1$. It is called density operator of the state ρ.

By analogy with classical probability theory we may now introduce observables (this term replaces classical "random variables") with finite number of values $\{a_i\}$ as

$$A = \sum_i a_i P_i$$

where $\{P_i\}$ is a finite orthogonal decomposition of the identity in H ; some more elaborated consideration [3] leads to definition of an observable as an arbitrary self-adjoint operator in H :

$$A = \int_{-\infty}^{\infty} a\,P(da), \tag{1}$$

where $P(\cdot)$ is the spectral decomposition of A.

We may define measurement corresponding to observable (1) as a map

$$F \longrightarrow P(F), \quad F \in \mathcal{A}, \tag{2}$$

where $\mathcal{A}$, in this case, is the σ-algebra of Borel sets of real line. If ρ is a state, then the probability distribution on $\mathcal{A}$, given by

$$P_A(F) = \rho(P(F)), \quad F \in \mathcal{A}, \tag{3}$$

is called the probability distribution of the measurement (2) relative to the state ρ. Thus the mean value of observable A relative to the state ρ (if it exists) may be defined as

$$\int a\,P_A(da) = \int a\rho(P(da)).$$

If, in particular, A is bounded, then it is equal to $Tr\rho A$, and putting

$$\rho(A) = Tr\rho A$$

we obtain an extension of a state ρ to $\mathcal{L}(H)$, the algebra of all bounded operators on H, which is a linear positive normal functional on $\mathcal{L}(H)$ such that $\rho(I) = 1$.

Now let $(\Omega, \mathcal{A})$ be an arbitrary measurable space; in view of the following it is natural to define measurements with values in $(\Omega, \mathcal{A})$ as orthogonal decompositions $\{P(F)\,;\ F \in \mathcal{A}\}$ of the identity in H; as before the probability distribution of such a measurement (relative to the state ρ) is given by (3). To avoid a confusion with more general measurements to be introduced below we shall distinguish the measurements considered here as simple ones.

§ 2. Measurements and statistical decisions

Some previous works (see, for example,[6]) have led to the following

Definition. Let $(\Omega, \mathcal{A})$ be a measurable space; a measurement with values in $(\Omega, \mathcal{A})$ is a map

$$F \longrightarrow X(F), \quad F \in \mathcal{A},$$

where

1. $X(F)$ is a Hermitean nonnegative operator in $\mathcal{L}(H)$ for each $F \in \mathcal{A}$;

2. for each countable measurable decomposition $\{F_i\}$ of Ω (i.e. $F_i \in \mathcal{A}$, $F_i F_j = \emptyset$ for $i \neq j$ and $\bigcup_i F_i = \Omega$)

$$\sum_i X(F_i) = I$$

in the sense of strong operator topology.

In other words, measurement is a decomposition, not necessarily orthogonal, of the identity in H , or a positive operator-valued measure on $(\Omega,\mathcal{A})$ such that $X(\Omega)=I$ [8].

If $FG=\varnothing$ implies $X(F)X(G)=0$, then we obtain an usual quantum measurement in the sense of § 1, that is, a simple measurement.

The next proposition, which is an easy consequence of Naimark's theorem [9], justifies the definition from quantum-mechanical point of view.

<u>Proposition 1. Let $X(\cdot)$ be a measurement. There exist a Hilbert space H_0, a state ρ_0 on $\mathcal{B}(H_0)$ and a simple measurement $P(\cdot)$ in tensor-product space $H\otimes H_0$ such that</u>

$$\rho(X(F))=(\rho\otimes\rho_0)(P(F)),\quad F\in\mathcal{A},$$

<u>for each state ρ on $\mathcal{B}(H)$</u>.

Thus the probability distributions of measurements $X(\cdot)$ and $P(\cdot)$ relative to an arbitrary state ρ on $\mathcal{B}(H)$ coincide. In this sense, every measurement is statistically equivalent to a simple measurement over some extension of the initial system, which contains another "uncoupled" system in a fixed state ρ_0. We call triple $(H_0,\rho_0,P(\cdot))$ a realisation of a measurement $X(\cdot)$. It is by no means unique.

Now we give a general formulation for the problem of statistical decision. Suppose we are given a family of states $\{\rho_\theta\}$ on $\mathcal{B}(H)$ where parameter θ varies over a measurable space $(\Theta,\mathcal{T})$; a measurable space of decisions $(\Omega,\mathcal{A})$, and a nonnegative measurable function $W_\theta(\omega)$, $\theta\in\Theta$, $\omega\in\Omega$, which is to be interpreted as the loss corresponding to the decision ω when the true state is ρ_θ. Measurements with values in $(\Omega,\mathcal{A})$ will play the role of strategies.

The risk corresponding to parameter value θ and measurement

$X(\cdot)$ is defined as

$$R_\theta\{X(\cdot)\} = \int_\Omega W_\theta(\omega)\rho_\theta(X(d\omega)).$$

To be concrete consider the problem of minimizing the Bayes risk

$$R\{X(\cdot)\} = \int_\Theta R_\theta\{X(\cdot)\}\pi(d\theta),$$

where $\pi(\cdot)$ is a prior probability distribution of parameter θ. Introducing operators

$$K(\omega) = \int_\Theta W_\theta(\omega)\rho_\theta\pi(d\theta), \quad \omega\in\Omega,$$

we have

$$R\{X(\cdot)\} = \mathrm{Tr}\int_\Omega K(\omega)X(d\omega). \tag{4}$$

We will not dwell here on an exact definition of such an expression, but we point out that it is easily defined for the operator-valued functions of the form

$$K(\omega) = \sum_{i=1}^{n} K_i f_i(\omega), \tag{5}$$

where K_i are trace class operators, $f_i(\cdot)$ - measurable real-valued functions.

Denote by $M(\Omega,\mathcal{A})$ the set of all measurements with values in $(\Omega,\mathcal{A})$. It is evident that $M(\Omega,\mathcal{A})$ is a convex set, and $R\{X(\cdot)\}$ is an affine functional on $M(\Omega,\mathcal{A})$; therefore the knowledge of the extreme points of $M(\Omega,\mathcal{A})$ is important.

We have the next partial result.

Theorem 1. Every simple measurement is an extreme point of $M(\Omega,\mathcal{A})$; on the other hand, if $X(\cdot)$ is an extreme point of $M(\Omega,\mathcal{A})$, such that for some $F, G\in\mathcal{A}$, $F\neq G$

$$X(F)X(G) = X(G)X(F)$$

then $X(\cdot)$ is a simple measurement.

One may ask if there exist extreme points other than simple measurements; further we shall give an affirmative answer to this question.

The next theorem gives some conditions for a measurement to be optimal.

Theorem 2. If there exists $X(\cdot)$, minimizing functional (4) then there exists a Hermitean operator Λ such that $K(\omega)-\Lambda$ is zero almost everywhere with respect to the operator-valued measure $X(\cdot)$, that is

$$\int_F (K(\omega)-\Lambda)\,X(d\omega)=0$$

for any $F\in\mathcal{A}$. Putting $F=\Omega$, we see that

$$\Lambda=\int_\Omega K(\omega)X(d\omega).$$

On the other hand, if such Λ satisfies

$$K(\omega)-\Lambda\geq 0\,,\quad \omega\in\Omega\,,$$

then $X(\cdot)$ is an optimal measurement. The minimal risk is evidently equal to $Tr\,\Lambda$.

A necessary and sufficient condition could be given, but it is rather complicated and seems to be of little use, so that we omit it.

We illustrate the said above by the Bayes problem with finite number of decisions $\Omega=\{1,\dots n\}$. In this case a measurement is a collection of nonnegative operators $X(1),\dots X(n)$, such that

$$X(1)+\dots+X(n)=I$$

The Bayes risk is

$$R\{X(\cdot)\}=Tr\sum_{\omega=1}^{n}K(\omega)X(\omega).$$

The set of measurements $M(\Omega)$ is a compact set with respect to the topology induced by the weak operator topology in $\mathcal{B}(H)$; $R\{X(\cdot)\}$ is continuous with respect to this topology. Therefore in the case where the number of decisions is finite there always exists an optimal measurement. From theorem 2 we deduce a necessary condition for a measurement $X(\cdot)$ to be optimal: operator $\Lambda = \sum_{\omega} K(\omega) X(\omega)$ must be Hermitean trace class and satisfy the equalities

$$(K(\omega) - \Lambda) X(\omega) = 0, \quad \omega = 1, \dots n.$$

If Λ is Hermitean and satisfies the equalities $K(\omega) - \Lambda \geq 0$, $\omega = 1, \dots n$, then the measurement $\{X(\cdot)\}$ is optimal.

Consider an example where H is a two-dimensional Euclidean space and P_1, P_2, P_3 are projections on three vectors with angles between them equal to $2\pi/3$. Notice that

$$(6) \qquad 2/3\, P_1 + 2/3\, P_2 + 2/3\, P_3 = I$$

Introduce the states ρ_θ with density operators P_θ and put $\pi_\theta = 1/3$, $\theta = 1,2,3$. Let $\Omega = \{1,2,3\}$ and $W_\theta(\omega) = 1 - \delta_{\theta\omega}$. Then, taking into account (6), we have

$$R\{X(\cdot)\} = Tr \sum_{\omega=1}^{3} (1/2\, I - 1/3\, P_\omega) X(\omega).$$

If we put $\Lambda = 1/6\, I$ then the sufficient condition is satisfied for $X(\omega) = 2/3\, P_\omega$, $\omega = 1,2,3$, which is thus the optimal measurement. The minimal risk is equal to $Tr\Lambda = 1/3$. On the other hand, it is easy to show, that

$$\min_{M_0(\Omega)} R\{X(\cdot)\} = \frac{2 - \sqrt{3}/2}{3} > 1/3,$$

where $M_0(\Omega)$ is the collection of all simple measurements with values in Ω . The measurement $\{2/3\, P(\omega);\ \omega \in \Omega\}$ is not a simple one but it is an extreme point of $M(\Omega)$ which gives an answer to the question raised after theorem 1.

§ 3. Randomized measurements. Optimal testing a simple hypotesis

We consider measurements $X(\cdot)$, satisfying

(7) $$X(F)X(G) = X(G)X(F), \quad F, G \in \mathcal{A}.$$

Let $(\mathcal{P}_\Omega, \mathcal{A}_\Omega)$ be the space of all probability measures on $(\Omega, \mathcal{A})$ with the σ-algebra generated by sets of the form

$$\{\mu : \mu \in \mathcal{P}_\Omega, \mu(F_1) < c_1, \dots, \mu(F_n) < c_n\},$$

for arbitrary $F_1, \dots, F_n \in \mathcal{A}$ and arbitrary real numbers $c_1, \dots, c_n$.

Theorem 3 [6]. For each measurement $X(\cdot)$ with values in $(\Omega, \mathcal{A})$ satisfying (7), there exists a unique simple measurement $P(\cdot)$ with values in $(\mathcal{P}_\Omega, \mathcal{A}_\Omega)$, such that

$$X(F) = \int_{\mathcal{P}_\Omega} \mu(F) P(d\mu), \quad F \in \mathcal{A}.$$

In the case when $\Omega = \{1, \dots n\}$, the space $(\mathcal{P}_\Omega, \mathcal{A}_\Omega)$ is the simplex

$$\mathcal{P}_\Omega = \{(\mu_1, \dots \mu_n) : \mu_i \geq 0, \sum_i \mu_i = 1\}$$

with the σ-algebra of Borel sets. If $\{X(1), \dots, X(n)\}$ is a measurement such that $X(i)X(j) = X(j)X(i)$, then from the theorem we have

$$X_\omega = \int_{\mathcal{P}_\Omega} \mu_\omega P(d\mu_1, \dots, d\mu_n)$$

Using theorem 3, we can construct a realization of such a measurement, which roughly speaking consists, first, of a simple measurement $P(\cdot)$ over a given system, giving us some probability distribution $\{\mu_\omega\}$, and second, of a measurement over an auxiliary system, which gives a decision ω with probability μ_ω. Therefore measurements satisfying commutativity condition (7) may be interpreted as randomized measurements (see [6]).

In the case of two decisions $\Omega = \{1,2\}$ we have $X(2) = I - X(1)$ and every measurement automatically is a randomized one. The measurement is uniquely defined by the test $X = X(2)$ which is a Hermitean operator such that $0 \le X \le I$. Suppose that there exist two states ρ_0, ρ_1 and consider the problem of finding the most powerful test of size ε, that is an operator X, $0 \le X \le I$, which maximizes $\rho_1(X)$ subject to the condition $\rho_0(X) \le \varepsilon$. We have the following result.

Theorem 4 [6]. Let $\varepsilon > 0$. There exists the most powerful test of size ε; for a test X to be the most powerful it is necessary and sufficient, that

(i) if $\rho_0(X) < \varepsilon$, then $\rho_1(X) = 1$;

(ii) if $\rho_0(X) = \varepsilon$, then there exist nonnegative number $c = c(\varepsilon)$ such that X maximizes $\rho_1(X) - c\rho_0(X)$, $0 \le X \le I$.

§ 4. Information carried by quantum-mechanical measurements

Let ρ_θ, $\theta = 1, \dots n$ be the states on $\mathcal{B}(H)$, $\{\pi_\theta\}$ a prior probability distribution. Proceeding from the Shannon's formula, we define information (about the state of the system) carried by measurement $X(\cdot)$ as

$$\mathcal{J}\{X(\cdot)\} = \sum_\theta \pi_\theta \int \ln\left[\frac{\rho_\theta(X(d\omega))}{\sum_\alpha \pi_\alpha \rho_\alpha(X(d\omega))}\right] \rho_\theta(X(d\omega)).$$

Denote by M the set of all measurements (with values in an arbitrary measurable space), and by M_0 the subset of all simple measurements.

We shall assume that the entropy

$$\mathcal{H}(\rho) = -Tr\rho \ln\rho$$

is finite for each state ρ_θ.

Proposition 2. If the density operators $\{\rho_\theta\}$ all commute, then

$$(8)\quad \sup_{M_0} \mathcal{J}\{X(\cdot)\} = \sup_{M} \mathcal{J}\{X(\cdot)\} = \mathcal{H}(\sum_\theta \pi_\theta \rho_\theta) - \sum_\theta \pi_\theta \mathcal{H}(\rho_\theta).$$

Thus in the "classical" case measurements in our sense give the same information about the state of a system as simple measurements. On the other hand, it can be shown that in the example considered at the end of §2:

$$\sup_{M_0} \mathcal{J}\{X(\cdot)\} < \sup_{M} \mathcal{J}\{X(\cdot)\}$$

This quantity $\sup_{M} \mathcal{J}\{X(\cdot)\}$ may be considered as information rate for a simple quantum channel. It can be shown that this quantity is additive, whereas $\sup_{M_0} \mathcal{J}\{X(\cdot)\}$ is not.

In the earliest applied works on quantum channels the quantity in the right hand side of (8) was considered as information rate. The next result together with proposition 2 shows that, strictly speaking, such an identification is permissible only in the "classical" case.

Theorem 5. If there are noncommuting operators among $\{\rho_\theta\}$ then

$$\sup_{M} \mathcal{J}\{X(\cdot)\} < \mathcal{H}(\sum_\theta \pi_\theta \rho_\theta) - \sum_\theta \pi_\theta \mathcal{H}(\rho_\theta).$$

In the case of two states ρ_0, ρ_ε, one of which is "close" to the other, we can find a measurement which asymptotically (as $\varepsilon \to 0$) maximizes the information $\mathcal{J}\{X(\cdot)\}$.

Proposition 3. Let ρ_0 be a density operator, σ be a Hermitean operator such that

(i) for some number k, $-k\rho_0 \leq \sigma \leq k\rho_0$;

(ii) $Tr\,\sigma = 0$.

Then there exist a Hermitean L satisfying the equation

$$1/2\,(L\rho_0 + \rho_0 L) = \sigma. \tag{9}$$

If $|\varepsilon| \leq k$ then $\rho_\varepsilon = \rho_0 + \varepsilon\sigma$ is evidently a density operator. Denote by $\mathcal{J}_\varepsilon\{X(\cdot)\}$ the information carried by the measurement $X(\cdot)$ if the states are ρ_0, ρ_ε with some probabilities π_0, π_1.

Theorem 6. Let $P(\cdot)$ be a spectral decomposition of the operator L ; this simple measurement is asymptotically optimal in the sense that

$$\lim_{\varepsilon\to 0} \mathcal{J}_\varepsilon\{P(\cdot)\} / \sup_M \mathcal{J}_\varepsilon\{X(\cdot)\} = 1.$$

Moreover

$$\mathcal{J}_\varepsilon\{P(\cdot)\} \sim \frac{\varepsilon^2}{2}\pi_0\pi_1\rho_0(L^2).$$

Operator L satisfying (9) is, to some extent, an analogue of the classical likelihood ratio [1] .

§ 5. Optimal joint measurement of several observables

Let $A_1, \ldots, A_n$ be arbitrary observables (i.e. self-adjoint operators in H). The observables are called compatible if the operators $A_1, \ldots, A_n$ all commute. It is only in this case when the joint spectral decomposition

$$A_i = \int \ldots \int a_i P(da_1, \ldots, da_n), \quad i = 1, \ldots, n,$$

exists, defining joint measurement of these observables. In classical probability theory all the observables are compatible and can be jointly measured with arbitrary precision; however, for a quantum system a nontrivial question of optimal joint measurement of several (incompatible) observables arises. We must find compatible observables $\widetilde{A}_1, \dots, \widetilde{A}_n$, generally, in some extension of a given system, which would give an optimal approximation to $A_1, \dots, A_n$. Let H_0 be another Hilbert space, ρ_0 be a state on $\mathcal{L}(H_0)$, $\widetilde{A}_1, \dots, \widetilde{A}_n$ be self-adjoint commuting operators in $H \otimes H_0$. Let ρ be a state on $\mathcal{L}(H)$ and we suppose that the second moments of observables $A_1, \dots, A_n$ are finite. We write total mean-square error as

$$\Sigma(\rho \otimes \rho_0)\{(A_i \otimes I - \widetilde{A}_i)^2\} = Tr(\rho \otimes \rho_0) \int \Sigma (a_i I - A_i)^2 P(da_1, \dots da_n)$$

where $P(\cdot)$ is the joint spectral decomposition of $\widetilde{A}_1, \dots, \widetilde{A}_n$

Now there exists a unique positive linear normal map $\mathcal{E}$ from $\mathcal{L}(H) \otimes \mathcal{L}(H_0)$ to $\mathcal{L}(H)$ such that

$$\mathcal{E}(X \otimes Y) = \rho_0(Y) X; \quad X \in \mathcal{L}(H), \ Y \in \mathcal{L}(H_0),$$

(this is closely connected with the notion of expectation, studied in [7]). If we put $X(F) = \mathcal{E}(P(F))$, $F \in \mathcal{A}$, then we obtain a measurement $X(\cdot)$ over the initial system. The mean-square error may be rewritten as

$$(10) \qquad R\{X(\cdot)\} = Tr \int_{R^n} K(a_1, \dots, a_n) X(da_1, \dots, da_n),$$

where

$$K(a_1, \dots, a_n) = \sum_i (A_i - a_i I) \rho (A_i - a_i I).$$

Thus, finally the problem of optimal joint measurement of observables $A_1, \dots, A_n$ may be formulated as follows: in the set $M(R^n, \mathcal{B}(R^n))$ of measurements with values in $(R^n, \mathcal{B}(R^n))$, where $\mathcal{B}(R^n)$ is the σ-algebra of Borel sets of a real n-dimensional space R^n, find an optimal one, minimizing the functional (10).

Theorem 7. Let the observables $A_1, \dots, A_n$ have finite second moments. In $M(R^n, \mathcal{B}(R^n))$ there exist an optimal joint measurement of $A_1, \dots, A_n$.

Notice that mathematically the problem is quite similar to one considered in §2. Now we apply theorem 2 to obtain a solution of this problem in an important case. For a more detailed account of related quantum-mechanical notions see, for example [3], [1]. Let p, q be regular irreducible representation of commutation relation $qp - pq \subseteq iI$. Thus p, q are noncompatible observables which are usually labelled as "the momentum" and "position" of a Bose particle. We consider the problem of optimal joint measurement of p and q for Gaussian states, which are defined as those having the generating function of the form

$$\rho(\exp(ixp + iyq)) = \exp\left(-\frac{\sigma^2}{2}(x^2 + y^2)\right).$$

We have $\rho(p^2) = \rho(q^2) = \sigma^2$, hence, taking into account the Heisenberg inequality

$$\text{(11)} \qquad \rho(p^2)\rho(q^2) \geq 1/4,$$

we see that $\sigma^2 \geq 1/2$. Put

$$R(\{X(\cdot)\} = Tr \iint K(a, b) X(da\,db),$$

where

$$K(a, b) = (p - aI)\rho(p - aI) + (q - bI)\rho(q - bI).$$

Theorem 8. The minimal mean-square error

$$\min R\{X(\cdot)\} = \frac{2\sigma^2}{2\sigma^2 + 1} \; (= c)$$

is attained for the measurement

$$X(F) = \frac{1}{2\pi c^2} \iint_F P(c^{-1}x, c^{-1}y)\,dx\,dy, \quad F \in \mathcal{B}(R^2),$$

where $P(x,y),\ (x,y)\in R^2$ are projections on the so called "coherent" states (cf. for example [1]). A realisation of this measurement is $(H_0, \rho_0, P(\cdot))$, where H_0 is a space of another irreducible representation p_0, q_0 of commution relation, ρ_0 is the vacuum state on $\mathcal{L}(H_0)$, and $P(\cdot)$ is the joint spectral decomposition of (commuting) operators

$$\tilde{p} = c(p\otimes I + I\otimes p_0), \qquad \tilde{q} = c(q\otimes I - I\otimes q_0).$$

This theorem follows from theorem 2 if we put

$$\Lambda = \rho p \rho + q\rho q - \frac{4\sigma^4}{4\sigma^4 - 1}(p^2+q^2-1)\rho .$$

Notice that in the case $\sigma^2 = 1/2$ we have

$$\rho(\tilde{p}-p)^2\rho(\tilde{q}-q)^2 = 1/16$$

so that Heisenberg inequality (11) should not be regarded as giving the lower bound for the error of a joint measurement of p and q.

§ 6. Conlusion and remarks

1. We have shown that various problems concerning optimal quantum measurements may be joint in the following general formulation: find the minimum of an affine functional of the form (4) over the convex set of positive operator-valued measures on a fixed $(\Omega,\mathcal{A})$. Such problems as parameter estimation are in fact included in considerations of §2 and we omit them here only for brevity. This is a new mathematical problem, to which we gave a solution, applicable to important concrete problems

2. In §2 we have introduced a generalization of the notion of a measurement, whose quantum-mechanical meaning was displayed in Proposition 1. The example considered in §2, §4 leads us to the essential observation that a measurement over an extension of a given quantum-mechanical system including another independent

"noise" system may substantially reduce the risk (or give more information about the initial system) as compared with any direct measurement over the initial system.

This seems somewhat paradoxal because it has no counterparts in the classical information and statistical decision theory, where randomized strategies play the role of measurements in the sense of our definition (cf. §3) (in the classical "commutative" case all the extreme points are simple measurements; cf. Theorem 1).

3. There is a range of statistical problems such as estimation of a single parameter etc. (see [1]), in which an optimal measurement is necessarily simple. In [10] we gave a mathematical treatment for some problems of such a kind, namely, extrapolation and mean estimation of a quasi-free (Gaussian) Boson field. It was shown that these lead to an integral equation which is a regularization of the corresponding classical one.

When the quantum system has infinitely many degrees of freedom (fields), the corresponding algebra of bounded observables may be much more complicated than $\mathcal{L}(H)$ [11]. Some of statistical problems for this more general case were considered in [6], [10], but there are still unsolved questions; in particular, it is not known whether Proposition 1 may be extended to an arbitrary von Neumann algebra (or even to a factor of type II or III).

References

[1] C.W.Helstrom et al: Quantum Mechanical Communication Theory, Proc. IEEE, 58 (1970), 1578-1598.

[2] J.V.Neumann: Mathematical foundations of quantum mechanics, Princeton University Press, 1955.

[3] G.W.Mackey: The Mathematical Foundations of Quantum Mechanics, Benjamin, 1963.

[4] V.S.Varadarajan: Geometry of Quantum Theory, Van Nostrand,1968.

[5] A.M.Gleason: Measures on closed subspaces of a Hilbert space, J. Rat. Mech. Anal. 6 (1957), 885-894

[6] А.С.Холево: Аналог теории статистических решений в некоммутативной теории вероятностей, Тр.Мос.мат.об-ва, 26 (I972), I33-I49.

[7] H.Umegaki: Conditional expectations in an operator algebra, Tohoku Math. J., 6, (1954), 177-181.

[8] S.K.Berberian, Notes on spectral theory, Van Nostrand, 1966.

[9] М.А.Наймарк: Спектральные функции симметричного оператора, Изв. АН СССР, 4 (I940), 277-3I8.

[I0] А.С.Холево: Некоторые статистические задачи для квантовых полей, Теория вероят. и ее примен., I7, I972, 360-365.

[11] H.Araki, E.J.Woods: Representations of the canonical commutation relations, describing a nonrelativistic infinite free Bose gas, J. Math. Phys. 4 (1963), 637-662.

Steklov Mathematical Institute
of the Academy of Sciences of the USSR
Moscow

OPTIMAL CODING IN WHITE GAUSSIAN CHANNEL WITH FEEDBACK

Shunsuke Ihara

Let us consider the coding scheme for the transmission of a Gaussian random variable θ through a white Gaussian channel with feedback. The model is formulated as follows: $d\xi_t = z(t,\xi_0^t,\theta)\,dt + N\,dw_t$, $t \geqq 0$. Shiryayev [1] constructed the "optimal" coding $z^*(t)$ in the linear codings: $z(t) = z(t,\xi_0^t,\theta) = A_0(t,\xi_0^t) + A_1(t)\,\theta$, conditioned by $Ez^2(t) \leqq P_0$, $t \geqq 0$ (Theorem 1). The "optimal" means that the coding minimizes the mean square error of estimating θ based on the observations ξ_s, $0 \leqq s \leqq t$. The purpose of this paper is to show that, in all codings z (not necessarily of linear type) conditioned by $Ez^2(t) \leqq P_0$, the coding z^* is optimal in the above sense and moreover z^* is optimal also in the sense of maximizing the information quantity between θ and $\xi_0^t = \{\xi_s,\ 0 \leqq s \leqq t\}$ (Theorem 2).

1. Let θ be a Gaussian random variable with the distribution $N(m,\gamma)$, which is the message to be transmitted. And let $w = \{w_t,\ t \geqq 0\}$ be a standard Wiener process independent of θ. The model for a white Gaussian noise channel with noise-free feedback is formulated as follows: let $\xi = \{\xi_t,\ t \geqq 0\}$ be the output, then

$$(1) \qquad \begin{cases} d\xi_t = z(t,\xi_0^t,\theta)\,dt + N\,dw_t, & t > 0, \\ \xi_0 = 0 \end{cases}$$

where ξ_0^t stands for the path ξ_s, $0 \leqq s \leqq t$ and $N > 0$ is a given constant. We discuss only on the codings z which satisfy the following assumptions.

Assumption (a). The equation (1) has the unique solution $\xi = \{\xi_t\}$.

(b). $Ez^2(t) \leqq P_0$, for each $t \geqq 0$, where $P_0 > 0$ is a constant.

Denote by Z the set of all codings satisfying the assumptions (a) and (b).

Our problem is to find the optimal codings z_1^* and z_2^* in the sense of (I) and of (II), respectively:

(I) Minimizing the mean square filtering error of estimating θ by the data ξ_0^t.

(II) Maximizing the information quantity $I(\theta,\xi_0^t)$ between θ and ξ_0^t.

Denote by $\sigma^2(t)$ the minimum of the mean square errors:

(2) $$\sigma^2(t) = \inf_{z \in Z} E(\theta - \hat{\theta}_t)^2,$$

where $\hat{\theta}_t = \hat{\theta}_t(z) = E[\theta|F_t^\xi]$ is the best estimate corresponding to $z \in Z$. *)

And denote by $I(t)$ the maximum of the information quantity:

(3) $$I(t) = \sup_{z \in Z} I(\theta,\xi_0^t).$$

2. Shiryayev [1] found out the optimal coding $z^*(t)$ in the linear codings:

(4) $$z(t) = A_0(t,\xi_0^t) + A_1(t)\,\theta,$$

in the sense of (I). It is given in the following manner. Let

$$A_1^*(t) = \frac{P_0}{\gamma} \exp\left(\frac{P_0}{2N^2}\,t\right)$$

and $$A_0^*(t,\xi^*) = -\,\hat{\theta}_t^*\,A_1^*(t),$$

where

$$d\xi_t^* = [-\,A_1^*(t)\,\hat{\theta}_t^* + A_1^*(t)\,\theta]\,dt + N\,dw_t, \quad \xi_0^* = 0,$$

and $$\hat{\theta}_t^* = E[\theta|F_t^{\xi^*}].$$

Then the coding $z^*(t)$ is defined by

(5) $$z^*(t) = A_0^*(t,\xi^*) + A_1^*(t)\,\theta.$$

Denote by Z_0 the set of codings $z \in Z$ of type (4), and put

(6) $$\sigma_0^2(t) = \inf_{z \in Z_0} E(\theta - \hat{\theta}_t)^2.$$

Then the Shiryayev's result can be stated as follows.

---

*) F_t^ξ is the σ-algebra generated by ξ_s, $0 \leq s \leq t$.

THEOREM 1. The coding z^* is optimal in Z_0 in the sense of (I), that is,

(7) $$\sigma_0^2(t) = E(\theta - \hat{\theta}_t^*)^2 \quad (= \gamma \exp (- \frac{P_0}{N^2} t)).$$

3. Now the solution of our problem is given by

THEOREM 2. The coding $z^*(t)$ given by (5) is optimal in Z in the sense of (I) and also of (II), that is,

(8) $$\sigma^2(t) = E(\theta - \hat{\theta}_t^*)^2 \quad (= \gamma \exp (- \frac{P_0}{N^2} t)),$$

and

(9) $$I(t) = I(\theta, \xi_0^{*t}) \quad (= \frac{P_0}{2N^2} t).$$

Proof. At first, we give an inequality related to the information quantity and the mean square error. Define the ε-entropy $H_\varepsilon(\theta)$ of the Gaussian random variable θ by the quantity:

$$H_\varepsilon(\theta) = \inf \{I(\theta,\tilde{\theta}); E(\theta - \tilde{\theta})^2 \leqq \varepsilon^2\}.$$

Then we have the following well known formula:

$$H_\varepsilon(\theta) = \frac{1}{2} \log \max (\frac{\gamma}{\varepsilon^2}, 1).$$

Therefore the following ineqality holds for any random variable $\tilde{\theta}$,

(10) $$I(\theta,\tilde{\theta}) \geqq \frac{1}{2} \log \max (\frac{\gamma}{E(\theta - \tilde{\theta})^2}, 1).$$

On the other hand, Kadota, Zakai and Ziv [2] proved that the following inequality holds for any output $\xi = \{\xi_t\}$ corresponding to $z \in Z$.

$$\frac{P_0}{2N^2} t \geqq I(\theta, \xi_0^t)$$

$$(= \frac{1}{2N^2} \int_0^t [Ez^2(t) - E\hat{z}^2(t)]\, dt, \text{ where } \hat{z}(t) = E[z(t) | F_t^\xi]).$$

And it follows from (10) that

$$I(\theta,\xi_0^t) \geq I(\theta,\hat{\theta}_t) \geqq \frac{1}{2} \log \max \left(\frac{\gamma}{E(\theta - \hat{\theta}_t)^2} , 1 \right).$$

Taking into account the relations

$$\sigma^2(t) = \inf_{z \in Z} E(\theta - \hat{\theta}_t)^2 \leqq \sigma_0^2(t) = E(\theta - \hat{\theta}_t^*)^2 = \gamma \exp\left(- \frac{P_0}{N^2} t \right),$$

which are easily shown from (2), (6) and (7), we can derive

$$(11) \qquad \frac{P_0}{2N^2} t \geqq \sup_{z \in Z} I(\theta,\xi_0^t) \geqq \frac{1}{2} \log \max \left(\frac{\gamma}{\inf_{z \in Z} E(\theta - \hat{\theta}_t)^2} , 1 \right)$$

$$= \frac{1}{2} \log \frac{\gamma}{\sigma^2(t)} \geqq \frac{1}{2} \log \frac{\gamma}{\sigma_0^2(t)} = \frac{P_0}{2N^2} t.$$

Thus, all quantities in (11) must be equal each other. And (8) and (9) are proved simultaneously.

Remark. The relation (11) implies that if $z_1^* \in Z$ is optimal in the sense of (I), then z_1^* is optimal also in the sense of (II).

References

[1] A.N. Shiryayev; Statistics of diffusion type processes, Proceedings of the Second Japan-USSR Symposium on Probability Theory.

[2] T.T. Kadota, M. Zakai and J. Ziv; Mutual information of white Gaussian channel with and without feedback, IEEE Trans. Inform. Theory, IT-17 (1971), 368-371.

Faculty of General Education

Nagoya City University

Nagoya, Japan.

THE LOCAL STRUCTURE OF A CLASS OF DIFFUSIONS AND RELATED PROBLEMS

Nobuyuki Ikeda and Shinzo Watanabe

§ 0 Introduction. It is a difficult problem to discribe completely the structure of diffusion processes except the one-dimensional case where an almost complete theory is established mainly by Feller, Itô, McKean and Dynkin. For multi-dimensional case, we usually consider diffusions which have as its infinitesimal generator a differential operator. We can construct such diffusions with help of the theory of partial differential equations. Also, by solving stochastic differential equations, we can construct path functions of diffusions for which corresponding differential operators may degenerate.

Purpose of this paper is to investigate a class of multi-dimensional diffusions which are not in the framework of the classical diffusions i.e. diffusions whose infinitesimal generators are not necessarily differential operators. Of course, some classes of such diffusions are already considered: e.g., diffusions with Brownian hitting probabilities, rotation invariant diffusions, diffusions with boundary conditions without distinguishing the boundary and the interior. Also, there are several works discussing the general structure of such diffusions, (cf. e.g. Skorohod [22], Knight [15]). Here, we would like to discuss the local structure of such diffusions. For this, we must first know the quantities which characterize a diffusion process. In § 1, under the assumption of a symmetry, we define a system of measures called a system of generators which characterizes a given diffusion. Also we study some properties of such system of measures though we must say that our results on these lines are still quite unsatisfactory.

In § 2, we shall investigate some typical examples of diffusions in the framework of § 1. We shall discuss mainly the construction of sample functions. Here, for example, the method of skew product and construction of the excursions as a Poisson point process are used effectively. Also we are interested in the local property of sample functions and corresponding problem in analysis. In particular, a space of harmonic functions is studied in connection with the property of sample functions (e.g. Theorem 2.5). We note that such a study has been done extensively for diffusions with oblique reflections (cf. e.g. [5],[18],[20]).

In § 3, we study, in connection with applications for diffusions in § 2, a local property of sample functions of one-dimensional Lévy processes. The possibility of hitting a single point for such processes has been studied extensively by Kesten [14] (cf. also [3]) and purpose of § 3 is to analyse various aspects of such hitting.

§ 1. The system of generators of a class of diffusions.

Let D be a domain in R^n and $X = \{X_t, P_x, x \in D\}$ be a diffusion process on D. Let $\mathscr{D}(D)$ be the class of C^∞-functions with compact support. Let $m(dx)$ be an everywhere dense positive measure on D. We assume that X satisfies the following conditions:

(A.1) X is m-symmetric.

Under this condition, the resolvent operator G_α of X defines a bounded operator on $\mathcal{L}^2(D, m(dx))$ and we have an $\mathcal{L}^2$-Dirichlet space $F_\alpha = \{F, \varepsilon_\alpha\}$ by Fukushima's theory [6]. Let $\varepsilon[u, v] = \varepsilon_\alpha[u, v] - \alpha(u, v)_{\mathcal{L}^2}$. Then ε is independent of α. We assume further

(A.2) As a linear topological space, $\mathcal{D}(D) \underset{\text{dense}}{\hookrightarrow} F$ i.e., $\mathcal{D}(D) \subset F$ and the injection is continuous with dense range in F.

(A.3) ε is local in the sense that, if $u, v \in \mathcal{D}(D)$ and $v = 0$ on $\mathrm{Supp}(u)$, then $\varepsilon[u, v] = 0$.

<u>Theorem 1.1</u>. (Beurling - Deny [1]). <u>Let</u> X <u>satisfy the above conditions</u> (A.1), (A.2) <u>and</u> (A.3). <u>Then, there exist a symmetric and non-negative definite</u> $n \times n$-<u>matrix of signed Radon measures</u> (ν_{ij}) <u>on</u> D <u>and a non-negative Radon measure</u> k <u>on</u> D <u>such that</u>

$$\varepsilon[u,v] = \sum_{i,j=1}^{n} \int_D \frac{\partial u}{\partial x_i}(x) \frac{\partial v}{\partial x_j}(x) \nu_{ij}(dx) + \int_D u(x)v(x)k(dx).$$

(ν_{ij}) <u>and</u> k <u>are uniquely determined</u>.

Proof of the theorem can be found in [10].

<u>Definition 1.1</u>. The system of measures (m, ν_{ij}, k) is called the <u>system of generators</u> of the diffusion X. m is called the <u>speed measure</u>, $\{\nu_{ij}\}_{i,j=1}^{n}$ the <u>system of energy measures</u> and k the <u>killing measure</u>.

It is clear that the process X is uniquely determined from the system of generators. Of course, the system of generators cannot be given arbitrarily and it is a difficult problem to give its necessary and sufficient condition. In this section, we shall obtain some conditions satisfied by a system of generators and give a few examples. Some typical examples of such diffusions will be studied, in detail, in the next section.

Definition 1.2. A function $s(x)$ defined on a domain $D_1 \subset D$ is called (F,ε)-harmonic if, for each open subset $G \subset D_1$ such that $\bar{G} \subset D_1$, there exists $s^* \in F$ such that $s^* = s$ on G and for every $v \in F$, with $\mathrm{Supp}(v) \subset G$, $\varepsilon(s^*,v) = 0$. If in a neighborhood of a point $x \in D$, a coordinate system $S = (s_1(x), s_2(x), \cdots, s_n(x))$ exists such that each $s_i(x)$ is (F,ε)-harmonic, then S is called a harmonic coordinate.

We set further assumptions:

(A.4) $$k = 0.$$

(A.5) The Euclidean coordinate $x = (x_1, x_2, \cdots, x_n)$ is a harmonic coordinate.

It is easy to see that (A.4) is satisfied if and only if for every $u,v \in \mathcal{D}(D)$ such that

$$u = \text{const.} \quad \text{on} \quad \mathrm{Supp}(v),$$

we have $$\varepsilon[u, v] = 0.$$

Lemma 1.1. We assume that X satisfy (A.1)~(A.4). Then (A.5) is satisfied if and only if

(1.1) $$\int_D \sum_{k=1}^{n} \frac{\partial}{\partial x_k} u(x) \nu_{j,k}(dx) = 0, \quad \text{for every } u \in \mathcal{D}(D), \quad j = 1,2,\cdots, n.$$

Proof. Suppose (A.5) is satisfied. Let $u \in \mathcal{D}(D)$ and let G be an open set such that $\mathrm{Supp}(u) \subset G \subset \bar{G} \subset D$. Then, there exist $s_j(x) \in F$, $j = 1,2,\cdots, n$ such that $s_j(x) = x_j$ on G and $\varepsilon[s_j, u] = 0$. Thus

$$0 = \varepsilon[s_j,u] = \sum_{i,k=1}^{n} \int_D \frac{\partial s_j}{\partial x_i}(x) \frac{\partial u}{\partial x_k}(x) \nu_{ik}(dx)$$

$$= \sum_{k=1}^{n} \int_D \frac{\partial u}{\partial x_k}(x)\,\nu_{j,k}(dx), \qquad j = 1,2,\cdots, n.$$

Conversely, if $s_j(x) \in \mathcal{D}(D)$ is such that $s_j(x) = x_j$ on G and if $u \in \mathcal{D}(D)$ is such that $\mathrm{Supp}(u) \subset G$, then, as above, $\varepsilon[s_j,u] = \sum_{k=1}^{n} \int_D \frac{\partial u}{\partial x_k}(x)\,\nu_{jk}(dx)$. Thus, if (1.1) is satisfied, $x = (x_1,\cdots, x_n)$ is a harmonic coordinate.

Corollary 1. *If* X *satisfies* (A.1)$\sim$(A.5), *then*

(1.2) $$\varepsilon[u,v] = -\sum_{i,j=1}^{n} \int_D v(x)\frac{\partial^2 u}{\partial x_i \partial x_j}(x)\,\nu_{ij}(dx),$$

for every $u,v \in \mathcal{D}(D)$.

Proof. Let $u,v \in \mathcal{D}(D)$. Applying (1.1) for $w_j = v\,\frac{\partial u}{\partial x_j}$, we have

$$\varepsilon[u,v] = \sum_{i,j=1}^{n} \int \frac{\partial}{\partial x_i}\Big[v\,\frac{\partial u}{\partial x_j}\Big]\nu_{ij}(dx) - \sum_{i,j=1}^{n} \int v(x)\frac{\partial^2 u}{\partial x_i \partial x_j}\nu_{ij}(dx)$$

$$= -\sum_{i,j=1}^{n} \int v(x)\frac{\partial^2 u}{\partial x_i \partial x_j}\nu_{ij}(dx),$$

which completes the proof.

Proposition 1.1. *If* X *satisfies* (A.1)$\sim$(A.5), *then for every* $x \in D$, $\nu_{ij}(\{x\}) = 0$, $i,j=1,2,\cdots, n$.

Proof. Let $x_0 \in D$ be fixed. Set $\nu^*_{j,k}(dx) = \nu_{j,k}(dx) - \nu_{j,k}(\{x_0\})\delta_{\{x_0\}}(dx)$, $j,k=1,2,\cdots, n$. Choose $g \in \mathcal{D}(D)$ with the following properties:

(1.3) $g(x) = 0$, $|x - x_0| \geqq 1$,

(1.4) $g_i = \frac{\partial g}{\partial x_i}$ satisfies $g_i(x_0) > 0$ and $|g_i| < M$ for some constant M, $i = 1,2,\cdots, n$.

Set $u_\varepsilon(x) = \varepsilon\, g(\frac{x-x_0}{\varepsilon} + x_0)$, then $u_\varepsilon \in \mathcal{D}(D)$ and $\frac{\partial}{\partial x_i} u_\varepsilon(x) = g_i(x_0 + \frac{x-x_0}{\varepsilon})$. By Lemma 1.1, we have

$$\sum_{k=1}^{n} \int_D \frac{\partial}{\partial x_k} u_\varepsilon(x) \nu_{j,k}(dx) = 0, \quad j = 1,2,\cdots, n,$$

implying

$$-\sum_{k=1}^{n} \int_{|x-x_0| \leq \varepsilon} g_k(\frac{x-x_0}{\varepsilon} + x_0) \nu^*_{j,k}(dx) = \sum_{k=1}^{n} g_k(x_0) \nu_{j,k}(\{x_0\}), \; j=1,2,\cdots, n.$$

Letting $\varepsilon \downarrow 0$, we have

$$\sum_{k=1}^{n} g_k(x_0) \nu_{j,k}(\{x_0\}) = 0, \quad j = 1,2,\cdots, n.$$

Since this holds for any choice of above g, we have

$$\nu_{j,k}(\{x_0\}) = 0, \qquad j,k = 1,2,\cdots, n.$$

Theorem 1.2. Let $D = R^n$ and suppose X satisfies (A.1) $\sim$ (A.5). Further, we assume that

(A.6) $\quad m(dx) = dx \quad$ (: the Lebesgue measure),

and

(A.7) $\quad \{\nu_{ij}\}$ is of the form

(1.5) $\quad \nu_{ij}(dx) = \frac{1}{2}\delta_{ij} dx + \nu^0_{ij}(dx), \qquad i,j=1,2,\cdots, n,$

where ν^0_{ij} is singular to the Lebesgue measure. Then, if $\{\nu^0_{ij}\}$ is not identically 0 and if a Borel subset $A \subset R^n$ satisfies $\nu^0_{ij}(D \setminus A) = 0$ for all $i,j = 1,2,\cdots, n$, then A has a positive n-dimensional Newtonian capacity; i.e., A is non-polar with respect to n-dimensional Brownian motion.

<u>Proof</u>. Without loss of generality, we may assume that ν^0_{ij} is supported on $K = [0, 1]^n$, $i,j=1,2,\cdots, n$. Let $X^B = (X_t, P_x{}^B)$ be an n-dimensional Brownian motion (i.e. a diffusion with system of generators given by $m(dx) = dx$, $\nu_{ij}(dx) = \frac{1}{2}\delta_{ij}dx$ and $k = 0$.). We will show that, if $A \subset K$ satisfies $\nu^0_{ij}(K \setminus A) = 0$, $i,j=1,2,\cdots, n$ and A is polar with respect to X^B, then $\{\nu^0_{ij}\}$ is identically 0. First, we note that A is also polar with respect to X. In fact, let G_m be a sequence of open sets such that $G_m \supset A$ and $E^B_x(e^{-\lambda\sigma_{G_m}}) \searrow 0$ almost everywhere. Since the diffusion X is an n-dimensional Brownian motion up to the time σ_{G_m}, we have $E_x(e^{-\lambda\sigma_{G_m}}) = E^B_x(e^{-\lambda\sigma_{G_m}})$ and hence $E_x(e^{-\lambda\sigma_A}) = 0$ almost everywhere. Thus, A is polar with respect to X by Theorem 3.12 of [8]. Let the set of Radon measures M^+_0 be defined as in [8]. By Corollary 1, we have

$$\varepsilon_1[u,v] = \varepsilon[u,v] + \int_{R^n} u(x)v(x)dx$$

(1.6)
$$= - \int_{R^n} v(x) \sum_{i,j=1}^{n} \frac{\partial^2 u}{\partial x_i \partial x_j}(x)\nu_{ij}(dx) + \int_{R^n} u(x)v(x)dx.$$

Generally, if we set $g^u_\beta(x) = \beta(u - \beta G_{\beta+1}\, u)(x)$, we have

$$\varepsilon_1[u,v] = \lim_{\beta\uparrow\infty} \int_{R^n} v(x)g^u_\beta(x)dx, \qquad u,v \in F.$$

Combining this with (1.6), we see that the measure $g^u_\beta(x)dx$ converges vaguely to $\mu^u(dx) = -\sum_{i,j=1}^{n} \frac{\partial^2 u}{\partial x_i \partial x_j}(x)\nu_{ij}(dx) + u(x)dx$. In particular, (1.6) holds for every $v \in F \cap C(R^n)$ and $u \in \mathcal{D}(R^n)$. Choose $w,u \in \mathcal{D}(R^n)$ such that, $w = 1$ on K and $0 \leq w \leq 1$, and $\frac{\partial^2 u}{\partial x_1^2} = -1$, $\frac{\partial u}{\partial x_i} = 0$ $i=2,\cdots, n$ and $u > 0$ on $\mathrm{Supp}(w)$. Then,

$$\varepsilon_1[u,v] = \int_{R^n} v(x)(1-w(x))(u(x)-\tfrac{1}{2}\Delta u(x))dx + \int_{R^n} v(x)\mu(dx)$$

where $\mu(dx)=w\{\nu_{11}(dx) + u(x)dx\}$. Set $f = (1-w)(u-\frac{1}{2}\Delta u)$. Then, we have for every $v \in C(R^n)\cap F$,

$$\int_{R^n} v(x)\mu(dx) = \varepsilon_1[u, v - G_1 f],$$

and hence $\mu \in M_0^+$. Then, by Theorem 1.5 of [8], we have $\mu(A) = 0$ which implies that $\nu_{11}^0(A) = 0$. Similarly, we can prove that $\nu_{ii}^0(A) = 0$ and thus $\nu_{ij}^0 = 0$.

Remark 1.1. Under the assumption of Theorem 1.2, it can happen that some ν_{ii} has positive mass on a polar set A with respect to n-dimensional Brownian motion as is seen in the following example.

Example 1.1. Let $D = R^3$. There exists a diffusion X which satisfies (A.1)$\sim$(A.7) whose system of generators is given by

$$(1.7)\quad \begin{cases} m(dx) = dx (\equiv dx_1 dx_2 dx_3), \\ \nu_{11}(dx) = \frac{1}{2}dx, \\ \nu_{22}(dx) = \frac{1}{2}dx + I_{\{x_1=0\}}dx_2 dx_3, \\ \nu_{33}(dx) = \frac{1}{2}dx + I_{\{x_1=0\}}dx_2 dx_3 + I_{\{x_1=0, x_2=0\}}dx_3, \\ \nu_{ij} = 0, \quad i \neq j, \quad k(dx) = 0. \end{cases}$$

In fact, X can be constructed by the method of skew-product as follows: Let B_t^1, B_t^2, B_t^3 be three mutually independent one-dimensional Brownian motions starting at 0. Let $x_t^1 = x_1 + B_t^1$ and let ψ_t be the local time at 0 of x_t^1. Let $x_t^2 = x_2 + B_{t+\psi_t}^2$. Then (x_t^1, x_t^2) is a two-dimensional diffusion process for which the origin (0, 0) is regular for itself (cf. Example 2.1 of § 2). Thus the local time at (0, 0) of (x_t^1, x_t^2) exists which is denoted by $\tilde{\psi}_t$. Set $x_t^3 = x_3 + B^3_{t+\psi_t+\tilde{\psi}_t}$. Then $X_t = (x_t^1, x_t^2, x_t^3)$ defines a sample path of X starting at $x = (x_1, x_2, x_3)$.

Proposition 1.2. Let $D = R^2$ and suppose X satisfy (A.1)$\sim$(A.5).

(i) If $\nu_{12} = \nu_{21} = 0$, then ν_{11} and ν_{22} must be of the form

$$(1.8)\qquad \begin{aligned} \nu_{11}(dx_1 dx_2) &= dx_1 \nu_1(dx_2), \\ \nu_{22}(dx_1 dx_2) &= \nu_2(dx_1) dx_2, \end{aligned}$$

where ν_1 and ν_2 are measures on R^1.

(ii) If $\nu_{ij}(dx) = \frac{1}{2}\delta_{ij} dx + \nu_{ij}^0(dx)$, where $\nu_{ij}^0(dx)$ is singular to the Lebesgue measure dx and if ν_{11}^0 and ν_{22}^0 are singular each other then ν_{11}^0 and ν_{22}^0 must be of the form (1.8) and $\nu_{12}^0 = \nu_{21}^0 = 0$.

(iii) If $\nu_{ij}(dx) = \frac{1}{2}\delta_{ij} dx + \nu_{ij}^0(dx)$, where $\nu_{ij}^0(dx)$ is singular with respect to the Lebesgue measure dx and ν_{11}^0 is of the form (1.8), then ν_{22}^0 must be of the form (1.8) and $\nu_{12}^0 = \nu_{21}^0 = 0$.

We can prove this theorem easily by using Lemma 1.1 and the detail is omitted.

Example 1.2. Let X be a diffusion on R^2 which satisfies (A.1)$\sim$(A.7) such that

$$(1.9)\qquad \begin{cases} \nu_{11}(dx) = \frac{1}{2} dx_1 (dx_2 + \nu_1^0(dx_2)), \\ \nu_{22}(dx) = \frac{1}{2}(dx_1 + \nu_2^0(dx_1)) dx_2, \\ \nu_{12} = \nu_{21} = 0, \end{cases}$$

where ν_1^0 and ν_2^0 are measures on R^1 singular to the Lebesgue measure. This diffusion X can be constructed in the following way: Let B_1 and B_2 be the Borel sets of the Lebesgue measure 0 such that $\nu_2^0(R^1 \setminus B_1) = 0$ and $\nu_1^0(R^1 \setminus B_2) = 0$. Let X_t^1 and X_t^2 be the two mutually independent one-dimensional diffusion processes with infinitesimal generators $\frac{d}{2d\nu_2}\frac{d}{dx}$ and $\frac{d}{2d\nu_1}\frac{d}{dx}$ respectively where $\nu_i = dx + \nu_i^0$, $i = 1,2$. Let $\tilde{X}_t = (X_t^1, X_t^2)$ and $A_t = \int_0^t I_{(R^1\setminus B_1)\times(R^1\setminus B_2)}(\tilde{X}_s)ds$. Then $X_t = \tilde{X}_{A_t^{-1}}$, where A_t^{-1} is the inverse function of $t \longrightarrow A_t$, defines a sample function of X.

§2. Some examples of two dimensional diffusions.

a) Example 2.1. Let $D = R^2$ and consider the diffusion process X determined by the following system of generators:

$$(2.1)\quad \begin{cases} m(dx) = dx, \\ \nu_{11}(dx) = \frac{1}{2}dx + \delta_0(x_2)dx, \\ \qquad\quad = \frac{1}{2}dx + I_{\{x_2=0\}}dx_1, \\ \nu_{22}(dx) = \frac{1}{2}dx, \\ \nu_{12}(dx) = \nu_{21}(dx) = 0, \\ k(dx) = 0. \end{cases}$$

Sample functions of this diffusion X can be constructed by the method of skew product: Let $B^1(t)$ and $B^2(t)$ be two mutually independent one-dimensional Brownian motions. Let $x^2(t) = x_2 + B^2(t)$ and let ψ_t be the local time at 0 of x^2-process;

$$\psi_t = \lim_{\varepsilon\downarrow 0} \frac{1}{\varepsilon}\int_0^t I_{(-\varepsilon,\varepsilon)}(x_s^{\,2})ds.$$

Set

$$x^1(t) = x_1 + B^1(t + \psi(t)).$$

Proposition 2.1. $X_t = (x_t^1, x_t^2)$ defines a sample function starting at $x = (x_1, x_2)$ of the diffusion process X with the system of generators given by (2.1).

Proof. It is known (cf. [17],[12]) that

$$P_0[x_t^2 \in da, \psi_t \in db] = \frac{|a|+\frac{b}{2}}{\sqrt{2\pi t^3}} e^{-\frac{(|a|+\frac{b}{2})^2}{2t}} \frac{da db}{2}, \quad a \in R^1, \ b > 0, \tag{2.2}$$

and

$$P_x[\sigma_{L_1} \in d\theta] = \frac{|x_2|}{\sqrt{2\pi\theta^3}} \exp\{-\frac{|x_2|^2}{2\theta}\} d\theta, \tag{2.3}$$

where σ_{L_1} is the first hitting time of X_t to the x_1-axis L_1;

$$\sigma_{L_1} = \inf\{t \ ; \ x_t^2 = 0\}. \tag{2.4}$$

By (2.2) and (2.3), the transition probability of X_t-process is given by $P(t,x,dy) = p(t,x,y)dy$ where

$$\begin{aligned} & p(t,x,y) \\ & = I(x_2,y_2)\frac{1}{\sqrt{2\pi t}} e^{-\frac{(x_1-y_1)^2}{2t}} \frac{1}{\sqrt{2\pi t}} \times (e^{-\frac{(x_2-y_2)^2}{2t}} - e^{-\frac{(x_2+y_2)^2}{2t}}) \\ & + \int_0^\infty \frac{1}{\sqrt{2\pi(t+s)}} \exp\{-\frac{(x_1-y_1)^2}{2(t+s)}\} \times \frac{|x_2|+|y_2|+\frac{s}{2}}{2\sqrt{2\pi t^3}} \exp\{-\frac{(|x_2|+|y_2|+\frac{s}{2})^2}{2t}\} ds, \end{aligned} \tag{2.5}$$

and

$$I(\xi, \eta) = \begin{cases} 1, & \xi > 0, \ \eta > 0 \ \text{ or } \ \xi < 0, \ \eta < 0, \\ 0, & \text{otherwise.} \end{cases}$$

By Itô's formula (cf. [16]), we have, for $u \in \mathcal{D}(R^2)$,

$$u(X_t) - u(X_0) = \sum_{i=1}^{2} \int_0^t \frac{\partial u}{\partial x_i}(X_s) dA_s^{(i)} + \frac{1}{2} \int_0^t \sum_{i=1}^{2} \frac{\partial^2 u}{\partial x_i^2}(X_s) ds$$

$$+ \frac{1}{2} \int_0^t \frac{\partial^2 u}{\partial x_1^2}(X_s) d\psi_s,$$

where $\int dA_s^{(i)}$ (i=1,2) are stochastic integrals with respect to martingales $A_s^{(i)} = X^i(s) - X^i(0)$. Hence

$$\alpha G_\alpha u(x) - u(x) = \frac{1}{2}\sum_{i=1}^{2}\int_{R^2} g_\alpha(x,y)\frac{\partial^2 u}{\partial x_i^2}(y)dy + \int_{-\infty}^{\infty} g_\alpha(x,(y_1,0))\frac{\partial^2 u}{\partial x_1^2}((y_1,0))dy_1,$$

where $G_\alpha u(x) = \int_0^\infty e^{-\alpha t}E_x(u(X_t))dt = \int_{R^2} g_\alpha(x,y)u(y)dy$ with $g_\alpha(x,y) = \int_0^\infty e^{-\alpha t}p(t,x,y)dt$. Noting that $g_\alpha(x,y) = g_\alpha(y,x)$ and $\alpha G_\alpha v(x) \longrightarrow v(x)$, $v \in \mathcal{D}(R^2)$, we have

$$\lim_{\beta\uparrow\infty} \beta(u-\beta G_\beta u, v)_{L^2} = \sum_{i,j=1}^{2}\int_{R^2}\frac{\partial u}{\partial x_i}\frac{\partial v}{\partial x_j}\cdot \nu_{ij}(dx)$$

where $\{\nu_{ij}\}$ is given by (2.1). q.e.d.

Let $x \in R^2$ and σ_x be the hitting time to the point x: $\sigma_x = \inf\{t > 0 ; X_t = x\}$. Let $x^{(0)} = (x_1^{(0)}, x_2^{(0)})$ and L^1 be the x_1-axis: $L^1 = \{x = (x_1,x_2) ; x_2 = 0\}$.

<u>Proposition 2.2</u>. <u>If</u> $x^{(0)} \in L^1$, <u>then</u> $P_x(\sigma_{x^{(0)}} < \infty) = 1$ <u>for every</u> $x \in R^2$. <u>If</u> $x^{(0)} \notin L^1$, <u>then</u> $P_x(\sigma_{x^{(0)}} = \infty) = 1$ <u>for every</u> $x \in R^2$.

<u>Proof</u>. Using (2.5), we can show that if $x^{(0)} \in L^1$, $x \longmapsto g_\alpha(x,x^{(0)})$ is bounded and continuous. This implies that $P_x(\sigma_{x^{(0)}} < \infty) > 0$, cf. [2]. Also, we have $\lim_{\alpha\downarrow 0} g_\alpha(x,y) = \infty$ and hence X is recurrent. Thus, $P_x(\sigma_{x^{(0)}} < \infty) = 1$.

If $x^{(0)} \notin L^1$, then $P_x(\sigma_{x^{(0)}} = \infty) = 1$ by a well known property of two-dimensional Brownian motion.

Proposition 2.2 shows that each point of the x_1-axis is non-polar. We shall now discuss, in more detail, the behaveir of sample functions near x_1-axis.

Definition 2.1. Let $\mathcal{H}(0)$ be the class of all X-harmonic functions on $R^2 \setminus \{0\}$. More concretely, $u \in \mathcal{H}(0)$ if and only if u is a function defined on $R^2 \setminus \{0\}$ with the following properties:

(i) u is continuous on $R^2 \setminus \{0\}$,

(ii) u is harmonic in $R^2 \setminus L^1$ (i.e. $\Delta u = 0$ in $R^2 \setminus L^1$),

(iii) $\bar{u}(x_1) = u(x_1, 0) \in C^2(R^1 \setminus \{0\})$ and

$$\text{(2.6)} \qquad \frac{\partial u}{\partial x_2}(x_1, 0+) - \frac{\partial u}{\partial x_2}(x_1, 0-) + 2\bar{u}''(x_1) = 0, \quad x_1 \in R^1 \setminus \{0\}.$$

Theorem 2.1. Let $\mathcal{H}_b(0) = \{u \in \mathcal{H}(0);$ bounded$\}$. Then $\mathcal{H}_b(0)$ is two dimensional: i.e. $\exists\, u_1, u_2 \in \mathcal{H}_b(0)$ such that every $u \in \mathcal{H}_b(0)$ is expressed uniquely as $u = c_1 u_1 + c_2 u_2$, (c_1, c_2: constants).

Proof. $u \in \mathcal{H}_b(0)$ is expressed as

$$u(x) = u(x_1, x_2) = \frac{1}{\pi} \int_{-\infty}^{\infty} \frac{|x_2|}{(x_1 - \xi)^2 + x_2^2} \bar{u}(\xi)\, d\xi$$

with $\bar{u}(\xi) = u(\xi, 0) \in C^2(R^1 \setminus \{0\})$. By (2.6)

$$\text{(2.7)} \qquad 2\bar{u}''(\xi) + \lim_{\varepsilon \downarrow 0} \frac{2}{\pi} \int_{-\infty}^{\infty} \frac{(\xi - \eta)^2 - \varepsilon^2}{[(\xi - \eta)^2 + \varepsilon^2]^2} \bar{u}(\eta)\, d\eta = 0, \quad \xi \in R^1 \setminus \{0\}.$$

Since $\bar{u}$ is bounded, $\bar{u} \in \mathcal{D}'_{L^\infty}(R^1)$ and the distribution $\frac{1}{|\xi|^2} \in \mathcal{D}'_{L^1}(R^1)$. Hence $\frac{1}{|\xi|^2} * \bar{u} \in \mathcal{D}'_{L^\infty}$, (cf. [21]). Now (2.7) is equivalent to

$$\text{(2.8)} \qquad \operatorname{Supp}(T) \subset \{0\}$$

where T is the distribution given by

(2.9) $$T = \bar{u}'' + \frac{1}{\pi}\frac{1}{|\xi|^2} * \bar{u}.$$

Thus, $T = \sum_{\ell=0}^{p} a_\ell \, \delta^{(\ell)}_{\{0\}}$, ($a_\ell$: constants). Taking the Fourier transform of the both sides, we have

(2.10) $$-|\lambda|^2 \tilde{u}(\lambda) - |\lambda|\tilde{u}(\lambda) = \sum_{\ell=0}^{p}(-i)^\ell a_\ell \lambda^\ell$$

where $\tilde{u}(\lambda) = \int_{-\infty}^{\infty} e^{i\lambda\xi}\bar{u}(\xi)d\xi$. Then, noting that $\bar{u}(\xi)$ is bounded, it is not difficult to see that the distribution $\tilde{u}(\lambda)$ must be given in the form

$$\tilde{u}(\lambda) = \frac{a_0 - i\lambda a_1}{|\lambda|+1} - \frac{a_0 - i\lambda a_1}{|\lambda|} + b_0\delta_{\{0\}}.$$

Thus $\tilde{u}(\lambda)$ must be a linear combination of the following distributions whose Fourier transforms are given by

$$\tilde{f}_1(\lambda) = \delta_{\{0\}},$$
$$\tilde{f}_2(\lambda) = \frac{1}{|\lambda|+1} - \frac{1}{|\lambda|},$$
$$\tilde{f}_3(\lambda) = \frac{-i\lambda}{|\lambda|+1} + \frac{i\lambda}{|\lambda|}.$$

Inverting the Fourier transforms,

(2.11) $$f_1(\xi) = \frac{1}{2\pi} : \text{constant},$$

(2.12) $$f_2(\xi) = \frac{1}{\pi}\int_1^\infty \frac{e^{-t}t}{t^2+\xi^2}dt + \frac{1}{2\pi}\log(1+\xi^2) + \frac{1}{\pi}\int_0^1 \frac{(e^{-t}-1)t}{t^2+\xi^2}dt,$$

(2.13) $$f_3(\xi) = \frac{1}{\pi}\frac{\xi}{1+\xi^2} - \frac{2\xi}{\pi}\int_1^\infty \frac{e^{-t}t}{(t^2+\xi^2)^2}dt - \frac{2\xi}{\pi}\int_0^1 \frac{(e^{-t}-1+t)t}{(t^2+\xi^2)^2}dt$$
$$+ \frac{2}{\pi}\operatorname{sgn}\xi \int_0^{\frac{1}{|\xi|}} \frac{t^2}{(t^2+1)^2}dt.$$

$f_2(\xi)$ is not bounded but $f_1(\xi)$ and $f_3(\xi)$ are bounded. Thus, $\bar{u}$ must be of the form

(2.14) $$\bar{u}(\xi) = af_3(\xi) + b, \quad (a,b : \text{constants}).$$

Conversely, any $\bar{u}$ in the form (2.14) satisfies (2.8); in fact, $f_3'' + \frac{1}{\pi}\frac{1}{|\xi|^2} * f_3 = \delta'_{\{0\}}$. Thus

$$\mathcal{H}_b\{0\} = \{u(x) = \frac{1}{\pi}\int_{-\infty}^{\infty} \frac{|x_2|}{(x_1-\xi)^2+x_2^2}\,\bar{u}(\xi)\,d\xi \ ;\ \bar{u}(\xi) = af_3(\xi) + b\}$$

proving the theorem.

We can choose a and b so that $\bar{u}_1(\xi) = af_3(\xi) + b$ has the property

(2.15) $$\lim_{\xi\to 0-} \bar{u}_1(\xi) = 0, \qquad \lim_{\xi\to 0+} \bar{u}_1(\xi) = 1.$$

Let

(2.16) $$u_1(x) = \frac{1}{\pi}\int_{-\infty}^{\infty} \frac{|x_2|}{(x_1-\xi)^2+x_2^2}\,\bar{u}_1(\xi)\,d\xi.$$

Lemma 2.1.

(i) $u_1(x) \geqq 0,$

(ii) $\lim_{|x|\to\infty} u_1(x) = \frac{1}{2},$

(iii) $\lim_{x_1\to 0+} u_1(x_1, \alpha x_1) = \frac{1}{\pi}\int_{-\frac{1}{|\alpha|}}^{\infty} \frac{d\eta}{1+\eta^2}.$

We omit the proof. By (iii), if L is a half line in R^2 starting at 0, which does not coincide with the positive half line of x_1-axis,

(2.17) $$\sup_{x\in L} u_1(x) < 1.$$

Let $\omega(t)$, $0 \leqslant t < \sigma_0$, be a continuous curve in $R^2 \setminus \{0\}$

such that $\lim_{t\uparrow\sigma_0} \omega(t) = 0$. Let L be a half-line in R^2 starting at 0.

Definition 2.2. We shall say that $\omega(t)$ approaches 0 tangentially on L if, for every $\varepsilon > 0$, there exists $\delta > 0$ such that $\omega(t) \in U_\varepsilon^L$ for every $\sigma_0 - \delta \leqslant t < \sigma_0$ where U_ε^L is a domain as in Fig.1.

Fig.1 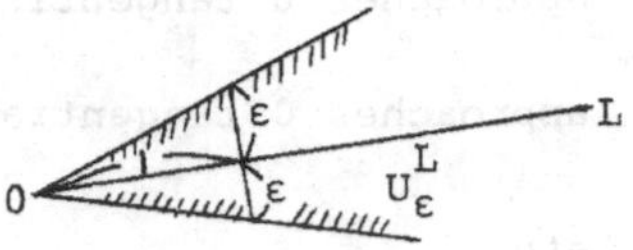

Theorem 2.2. Let L_+ and L_- be the positive and negative hafl-lines of x_1-axis respectively. Then for every $x \in R^2 \setminus \{0\}$

(2.18) $$P_x(X_t \text{ approaches } 0 \text{ tangentially on } L_+ \text{ or } L_-) = 1$$

and

(2.19) $$u_1(x) = P_x(X_t \text{ approaches } 0 \text{ tangentially on } L_+).$$

Proof. First we note that

(2.20) $$\xi_\infty = \lim_{t\uparrow\sigma_0} u_1(X_t)$$

exists almost surely. In fact, by Itô's formula, if $t < \sigma_0$,

$$u_1(X_t) - u_1(X_0) = \sum_{i=1}^{2}\int_0^t \frac{\partial u_1}{\partial x_i}(X_s)dA_s^i + \frac{1}{2}\int_0^t \Delta u_1(X_s)I_{R^2\setminus L^1}(X_s)ds$$

$$+ \int_0^t \left[\frac{\partial^2 u_1}{\partial x_1^2}(X_s) + \frac{1}{2}\frac{\partial u_1}{\partial x_2}\Bigg|_{x_2=0-}^{x_2=0+}(X_s)\right]d\psi_s$$

$$= \sum_{i=1}^{2}\int_0^t \frac{\partial u_1}{\partial x_i}(X_s)dA_s^i, \qquad (A_t^i = x^i(t) - x^i(0)),$$

and hence $\{u_1(X_t)\}$, $0 \leqq t < \sigma_0$, is a part of bounded martingale.

In particular, (2.20) exists almost surely and $u_1(x) = E_x(\xi_\infty)$. Note also that $P_x[\exists\varepsilon>0,\ \sigma_0 - \varepsilon \leq t < \sigma_0 \Rightarrow X_t \in R^2\setminus L^1] = 0$ since X_t is a Brownian motion before it hits the x_1-axis L^1. Thus, it is clear that $P_x[\xi_\infty = 0$ or $\xi_\infty = 1] = 1$ and $u_1(x) = E_x(\xi_\infty) = P_x(\xi_\infty = 1)$. By (2.15) and (2.17), we have

$$\{\xi_\infty = 1\} \underset{\text{a.s.}}{=} \{X_t \text{ approaches } 0 \text{ tangentially on } L_+\},$$

$$\{\xi_\infty = 0\} \underset{\text{a.s.}}{=} \{X_t \text{ approaches } 0 \text{ tangentially on } L_-\}.$$

The proof is now complete.

b) Example 2.2. Let $D = R^2$ and consider the diffusion process X determined by the following system of generators:

$$(2.21)\qquad \begin{cases} m(dx) = dx, \\ \nu_{11}(dx) = \psi(x_2)dx_1dx_2, \\ \nu_{22}(dx) = \frac{1}{2}\, dx, \\ \nu_{12}(dx) = \nu_{21}(dx) = 0, \\ k(dx) = 0. \end{cases}$$

Here, we assume that the density $\psi(\eta)$ satisfies $\psi(\eta) = \psi(-\eta)$, $\psi(\eta) > 0$ for $\eta > 0$, $\int_{-1}^{1} \psi(\eta)d\eta < \infty$ and continuous in $\eta \in (0,\infty)$. Sample functions of this diffusion X can be constructed by skew product as follows: Let $B^1(t)$ and $B^2(t)$ be two mutually independent one-dimensional Brownian motions. Let $x^2(t) = x_2 + B^2(t)$ and $\alpha(t) = 2\int_0^t \psi(x^2(s))ds$. Set $x^1(t) = x_1 + B^1(\alpha(t))$. Then $X_t = (x^1(t), x^2(t))$ defines a sample function starting at $x = (x_1, x_2)$ of the diffusion X with the system of generators given by (2.21).

Lemma 2.2. Let L^1 be the x_1-axis. If $x^0 \notin L^1$, then

$$P_x(\sigma_{x^0} = \infty) = 1.$$

If $x^0 \in L^1$, then either

(A) $$P_x(\sigma_{x^0} = \infty) = 1, \quad \text{for all} \quad x \in R^2,$$

or

(B) $$P_x(\sigma_{x^0} < \infty) > 0, \quad \text{for all} \quad x \in R^2.$$

Proof is easy and so it is omitted. We write $\psi \in (A)$ if (A) holds and $\psi \in (B)$ if (B) holds.

Lemma 2.3. (A comparison theorem)

(i) If $\psi_1 \in (A)$ and $\psi \leq \psi_1$, then $\psi \in (A)$.

(ii) If $\psi_1 \in (B)$ and $\psi \geq \psi_1$, then $\psi \in (B)$.

Proof. Without loss of generality, we may take $x^0 = 0$ in the classification of (A) and (B). Let $X_t = (x^1(t), x^2(t))$ be given as above. Then $\xi(t) = x^2(\alpha^{-1}(t))$ is a one-dimensional diffusion process with the generator $(4\psi(\xi))^{-1}\frac{d^2}{d\xi^2}$. Since its speed measure and canonical measure are given by $dm(\xi) = 4\psi(\xi)d\xi$ and $ds(\xi) = d\xi$ respectively, $\xi = 0$ is a regular point. Let $\eta(t)$ be the local time at 0 of $\xi(t)$. Then $a(t) = x^1(\eta^{-1}(t))$ is the trace on L^1 of X_t. $a(t)$ is a symmetric Lévy process and it is clear that $\psi \in (A)$ or $\psi \in (B)$ according as $a(t)$ never hits 0 or hits 0 with positive probability. Then, by a result of Kesten [14], $\psi \in (A)$ or $\psi \in (B)$ according as

$$(2.22) \qquad \int_{-\infty}^{\infty} \frac{1}{1-\Psi(\lambda)} d\lambda = \infty \quad \text{or} \quad < \infty,$$

where $\Psi(\lambda)$ is the exponent of $a(t)$:

(2.23) $$E(e^{i\lambda a(t)}) = e^{t\Psi(\lambda)}.$$

Let

(2.24) $$E(e^{-\lambda\eta^{-1}(t)}) = e^{-t\theta(\lambda)}.$$

Then, clearly $\Psi(\lambda) = -\frac{1}{2}\theta^2(\lambda)$. On the otherhand,

(2.25) $$\eta^{-1}(t) = \int_{-\infty}^{\infty} 4T(t,y)\psi(y)dy$$

where $T(t,y) = \underline{\underline{t}}(\underline{\underline{t}}^{-1}(t,0), y)$ and $\underline{\underline{t}}(t,y)$ is the local time at $y \in R^1$ of the Brownian motion $x^2(t)$, cf.[12]. Then it is clear that, η^{-1} and hence θ and $-\Psi$ are monotone increasing in ψ. Hence, if $\psi_1 \in (A)$ and $\psi \leqq \psi_1$ then, if Ψ_1 corresponds to ψ_1,

$$\int_{-\infty}^{\infty} \frac{1}{1-\Psi(\lambda)} d\lambda \geqq \int_{-\infty}^{\infty} \frac{1}{1-\Psi_1(\lambda)} d\lambda = \infty$$

implying $\psi \in (A)$. This proves (i) and (ii) can be proved similarly.

<u>Theorem 2.3.</u> <u>Let</u> K <u>and</u> β <u>be any constants such that</u> $K > 0$ <u>and</u> $0 < \beta < 1$.

(i) <u>If</u> $\psi(x_2) \leqq K$ <u>on a neighborhood of</u> $x_2 = 0$, <u>then</u> $\psi \in (A)$.

(ii) <u>If</u> $\psi(x_2) \geqq K|x_2|^{-\beta}$ <u>on a neighborhood of</u> $x_2 = 0$, <u>then</u> $\psi \in (B)$.

<u>Proof</u>. Clearly, we may assume that $\psi(x_2) \leqq K$ or $\psi(x_2) \geqq K|x_2|^{-\beta}$ holds everywhere. Then (i) follows from Lemma 2.3 (i) by taking $\psi_1 = K$. For the proof of (ii), let $\psi_1(\xi) = K|x_2|^{-\beta}$. Then the trace $a(t)$ of X_t on x_1-axis L^1 is a symmetric stable process with the exponent $\frac{2}{2-\beta}(> 1)$ (cf.[12], [19]). Then, $\psi_1 \in (B)$

and by Lemma 2.3 (ii), $\psi \in (B)$.

Thus, if $\psi(x_2) \geqq K|x_2|^{-\beta}$ near $x_2 = 0$ for some $K > 0$ and $0 < \beta < 1$, every point on x_1-axis is non-polar but sample functions approach a given point in a quite different way from those of diffusion of Example 2.1. (The diffusion of Example 2.1 is the case when $\psi(x_2) = \frac{1}{2} + \delta_0(x_2)$, where δ_0 is the delta function.)

Theorem 2.4. Let $\psi \in (B)$. Then

$$P_x(\ X_t \text{ approaches } 0 \text{ tangentially on } L_+ \text{ or } L_- / \sigma_0 < \infty) = 0$$

for all $x \in R^2 \setminus \{0\}$. More precisely, let L be any half-line starting at 0. Then

$$P_x(\exists t_n \uparrow \sigma_0,\ X_{t_n} \in L / \sigma_0 < \infty) = 1.$$

Proof. Let $a(t)$ be, as above, the trace on L^1 of the diffusion X_t. Then $a(t)$ is a symmetric Lévy process with does not possess the Brownian motion part; i.e. in the Lévy-Kchinchin canonical form of the exponent $\Psi(\lambda)$ of $a(t)$, the term $\frac{\sigma^2}{2}\lambda^2$ does not appear. Also $-\Psi(\lambda)$ is increasing in λ. Then by Example 3.3 in §3, it holds that, if τ_0 is the hitting time to 0 of $a(t)$-process,

$$P_a(\exists t_n \uparrow \tau_0,\ a(t_{2n-1}) < 0 < a(t_{2n}) / \tau_0 < \infty) = 1,\ a \in R^1 \setminus \{0\},$$

and the theorem follows easily from this.

c) Example 2.3. Let $D = R^2$ and consider the diffusion process X determined by the following system of generators:

$$(2.26)\quad \begin{cases} m(dx) = dx, \\ \nu_{ij} = \frac{1}{2}\delta_{ij}\,dx + \sum_{L_\lambda\in\Lambda} \ell_i^\lambda \ell_j^\lambda\, d\ell_\lambda, \\ k(dx) = 0, \end{cases}$$

where Λ is a locally finite family of line segments on R^2 (i.e. the number of segments which meet a bounded set is finite) and, for $L_\lambda \in \Lambda$, $(\ell_1^\lambda, \ell_2^\lambda)$ is the unit direction vector and $d\ell_\lambda$ is the line element on L_λ. Clearly, the diffusion discussed in Example 2.1 is a special case when $\Lambda = \{L^1\}$ where L^1 is the x_1-axis. In the following, we consider only the case

$$(2.27)\qquad \Lambda = \{L_1, L_2, \cdots, L_k\},$$

where $L_1, L_2, \cdots, L_k$ are (open) half-lines starting at the origin (cf. Fig.2). Let $X^0 = (X_t^0, P_x^0)$ be a diffusion on $R^2\setminus\{0\}$ with the following property; in a neighborhood of $x \in R^2\setminus(L_1 \cup L_2 \cup \cdots \cup L_k \cup \{0\})$ it coincides with the two-dimensional Brownian motion and in some neighborhood of $x \in L_i$, $i=1,2,\cdots,k$, it coincides with the diffusion process defined as in Example 2.1 by identifying L_i with one of half-lines of x_1-axis. By a standard argument of "recollement" (cf.[4]), such a diffusion exists and is unique: its sample functions are defined up to the time of approaching the origin. We will now construct the sample functions of the diffusion X on R^2 given by (2.26) which is clearly an extension to R^2 of the diffusion X^0 on $R^2\setminus\{0\}$. For this, we shall study the diffusion X^0 in detail.

Fig.2

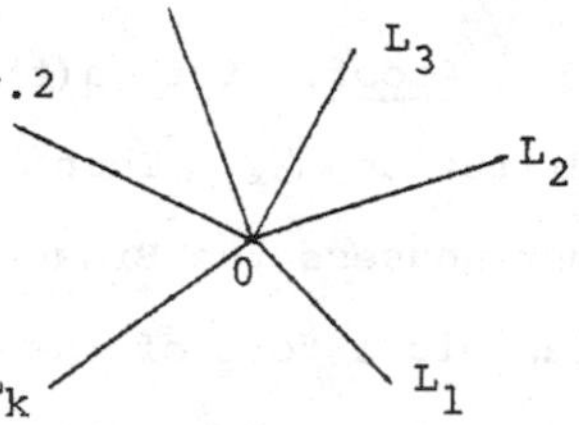

First, we extend Theorem 2.1. The following theorem is purely analytical but its proof without using the process X^0 would be much more complicated.

Definition 2.3. Let $\mathcal{H}(0)$ be the class of all X^0-harmonic functions on $R^2 \setminus \{0\}$. As in Example 2.1., $u \in \mathcal{H}(0)$ if and only if u is a function defined on $R^2 \setminus \{0\}$ with the following properties:

(i) u is continuous on $R^2 \setminus \{0\}$.

(ii) u is harmonic in $R^2 \setminus (L_1 \cup L_2 \cup \cdots \cup L_k \cup \{0\})$.

(iii) $u|_{L_i} \equiv \bar{u}_i \in C^2(L_i)$ $(i=1,2,\cdots,k)$ and if we introduce a local coordinate $\xi = (\xi_1, \xi_2)$ in a neighborhood of $x \in L_i$ such that $\det \frac{dx}{d\xi} = 1$ and $\{\xi ; \xi = (\xi_1, 0), \xi_1 > 0\}$ coincides with L_i, then

$$\frac{\partial u}{\partial \xi_2}(\xi_1, 0+) - \frac{\partial u}{\partial \xi_2}(\xi_1, 0-) + 2\bar{u}_i''(\xi_1) = 0, \quad \forall \xi_1 > 0.$$

Theorem 2.5. Let $\mathcal{H}_b(0) = \{u \in \mathcal{H}(0) ; \text{bounded}\}$. Then $\mathcal{H}_b(0)$ is k-dimensional.

Proof. 1°) We supplement some properties of the diffusion $X^1 = (X^1(t), P_x^1)$ of Example 2.1. Let $L_1 = \{x = (x_1, 0), x_1 > 0\}$ and L_1^* and L_1^{**} be open half-lines starting at 0, L_1^* being in the upper half-plane and L_1^{**} in the lower half-plane respectively (cf. Fig.3). Let D be the open domain surrounded by L_1^* and L_1^{**} which contains L_1. Let $\tau = \inf\{t ; X_t^1 \in L_1^* \cup L_1^{**}\}$.

Fig.3

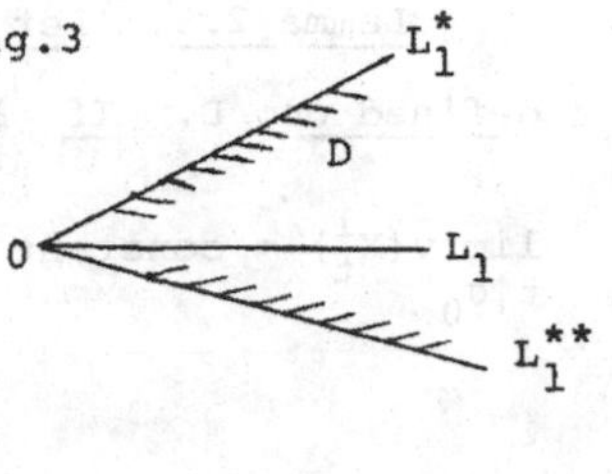

Lemma 2.4. (i) $P^1_x(\sigma_0 < \tau) > 0$. for every $x \in D$ and (ii) $\lim_{\substack{x \in L_1 \\ x \to 0}} P^1_x(\sigma_0 < \tau) = 1$.

Proof. Set $v(x) = P^1_x(\sigma_0 < \tau) = 0$. Since this function is X^1-harmonic in D, it is easy to see that it is positive everywhere in D or identically 0. If $v(x) = 0$, then, using the same notation as in Example 2.1, we have for $x \in D$,

$$u_1(x) = P^1_x(\xi_\infty = 1) = E^1_x(u_1(X^1_\tau)) \leqq \beta,$$

where $\beta = \sup_{x \in L_1^* \cup L_1^{**}} u_1(x)$ and $\beta < 1$ by (2.17). This contradicts $\lim_{\substack{x \in L_1 \\ x \to 0}} u_1(x) = 1$. Thus, $P_x(\sigma_0 < \tau) > 0$ everywhere in D. Also

$$\begin{aligned} u_1(x) &= P^1_x(\xi_\infty = 1,\ \sigma_0 < \tau) + E^1_x[P^1_{X^1_\tau}(\xi_\infty = 1) \ ;\ \tau < \sigma_0] \\ &\leqq P^1_x(\xi_\infty = 1,\ \sigma_0 < \tau) + \beta P^1_x(\tau < \sigma_0) \\ &\leqq P^1_x(\xi_\infty = 1,\ \sigma_0 < \tau) + \beta P^1_x(\xi_\infty = 1,\ \tau < \sigma_0) + \beta P^1_x(\xi_\infty = 0) \\ &\leqq \beta P^1_x(\xi_\infty = 1) + (1-\beta) P^1_x(\xi_\infty = 1,\ \sigma_0 < \tau) + \beta P^1_x(\xi_\infty = 0). \end{aligned}$$

Hence if $u_1(x) > 1-\varepsilon$, then $P^1_x(\xi_\infty = 0) < \varepsilon$ and hence,

$$\begin{aligned} P^1_x(\sigma_0 < \tau) &\geqq P^1_x(\xi_\infty = 1) - \frac{\beta}{1-\beta} P^1_x(\xi_\infty = 0) \\ &\geqq 1 - \varepsilon - \frac{\beta\varepsilon}{1-\beta} = 1 - \frac{\varepsilon}{1-\beta}. \end{aligned}$$

Since $\lim_{\substack{x \in L_1 \\ x \to 0}} u_1(x) = 1$, this proves (ii).

Lemma 2.5. Let $v(x)$ be a bounded measurable function defined on D. If $\lim_{t \uparrow \sigma_0} v(X^1_t)$ exists a.s. on $\{\sigma_0 < \tau\}$, then $\lim_{t \uparrow \sigma_0} v(X^1_t) = $ constant a.s. on $\{\sigma_0 < \tau\}$.

Proof. If there exist a constant a and $x \in D$ such that

$$P_x^1[\lim_{t\uparrow\sigma_0} v(X_t^1) > a \; ; \; \sigma_0 < \tau] > 0,$$

and

$$P_x^1[\lim_{t\uparrow\sigma_0} v(X_t) \leqq a \; ; \; \sigma_0 < \tau] > 0,$$

then, this would imply that, for every $x \in R^2 \setminus \{0\}$,

$$w_1(x) = P_x^1[\overline{\lim_{t\uparrow\sigma_0}} \tilde{v}(X_t^1) > a \; ; \; \xi_\infty = 1] > 0,$$

and

$$w_2(x) = P_x^1[\overline{\lim_{t\uparrow\sigma_0}} \tilde{v}(X_t^1) \leqq a \; ; \; \xi_\infty = 1] > 0,$$

where $\tilde{v}$ is an extension of v on $R^2 \setminus \{0\}$. Now $u_1 = w_1 + w_2$ and $w_1, w_2 \in \mathcal{H}(0)$. Then, by Theorem 2.1, we have $w_1 = cw_2$ for some constant $c > 0$. But this is a contraction since

$$\lim_{t\uparrow\sigma_0} w_1(X_t^1) = I_{[\overline{\lim}_{t\uparrow\sigma_0} \tilde{v}(X_t^1) > a \; ; \; \xi_\infty = 1]},$$

and

$$\lim_{t\uparrow\sigma_0} w_2(X_t^1) = I_{[\overline{\lim}_{t\uparrow\sigma_0} \tilde{v}(X_t^1) \leqq a \; ; \; \xi_\infty = 1]}.$$

2°) Consider the diffusion X^0 on $R^2 \setminus \{0\}$ defined above. Let $\sigma_0 = \lim_{\varepsilon\downarrow 0} \sigma_\varepsilon$ where $\sigma_\varepsilon = \inf\{t \; ; \; |X_t^0| = \varepsilon\}$ and η_i (i=1,2, ..., k) be the hitting time to the set $L_1 \cup L_2 \cup \cdots \cup L_{i-1} \cup L_{i+1} \cup \cdots \cup L_k$. Set

(2.28) $u_i(x) = P_x^0[\{\sigma_0 < \infty\} \cap \{X_t^0 \to 0, \text{ tangentially on } L_i \text{ when } t\uparrow\sigma_0\}]$,

(cf. Definition 2.2). Then $u_i \in \mathcal{H}_b(0)$, $i = 1,2,\cdots,k$. By Lemma 2.4,

(2.29) $u_i(x) \geqq P_x^0(\sigma_0 < \eta_i) \longrightarrow 1$, if $x \in L_i$ and $x \longrightarrow 0$.

Then $i \neq j$,

(2.30) $$u_i(x) \leqq 1-P^0_x(\sigma_0 < \eta_j) \longrightarrow 0, \qquad \text{if} \quad x \in L_j \quad \text{and} \quad x \longrightarrow 0.$$

If we set, $u(x) = P^0_x(\sigma_0 < \infty)$ then

$$u(x) \geqq \sum_{i=1}^{k} u_i(x)$$

and by (2.29)

(2.31) $$\lim_{\substack{x\to 0 \\ x \in \bigcup_{i=1}^{k} L_i}} u(x) = 1.$$

Since u is harmonic in $R^2 \setminus (\bigcup_{i=1}^{k} L_i \cup \{0\})$, this implies

(2.32) $$\lim_{x\to 0} u(x) = 1.$$

We will show that

(2.33) $$u(x) \equiv P^0_x(\sigma_0 < \infty) = 1, \qquad \text{for every} \quad x \in R^2 \setminus \{0\}.$$

For this, noting (2.32), it is sufficient to show

(2.34) $$P^0_x(\sigma_B < \infty) = 1, \qquad x \in R^2 \setminus B,$$

for every disk B with center at the origin and σ_B is the first hitting time to B. Let

$$u_\lambda(x) = E^0_x(e^{-\lambda\sigma_B}).$$

Let $(\tilde{X}_t, \tilde{P}_x)$ be a two-dimensional Brownian motion and set

$$\tilde{u}_\lambda(x) = \tilde{E}_x(e^{-\lambda\sigma_B}).$$

As is well known, $\tilde{u}_\lambda$ satisfies

$$\lambda\tilde{u}_\lambda - \frac{1}{2}\Delta\tilde{u}_\lambda = 0, \qquad \text{on} \quad R^2 \setminus B.$$

Also, $\tilde{u}_\lambda(x) = w_\lambda(|x|)$ and w_λ satisfies

$$(2.35)\qquad \lambda w_\lambda - \frac{1}{2}(w''_\lambda + \frac{w'_\lambda}{r}) = 0, \quad \text{for } r \geqq r_0 = \text{radius of } B.$$

By Itô's formula,

$$e^{-\lambda t}\,\tilde{u}(X_t^0) - \tilde{u}_\lambda(X_t^0) = \text{a stochastic integral}$$
$$+ \int_0^t e^{-\lambda s}(\lambda\tilde{u}_\lambda - \frac{1}{2}\Delta\tilde{u}_\lambda)(X_s^0)ds + \frac{1}{2}\sum_{i=1}^{k}\int_0^t e^{-\lambda s}\nabla_{\ell_i}^2\,\tilde{u}_\lambda(X_s^0)d\psi_s^i,$$

where $\psi^i = \langle M^i\rangle$ is the quadratic variational process of the martingale $M^i = \int_0^t I_{L_i}(X_s^{(0)})dX_s^{(i)}$ and $X_t^{(i)} = (\ell_i, X_t^0)$ (ℓ_i is the direction vector of the line L_i). Since $\nabla_{\ell_i}^2\tilde{u}_\lambda(x) = w''_\lambda(|x|) \geqq 0$ by (2.35), $E_x^0(e^{-\lambda\sigma_B\wedge n}) \geqq E_x^0(e^{-\lambda\sigma_B\wedge n}\,\tilde{u}_\lambda(X^0_{\sigma_B\wedge n})) \geqq \tilde{u}_\lambda(x) = \tilde{E}_x(e^{-\lambda\sigma_B})$. Letting $n \to \infty$ and $\lambda \downarrow 0$,

$$P_x^0(\sigma_B < \infty) \geqq \tilde{P}_x(\sigma_B < \infty) = 1.$$

By (2.33), we have

$$(2.36)\qquad \sum_{i=1}^{k} u_i(x) = P_x^0(X_t^0 \to 0 \text{ tangentially on some } L_i,\ i = 1,2,\cdots,k \text{ when } t\uparrow\sigma_0) = 1,$$

since $\bar{u} = \sum_{i=1}^{k} u_i(x)$ satisfies $\lim_{x\to 0}\bar{u}(x) = 1$. Finally, we shall prove that every $u \in \mathcal{H}_b(0)$ is a linear combination of $u_1, u_2, \cdots, u_k$. Let $u \in \mathcal{H}_b(0)$. Then

$$Y_t = \begin{cases} u(X_t^0), & t < \sigma_0, \\ \lim_{t\uparrow\sigma_0} u(X_t^0), & t \geqq \sigma_0, \end{cases}$$

is a bounded martingale. By Lemma 2.5, there exist constants $c_1, c_2, \cdots, c_k$ such that

$$\lim_{t\uparrow\sigma_0} u(X_t^0) = c_i \quad \text{a.s. on } \{X_t^0 \to 0 \text{ tangentially on } L_i\}.$$

Since $u(x) = E_x^0(\lim_{t\uparrow\sigma_0} u(X_t^0))$, we have $u(x) = \sum_{i=1}^{k} c_i u_i(x)$. q.e.d.

<u>Corollary</u>. <u>For every</u> $x \in R^2 \setminus \{0\}$,

$P_x^0\{X_t^0 \longrightarrow 0$ tangentially on L_i for some $i=1,2,\cdots, k\} = 1$.

Now, we will discuss a construction of sample functions of X. More generally, we will determine all possible extensions as conservative diffusion process on R^2 of X^0. Let

$$u_i^\alpha = E_x^0(e^{-\alpha\sigma_0} \; ; \; X_t^0 \longrightarrow 0 \text{ tangentially on } L_i \text{ when } t\uparrow\sigma_0).$$

Let
$$a_i = \int_{R^2\setminus\{0\}} u_i^1(x)dx, \quad i = 1,2,\cdots, k,$$

and $w_i(x) = a_i^{-1}u_i(x)$. By the symmetry of X^0 with respect to the Lebesgue measure dx,

$$\eta_i(dx) = w_i(x)dx, \quad i = 1,2,\cdots, k,$$

is an excessive measure for X^0. Now, we construct approximate process $X^{(i)} = \{\omega(t), \alpha\equiv 0, \beta, E^i\}$ of X^0 with potential measure η_i (i.e., $E^i(\int_\alpha^\beta I_A(X_t)dt) = \eta_i(A)$) (cf. M.Weil [24]). E^i is a measure on continuous functions:

$$\Omega = \{\omega : [0, \infty) \ni t \longmapsto \omega(t) \in R^2 : \text{continuous},$$

$$\exists_{\beta(\omega)} : \begin{array}{l} 0 < t < \beta(\omega) \Rightarrow \omega(t) \in R^2\setminus\{0\} \\ t \geqq \beta(\omega) \Rightarrow \omega(t) = 0\}. \end{array}$$

<u>Theorem 2.6</u>. <u>An extension of</u> X^0 <u>as a conservative diffusion on</u> R^2 <u>is determined by the parameters</u> p_i, $i = 1,2,\cdots, k$ <u>and</u> m such that

$$p_i \geqq 0,\ m \geqq 0 \ \underline{\text{and}} \ \sum_{i=1}^{k} p_i + m = 1.$$

Proof. We shall prove this theorem by constructing sample functions directly following K.Itô [11]. Let p_i, $i = 1,2,\cdots, k$ and m be given as above. Let $\mathcal{N} = \sum_{i=1}^{k} p_i E^i$. Then $\mathcal{N}$ is a measure on Ω and we can construct Poisson point process Y on Ω with characteristic measure $\mathcal{N}$, cf. [11]. Note that Y can be identified with a probability measure on the space of Ω-valued point functions p on $(0, \infty)$:

$$p : D_p \ni t \longmapsto \omega = p_t \in \Omega,$$

where the domain D_p of p is a countable subset of $(0, \infty)$. For each point function p and $s \in [0, \infty)$, we set

$$S(s) = ms + \sum_{\substack{t \leqq s \\ t \in D_p}} \beta(p_t).$$

Let, for $u \in [0, \infty)$

$$X(u) = \begin{cases} p_t(u - S(t-)), & S(t-) \leqq u < S(t), \\ 0 & , \quad u = S(t-) = S(t). \end{cases}$$

Then $X(u) = X(u, p)$, with a point function $p : D_p \longrightarrow \Omega$ as the stochastic parameter, defines a sample function starting at 0 of a diffusion which is a conservative extension on R^2 of X^0.

The diffusion X corresponding to (2.26) is a special case when $m = 0$ and $p_1 = p_2 = \cdots = p_k = \frac{1}{k}$. This diffusion satisfies (A.1)$\sim$(A.7) except (A.5). (A.5) is satisfied if and only if the unit direction ℓ_i $(i=1,2,\cdots, k)$ vector (starting at 0) of L_i satisfies

$$\ell_1 + \ell_2 + \cdots + \ell_k = 0.$$

b) Example 2.4. (Rotation invariant diffusions). Let $D = R^2 \setminus \{0\}$ and let X be a diffusion on D which is invariant under all rotations around the origin. This class of diffusions has been studied by Wentzell [25] and Galmarino [9]. In particular, they showed that sample functions of X can be given by the skew product as follows: let $r(t)$ be a one-dimensional diffusion process on $(0, \infty)$, $\psi(t)$ be a non-negative additive functional of $r(t)$ and $\theta(t)$ be a Brownian motion on the unit circle S^1. Then $X(t) = (r(t), \theta(\psi(t)))$ gives, in the polar coordinate representation, a sample function of a rotation invariant diffusion X. Let the local generator of $r(t)$ be $\frac{d}{dm}\frac{d}{ds}$ and the potential measure of $\psi(t)$ be $k(dr)$. Then, introducing the polar coordinate (r, θ), we have, for every $v, u \in \mathcal{D}(G)$, (G: a domain in $R^2 \setminus \{0\}$),

$$\varepsilon[u, v] = \int \frac{\partial u}{\partial s}\frac{\partial v}{\partial s}\, ds d\theta + \frac{1}{2}\int \frac{\partial u}{\partial \theta}\frac{\partial v}{\partial \theta}\, k(ds) d\theta,$$

Now, we shall summarize some results on a local property of sample functions near the origin. We consider the following three cases:

$$\text{(i)} \qquad \int_{0+} ds = \infty,$$

$$\text{(ii)} \qquad \int_{0+} ds < \infty \quad \text{and} \quad \int_{0+}^{1} \left(\int_{0+}^{r} ds(u) \right) dk(r) = \infty,$$

$$\text{(iii)} \qquad \int_{0+} ds < \infty \quad \text{and} \quad \int_{0+}^{1}\int_{0+}^{r} ds(u)\, dk(r) < \infty.$$

In the case (i), $\sigma_0 = \infty$ and X_t does not approach 0 when $t \uparrow \infty$, while in the cases (ii) and (iii), X_t approaches 0 when $t \uparrow \sigma_0$. (ii) or (iii) occurs according as

$$\eta = \lim_{t \uparrow \sigma_0} \psi(t) = \infty \text{ a.s.} \quad \text{or} \quad \eta = \lim_{t \uparrow \sigma_0} \psi(t) < \infty \text{ a.s.}.$$

Let $D = \{x \,;\, 0 < |x| < r_0,\ x \in R^2\}$ and $\mathcal{H}_b(D)$ be the set of all bounded X-harmonic functions $u(x)$ in D such that $\lim_{|x| \to r_0} u(x) = 0$. Then, we

$$\dim \mathcal{H}_b(D) = 0, \qquad \text{if (i) occurs,}$$

$$\dim \mathcal{H}_b(D) = 1, \qquad \text{if (ii) occurs,}$$

$$\dim \mathcal{H}_b(D) = \infty, \qquad \text{if (iii) occurs.}$$

In the case of (iii), we have an independent system $\{v_n(x)\}_{n=-\infty}^{\infty}$ in $\mathcal{H}_b(D)$ as follows. Under the assumption of $\int_{0+} ds(u) < \infty$, $\int_{0+}^{1} (\int_{0+}^{r} ds(u)) dk(r) < \infty$, there exists a unique solution of $\frac{d}{dk}\frac{du}{ds} - n^2 u = 0$, $u(r_0) = 0$, $u(0+) = 1$, for each $n = 1, 2, \cdots$. Then if we set,

$$v_n(x) = u_n(r) \cos n\theta, \qquad v_{-n}(x) = u_n(r) \sin n\theta, \qquad x = (r\cos\theta,\ r\sin\theta),$$

$\{v_n(x)\}_{n=-\infty}^{\infty}$ defines an independent system of $\mathcal{H}_b(D)$.

If the local generator L of X is given as

$$Lu = \sum_{i,j=1}^{2} a_{ij} \frac{\partial^2 u}{\partial x_i \partial x_j},$$

then a_{ij} is expressed in the following form

$$a_{ij}(x) = \delta_{ij}\, b(|x|) + \{a(|x|) - b(|x|)\} \frac{x_i x_j}{|x|^2}$$

and hence

$$ds = e^{-c(r)} dr, \quad dm = \frac{e^{c(r)}}{a(r)} dr \quad \text{and} \quad dk = \frac{b(r)}{a(r)} \frac{e^{c(r)}}{r^2} dr,$$

where $c(r) = \int^r \frac{1}{\xi}\frac{b(\xi)}{a(\xi)}\,d\xi$. a_{ij} is continuous and uniformly elliptic if and only if $a(r)$ and $b(r)$ are continuous, $c_1 \leqq a(r)$, $b(r) \leqq c_2$ for some positive constants c_1 and c_2 and $\lim_{r\downarrow 0}(a(r) - b(r)) = 0$. In this case, it is possible to have $\int_{0+} ds < \infty$ and $\int_0 dm < \infty$ as is seen in the following example of Kanda ([13]);

$$a(r) = \frac{\log r}{2+\log r}, \quad b(r) = 1.$$

But, $\int_0^1 (\int_0^r ds(u))dk = \infty$ holds always and thus, the case (iii) never occurs. If a_{ij} degenerates, then the case (iii) occurs; e.g., $a(r) = r$ and $b(r) = r^2$.

Finally, we give an example of rotation invariant diffusion on R^2 which is not strong Markov. This diffusion X has the following interesting property: there exists a martingale additive functional $\alpha(t) \not\equiv 0$ such that

$$\alpha_t = \int_0^t I_{\{0\}}(X_s)d\alpha_s, \quad (\int d\alpha \text{ is a stochastic integral}),$$

i.e. $\alpha(t)$ is a non-zero martingale which varies only when the path is at the origin. Let $r(t)$ be a Brownian motion on $[0, \infty)$ with Feller's boundary condition

$$\lim_{r\downarrow 0} \tfrac{1}{2}\psi''(r) - \psi'(0) = 0.$$

Let $k(r)$ be a continuous and positive functions on $[0, \infty)$ and $\psi(t)$ be the corresponding non-negative additive functional of $r(t)$. Let $\theta(t)$ be a Brownian motion on the unit circle S^1 which is independent of $r(t)$ and set $\underline{\underline{t}}(t) = t - \int_0^t I_{\{0\}}(r_s)ds$.

Construct the diffusion process on $[0, \infty) \times S^1$ by skew product:

$$X(t) = (r(\underline{\underline{t}}^{-1}(t)),\ \theta(\psi(\underline{\underline{t}}^{-1}(t)))),$$

where $\underline{\underline{t}}^{-1}(t)$ is the inverse function of $\underline{\underline{t}}(t)$. Let $P_{(r,\theta)}$, $((r,\theta) \in [0,\infty) \times S^1)$ be the probability law of $X(t)$ starting at (r,θ). Let τ be the mapping $\tau : [0,\infty) \times S^1 \longrightarrow R^2$ given by

$$\tau(r,\theta) = (r\cos\theta,\ r\sin\theta),\ r \geqq 0,\ \theta \in S^1.$$

Let $P^*_{\tau(r,\theta)} = P_{(r,\theta)}$, if $r > 0$ and $P^*_0 = \frac{1}{2\pi}\int_{S^1} P_{(0,\theta)}d\theta$. Then it is not difficult to see that $X^* = (X^*_t = \tau(X(t)),\ P^*_x,\ x \in R^2)$ is a Markov process with continuous sample functions. It is not strong Markov. The resolvent operator of X^* is symmetric with respect to $dm = \frac{1}{2\pi}\,drd\theta$ and hence, we have the corresponding $\mathcal{L}^2$-Dirichlet space. If $u(x) \in C^\infty(R^2 \setminus \{0\})$ and, setting $u(x) = \tilde{u}(r,\theta)$ in the polar coordinate, $\tilde{u}(r,\theta)$ is extended to a function in $\mathcal{D}([0,\infty) \times S^1)$, then $u \in F$. For such functions u and v, we have

$$\varepsilon[u,v] = \int_{R^2\setminus\{0\}} \{\frac{\partial\tilde{u}}{\partial r}(r,\theta)\frac{\partial\tilde{v}}{\partial r}(r,\theta) + k(r)\frac{\partial\tilde{u}}{\partial\theta}(r,\theta)\frac{\partial\tilde{v}}{\partial\theta}(r,\theta)\}dm$$
$$+ \frac{1}{2}\int_{S^1} \frac{\partial\tilde{u}}{\partial\theta}(0+,\theta)\frac{\partial\tilde{v}}{\partial\theta}(0+,\theta)\frac{d\theta}{2\pi}.$$

If $u(x) \in \mathcal{D}(R^2)$, then the second term disappears.

§ 3. Hitting a single point of Lévy processes.

In this section, we shall study a local property of sample functions of one-dimensional Lévy processes. As we have seen in § 2, the results obtained here have some applications to the diffusion processes.

Let (X_t, P_x) be a one-dimensional Lévy process (P_x: the probability law for sample functions starting at x) i.e., a process with stationary independent increments whose sample functions are right continuous. Problem of hitting a single point for such processes has been studied extensively by H.Kesten [14], (cf. also Bretagnolle [3]). Our purpose is to analyse various aspects of hitting, when it occurs with positive probability. Let $\Psi(\xi)$ be the exponent of the process:

$$E_0(e^{i\xi X_t}) = e^{t\Psi(\xi)},$$

given in the canonical form

$$\Psi(\xi) = ia\xi - \frac{1}{2}\sigma^2\xi^2 + \int_{-\infty}^{\infty}(e^{i\xi u} - 1 - \frac{i\xi u}{1+u^2})n(du). \tag{3.1}$$

We assume that

$$P_x(\sigma_0 < \infty) > 0, \qquad \text{for every } x \in R^1, \tag{3.2}$$

where $\sigma_0 = \inf\{t ; X_t = 0\}$. Set

$$\Omega_1 = \{\sigma_0 < \infty\}, \tag{3.3}$$

$$\Omega_1^+ = \Omega_1 \cap \{\exists \varepsilon > 0 \text{ such that if } t \in [\sigma_0 - \varepsilon, \sigma_0), \text{ then } X_t > 0\}, \tag{3.4}$$

$$\Omega_1^- = \Omega_1 \cap \{\exists \varepsilon > 0 \text{ such that if } t \in [\sigma_0 - \varepsilon, \sigma_0), \text{ then } X_t < 0\} \tag{3.5}$$

and

$$\Omega_1^\pm = \Omega_1 \cap \{\exists t_n \uparrow \sigma_0 \text{ such that } X_{t_{2n-1}} < 0 < X_{t_{2n}}\}. \tag{3.6}$$

Clearly, $\Omega_1 = \Omega_1^+ \cup \Omega_1^- \cup \Omega_1^\pm$ (disjoint union).

<u>Theorem 3.1</u>. <u>Assume</u> $\sigma^2 > 0$. <u>Then</u>

$$P_x(\Omega_1^+ \cup \Omega_1^- / \Omega_1) = 1, \qquad \underline{\text{for all}} \;\; x \in R^1\setminus\{0\}.$$

<u>If further,</u> $n_+ \neq 0$ (n_+ <u>and</u> n_- <u>being the positive and negative parts of the Lévy measure</u> n <u>respectively</u>), <u>then</u>

$$P_x(\Omega_1^+/\Omega_1) > 0 \quad \underline{\text{and}} \quad P_x(\Omega_1^-/\Omega_1) > 0, \qquad \underline{\text{for all}} \quad x < 0.$$

<u>If further,</u> $n_+ \neq 0$ <u>and</u> $n_- \neq 0$, <u>then</u>

$$P_x(\Omega_1^+/\Omega_1) > 0 \quad \underline{\text{and}} \quad P_x(\Omega_1^-/\Omega_1) > 0 \qquad \underline{\text{for all}} \quad x \in R^1\setminus\{0\}.$$

The proof is based on the following property of the resolvent density; let $u_\lambda(x)$ be the resolvent density i.e., $\int_E u_\lambda(x)dx = E_0(\int_0^\infty e^{-\lambda t} I_E(x_t)dt)$, $E \in \mathcal{B}(R^1)$. If $\sigma^2 > 0$ then $u_\lambda(x)$ is continuous. Further, if the Lévy measure has a compact support, $u_\lambda(x)$ is continuously differentiable on $R^1\setminus\{0\}$ such that both

$$\lim_{x\uparrow 0} u_\lambda'(x) = u_\lambda'(0-) \quad \text{and} \quad \lim_{x\downarrow 0} u_\lambda'(x) = u_\lambda'(0+)$$

exist finitely and

$$(3.7) \qquad u_\lambda'(0-) - u_\lambda'(0+) = \frac{2}{\sigma^2} \qquad \text{(cf. [23])}.$$

Then, we have

$$(3.8) \qquad E_x(e^{-\lambda\sigma_0}\ ;\ \Omega_1^+) = \frac{u_\lambda(0)u_\lambda'(-x)-u_\lambda(-x)u_\lambda'(0+)}{u_\lambda(0)(u_\lambda'(0-)-u_\lambda'(0+))},$$

$$(3.9) \qquad E_x(e^{-\lambda\sigma_0}\ ;\ \Omega_1^-) = \frac{-u_\lambda(0)u_\lambda'(-x)+u_\lambda(-x)u_\lambda'(0-)}{u_\lambda(0)(u_\lambda'(0-)-u_\lambda'(0+))}.$$

We omit the details.

When $\sigma^2 = 0$, the situation is completely different. It may be reasonably conjectured that only the following three cases are possible:

$$P_x(\Omega_1^+/\Omega_1) \equiv 1 \quad \text{on} \quad R^1\setminus\{0\}, \tag{3.10}$$

or

$$P_x(\Omega_1^-/\Omega_1) \equiv 1 \quad \text{on} \quad R^1\setminus\{0\}, \tag{3.11}$$

or

$$P_x(\Omega_1^\pm/\Omega_1) \equiv 1 \quad \text{on} \quad R^1\setminus\{0\}. \tag{3.12}$$

Though our result is incomplete in this respect, we give several sufficient conditions. In the following, we assume $\sigma^2 = 0$ and $\int_{-1}^{1} |u| n(du) = +\infty$. Combining with (3.2), it is known that the continuous resolvent density $u_\lambda(x) > 0$ exists.

Theorem 3.2. *Suppose one of the following conditions is satisfied:*

(A) $u_\lambda(x)$ *is continuously differentiable in* $R^1\setminus\{0\}$, $\lim_{x\uparrow 0} u'_\lambda(x) = u'_\lambda(0-)$ *exists finitely and* $\lim_{x\downarrow 0} u'_\lambda(x) = -\infty$.

(B) $u_\lambda(x)$ *is continuously differentiable in* $R^1\setminus\{0\}$ *and there exist positive constants* c_1, c_2, α, β *and* ε_0 *such that* $0 < \alpha < \beta < 1$, $c_1 > 1$, $c_2 > 1$ *and*

$$\frac{1}{c_1}\varepsilon^\beta < u_\lambda(0) - u_\lambda(-\varepsilon) < c_1\varepsilon^\beta, \tag{3.13}$$

$$\frac{1}{c_2}\varepsilon^\alpha < u_\lambda(0) - u_\lambda(\varepsilon) < c_2\varepsilon^\alpha, \tag{3.14}$$

for all $\varepsilon \in [0, \varepsilon_0]$. *Then*

$$P_x(\Omega_1^+/\Omega_1) \equiv 1 \quad \text{on} \quad R^1\setminus\{0\}.$$

Proof. Let (X_t^0, P_x) be the absorbing barrier process of X_+ at $x = 0$:

$$X_t^0 = \begin{cases} X_t, & t < \sigma_0, \\ 0, & t \geqq \sigma_0. \end{cases}$$

Its resolvent density $g_\lambda^0(x, y)$ (i.e. $\int_E g_\lambda^0(x, y)dy = E_x(\int_0^{\sigma_0} e^{-\lambda t} I_E(X_t)dt)$, $E \in \mathcal{B}(R^1 \setminus \{0\})$) is given by

$$g_\lambda^0(x, y) = u_\lambda(y-x) - \frac{u_\lambda(-x)u_\lambda(y)}{u_\lambda(0)}.$$

Now, suppose that the condition (A) is satisfied. It is easy to see that there exists $c \in R^1\setminus\{0\}$ such that

$$-u_\lambda(0)u_\lambda'(-c) + u_\lambda(-c)u_\lambda'(0-) \neq 0.$$

Then,

$$(3.15) \qquad \frac{g_\lambda^0(x,y)}{g_\lambda^0(c,y)} = \frac{u_\lambda(0)[u_\lambda(y-x)-u_\lambda(-x)]+u_\lambda(-x)[u_\lambda(0)-u_\lambda(y)]}{u_\lambda(0)[u_\lambda(y-c)-u_\lambda(-c)]+u_\lambda(-c)[u_\lambda(0)-u_\lambda(y)]}$$

and, if we set $K_\lambda(x) = \lim_{y\uparrow 0} \frac{g_\lambda^0(x,y)}{g_\lambda^0(c,y)}$, $x \in R^1\setminus\{0\}$,

$$(3.16) \qquad K_\lambda(x) = \frac{-u_\lambda(0)u_\lambda'(-x)+u_\lambda(-x)u_\lambda'(0-)}{-u_\lambda(0)u_\lambda'(-c)+u_\lambda(-c)u_\lambda'(0-)}.$$

It is easy to see that

$$\lim_{x\uparrow 0} K_\lambda(x) = \infty \quad \text{and} \quad \lim_{x\downarrow 0} K_\lambda(x) = 0.$$

Now, $y_t = K_\lambda(X_t^0)$ is a non-negative super-martingale for each P_x, $x \in R^1\setminus\{0\}$ and hence, by a standard argument,

$$P_x\{\exists t_n \uparrow \sigma_0 : X(t_n) < 0\} = 0$$

implying

$$P_x(\Omega_1^+ | \Omega_1) = 1, \qquad \text{for all } x \in R^1\setminus\{0\}.$$

Now, suppose that the condition (B) is satisfied. We will construct a finite measure μ on $(-\infty, 0)$ such that

$$\mho^{\mu}(x) = \int_{-\infty}^{0} \frac{g_\lambda^0(x,y)}{g_\lambda^0(c,y)} \mu(dy)$$

has the property

$$\mho^{\mu}(x) < \infty \quad \text{for} \quad x \in R^1 \setminus \{0\} \quad \text{and} \quad \lim_{x \uparrow 0} \mho^{\mu}(x) = \infty.$$

Then the proof of $P_x(\Omega_1^+/\Omega_1) \equiv 1$ proceeds just as above.

Let $c \in R^1 \setminus \{0\}$ be fixed. We have

$$(3.17) \qquad \frac{g_\lambda^0(x,-\varepsilon)}{g_\lambda^0(c,-\varepsilon)} = \frac{u_\lambda(0)[u_\lambda(-\varepsilon-x)-u_\lambda(-x)]+u_\lambda(-x)[u_\lambda(0)-u_\lambda(-\varepsilon)]}{u_\lambda(0)[u_\lambda(-\varepsilon-c)-u_\lambda(-c)]+u_\lambda(-c)[u_\lambda(0)-u_\lambda(-\varepsilon)]}$$

and

$$u_\lambda(-\varepsilon-x)-u_\lambda(-x)=u_\lambda(-y)-u_\lambda(\varepsilon-y)=[u_\lambda(0)-u_\lambda(\varepsilon-y)]-[u_\lambda(0)-u_\lambda(-y)]$$

if we set $y = x + \varepsilon$. By (3.13) and (3.14), it is easy to see that there exist $0 < a < 1$ and $K > 0$ such that if $0 \leq y \leq a\varepsilon$ $(0 \leq \varepsilon \leq \varepsilon_0)$ i.e. if $-\varepsilon \leq x \leq -(1-a)\varepsilon$ $(0 \leq \varepsilon \leq \varepsilon_0)$, then

$$u_\lambda(-\varepsilon-x) - u_\lambda(-x) \geq K\varepsilon^{\alpha}.$$

Hence, there exists $K' > 0$ such that if $-\varepsilon \leq x \leq -(1-a)\varepsilon$ and $0 \leq \varepsilon \leq \varepsilon_0$,

$$\frac{g_\lambda^0(x,-\varepsilon)}{g_\lambda^0(c,-\varepsilon)} > K'\varepsilon^{\alpha-\beta}.$$

Now, let $x_n = -\varepsilon_0(1-a)^n$. Then there exists $K'' > 0$ such that

$$\frac{g_\lambda^0(x,x_n)}{g_\lambda^0(c,x_n)} > K''(1-a)^{n(\alpha-\beta)} \qquad \text{if} \quad x_n \leq x \leq x_{n+1}, \ n = 1,2,\cdots.$$

Take $0 < b < 1$ such that $(1-a)^{(\alpha-\beta)}b > 1$. Then

$$\mu = \sum_{n=1}^{\infty} b^n \delta_{\{x_n\}}$$

is a finite measure on $(-\infty, 0)$. Since, for each $x \in R^1\setminus\{0\}$,

$$\lim_{\varepsilon\to 0} \frac{g_\lambda^0(x,-\varepsilon)}{g_\lambda^0(c,-\varepsilon)} = \frac{u_\lambda(-x)}{u_\lambda(-c)} ,$$

we have $\mho^\mu(x) < \infty$. Also,

$$\mho^\mu(x) = \int_{-\infty}^{0} \frac{g_\lambda^0(x,y)}{g_\lambda^0(c,y)} \mu(dy) \geqq \frac{g_\lambda^0(x,x_n)}{g_\lambda^0(c,x_n)} b^n \geqq K''[(1-a)^{(\alpha-\beta)}b]^n$$

if $x \in [x_n, x_{n+1}]$, $n = 1,2,\cdots$. Thus, $\lim_{x\uparrow 0} \mho^\mu(x) = \infty$.

<u>Theorem 3.3</u>. <u>Assume that</u> $u_\lambda(x)$ <u>satisfies</u>

(i) $$0 < \varliminf_{\varepsilon\downarrow 0} \frac{u_\lambda(0)-u_\lambda(-\varepsilon)}{u_\lambda(0)-u_\lambda(\varepsilon)} \leqq \varlimsup_{\varepsilon\downarrow 0} \frac{u_\lambda(0)-u_\lambda(-\varepsilon)}{u_\lambda(0)-u_\lambda(\varepsilon)} < \infty$$

(ii) <u>for</u> $x \in R^1\setminus\{0\}$, $$\lim_{\varepsilon\downarrow 0} \frac{u_\lambda(x)-u_\lambda(x\pm\varepsilon)}{u_\lambda(0)-u_\lambda(\varepsilon)} = 0.$$

<u>Then</u>, $P_x(\Omega_1^\pm|\Omega_1) \equiv 1$ <u>on</u> $R^1\setminus\{0\}$.

<u>Proof</u>. By the expression (3.17) and the assumptions (i), (ii), we have

$$\lim_{\varepsilon\to 0} \frac{g_\lambda^0(x,\varepsilon)}{g_\lambda^0(c,\varepsilon)} = \frac{u_\lambda(-x)}{u_\lambda(-c)} .$$

This implies, by a standard argument involving the theory of Martin boundaries, that every bounded λ-harmonic function with respect to X_t^0 is a constant multiple of $u_\lambda(-x)$. In particular, it implies

$$E_x(e^{-\lambda\sigma_0}\ ;\ \Omega_1^+) = \frac{u_\lambda(-x)}{u_\lambda(0)} , \quad \text{i.e.} \quad P_x(\Omega_1^+/\Omega_1) \equiv 1,$$

or

$$E_x(e^{-\lambda\sigma_0} ; \Omega_1^-) = \frac{u_\lambda(-x)}{u_\lambda(0)} , \quad \text{i.e.} \quad P_x(\Omega_1^-/\Omega_1) \equiv 1,$$

or

$$E_x(e^{-\lambda\sigma_0} ; \Omega_1^\pm) = \frac{u_\lambda(-x)}{u_\lambda(0)} , \quad \text{i.e.} \quad P_x(\Omega_1^\pm/\Omega_1) \equiv 1.$$

Thus, it is sufficient to show that the first two possibilities can not occur.

Let $\sigma_{-\varepsilon} = \inf\{t ; X_t = -\varepsilon\}$. Then, for $x \in R^1\setminus\{0\}$,

$$P_x(\sigma_{-\varepsilon} < \sigma_0) \geqq E_x(e^{-\lambda\sigma_{-\varepsilon}} ; \sigma_{-\varepsilon} < \sigma_0) = \frac{g_\lambda^0(x,-\varepsilon)}{g_\lambda^0(-\varepsilon,-\varepsilon)}$$

$$= \frac{u_\lambda(-x)[u_\lambda(0)-u_\lambda(-\varepsilon)]+u_\lambda(0)[u_\lambda(-\varepsilon-x)-u_\lambda(-x)]}{u_\lambda(\varepsilon)[u_\lambda(0)-u_\lambda(-\varepsilon)]+u_\lambda(0)[u_\lambda(0)-u_\lambda(\varepsilon)]}$$

and hence

$$\varliminf_{\varepsilon\downarrow 0} P_x(\sigma_{-\varepsilon} < \sigma_0) > 0$$

implying that $\varepsilon_n \downarrow 0$ exists such that

$$P_x(\varlimsup_{n\to\infty}\{\sigma_{-\varepsilon_n} < \sigma_0\}) > 0.$$

Let $T_n = \sigma_{-\varepsilon_n} \wedge \sigma_{-\varepsilon_{n+1}} \wedge \cdots$, $n = 1,2,\cdots$. Then T_n increases and the quasi-left continuity of X_t implies that

$$\sigma_0 \wedge T_n \uparrow \sigma_0 \qquad \text{a.s. on } \{\sigma_0 < \infty\}.$$

Note also that

$$\varlimsup_{n\to\infty}\{\sigma_{-\varepsilon_n} < \sigma_0\} = \{T_n < \sigma_0 \quad \text{for all} \quad n\}.$$

Thus,

$$P_x(T_n < \sigma_0 \quad \text{and} \quad T_n \uparrow \sigma_0) > 0$$

implying

$$P_x(\exists t_n \uparrow \sigma_0, X_{t_n} < 0) > 0$$

and hence it is impossible to have

$$P_x(\Omega_1^+/\Omega_1) = 1.$$

Similarly, $P_x(\Omega_1^-/\Omega_1) = 1$ is impossible.

Example 3.1. (Stable processes). Let

$$(3.18) \quad \Psi(\xi) = -|\xi|^\alpha(1-i\beta \tan \frac{\pi\alpha}{2} \operatorname{sgn}(\xi)), \quad |\beta| \leqq 1, \; 2 > \alpha > 1.$$

Then, it can be proved easily that the derivative $u_\lambda'(x)$ of the resolvent density has an expression

$$(3.19) \qquad u_\lambda'(x) = -\frac{\Gamma(2-\alpha)\sin\frac{\pi\alpha}{2}}{\pi(1+h^2)}(\operatorname{sgn}(x)+\beta)|x|^{\alpha-2} + v_\lambda(x)$$

where $h = \beta \tan \frac{\pi\alpha}{2}$ and $v_\lambda(x)$ is bounded and continuous. Hence, by Theorem 3.2 and Theorem 3.3,

$$P_x(\Omega_1^\pm|\Omega_1) \equiv 1, \qquad \text{if} \qquad |\beta| < 1,$$

$$P_x(\Omega_1^+|\Omega_1) \equiv 1, \qquad \text{if} \qquad \beta = 1,$$

$$P_x(\Omega_1^-|\Omega_1) \equiv 1, \qquad \text{if} \qquad \beta = -1.$$

Example 3.2. Let

$$(3.20) \quad \Psi(\xi) = -|\xi|^{\alpha_1}(1-i\beta_1\tan \frac{\pi\alpha_1}{2} \operatorname{sgn}(\xi))-|\xi|^{\alpha_2}(1-i\beta_2\tan\frac{\pi\alpha_2}{2}\operatorname{sgn}(\xi)),$$

where $\alpha_1 > 1$ and $\alpha_1 > \alpha_2 > 0$. Then $u_\lambda'(x)$ has an expression

$$u_\lambda'(x) = -\frac{\Gamma(2-\alpha_1)\sin\frac{\pi\alpha_1}{2}}{\pi(1+h_1^2)}(\mathrm{sgn}(x) + \beta_1)|x|^{\alpha_1-2}$$

$$+ (a_1\ \mathrm{sgn}(x) + b_1)|x|^{2\alpha_1-\alpha_2-2}$$

$$+ (a_2\ \mathrm{sgn}(x) + b_2)|x|^{3\alpha_1-2\alpha_2-2}$$

$$\vdots$$

$$+ (a_n\ \mathrm{sgn}(x) + b_n)|x|^{\alpha_1-2+n(\alpha_1-\alpha_2)}$$

$$+ v_\lambda(x),$$

where $v_\lambda(x)$ is bounded and continuous and n is the largest integer satisfying $\alpha_1 - 2 + n(\alpha_1-\alpha_2) \leqq 0$. ($a_i, b_i$ depend on $\alpha_1, \alpha_2, \beta_1, \beta_2$ but not on λ). When $\alpha_1 - 2 + n(\alpha_1-\alpha_2) = 0$, the term $(a_n\ \mathrm{sgn}(x)+b_n)$ should be modified as $(a_n\ \mathrm{sgn}(x) + b_n \log(\frac{1}{|x|}\vee 1))$. From this, we see by Theorems 3.2 and 3.3 that if we have near $u = 0$,

$$\text{(3.22)}\qquad \begin{aligned} n_+(du) &\sim c_+ \frac{du}{u^{\alpha_1+1}}, \\ n_-(du) &\sim c_- \frac{du}{|u|^{\alpha_2+1}} \end{aligned} \qquad c_+ > 0,\ c_- > 0,\ \alpha_1 \geqq \alpha_2,\ \alpha_1 > 1,$$

then $P_x(\Omega_1^+/\Omega_1) \equiv 1$, if $\alpha_1 > \alpha_2$, and $P_x(\Omega_1^\pm/\Omega_1) \equiv 1$, if $\alpha_1 = \alpha_2$.

Proof of (3.21). By (3.20),

$$u_\lambda(x) = \frac{1}{2\pi}\int_{-\infty}^{\infty} e^{-i\xi x}\frac{1}{\lambda-\Psi(\xi)}\,d\xi$$

$$= \frac{1}{\pi}\int_0^\infty \frac{\cos \xi x(\lambda+\xi^{\alpha_1}+\xi^{\alpha_2})}{F(\xi\ ;\ \lambda)}\,d\xi + \frac{1}{\pi}\int_0^\infty \frac{\sin \xi x(h_1\xi^{\alpha_1}+h_2\xi^{\alpha_2})}{F(\xi\ ;\ \lambda)}\,d\xi,$$

where we set

$$F(\xi\ ;\ \lambda) = (\lambda + \xi^{\alpha_1} + \xi^{\alpha_2})^2 + (h_1\xi^{\alpha_1} + h_2\xi^{\alpha_2})^2$$

with $h_i = \beta_i \tan\frac{\pi\alpha_i}{2}$, $i = 1,2$. Hence

$$u_\lambda'(x) = -\frac{1}{\pi}\int_0^\infty \frac{\sin \xi x\cdot\xi(\lambda+\xi^{\alpha_1}+\xi^{\alpha_2})}{F(\xi\ ;\ \lambda)}\,d\xi + \frac{1}{\pi}\int_0^\infty \frac{\cos \xi x\cdot\xi(h_1\xi^{\alpha_1}+h_2\xi^{\alpha_2})}{F(\xi\ ;\ \lambda)}\,d\xi$$

$$\equiv I_1(x) + I_2(x).$$

Now, we shall prove

$$(3.23)\quad I_1(x) = -\frac{\Gamma(2-\alpha)\sin\frac{\pi\alpha}{2}}{\pi(1+h_1^2)}\cdot \mathrm{sgn}(x)|x|^{\alpha_1-2}+a_1\mathrm{sgn}(x)|x|^{\alpha_1-2+(\alpha_1-\alpha_2)} + \cdots + a_n\,\mathrm{sgn}(x)|x|^{\alpha_1-2+n(\alpha_1-\alpha_2)} + v_\lambda(x),$$

where $v_\lambda(x)$ is bounded continuous and n is the largest integer satisfying $\alpha_1 - 2 + n(\alpha_1-\alpha_2) \leqq 0$. For this, we need the following

<u>Lemma 3.1.</u> <u>If</u> $\alpha_1 > \alpha_2 > 0$, $\alpha_1 > 1$ <u>and</u> $1 \leqq k \leqq n+1$,

$$\int_0^\infty \sin \xi x\,\frac{A_k\xi^{1+\alpha_1-(k-1)(\alpha_1-\alpha_2)}+B_k\xi^{1+\alpha_1-k(\alpha_1-\alpha_2)}}{F(\xi\ ;\ \lambda)}\,d\xi - \frac{A_k}{(1+h_1^2)}\int_0^\infty \sin \xi x\cdot\xi^{1-\alpha_1-(k-1)(\alpha_1-\alpha_2)}\,d\xi$$

$$= \int_0^\infty \sin \xi x\,\frac{A_{k+1}\xi^{1+\alpha_1-k(\alpha_1-\alpha_2)}+B_{k+1}\xi^{1+\alpha_1-(k+1)(\alpha_1-\alpha_2)}}{F(\xi\ ;\ \lambda)}\,d\xi$$

+ a bounded continuous function,

<u>where</u> A_{k+1} <u>and</u> B_{k+1} <u>are determined by</u> A_k <u>and</u> B_k <u>as follows:</u>

$$A_{k+1} = B_k - 2\,\frac{1+h_1h_2}{1+h_1^2}A_k \quad\text{and}\quad B_{k+1} = -\frac{1+h_2^2}{1+h_1^2}A_k.$$

We can prove this lemma by a direct calculation. Using this lemma, the proof of (3.23) proceeds as follows. First, we note the following formula:

$$\int_0^\infty \sin \xi x \cdot \xi^{1-\alpha_1-k(\alpha_1-\alpha_2)} d\xi = \Gamma(2-\alpha_1-k(\alpha_1-\alpha_2)) \sin(\frac{\alpha_1+k(\alpha_1-\alpha_2)}{2}\pi)$$
$$\times |x|^{\alpha_1-2+k(\alpha_1-\alpha_2)} \mathrm{sgn}(x).$$

$I_1(x)$ can be written as

$$-\frac{1}{\pi}\int_0^\infty \sin \xi x \frac{\xi^{1+\alpha_1}+\xi^{1+\alpha_1-(\alpha_1-\alpha_2)}}{F(\xi\ ;\ \lambda)} d\xi + \text{a bounded continuous function}$$

Then, applying the lemma successively and noting that

$$\int_0^\infty \sin \xi x \frac{\xi^{1+\alpha_1-k(\alpha_1-\alpha_2)}}{F(\xi\ ;\ \lambda)} d\xi$$

is bounded continuous if $k = n + 1$ and $k = n + 2$, we obtain (3.23).

Similarly, we can calculate $I_2(x)$ and obtain (3.21).

<u>Example 3.3</u>. (Symmetric processes) Let X_t be a symmetric Lévy process. We assume (as we may without losing the generality) that the Lévy measure has compact support. The condition (i) of Theorem 3.3 is trivially satisfied and (3.2) is satisfied if and only if $\int_0^\infty \frac{1}{1-\Psi(\xi)} d\xi < \infty$. Assume further that $\sigma^2 = 0$ and

(3.24) $$\lim_{\xi\uparrow\infty} \frac{-\Psi(\xi)}{\xi} = \infty.$$

(3.24) holds, for instance, if $-\Psi(\xi)$ is monotone increasing in $\xi > 0$. Then $u_\lambda(x)$ is continuously differentiable on $R^1\setminus\{0\}$ since the integral

$$u_\lambda'(x) = -\frac{1}{\pi}\int_0^\infty \frac{\xi \sin \xi x}{\lambda-\Psi(\xi)} d\xi$$

is uniformly convergent in $x \in R^1 \setminus [-\varepsilon, \varepsilon]$ by the second meanvalue theorem. Also

$$\frac{u_\lambda(0)-u_\lambda(\varepsilon)}{\varepsilon} = \frac{1}{\pi\varepsilon}\int_0^\infty \frac{1-\cos\xi\varepsilon}{\lambda-\Psi(\xi)}\,d\xi$$

$$= \frac{1}{\pi}\int_0^\infty \frac{1-\cos\xi}{\varepsilon^2[\lambda-\Psi(\frac{\xi}{\varepsilon})]}\,d\xi$$

$$> K\int_a^b \frac{d\xi}{-\frac{\varepsilon^2}{\xi^2}\,\Psi(\frac{\xi}{\varepsilon})} \longrightarrow \infty, \qquad (\varepsilon \downarrow 0),$$

since if $\sigma^2 = 0$, it is well known that

$$-\Psi(\xi) = o(|\xi|^2), \quad \text{when} \quad |\xi| \longrightarrow \infty.$$

Then, the condition (ii) of Theorem 3.3 is also satisfied and hence $P_x(\Omega_1^{\pm}|\Omega_1) = 1$ for all $x \in R^1 \setminus \{0\}$.

References

[1] A. Beurling and J. Deny: Dirichlet spaces, Proc. Nat. Acad. Sci. U.S.A. 45 (1959), 208-215.

[2] R.M. Blumenthal and R.K. Getoor: Markov processes and potential theory, Academic Press, New York and London, 1968.

[3] J. Bretagnolle: Resultats de Kesten sur les processus a accroissements independants, Séminaire de Probabilités V, Lecture note of Math. Vol.191, Springer, (1971), 21-36.

[4] Ph. Courrège and P. Priouret: Recollements de processus de Markov, Publ. Inst. Statist. Univ. Paris 14 (1965), 275-377.

[5] E.B. Dynkin: Martin boundary for nonnegative solutions of a boundary value problem with a directional derivative, Russian Math. Surveys 19(5), (1964), 1-48.

[6] M. Fukushima: On boundary conditions for multi-dimensional Brownian motions with symmetric resolvent densities, Jour. Math. Soc. Japan 21 (1969), 58-93.

[7] M. Fukushima: Regular representations of Dirichlet spaces, Trans. Amer. Math. Soc. 155 (1971), 455-473.

[8] M. Fukushima: Dirichlet spaces and strong Markov processes, Trans. Amer. Math. Soc. 162 (1971), 185-224.

[9] A.R. Galmarino: Representation of an isotropic diffusion as a skew product, Z. Wahrscheinlichkeitstheorie 1 (1963), 359-378.

[10] N. Ikeda and S. Watanabe: The local structure of diffusion processes, Seminar on Probability Vol.35 (1971) (Japanese).

[11] K. Itô: Poisson point processes attached to Markov processes, to appear in Proc. 6-th Berkeley Symp.

[12] K. Itô and H.P. McKean Jr. : Diffusion processes and their sample paths, Springer, Berlin, 1965.

[13] M. Kanda: Regular points and Green functions in Markov processes, Jour. Math. Soc. Japan 19 (1967), 46-69.

[14] H. Kesten: Hitting probabilities of single points for processes with stationary independent increments, Memoir 93. Amer. Math. Soc., (1969).

[15] F. Knight: An infinitesimal decomposition for a class of Markov processes, Ann. Math. Stat. (1970) 5, 1510-1529.

[16] H. Kunita and S. Watanabe: On square integrable martingales, Nagoya Math. J. 30 (1967), 209-245.

[17] P. Lévy: Processus stochastiques et mouvement brownien, Paris, Gauthier-Villars, 1948.

[18] H.P. McKean Jr. : Stochastic integrals, Academic Press, 1969.

[19] S.A. Molchanov and E.Ostrovskii: Symmetric stable processes as traces of degenerate diffusion processes, Theory of Prob. and its Appl. 14 (1969), 128-131.

[20] M. Motoo: The boundary condition with the discontinuous inclined derivative, Proc. U.S.S.R-Japan Symp. on Probability, Acad. Sci. USSR, Novosibirsk (1969), 247-256.

[21] L. Schwartz: Théorie des distributions, Paris Hermann, 1950-51.

[22] A.V. Skorohod: On the local structure of continuous Markov processes, Theory of Prob. and its Appl. 11 (1966), 336-372.

[23] T. Takada: Hitting probabilities of single points and local times for Lévy processes. Master Thesis, Kyoto Univ. (1972) (Japanese).

[24] M. Weil: Quasi-processus, Séminaire de Probabilités IV, Lecture notes in Math. 124 (1970), 216-239.

[25] A.D. Wentzell: On lateral conditions for multi-dimensional diffusion processes, Theory of Prob. and its Appl. 4 (1959), 164-177.

Nobuyuki Ikeda
Department of Mathematics
Osaka University
Toyonaka, Japan.

Shinzo Watanabe
Department of Mathematics
Kyoto University
Kyoto, Japan.

REMARKS ON MARKOV PROCESSES HAVING GREEN FUNCTIONS WITH ISOTROPIC SINGULARITY

Mamoru Kanda

[1] In this report we study the following three relations between two Markov processes(*) X_i, i=1,2, of a special type. Let us assume that, for i=1 and 2, X_i has Green function $G_i(x,y)$.

S)(**) For each compact set K, there exist positive constants M_i, i=1,2, such that $M_1G_2(x,y) \geqq G_1(x,y) \geqq M_2G_2(x,y)$, $(x,y) \in K \times K$.

P) X_i, i=1,2, have the same polar sets.

R) $K^r_{X_1} = K^r_{X_2}$ holds for each compact set K, where

$$K^r_{X_i} = \{x \in R^n,\ P^i_x(\sigma^i_K = 0) = 1\}. \quad (***)$$

In case both X_i, i=1 and 2, satisfy certain regularity conditions, P) and R) follow from S). For precise statement, see [3]. Under Hunt's condition (H) (i.e $K^r_{X_i} \neq \phi$ for any compact set K which is not polar), R) is sufficient for P). But P) is not sufficient for S) even for isotropic Lévy processes. Further P) is not sufficient for R) as is easily seen by considering a one-sided stable process X_1 and its dual process X_2. Note that it is open except in one-dimensional case whether (H) holds for general unsymmetric stable processes.

Throughout this report we impose the following conditions on our Markov process X without special mentioning.

A1) X has a Green function $G(x,y)$ such that

(*) We always assume that Markov processes are Hunt processes on R^n.
(**) We denote this property by writing $G_1(x,y) \approx G_2(x,y)$.
(***) $\sigma^i_K = \inf\{t>0,\ x^i_t \in K\}$.

$$G(x,y) \approx \phi(|x-y|),$$

where ϕ is a positive, finite and measurable function on $(0,+\infty)$.

A2) X satisfies the duality condition in Blumenthal-Getoor [1], P253 and $G(x,y)$ satisfies (2.1), (2.2), (4.1) and (4.2) in Chap. VI of [1].

Remark 1. For each Borel set B there exists a unique measure $\mu_B(dy)$ carried by $\overline{B}$ such that

$$P_x(\sigma_B<+\infty) = \int_{R^n} G(x,y)\mu_B(dy).$$

We denote the total mass of μ_B by $C(B)$ and call it the capacity of B as usual. Note that $C(\cdot)$ is a Choquet capacity.

Remark 2. For X_i, i=1,2, satisfying A1) and A2), we have S)$\Rightarrow$R)$\Rightarrow$P).

S)$\Rightarrow$R) follows from Corollary 1 in [3]. Now let X be X_1 or X_2 and $\hat{X}$ be the dual process of X. Then, we see from A1) and Corollary 1 in [3] that fine topologies induced by X and $\hat{X}$ are equivalent. Hence (H) holds for X, which implies R)$\Rightarrow$P).

[2] In the sequel we always assume that the dimension $n \geq 3$. An n-dimensional isotropic stable process of index α is denoted by X_α. Set

$$Q_r=\{|x|\leq r\},\ \tilde{Q}_r=\{{}^r/_2\leq|x|\leq r\},\ S_r=\{|x|=r\}.$$

We consider the following conditions.

B1)$_\alpha$ $\quad K^r_X \supset K^r_{X_\alpha}$ for each compact set K.

B2) $\quad \liminf_{r\to 0} P_0(\sigma_{S_r} < +\infty) > 0.$

B3) $\quad \liminf_{r\to 0} P_0(\sigma_{\tilde{Q}_r} < +\infty) > 0.$

B4) $\quad$ There exists a positive function ϕ_o such that $r^p\phi_o(r)$ is monotone increasing on $(0,\delta)$ for some $p,\delta > 0$ and

$$\phi_o(r) \asymp \phi(r)\ ^{(*)}.$$

B5) $\phi(r)$ is monotone decreasing.

<u>Remark 3</u>. $B1)_\alpha \Rightarrow B3)$ in case $\alpha>0$ and $B1)_\alpha \Rightarrow B2) \Rightarrow B4)$ in case $\alpha>1$.

<u>Proof</u>. By a simple computation we have

$$P_0^\alpha(\sigma_{\tilde{Q}_r} < +\infty) \geqq M > 0.$$

Hence $0 \in \{\bigcup_k \tilde{Q}_{r_k} \cup \{0\}\}^r_{X_\alpha}$ for each sequence $\{r_k\}$, $r_k \to 0$, which together with $B1)_\alpha$ implies B3) by Borel-Cantelli lemma. The proof of the latter is as follows. First note that for each sequence $\{r_k\}$, $r_k \to 0$,

$$0 \in (\bigcup_{k=1}^{+\infty} S_{r_k} \cup \{0\})^r_{X_\alpha} \quad , \quad \alpha > 1.$$

Hence $B1)_\alpha$ implies

$$0 \in (\bigcup_{k=1}^{+\infty} S_{r_k})^r_X \quad ,$$

from which B2) follows by Borel-Cantelli lemma. Set

$$\phi_o(r) = r^{1-n}\int_0^r \phi(2s)s^{n-2}(1-(\tfrac{s}{r})^2)^{\frac{n-3}{2}} ds \ .$$

Then, since

$$C(S_r)^{-1} \asymp \phi_o(r) \quad , \quad r \to 0$$

(see Lemma 4 in [2])$^{(**)}$, it holds that

$$\phi(r) \asymp \phi_o(r) \quad , \quad r \to 0$$

(*) This means that

$$0 < \liminf_{r\to o} \frac{\phi(r)}{\phi_o(r)} \leqq \liminf_{r\to o} \frac{\phi(r)}{\phi_o(r)} < +\infty.$$

(**) Note that the monotone property is not necessary for the proof of Lemma 4 in [2].

by B2). The above argument is essentially same as that in the proof of Lemma 6 in [3] and Theorem in [2].

<u>Remark 4</u>. Under the conditions B3) and B4), it holds that

(2.1) $$\phi(r) \asymp \frac{1}{C(Q_r)} \asymp r^{-n} \int_0^r \frac{1}{C(Q_s)} s^{n-1} ds, \quad r \to 0.$$

<u>Proof</u>. We denote absolute positive constants by M_k, k=1,2,··· . Since $r^p \phi_o(r)$ is monotone increasing, we have

(2.2) $$M_1 \phi(r) \geqq \sup_{r/2 \leqq t \leqq r} \phi(t) \geqq \inf_{r/2 \leqq t \leqq r} \phi(t) \geqq M_2 \phi(r/2).$$

Combining B3) with (2.2), we can show

(2.3) $$M_3 \frac{1}{C(\tilde{Q}_{2r})} \geqq \phi(r) \geqq M_4 \frac{1}{C(\tilde{Q}_r)} .$$

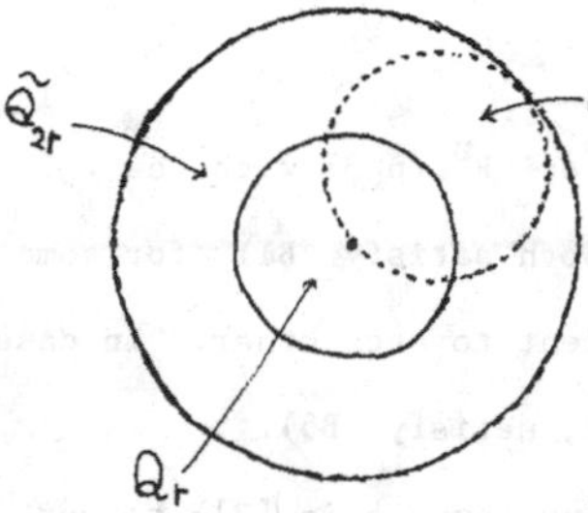

Noting $C(Q_r) \leqq M_5 C(Q'_r)$, we see that

(2.4) $$C(\tilde{Q}_{2r}) \geqq \frac{M_5}{2} C(Q_r) \geqq M_3 C(\tilde{Q}_r) .$$

Combining (2.3) with (2.4), we get

(2.5) $$\phi(r) \asymp \frac{1}{C(\tilde{Q}_r)} \asymp \frac{1}{C(\tilde{Q}_{2r})}, \quad r \to 0.$$

On the other hand, we have by (2.4)

$$2C(\tilde{Q}_{2r}) \geqq C(\tilde{Q}_{2r}) ,$$

because $C(Q_r) + C(\tilde{Q}_{2r}) \geqq C(Q_{2r})$. Noting $C(Q_r) \geqq C(\tilde{Q}_r)$, it follows from (2.5) that

$$\phi(r) \asymp \frac{1}{C(Q_r)} , \quad r \to 0.$$

Set $f(r) = C(Q_r)^{-1}$ and

$$\tilde{f}(r) = r^{-n} \int_0^r f(s) s^{n-1} ds.$$

Then, we have

$$\sup_{y\in Q_r}\int_{Q_r} f(|z-y|)dz = \int_{Q_r} f(|z|)dz \leqq 2n \inf_{y\in Q_r}\int_{Q_r} f(|z-y|)dz,$$

because f is monotone decreasing on $(0, +\infty)$. Hence

$$M_6 \check{f}(r)C(Q_r) \leqq \frac{1}{|Q_r|}\int_{Q_r} P_z(\sigma_{Q_r} < +\infty)dz \leqq M_7 \check{f}(r)C(Q_r),$$

where $|Q_r|$ is the volume of Q_r. Since the value of the middle term equals one, we have

$$\check{f}(r) \asymp \frac{1}{C(Q_r)}.$$

The proof is complete.

Remark 5. For any $\alpha>0$, $B1)_\alpha$ together with B5) implies (2.1). This is shown in Lemma 6 [3].

Now we can prove

PROPOSITION. Let X_i, $i=1,2$, be Markov processes on R^n ($n\geqq 3$) with the properties A1)(*) and A2). Suppose that X_i, $i=1,2$, both satisfy $B1)_\alpha$ for some α. In case $\alpha>1$, the relations S), P) and R) are equivalent to each other. In case $\alpha>0$, the same conclution holds provided that X_i, $i=1,2$, satisfy B5).

Indeed, we can apply S. J. Taylor's result (see also Lemma 5 in [3]) to our processes X_i, $i=1,2$ by Remark 4 or 5. Hence

$$\liminf_{r\to 0} \phi_1(r)/\phi_2(r) = 0$$

implies that for some compact set K

$$C_1(K) > 0, \quad C_2(K) = 0 \quad ,$$

(*) That is, $G_i(x,y) \approx \phi_i(|x-y|)$, $i=1,2$.

where $C_i(\cdot)$ denotes the capacity relative to X_i, which together with Remark 2 completes the proof.

COROLLARY. Fix a Markov process X_1 whose Green function has monotone singularity ϕ_1 such that $r^{n-\alpha}\phi_1(r)$ is monotone increasing for some $\alpha>1$. Let X_2 be a Markov process whose Green function has singularity ϕ_2. Then R) is equivalent to S).

Indeed, by Theorem 1 in [3], X_1 satisfies $B1)_\alpha$ for $\alpha>1$.

Remark 6. Remark 4 together with Remark 3 implies that a Green function of a Markov process with A1) and A2) has monotone singularity if it satisfies $B1)_\alpha$ for some $\alpha>1$. It is known that there exists an isotropic Lévy process whose Green function has not monotone singularity. See J. Zabczyk [4].

References

[1] R. M. Blumenthal and R. K. Getoor, Markov processes and potential theory. Academic Press, New York and London, 1968.

[2] M. Kanda, On the singularity of Green functions in Markov processes II. Nagoya Math. J., 37, 207-217 (1970).

[3] M. Kanda, Comparison theorems on regular points for multi-dimensional Markov processes of transient type. Nagoya Math. J., 44, 165-214(1971).

[4] J. Zabczyk, Sur la théorie semi-classique du potentiel pour les processus à accroissements indépendants. Studia Math., XXXV, 227-247 (1970).

Hiroshima University

ON THE VARIATION OF GAUSSIAN PROCESSES

Takayuki Kawada and Norio Kôno

1. Introduction. The problem treated in this paper is initiated by P. Lévy [5], who discovered two interesting facts concerning the variation of Brownian motion $B(t)$. Let $\{P_k\}$, $k=1,2,\cdots$ be a sequence of finite partitions of the interval $[0,1]$ such that P_{k+1} is a refinement of P_k and the collection of partition points of all $\{P_k\}$, $k=1,2,\cdots$, is dense in $[0,1]$. Denote by Π the family of all finite partitions of $[0,1]$ and by $Q(\delta)$ the family of all finite partitions $P = \{0 = t_0 < t_1 < \cdots < t_n = 1\}$ such that $\max|t_i - t_{i-1}| < \delta$. Then he has shown the following facts:

(a) $\quad P[\lim_{k\to\infty} V_\psi(B;P_k) = 1] = 1,$

(b) $\quad P[\lim_{\delta\to 0} \sup_{P\in Q(\delta)} V_\psi(B;P) = +\infty] = 1,$

(b)' $\quad P[\sup_{P\in\Pi} V_\psi(B;P) = +\infty] = 1,$

where

$$V_\psi(B;P) = \sum_{i=1}^{n} \psi(|B(t_i) - B(t_{i-1})|),$$

$$\psi(t) = t^2, \quad P = \{0 = t_0 < t_1 < \cdots < t_n = 1\}.$$

We notice that the function $\psi(t) = t^2$ is the inverse function of the square root of the incremental covariance (E[(B(t)

$- B(0))^2])^{1/2} = \sqrt{t}$. In this point of view, one of the authors [4] extended (a) to some class of Gaussian processes.

Recently, S.J. Taylor [7] has presented the exact asymptotic estimate of Brownian path variation of type (b). He has proved the following:

$$P[\lim_{\delta\to 0} \sup_{P\in Q(\delta)} V_\psi(B;P) = 1] = 1,$$

where

$$\psi(t) = t^2/(2\log\log 1/t).$$

We notice here that the function $\psi(t) = t^2/(2\log\log 1/t)$ becomes the asymptotically the inverse function of $\sqrt{2t \log\log 1/t}$ which is local modulus of continuity of Brownian motion paths.

In this paper we extend the result of S.J. Taylor to some class of Gaussian processes in the direction pointed above.

The authors are greatly in debt to Professor S.J. Taylor, whose pre-print has been communicated by professor S. Watanabe.

2. 0-1 law. Before describing our main theorems we point out that the path variations of type (b) and (b)' satisfy the 0-1 laws.

__Theorem 1.__ __Let__ $\{X(t,\omega)\ ;\ 0\le t\le 1\}$ __be a centered Gaussian process with continuous sample functions such that__

$$E[(X(t) - X(s))^2] \le \sigma^2(|t-s|),$$

__where__ $\sigma(t)$ __is continuous function with__ $\sigma(0) = 0$.

Let $\psi(t)$ *be a non-decreasing regular varying function with exponent* $\alpha > 0$ *satisfying*

$$\psi(\sigma(t)) \leq t\gamma(t) \quad \text{for} \quad t \geq 0 \quad \text{and} \quad \lim_{t\to 0} \gamma(t) = 0.$$

Then

$$\lim_{\delta\to 0} \sup_{P\in Q(\delta)} V_\psi(X;P) = \text{constant (including the infinity)}$$

holds with probability 1.

Theorem 2. *Let* $\{X(t,\omega)\ ;\ 0 \leq t \leq 1\}$ *be a centered Gaussian process with continuous sample functions. If a continuous function* $\psi(t)$ *satisfies the inequality*

$$\psi(at + bs) \leq C_1(a)\psi(t) + C_2(b)\psi(s),$$

for any constants a *and* b, *then either*

$$\sup_{P\in\Pi} V_\psi(X;P) = +\infty$$

with probability 1, *or*

$$\sup_{P\in\Pi} V_\psi(X;P) < +\infty$$

with probability 1 *holds*.

The proof of Theorem 1 is based on Kolmogorov's 0-1 law. We can show by an analogous method in the 0-1 law for type (a) ([4,

Theorem 1]) that the event in Theorem 1 belongs to the tail σ-field of an independent Gaussian random sequence.

Whereas the proof of Theorem 2 is based on Kallianpur's 0-1 law [2]. We could show that the event in Theorem 2 is a measurable r-module.

3. Main Theorems. Now we describe our main theorems. Let $\{X(t,\omega)\ ;\ 0\leq t\leq 1\}$ be a centered Gaussian process with continuous sample functions. We assume that there exists a continuous function with $\sigma(0) = 0$ such that

(i) $E[(X(s) - X(t))^2] \leq \sigma^2(|t-s|)$,

(ii) $\sigma(t)$ is a non-decreasing regular varying function,

(iii) $\sigma(t)\sqrt{2\log\log 1/t}$ is strictly increasing near the origin.

Set

$\psi(t)$ = the inverse function of $\sigma(t)\sqrt{2\log\log 1/t}$ near the origin. Then we have

Theorem 3. Under the assumptions (i), (ii), *and* (iii),

$$\lim_{\delta\to 0}\ \sup_{P\in Q(\delta)} V_\psi(X;P) \leq 1$$

holds with probability 1.

We assume the followings:

(iv) $E[(X(t) - X(s))^2] = \sigma^2(|t-s|)$.

(v) $\sigma^2(t) - \sigma^2(t-h) \leq L\sigma^2(h)$ for small t and $h>0$.

Then we have

Theorem 4. Under the assumptions (ii) - (v),

$$\lim_{\delta\to 0}\ \sup_{P\epsilon Q(\delta)} V_\psi(X;P) \geq 1$$

holds with probability 1.

4. Proofs of Theorem 3 and Theorem 4. To prove Theorem 3, we need some lemmas.

Lemma 1. Let (S,ρ) be a compact metric space and $\{X(s,\omega)\ ;\ s\epsilon S\}$ be a real valued centered Gaussian process with continuous sample functions. We assume that there exists a non-decreasing continuous function with $\sigma(0) = 0$ such that

$$E[(X(s) - X(t))^2] \leq \sigma^2(\rho(s,t)),$$

and

$$\int^\infty \sigma(e^{-x^2})dx < +\infty$$

In addition we assume that for any compact subset K the minimal number of ε-covering $N_\varepsilon(K)$ of K satisfies

$$N_\varepsilon(K) \leq C(d(K)/\varepsilon)^N,$$

where $d(K)$ is the diameter of K , C is an absolute constant and N is a positive integer.

Then

$$P[\sup_{s\in K}|X(s)| \geq x(|\Gamma|_K + 4\int_0^\infty \sigma(d(K)e^{-u^2})du/\sqrt{\log p})]$$

$$\leq Cp^{2N} \int_x^\infty \frac{1}{\sqrt{2\pi}} e^{-u^2/2} du$$

holds for any $x \geq \sqrt{1+4N\log p}$ and $p > 1$, where $|\Gamma|_K^2 = \sup_{s\in K} E[X(s)^2]$.

This lemma is essentially due to X. Fernique [1].

The following lemmas 2 and 3 are concerned with uniform and local modulus of sample function continuity, respectively.

Lemma 2. Under the assumptions (i) and (ii) of §3,

$$\overline{\lim_{|t-s|\to 0}} \frac{|X(t) - X(s)|}{\sigma(|t-s|)\sqrt{2\log 1/|t-s|}} \leq 1$$

holds with probability 1.

This lemma is essentially contained in Theorem 4 of M. Nisio [6]. We notice that it is not necessary to assume concavity of $\sigma^2(t)$ and stationarity in her Theorem 4 when we require an upper bound estimate. It is not difficult to prove Lemma 2 from Lemma 1 directly.

Lemma 3. Under the assumptions (i) and (ii) of §3,

$$\lim_{\delta\to 0} \sup_{\substack{0\leq u,v \\ 0<u+v\leq\delta}} \frac{|X(t-u) - X(t+v)|}{\sigma(u+v)\sqrt{2\log\log 1/(u+v)}} \leq 1$$

holds for any $t \in (0,1)$ with probability 1.

Proof. Set

$$d_n = \exp[-n^{1-\varepsilon}] \quad (1/2 > \varepsilon > 0),$$

$$S_n = \{(u,v) \ ; \ 0 \le u,v, \ \ 0 \le u+v \le d_n\},$$

$$\rho((u,v),(u',v')) = \sqrt{(u-u')^2 + (v-v')^2},$$

$$Y(u,v) = X(t-u) - X(t+v).$$

Then (S_1,ρ) is a compact metric space with

$$N_\varepsilon(K) \le d(d(K)/\varepsilon)^2,$$

and

$$d(S_n) \le \sqrt{2}\, d_n.$$

On the other hand we have

$$E[(Y(u,v) - Y(u',v'))^2]$$

$$\le 2E[(X(t-u) - X(t-u'))^2 + (X(t+v) - X(t+v'))^2]$$

$$\le 4\sigma^2(\rho((u,v),(u',v'))),$$

$$|\Gamma|_n^2 = \sup_{(u,v)\varepsilon S_n} E[Y(u,v)^2] \le \sup_{(u,v)\varepsilon S_n} \sigma^2(u+v) = \sigma^2(d_n).$$

Set

$$A_n = \{\ \omega\ ;\ \sup_{(u,v)\varepsilon S_n} |X(t-u) - X(t+v)| \geq \sqrt{(2+4\varepsilon)\log\log 1/d_n}\ \sigma(d_n)(1+\varepsilon)\}.$$

Since we can choose sufficiently large p such that

$$8 \int_0^\infty \sigma(\sqrt{2}d_n e^{-u^2})du \leq \varepsilon\sqrt{\log p}\ \sigma(d_n),$$

because of the assumption (ii), we have, setting $N=2$ in Lemma 1,

$$\sum_n^\infty P(A_n) \leq \sum_n^\infty 9cp^4 \exp[-(1+2\varepsilon)\log\log 1/d_n]$$

$$= \sum_n^\infty 9cp^4 n^{-(1-\varepsilon)(1+2\varepsilon)} < +\infty.$$

Hence there exists n_0 with probability 1 such that for any $n \geq n_0$

$$\sup_{\substack{0\leq u,v\\ 0<u+v\leq d_n}} |X(t-u) - X(t+v)| \leq \sqrt{(2+4\varepsilon)\log\log 1/d_n}\ \sigma(d_n)(1+\varepsilon)$$

holds. But for $d \leq d_n$ we have

$$\sup_{\substack{0\leq u,v\\ 0<u+v\leq d}} \frac{|X(t-u) - X(t+v)|}{\sigma(u+v)\sqrt{2\log\log 1/(u+v)}}$$

$$\leq \sup_{m\geq n} \sup_{\substack{0\leq u,v\\ d_{m+1}\leq u+v\leq d_m}} \frac{|X(t-u) - X(t+v)|}{\sigma(u+v)\sqrt{2\log\log 1/(u+v)}}$$

$$\le \sup_{m\ge n}\{(\sup_{\substack{0\le u,v\\ 0<u+v\le d_m}} |X(t-u) - X(t+v)|)/\sigma(d_{m+1})\sqrt{2\log\log 1/d_m}\}$$

$$\le \sup_{m\ge n} \frac{\sqrt{(2+4\varepsilon)\log\log 1/d_m}\ \sigma(d_m)(1+\varepsilon)}{\sigma(d_{m+1})\sqrt{2\log\log 1/d_m}}$$

$$= \sup_{m\ge n} \frac{\sigma(d_m)}{\sigma(d_{m+1})} \sqrt{1+2\varepsilon}(1+\varepsilon).$$

Since

$$\lim_{m\to\infty} \frac{\sigma(d_m)}{\sigma(d_{m+1})} = 1,$$

and $\varepsilon > 0$ is arbitrary, it completes the proof of Lemma 3.

Now we begin to prove Theorem 3 in the line of S.J. Taylor's method. At first we notice that since $\sigma(t)$ is a regular varying function, so is $\psi(t)$ and there exists a constant $\mu > 0$ such that

$$\psi(\sigma(t)\sqrt{\log 1/t}) \le t(\log 1/t)^{\mu}$$

holds near the origin.

For any $\varepsilon > 0$, set

$$A_\delta = \{(t,\omega)\ ;\ \sup_{\substack{0<u+v\le\delta\\ 0\le u,v}} \frac{|X(t-u) - X(t+v)|}{\sigma(u+v)\sqrt{2\log\log 1/(u+v)}} < 1+\varepsilon\}.$$

Then A_δ is measurable with respect to (t,ω).

Set

$$I_\delta(t,\omega) = 1, \quad \text{if} \quad (t,\omega) \in A_\delta,$$
$$= 0, \quad \text{if} \quad (t,\omega) \notin A_\delta.$$

Then by virtue of Lemma 3 it follows that for any t, with probability 1 $\lim_{\delta\to 0} I_\delta = 1$. Therefore we have

$$E[\underline{\lim}_{\delta\to 0} \int_0^1 I_\delta(t,\omega)dt] \geq E[\int_0^1 \underline{\lim}_{\delta\to 0} I_\delta(t,\omega)dt]$$
$$= \int_0^1 E[\underline{\lim}_{\delta\to 0} I_\delta(t,\omega)]dt = 1.$$

Hence we have

$$\lim_{\delta\to 0} \int_0^1 I_\delta(t,\omega)dt = 1,$$

with probability 1. This means that with probability 1 there exists $\delta_0 > 0$ such that for any $\delta < \delta_0$, $\int_0^1 I_\delta(t,\omega)dt > 1 - \varepsilon$ holds.

Let $P = \{0 = t_0 < t_1 <\dots< t_n = 1\}$ be a partition from $Q(\delta)$ of sufficiently small δ. If an interval $[t_{i-1},t_i]$ contains t such that $(t,\omega) \in A_\delta$, then we have

$$\psi(|X(t_i) - X(t_{i-1})|) \leq \psi((1+\varepsilon)\sigma(t_i-t_{i-1})\sqrt{2\log\log 1/t_i-t_{i-1}})$$
$$\leq \{(i+\varepsilon)^\alpha + \varepsilon\}\psi(\sigma(t_i-t_{i-1})\sqrt{2\log\log 1/t_i-t_{i-1}})$$
$$= \{(1+\varepsilon)^\alpha + \varepsilon\}(t_i-t_{i-1}),$$

where α is the exponent of the regular varying function $\psi(t)$.

Set

$$\Lambda = \{i \ ; \ \text{for any} \ t\varepsilon[t_{i-1},t_i] \ (t,\omega) \notin A_\delta\},$$

and for a fixed constant $A > \mu + 2$,

$$\Lambda' = \{i \ ; \ |X(t_i) - X(t_{i-1})| > \sqrt{2A\log\log 1/t_i-t_{i-1}}\,\sigma(t_i-t_{i-1})(1+\varepsilon)\}.$$

Then with probability 1 we have $\sum\limits_{i\varepsilon\Lambda} (t_i - t_{i-1}) < \varepsilon$, and

$$\sum_{\substack{i\varepsilon\Lambda \\ i\notin\Lambda'}} \psi(|X(t_i)-X(t_{i-1})|) \le \sum_{i\varepsilon\Lambda} \psi(\sqrt{2A\log\log 1/t_i-t_{i-1}}\,\sigma(t_i-t_{i-1})(1+\varepsilon))$$

$$\le \sum_{i\varepsilon\Lambda} \{(1+\varepsilon)^\alpha+\varepsilon\}\psi(\sigma(t_i-t_{i-1})\sqrt{2\log\log 1/t_i-t_{i-1}})$$

$$\le \{(1+\varepsilon)^\alpha + \varepsilon\} \sum_{i\varepsilon\Lambda} (t_i - t_{i-1})$$

$$\le \varepsilon\{(1+\varepsilon)^\alpha + \varepsilon\}.$$

Now to estimate $\sum\limits_{i\varepsilon\Lambda\wedge\Lambda'} \psi(|X(t_i) - X(t_{i-1})|)$, let us calculate the number

$$Z_n(\omega) = \#\{j \ ; \sup_{t,s\varepsilon J_{n,j}} |X(t) - X(s)| \ge \sqrt{2A\log\log 1/h_n}\,\sigma(h_n)(1+\varepsilon)\},$$

where $h_n = e^{-n}$, and

$$J_{n,j} = [\tfrac{j}{2} h_n, (\tfrac{j}{2} + 1)h_n], \quad 0 \le j \le 2e^n - 1.$$

Set

$$A_{n,j} = \{\omega \; ; \sup_{t,s\in J_{n,j}} |X(t) - X(s)| \geq \sqrt{2A\log\log 1/h_n}\,\sigma(h_n)(1+\varepsilon)\}.$$

Then by virtue of Lemma 1 and choosing sufficiently large p we have

$$P(A_{n,j}) \leq 9dp^4 n^{-A}.$$

Therefore we have $E[Z_n(\omega)] \leq 18de^n p^4 n^{-A}$, and $P[Z_n(\omega) > e^n n^{-\nu}] \leq 18dp^4 n^{-A+\nu}$. Choosing ν such that $A > \nu + 1$ and $\nu > \mu + 1$, we have

$$\sum_n P[Z_n(\omega) > e^n n^{-\nu}] < +\infty.$$

Hence, with probability 1 there exists $n_0(\omega)$ such that for any $n \geq n_0$, $Z_n(\omega) \leq e^n n^{-\nu}$ holds. Taking a partition $P = \{0 = t_0 < t_1 < \cdots < t_n = 1\}$ from $Q(\delta)$, set $m_0 = [\log 1/2\delta]$, and

$$\Lambda_m = \{i \; ; h_{m+1}/2 \leq t_i - t_{i-1} < h_m/2\}.$$

If $i \in \Lambda_m$, then there exists j such that $(t_{i-1}, t_i) \subset J_{m,j}$. Combining Lemma 2, we have

$$\sum_{i\in\Lambda\wedge\Lambda'} \psi(|X(t_i) - X(t_{i-1})|)$$

$$= \sum_{m=m_0}^{\infty} \sum_{i\in\Lambda_m\wedge\Lambda'} \psi(|X(t_i) - X(t_{i-1})|)$$

$$\leq \sum_{m=m_0}^{\infty} Z_m(\omega)\psi(\sigma(h_m/2)\sqrt{(2+\varepsilon)\log 2/h_{m+1}})$$

$$\leq C \sum_{m=m_0}^{\infty} e^m m^{-\nu}(\log 1/h_m)^{\mu} h_m$$

$$\leq C' \sum_{m=m_0}^{\infty} m^{-(\nu-\mu)} < +\infty.$$

Therefore we have

$$\lim_{\delta\to 0} \sup_{P\varepsilon Q(\delta)} \sum_{i\varepsilon\Lambda\frown\Lambda'} \psi(|X(t_i) - X(t_{i-1})|) = 0.$$

This completes the proof of Theorem 3.

The proof of Theorem 4. First we notice that by virtue of M. Nisio's result [6, Theorem 3], under the assumption of Theorem 4, we have

$$\overline{\lim_{h\downarrow 0}} \frac{|X(t+h) - X(t)|}{\sigma(h)\sqrt{2\log\log 1/h}} \geq 1 ,$$

with probability 1 for any $t \varepsilon (0,1)$. Therefore the proof could be completed just in the same way as that of S.J. Taylor's Theorem [7], so we omit it.

5. The relation between two variations of type (a) and type (b). Let $P(k)$ be the set of all partitions such that $\min|t_i - t_{i-1}| \geq 1/k$. Then we have

<u>Theorem 5</u>. <u>Under the same assumptions of Theorem 3</u>,

$$\overline{\lim_{k\to\infty}} \sup_{P\varepsilon P(k)} V_{\sigma^{-1}}(X;P)/\phi(1/k) \leq 1$$

<u>holds with probability</u> 1, <u>where</u>

$$\phi(t) = \sup_{t\leq s} \sigma^{-1}(\sigma(s)\sqrt{2\log\log 1/s})/s.$$

<u>Proof</u>. From the proof of Theorem 3 we have

$$|X(t_i) - X(t_{i-1})| \leq (1+\varepsilon)\sigma(t_i-t_{i-1})\sqrt{2\log\log 1/t_i-t_{i-1}},$$

for an interval $[t_{i-1},t_i]$ containing t such that $(t,\omega) \varepsilon A_\delta$. If

$$|X(t_i) - X(t_{i-1})| \geq (1+\varepsilon)\sigma(t_i-t_{i-1})\sqrt{2\log\log 1/t_i-t_{i-1}},$$

then we have

$$\psi(|X(t_i)-X(t_{i-1})|) \geq c\sigma^{-1}(|X(t_i)-X(t_{i-1})|)/\phi(t_i-t_{i-1}).$$

Since $\phi(1/k) \uparrow +\infty$ and $\sum_{i\varepsilon\Lambda} (t_i-t_{i-1}) < \varepsilon$, for sufficiently large k, we have

$$\sup_{P\varepsilon P(k)\wedge Q(\delta)} V_{\sigma^{-1}}(X;P)/\phi(1/k) \leq 1 + \varepsilon,$$

and

$$\sup_{P\varepsilon P(k),\ \notin Q(\delta)} V_{\sigma^{-1}}(X;P)/\phi(1/k) \leq \varepsilon.$$

These yield Theorem 5.

Theorem 6. Under the assumptions (ii), (iv) and (v),

$$\varlimsup_{k\to\infty} \sup_{P\in P(k)} V_{\sigma^{-1}}(X;P)/\phi(1/k) \geq 1$$

holds with probability 1.

Proof. Set

$$A_n = \{(t,\omega)\ ;\ \sup_{1\leq j\leq 2^n} (X(t+2^{-jp})-X(t))/\sigma(2^{-jp}) \geq \sqrt{2(1-\varepsilon)\log 2^n}(1-\varepsilon)\}.$$

According to the proof of Nisio's Theorem 2 [4], for any $\varepsilon > 0$, there exists a constant p and for any fixed t, we have

$$P[\ \varlimsup_{n\to\infty} A_n^c\ ] = 0.$$

Let $I(A)$ be the indicator function of a set A. Then the above equality means

$$\lim_{m\to\infty} I(B_m) = I(B) \quad \text{a.e.} \quad (t,\omega),$$

where $B_m = \bigcap_{n\geq m} A_n$ and $B = \bigcup_{m=1} B_m$. Then by virtue of Egoroff's theorem, there exists a (t,ω)-measurable set F such that

$$E[\ \int_0^1 I(F)dt\ ] \geq 1 - \varepsilon,$$

and $I(B_m)$ converges uniformly to $I(B)$ on F. It yields

$$P[\int_0^1 I(F)dt \geq 1-2\varepsilon\ ] \geq \varepsilon.$$

On the other hand, by Nisio's result we have

$$P[\int_0^1 I(B)dt = 1\ ] = 1.$$

Therefore we have

$$P[\int_0^1 I(B\wedge F)dt \geq 1-2\varepsilon\ ] \geq \varepsilon.$$

Since $B\wedge F \subset B_m$ for sufficiently large m, we have

$$P[\int_0^1 I(B_m)dt \geq 1-2\varepsilon\ ] \geq \varepsilon.$$

But for sufficiently large k we can choose n such that $2^{2^n p} \leq k \leq 2^{2^{n+1}p}$ and also we can choose disjoint intervals $[t_i, t_i+h_i]$ ($2^{-2^n p} \leq h_i \leq 2^{-p}$) such that

$$X(t_i+h_i) - X(t_i) \geq \sigma(h_i)\sqrt{2(1-\varepsilon)\log 2^n}(1-\varepsilon).$$

Moreover the total length of these intervals is more than $1-3\varepsilon$, so we conclude that

$$P[\varlimsup_{k\to\infty} \sup_{P\in P(k)} V_{\sigma^{-1}}(X;P)/\phi(1/k) \geq 1-2\varepsilon] > 0.$$

But we can easily see that the above event obeys 0-1 law. This completes the proof of Theorem 6.

These theorems improve the results in [3].

References

[1] X. Fernique : Continuité des processus gaussiens, C.R. Acad. Sc. Paris, t. 258 (1964), 6058-6060.

[2] G. Kallianpur : Zero-one laws for Gaussian processes, Trans. Amer. Math. Soc. 149 (1970), 199-211.

[3] T. Kawada : Oscillation des processus gaussiens, C.R. Acad. Sc. Paris, t. 274 (1972), 97-99.

[4] N. Kôno : Oscillation of sample functions in stationary Gaussian processes, Osaka J. Math. 6 (1969), 1-12.

[5] P. Lévy : Le mouvement brownien plan, Amer. J. Math. 62 (1940), 487-550.

[6] M. Nisio : On the extreme values of Gaussian processes, Osaka J. Math. 4 (1967), 313-326.

[7] S.J. Taylor : Exact asymptotic estimates of Brownian path variation, Duke Math. J. 39 (1972), 219-241.

Kobe College of Commerce,
Tarumi, Kobe, Japan

Yosida College,
Kyoto University,
Kyoto, Japan

ON A CLASS OF LINEAR PROCESSES

Tatsuo Kawata

1. Introduction.

Let $\xi(S,\omega)=\xi(S)$ be a random measure, that is,

$$\xi(S_1 \cup S_2) = \xi(S_1) + \xi(S_2),$$

for all disjoint Borel sets S_1 and S_2, $E|\xi(S)|^2 < \infty$ for all bounded Borel sets S and $\xi(S)$ is σ-additive in the sense that

$$\xi(S) = \underset{n\to\infty}{\text{l.i.m.}} \sum_{k=1}^{n} \xi(S_k)$$

for $S=\bigcup_{k=1}^{\infty} S_k$, S_k being disjoint Borel sets. $E|\xi(S)|^2$ is not necessarily finite for an unbounded Borel set S.

Define the set function $F(S)=E|\xi(S)|^2$ and define the nondecreasing function $F(t)$ by $F(t)-F(s)=F([s,t))$, $s<t$. $F(t)$ is then left continuous.

We define as usual the integral

$$(1.1) \qquad \int_a^b \phi(\lambda)\xi(d\lambda)$$

for any $\phi(\lambda)\varepsilon L^2(F)$ on (a,b). (See for example Rozanov[5])

If $\phi(\lambda)\varepsilon L^2(F)$ on $(-\infty,\infty)$, then we see that

$$(1.2) \qquad \underset{\substack{a\to-\infty \\ b\to\infty}}{\text{l.i.m.}} \int_a^b \phi(\lambda)\xi(d\lambda)$$

exists and is described by

$$(1.3) \qquad \int_{-\infty}^{\infty} \phi(\lambda)\xi(d\lambda).$$

Let a real valued random measure $\xi(S)$ defind above be given. Let $F(\lambda)$ be the corresponding nondecreasing function. $F(\lambda)$ generates a σ-finite measure on $(-\infty,\infty)$. We suppose that $E\xi(S,\omega)=0$ for every bounded Borel set S. For convenience we modify $F(\lambda)$ to satisfy

(1.4) $$F(\lambda) = \frac{1}{2}[F(\lambda+0) + F(\lambda-0)].$$

Let a real valued nonrandom function m(t) of bounded variation over every finite interval be given and let m(S) be the signed measure generated by m(t). We suppose

(1.5) $$m(t) = \frac{1}{2}[m(t+0) + m(t-0)].$$

Furthermore suppose that a real valued continuous function a(t) is given, which satisfies that, for all t,

(1.6) $${}_{\alpha}\int^{\beta} |a(t-\lambda)|^2 \, dF(\lambda) < \infty,$$

(1.7) $${}_{\alpha}\int^{\beta} |a(t-\lambda)||dm(\lambda)| < \infty,$$

for every finite (α,β).

Write

(1.8) $$\eta(S,\omega) = \xi(S,\omega) + m(S).$$

Then ${}_{\alpha}\int^{\beta} a(t-\lambda)\eta(d\lambda)$ exists for every t with the definition

(1.9) $${}_{\alpha}\int^{\beta} a(t-\lambda)\xi(d\lambda) + {}_{\alpha}\int^{\beta} a(t-\lambda)dm(\lambda).$$

Now suppose that there is a stochastic process $X(t,\omega)=X(t)$ with $EX^2(t) < \infty$ for every t, such that

(1.10) $$\int_I EX^2(t)dt < \infty,$$

(1,11) $$\int_I E|{}_{\alpha}\int^{\beta} a(t-\lambda)\eta(d\lambda) - X(t)|^2 dt \to 0,$$

for every finite interval I, when $\alpha\to-\infty$, $\beta\to\infty$.

We shall call a stochastic process $X(t)$ defined in this way a general linear process throughout this paper. This definition was motivated by Lugannani [4] who investigated a class of stochastic processes of pulse trains which appear in a variety of digital communication schemes. Actually he considered processes given formally by

$$(1.12) \qquad \sum_{n=-\infty}^{\infty} \alpha_n a(t-nT)$$

where $\{\alpha_n, n=0,\pm1,\pm2,\ldots\}$ is a sequence of uncorrelated random variables with identical first and second moments, $a(t)$ is a nonrandom function and T is a positive constant. He began with showing that if one denotes

$$a_N(t) = a(t), \quad \text{for } -(N+1/2)T \leq t < (N+1/2)T,$$
$$=0, \quad \text{otherwise}$$

and $X_N(t) = \sum_{n=-\infty}^{\infty} \alpha_n a_N(t-nT)$, then $X_N(t)$ converges to a stochastic process $X(t)$ in the sense that

$$(1.13) \qquad \int_I E|X_N(t)-X(t)|^2 dt \to 0,$$

for every finite interval I as $N\to\infty$. $X(t)$ can be written formally by (1.12). He then studied some properties of $X(t)$ including the weak and strong laws of large numbers.

(1.11) is a generalization of (1.13) which we have taken up as the definition of a class of stochastic processesthat we are going to study. In the case (1.12), each of $m(\lambda)$ and $F(\lambda)$ is a constant in each interval $(nT,(n+1)T)$, $n=0,1,2,\ldots$, $m(\lambda)$ has a constant saltus at nT and $F(\lambda)$ is nondecreasing with a positive constant saltus at nT.

We remark that if $m(\lambda)$ is constant for $-\infty<\lambda<\infty$ and $F(\lambda)$ is a linear function $c\lambda$ up to additive constants, then the existence of $X(t)$ in (1.11) implies the existence of $\int_{-\infty}^{\infty} a^2(t)dt$ and the definition (1.11) reduces to the ordinary one, namely

$$X(t) = \underset{\substack{\alpha \to -\infty \\ \beta \to \infty}}{\text{l.i.m.}} \int_\alpha^\beta a(t-\lambda)\xi(d\lambda)$$

which is a weakly stationary process.

We also remark that it is not difficult to see that for $X(t)$ defined by (1.11) with a fixed random measure $\xi(S)$ and $m(\lambda)$ a constant, to be a weakly stationary process for any $a(t)\varepsilon L^2(-\infty,\infty)$, it is necessary and sufficient that $F(\lambda)=c$ up to additive constants, c being a constant.

The main object of the present paper is to show the weak and strong laws of large numbers for a general linear process generalizing the Lugannani's results. But we will be concerned with the case in which $F(\lambda)$ does not differ very much from a linear function at least in the neighborhoods of $\pm\infty$.

2. Some lemmas

Let $X(t)$ be a general linear process defined in 1. From the definition we obviously have that

$$(2.1) \qquad E\int_\alpha^\beta a(t-\lambda)\eta(d\lambda) \to EX(t)$$

as $\alpha \to -\infty$ $\beta \to \infty$ in $L^2(I)$ for every finite interval I. Also we find that

$$(2.2) \qquad E\int_\alpha^\beta a(t-\lambda)\xi(d\lambda) = 0,$$

$$(2.3) \qquad E\int_\alpha^\beta a(t-\lambda)\eta(d\lambda) = \int_\alpha^\beta a(t-\lambda)dm(\lambda)$$

for every t, and

$$(2.4) \qquad \int_\alpha^\beta a(t-\lambda)dm(\lambda) \to EX(t)$$

in $L^2(I)$ for every finite interval I as $\alpha \to -\infty$, $\beta \to \infty$, for every t.

Because of frequent use we state

Lemma 2.1. Let $G(\lambda)$ be a function of locally bounded variation with normalization $G(\lambda) = \frac{1}{2}[G(\lambda+0) + G(\lambda-0)]$. We have, for every finite α, β, A, and B,

$$(2.5) \qquad \int_A^B dt \int_\alpha^\beta a(t-\lambda)\,dG(\lambda)$$

$$= \int_{A-\beta}^{B-\beta} a(u)\,du \int_{A-u}^{\beta} dG(\lambda) + \int_{A-\alpha}^{B-\alpha} a(u)\,du \int_{\alpha}^{B-u} dG(\lambda)$$

$$+ \int_{B-\beta}^{A-\alpha} a(u)\,du \int_{A-u}^{B-u} dG(\lambda).$$

In order to investigate the behavior of EX(t), we, in addition to the conditions in 1, impose the following condition (A) :

(A) $[m(\lambda + t) - m(t)] / \lambda$ converges as $|\lambda| \to \infty$ to a constant m_0 uniformly for t.

Lemma 2.2. If the condition (A) is satisfied and

$$(2.6) \qquad \int_\alpha^{\alpha+u} |a(t)|dt \to 0,$$

as $\alpha \to \pm\infty$, for every fixed u, then

$$(2.7) \qquad \int_{-\infty}^{+\infty} a(t)\,dt = \lim_{\substack{\alpha\to-\infty \\ \beta\to\infty}} \int_\alpha^\beta a(t)\,dt$$

converges.

Proof. Since $\int_\alpha^\beta a(t-\lambda)\,dm(\lambda)$ converges in $L^2(I)$ for every finite interval I, we see that for every fixed constant $u > 0$,

$$S = \int_0^u dt \int_\alpha^\beta a(t-\lambda)\,dm(\lambda)$$

converges as $\alpha \to -\infty$, $\beta \to \infty$. Using Lemma 2.1, we may write

$$S = \int_{-\beta}^{u-\beta} a(t)\,dt \int_{-t}^{\beta} dm(\lambda) + \int_{-\alpha}^{u-\alpha} a(t)\,dt \int_{\alpha}^{u-t} dm(\lambda)$$

$$+ \int_{u-\beta}^{-\alpha} a(t)\,dt \int_{-t}^{u-t} dm(\lambda) = S_1 + S_2 + S_3,$$

say. Here we have, using the condition (A),

$$|S_1| \leq \int_{-\beta}^{u-\beta} |a(t)|\,|m(\beta) - m(-t)|\,dt$$

$$(2.8) \quad \leq \int_{-\beta}^{u-\beta} |a(t)|\,(M_1(\beta + t) + M_2)\,dt,$$

where M_1, M_2 are some constants. The right member of (2.8) is

$$\leq M_3 u \int_{-\beta}^{u-\beta} |a(t)|\,dt$$

for some constant M_3, which, by (2.6), converges to zero as $\beta \to \infty$.

Similarly we may show that

$$S_2 \to 0 \quad \text{as } \alpha \to -\infty.$$

Hence S_3 should converge as $\alpha \to -\infty$, $\beta \to \infty$. In other words

$$(2.9) \qquad \int_{A}^{B} a(t)\,dt \int_{-t}^{u-t} dm(\lambda)$$

should converge to zero as $A, B \to \infty$, or as $A, B \to -\infty$.

Take u_o so large that, for some $\eta > 0$,

$$0 < \eta < \left|\int_{-t}^{u_o - t} dm(\lambda)\right| \leq M_4 u_o,$$

for all t. Writing $K(t) = \int_{-t}^{u_o - t} dm(\lambda)$, we have, using the second mean value theorem,

$$\int_{A}^{B} a(t)\,dt = \int_{A}^{B} a(t)K(t)\frac{dt}{K(t)}$$

$$\leq \frac{1}{K(A)} \int_A^{\xi} a(t)K(t)\,dt + \frac{1}{K(B)} \int_{\xi}^{B} a(t)K(t)\,dt$$

for some $A \leq \xi \leq B$, which is

$$\leq \frac{1}{\eta} \left[\, \left|\int_A^{\xi} a(t)\,dt \int_{-t}^{u_0 - t} dm(\lambda)\right| + \left|\int_{\xi}^{B} a(t)\,dt \int_{-t}^{u_0 - t} dm(\lambda)\right| \,\right].$$

This converges to zero as $A, B \to \infty$, or as $A, B \to -\infty$, by the convergence of (2.9) to zero. This completes the proof.

Since, if $a(t)$ is supposed to belong to $L^2(-\infty,\infty)$, (2.6) is satisfied, we have

<u>Lemma 2.3</u>. *If the condition (A) is satisfied and* $a(t)\varepsilon L^2(-\infty,\infty)$, *then* $\int_{\to-\infty}^{\to\infty} a(t)\,dt$ *converges.*

3. Average of EX(t)

Let X(t) be a given general linear process. We shall show

<u>Theorem 3.1</u>. *If (2.6) and the condition (A) are satisfied, then*

$$(3.1) \qquad \lim_{A \to +\infty} \frac{1}{2A} \int_{-A}^{A} EX(t)\,dt = m_0 \int_{\to-\infty}^{\to\infty} a(t)\,dt.$$

<u>Proof</u>.

$$I = \frac{1}{2A} \int_{-A}^{A} EX(t)\,dt$$

$$(3.2) \qquad = \lim_{\substack{\alpha \to -\infty \\ \beta \to \infty}} \frac{1}{2A} \int_{-A}^{A} dt \int_{\alpha}^{\beta} a(t-\lambda)\,dm(\lambda)$$

which is, by Lemma 2.1, equal to

$$\lim_{\substack{\alpha \to -\infty \\ \beta \to \infty}} \left[\frac{1}{2A} \int_{-A-\beta}^{A-\beta} a(u)\,du \int_{-A-u}^{\beta} dm(\lambda) + \frac{1}{2A} \int_{-A-\alpha}^{A-\alpha} a(u)\,du \int_{\alpha}^{A+u} dm(\lambda) + \frac{1}{2A} \int_{A-\beta}^{-A-\alpha} a(u)\,du \int_{-A-u}^{A-u} dm(\lambda) \right]$$

$$= \lim_{\substack{\alpha \to -\infty \\ \beta \to \infty}} (I_1+I_2+I_3),$$

say. I_1, $I_2 \to 0$ as $\alpha \to -\infty$, $\beta \to \infty$ under the condition (A) as in the proof of Lemma 2.2. Write

$$H_A(u) = [m(A-u) - m(-A-u)]/2A.$$

For any $\varepsilon > 0$, we may choose A_0 independently of u so large that

$$|H_A(u) - m_0| < \varepsilon \text{ , for } A > A_0.$$

Let B, B' be any real numbers. Since $H_A(u) - m_0$ is of bounded variation over [B, B'] (B B'), we have, using the second mean value theorem,

$$\left|\frac{1}{2A} \int_B^{B'} a(u)\, du \int_{-A-u}^{A-u} dm(\lambda) - m_0 \int_B^{B'} a(u)\, du\right|$$

$$= \left|\int_B^{B'} a(u)[H_A(u) - m_0]\, du\right|$$

$$\leq |H_A(B) - m_0| \left|\int_B^{\xi} a(u)\, du\right| + |H_A(B') - m_0| \left|\int_{\xi}^{B'} a(u)\, du\right|$$

$$(3.3) \quad < \varepsilon\left[\left|\int_B^{\xi} a(u)\, du\right| + \left|\int_{\xi}^{B'} a(u)\, du\right|\right]$$

for any A larger than A_o.

Letting $B' \to \infty$, $B \to \infty$ or $B \to -\infty$, $B' \to -\infty$, we see that $\frac{1}{2A} \int_{\to -\infty}^{\to \infty} a(u)\, du \int_{-A-u}^{A-u} dm(\lambda)$ exists. Since $\int_{\to -\infty}^{\to \infty} a(u)\, du$ exists, (3.3) also shows that

$$\left|\frac{1}{2A} \int_B^{B'} a(u)\, du \int_{-A-u}^{A-u} dm(\lambda) - m_o \int_B^{B'} a(u)\, du\right| \leq \varepsilon M_3$$

for some constant M_3 in dependent of A, B and B'. Hence keeping A fixed and letting $\alpha \to -\infty$, $\beta \to \infty$, we get

$$\lim_{\substack{\alpha \to -\infty \\ \beta \to \infty}} \left| I_3 - m_o \int_{\to -\infty}^{\to \infty} a(u)\, du \right| \leq \varepsilon M_3 .$$

This shows that I coverges to $m_o \int_{\to -\infty}^{\to \infty} a(u)\, du$, as $A \to \infty$. The proof is now complete.

If $m(\lambda) = m_o\lambda + c$, $-\infty < \lambda < \infty$, m_o, c being constants, then (3.2) reduces to $\frac{1}{2A} \int_{-A}^{A} EX(t) = m_o \int_{\to -\infty}^{\to \infty} a(u)\, du$, for every $A > 0$. Hence

$$(3.4) \qquad EX(t) = m_o \int_{\to -\infty}^{\to \infty} a(u)\, du, \text{ a.e.}$$

Also we can show that if m(t) is a step function on $(-\infty,\infty)$ with constant jump m_o at nT, $T > 0$, $n = 0, 1, 2, \ldots$, then EX(t) is periodic with period T and

$$(3.5) \qquad \frac{1}{T} \int_{-T/2}^{T/2} EX(t)\, dt = m_o \int_{\to -\infty}^{\to \infty} a(u)\, du.$$

This result was given by Lugannani.(loc. cit.)

4. Covariance function of X(t)

Let X(t) be a general linear process. We suppose that

$$(4.1) \qquad a(t) \in L^2(-\infty,\infty).$$

Let the covariance function of X(t) be

$$(4.2) \qquad \rho(s,t) = E[X(s) - EX(s)][X(t) - EX(t)].$$

For $F(\lambda)$, the nondecreasing function corresponding to the random measure ξ in the definition of X(t), we frequently assume the condition (B):

(B) $[F(\lambda + t) - F(t)] / \lambda$ is bounded for all t and λ, and converges to a constant v_o, as $\lambda \to \infty$ uniformly for $-\infty < t < \infty$.

We now prove the following

Theorem 4.1. Suppose (4.1) and the condition (B). Then

$$\rho(\tau) = \lim_{A \to \infty} \frac{1}{2A} \int_{-A}^{A} \rho(t+\tau, t)dt$$

exists and is equal to

$$(4.3) \qquad v_o \int_{-\infty}^{\infty} a(t+\tau)a(t)\,dt$$

which is written also by

$$(4.4) \qquad v_o \int_{-\infty}^{\infty} |\hat{a}(u)|^2 e^{iu\tau}\,du,$$

where $\hat{a}(u)$ is the Fourier transform in L^2 of $a(t)$,

$$(4.5) \qquad \hat{a}(u) = \frac{1}{\sqrt{2\pi}} \int_{-\infty}^{\infty} a(t)\, e^{-itu}\,dt \quad (\text{in } L^2).$$

Note that $\rho(\tau)$ is an even function.

Proof. We see that

$$\frac{1}{2A} \int_{-A}^{A} \rho(t+\tau, t)\,dt$$

$$= \frac{1}{2A} \lim_{\substack{\alpha \to -\infty \\ \beta \to \infty}} \lim_{\substack{\alpha' \to -\infty \\ \beta' \to \infty}} \int_{-A}^{A} E\Big[\int_{\alpha}^{\beta} a(t+\tau-\lambda)\xi(d\lambda) \cdot \int_{\alpha'}^{\beta'} a(t-\lambda)\xi(d\lambda)\Big]\,dt$$

$$= \frac{1}{2A} \lim_{\substack{\alpha \to -\infty \\ \beta \to \infty}} \int_{-A}^{A} dt \int_{\alpha}^{\beta} a(t+\tau-\lambda)a(t-\lambda)\,dF(\lambda).$$

Now using Lemma 2.1 we have that the expression under the lim sign is

$$\frac{1}{2A} \int_{\alpha}^{\beta} dF(\lambda) \int_{-A-\lambda}^{A-\lambda} a(v+\tau)a(v)\,dv$$

$$= \frac{1}{2A} \int_{-A-\beta}^{A-\beta} a(v+\tau)a(v)\,dv \int_{-A-v}^{\beta} dF(\lambda)$$

$$+ \frac{1}{2A} \int_{-A-\alpha}^{A-\alpha} a(v+\tau)a(v) \int_{\alpha}^{A-v} dF(\lambda)$$

$$+ \frac{1}{2A} \int_{-A-\beta}^{A-\alpha} a(v+\tau)a(v)dv \int_{-A-v}^{A-v} dF(\lambda)$$

$$= J_1 + J_2 + J_3,$$

say.

Since $|F(\beta) - F(-A - v)| \leq M_5(\beta + A + v) \leq 2AM_5$ for some constant M_5, we have

$$|J_1| \leq M_5 \int_{-A-\beta}^{A-\beta} |a(v+\tau)a(v)|dv$$

which converges to zero as $\beta \to \infty$ because of (4.1).

In a similar way we have that J_2 converges to zero as $\alpha \to -\infty$.

Hence letting $\alpha \to -\infty$, $\beta \to \infty$ in J_3, we get

$$\frac{1}{2A} \int_{-A}^{A} \rho(t+\tau,t)\, dt$$

$$(4.6) \qquad = \frac{1}{2A} \int_{-\infty}^{\infty} a(v+\tau)a(v)[F(A - v) - F(-A - v)]\, dv.$$

Since $[F(A - v) - F(-A - v)] / 2A$ is uniformly bounded and converges to v_0 as $A \to \infty$, (4.6) converges, as $A \to \infty$, to $v_0 \int_{-\infty}^{\infty} a(v+\tau)a(v)\, dv$. This shows (4.4), (4.5) is obvious from the Parseval relation for Fourier transform. The proof is now complete.

We agree to call $\rho(\tau)$ in the above theorem the <u>mean covariance function</u> of X(t).

<u>5. Weak law of large numbers</u>

The weak law of large numbers for a general not necessarily stationary process of the second order is known (for ex. see Kawata [3]) and could be applied to obtain the weak law of large numbers for a general linear process. But we shall treat the problem by a direct method. In doing this, we do not have to assume the condition such as (B), but the more general condition (5.1) is sufficient.

<u>Theorem 5.1</u>. <u>Let</u> X(t) <u>be a general linear process defined by</u> (1.11). <u>If</u> F(λ) <u>satisfies that</u>

$$F(t+\lambda) - F(t) = O(\lambda) \tag{5.1}$$

<u>for large</u> |λ| <u>uniformly for</u> t, <u>and if</u> (4.1) <u>is satisfied, then</u>

$$E\{\frac{1}{2A}\int_{-A}^{A} [X(t) - EX(t)]\, dt\}^2 \to 0 \tag{5.2}$$

<u>as</u> $A \to \infty$.

<u>Proof</u>. Write

$$L = E\{\frac{1}{2A}\int_{-A}^{A} [X(t) - EX(t)]\, dt\}^2$$

$$= \frac{1}{(2A)^2} E\int_{-A}^{A}\int_{-A}^{A} [X(s) - EX(s)][X(t) - EX(t)]\, dsdt$$

$$= \lim_{\substack{\alpha\to-\infty\\ \beta\to\infty}}\ \lim_{\substack{\alpha'\to-\infty\\ \beta'\to\infty}} \frac{1}{(2A)^2}\int_{-A}^{A}\int_{-A}^{A} E\int_{\alpha}^{\beta} a(s-\lambda)\xi(d\lambda) \int_{\alpha'}^{\beta'} a(t-\mu)\xi(d\mu)\, dsdt$$

$$= \lim_{\substack{\alpha\to-\infty\\ \beta\to\infty}} \frac{1}{(2A)^2}\int_{-A}^{A} ds\int_{-A}^{A} dt\int_{\alpha}^{\beta} a(s-\lambda)a(t-\lambda)\, dF(\lambda). \tag{5.3}$$

Setting $g(\lambda) = \int_{-A}^{A} a(s-\lambda)\, ds$ and applying Lemma 2.1 to (5.3), we have that the integral in (5.3) is

$$\frac{1}{(2A)^2}\int_{-A}^{A} dt\int_{\alpha}^{\beta} a(t-\lambda)g(\lambda)\, dF(\lambda)$$

$$= \frac{1}{(2A)^2}\int_{-A-\beta}^{A-\beta} a(u)\, du\int_{-A-u}^{\beta} g(\lambda)\, dF(\lambda)$$

$$+ \frac{1}{(2A)^2}\int_{-A-\alpha}^{A-\alpha}\int_{\alpha}^{A-u} + \frac{1}{(2A)^2}\int_{A-\beta}^{-A-\alpha}\int_{-A-u}^{A-u}$$

$$= L_1 + L_2 + L_3,$$

say. The integrands in L_2 and L_3 are the same as in L_1.

In L_1

$$\int_{-A-u}^{\beta} g(\lambda)\, dF(\lambda) = \int_{-A}^{A} ds \int_{-A-u}^{\beta} a(s-\lambda)\, dF(\lambda)$$

$$= \int_{-A-u}^{\beta} dF(\lambda) \int_{-A-\lambda}^{A-\lambda} a(v)\, dv.$$

Take β so large that $\beta > 3A$. We then see that the last integral can be written

$$\int_{-A-\beta}^{u} a(v)\, dv \int_{-A-v}^{\beta} dF(\lambda) + \int_{u}^{A-\beta} \int_{-A-u}^{\beta} + \int_{A-\beta}^{u+2A} \int_{-A-u}^{A-v}$$

Then L_1 turns out to be

$$L_1 = \frac{1}{(2A)^2} \int_{-A-\beta}^{A-\beta} a(u)\, du \int_{-A-\beta}^{u} a(v)\, dv \int_{-A-v}^{\beta} dF(\lambda)$$

$$+ \frac{1}{(2A)^2} \int_{-A-\beta}^{A-\beta} \int_{u}^{A-\beta} \int_{-A-u}^{\beta} + \frac{1}{(2A)^2} \int_{-A-\beta}^{A-\beta} \int_{A-\beta}^{u+2A} \int_{-A-u}^{A-v}$$

$$= L_{11} + L_{12} + L_{13},$$

say.

Let $|F(\lambda) - F(\mu)| \leq M_6 |\lambda - \mu|$, M_6 being a constant.

Then

$$|L_{11}| \leq \frac{1}{(2A)^2} M_6 \int_{-A-\beta}^{A-\beta} |a(u)|\, du \int_{-A-\beta}^{u} |a(v)|\, (\beta+A+v)\, dv$$

$$\leq \frac{1}{2A} M_6 \left[\int_{-A-\beta}^{A-\beta} |a(u)|\, du \right]^2 \leq M_6 \int_{-A-\beta}^{A-\beta} |a(u)|^2\, du,$$

from which it follows that

(5.4) $\qquad L_{11} \to 0$ as $\beta \to \infty$.

(5.5) $\qquad L_{12} \to 0$ as $\beta \to \infty$

follows in a similar way. We may handle L_{13} in a similar way and it is not difficult either to show that L_{13} converges to zero as $\beta \to \infty$. This together with (5.4) and (5.5) shows that

$$(5.6)\qquad L_1 \to 0 \text{ as } \beta \to \infty.$$

We may also show in a similar way that

$$(5.7)\qquad L_2 \to 0 \text{ as } \alpha \to -\infty.$$

Therefore what we have shown is that

$$L = \lim_{\substack{\alpha \to -\infty \\ \beta \to \infty}} L_3 = \frac{1}{(2A)^2} \int_{-\infty}^{\infty} a(u)\, du \int_{-A-u}^{A-u} dF(\lambda) \int_{-A}^{A} a(s-\lambda)\, ds,$$

in which

$$\int_{-A-u}^{A-u} dF(\lambda) \int_{-A}^{A} a(s-\lambda)\, ds = \int_{-A-u}^{A-u} dF(\lambda) \int_{-A-}^{A-} a(v)\, dv$$

$$= \int_{-2A+u}^{u} a(v)\, dv \int_{-A-v}^{A-u} dF(\lambda) + \int_{u}^{2A+u} \int_{-A-u}^{A-v} .$$

Hence

$$L = \frac{1}{(2A)^2} \int_{-\infty}^{\infty} a(u)\, du \int_{-2A+u}^{u} a(v)[F(A-v) - F(-A-v)]\, dv$$

$$+ \frac{1}{(2A)^2} \int_{-\infty}^{\infty} \int_{u}^{2A+u}$$

$$= L' + L'' ,$$

say. We now have

$$|L'| = \left| \frac{1}{(2A)^2} \int_{-\infty}^{\infty} a(u)\, du \int_{-2A}^{0} a(w+u)[F(A-u) - F(-A-u-w)]\, dw \right|$$

$$\leq \frac{M_6}{2A} \int_{-\infty}^{\infty} |a(u)|\, du \int_{-2A}^{0} |a(w+u)|\, dw$$

$$= \frac{M_6}{2A} \int_{-2A}^{0} dw \int_{-\infty}^{\infty} |a(u)a(w+u)|\, du$$

$$\frac{M_6}{2A} \int_{-2A}^{0} dw \int_{-\infty}^{\infty} |\widehat{a}_1(t)|^2\, e^{-itw}\, dt,$$

where $a_1(t)$ is the Fourier transform of $|a(u)|$ in L^2. Since the inner integral of the last one, as the Fourier transform of $|a_1(t)|^2$, converges to zero as $w \to \infty$. Hence L' converges as $A \to \infty$ to zero.

In the same way we may show that L" also converges to zero as $A \to \infty$. Therefore altogether L converges to zero as $A \to \infty$. Thus the proof of the theorem is complete.

6. Strong law of large numbers for nonstationary processes.

Generalizing some known results on the strong law of large numbers (Blanc-Lapierre and Brard [1], Kac, Salem and Zygmund [2]) Verbitskaya [6] has shown the following theorem:

Theorem A. *Let $X(t)$, $-\infty < t < \infty$ be a real valued weakly stationary process with $EX(t) = 0$ and the continuous covariance funtion $\rho(u)$. If*

$$(6.1) \qquad \int_1^\infty \frac{\log^2 u}{u^2} \left| \int_0^u \rho(s)\, ds \right| du < \infty,$$

then

$$(6.2) \qquad \lim_{T \to \infty} \frac{1}{T} \int_0^T X(t)\, dt = 0$$

almost surely.

In her succeeding paper [7], she sharpened the above theorem.

Theorem B. *In Theorem A the condition (6.1) can be replaced by the condition that the integral*

$$(6.3) \qquad \int_1^\infty \frac{\log^2 u}{u^2}\, du \int_0^u \rho(s)\, ds$$

converges.

We remark that her method of proving these theorems also applies even to the case of nonstationary processes to get the following theorems 6.1 and 6.2 below. Actually no substantial change of proofs is necessary. The condition (6.4) substitutes for the stationarity condition that the covariance function is a function of a single variable and the function h(u) plays a role of the covariance function of a stationary process. We therefore do not reproduce the proofs of the theorems here.

Theorem 6.1. *Let* $X(t)$ *be a real valued stochastic process of the second order with continuous covariance function* $\rho(s,t) = EX(s)X(t)$, $EX(t) = 0$, $-\infty < t < \infty$, *being assumed. If*

(i) $$\left|\int_n^{n+1} E[X(t)]^2\, dt\right| \leq K,$$

for all $-\infty < n < \infty$, K *being a constant independent of* n, *and*

(ii) *there is a nonnegative function* $g(u)$, $-\infty < u < \infty$, *such that*

$$\left|\int_0^x \int_0^x \rho(u+\tau, v+\tau)\, du dv\right| \leq \int_0^x g(u)\, du \tag{6.4}$$

for every $-\infty < \tau < \infty$, $0 < x < \infty$, *where*

$$\int_1^\infty \frac{\log^2 x}{x^2}\, g(x)\, dx < \infty, \tag{6.5}$$

then

$$\frac{1}{A} \int_0^A x(t)\, dt \to 0, \text{ as } A \to \infty, \tag{6.6}$$

almost surely.

Note that $\int_0^x \int_0^x \rho(u+\tau, v+\tau)\, du dv$ is nonnegative for $x > 0$.

Theorem 6.2. *Let* $X(t)$ *be a real valued stochastic process of the second order with continuous covariance function and* $EX(t) = 0$. *If* (i) *in Theorem* 6.1 *holds and*

(iii) *there is a function* $h(u)$, $-\infty < u < \infty$, *such that*

$$\left|\int_0^x \int_0^x \rho(u+\tau, v+\tau)\, du dv\right| \leq \left|\int_0^x dt \int_0^t h(u)\, du\right| \tag{6.7}$$

for every x, *where* $h(u)$ *is the Fourier-Stieltjes transform* $\int_{-\infty}^{\infty} e^{iu\lambda}\, dH(\lambda)$ *of some bounded nondecreasing function* $H(\lambda)$, *with the property that*

$$(6.8) \qquad \int_1^\infty \frac{\log^2 t}{t^2}\, dt \int_0^t h(u)\, du$$

<u>converges, then</u> (6.6) <u>holds almost surely.</u>

Combination of above two theorems gives us

<u>Theorem 6.3. Suppose</u> X(t) <u>is a process in Theorem</u> 6.1 <u>satisfying</u> (i). <u>If there exist</u> g(u) <u>and</u> h(u) <u>in Theorems</u> 6.1 <u>and</u> 6.2 <u>respectively with</u>

$$(6.9) \qquad \left|\int_0^x \int_0^x \rho(u+\tau, v+\tau)\, du dv\right| \leq \int_0^x g(u)\, du + \left|\int_0^x dt \int_0^t h(u)\, du\right|$$

<u>in place of</u> (6.4) <u>and</u> (6.7), <u>then</u> (6.6) <u>holds almost surely.</u>

7. Strong law of large numbers for linear processes.

Let X(t) be a general linear precess defined by (1.11). We shall prove the following law of large numbers for X(t).

<u>Theorem 7.1. Suppose that, for large</u> ,

$$(7.1) \qquad F(\lambda+t) - F(t) = v_o\lambda + O(1)$$

<u>holds uniformly for</u> $-\infty < t < \infty$. <u>If the mean covariance</u> $\rho(\tau)$ <u>satisfies that the integral</u>

$$\int_1^\infty \frac{\log^2 x}{x^2}\, dx \int_0^x \rho(\tau)\, d\tau$$

<u>converges, then</u>

$$(7.2) \qquad \frac{1}{A}\int_0^A [X(t) - EX(t)]\, dt \to 0$$

<u>as</u> $A \to \infty$, <u>almost surely.</u>

The existence of the mean covariance function is guaranteed by Theorem 4.1.

<u>Proof.</u> Let Y(t) = X(t) - EX(t). Obviously EY(t) = 0.

As in the proof of Theorem 4.1,

$$\left|\int_n^{n+1} EY^2(t)\,dt\right| = \left|\lim_{\substack{\alpha\to-\infty\\ \beta\to\infty}} \int_n^{n+1} dt \int_\alpha^\beta a^2(t-\lambda)\,dF(\lambda)\right|$$

$$= \int_{-\infty}^{\infty} a^2(v)\,|F((n+1)-v) - F(n-v)|\,dv.$$

Since from (7.1) there is a constant M_7 such that $F(b) - F(a) \leq M_7$ for all $a < b$, the last integral is not less than $M_7 \int_{-\infty}^{\infty} a^2(v)\,dv$ which is a constant. Thus the condition (1) of Theorem 6.1 is satisfied.

Next we verify the condition (6.9) of Theorem 3.

$$R = \int_0^x \int_0^x \rho(u+\tau, v+\tau)\,dudv$$

$$= \lim_{\substack{\alpha\to-\infty\\ \beta\to\infty}} \int_0^x \int_0^x dudv \int_\alpha^\beta a(u+\tau-\lambda)a(v+\tau-\lambda)\,dF(\lambda)$$

$$= \int_{-\infty}^{\infty} dF(\lambda) \int_0^x \int_0^x a(u+\tau-\lambda)a(v+\tau-\lambda)\,dudv$$

$$= \int_{-\infty}^{\infty} dF(\lambda) \int_{-\lambda}^{x-\lambda} \left[\int_0^x a(v+\tau-\lambda)\,dv\right] a(u+\tau)\,du$$

$$= \int_{-\infty}^{\infty} a(u+\tau)\,du \int_{-u}^{x-u} \left[\int_0^x a(v+\tau-\lambda)\,dv\right] dF(\lambda)$$

$$= \int_{-\infty}^{\infty} a(u+\tau)\,du \int_{-u}^{x-u} dF(\lambda) \int_{-\lambda}^{x-\lambda} a(v+\tau)\,dv$$

$$= \int_{-\infty}^{\infty} a(u+\tau)\,du \int_{-x+u}^{u} a(v+\tau)\,dv \int_{-v}^{x-v} dF(\lambda)$$

$$+ \int_{-\infty}^{\infty} a(u+\tau)\,du \int_{u}^{x+u} a(v+\tau)\,dv \int_{-u}^{x-v} dF(\lambda)$$

$$= R_1 + R_2,$$

say.

$$R_1 = \int_{-\infty}^{\infty} a(u+\tau)\,du \int_{-x}^{0} a(w+u+\tau)[F(x-u) - F(-w-u)]\,dw$$

$$= \int_0^x dv \int_{-\infty}^{\infty} a(u+\tau)a(u-v+\tau)[F(x-u) - F(v-u)]\, du$$

$$= \int_0^x dv \int_{-\infty}^{\infty} a(t)a(t-v)[F(x-t+\tau) - F(v-t+\tau)\, dt$$

which is, because of (7.1),

$$v_o \int_0^x (x-v)\, dv \int_{-\infty}^{\infty} a(t)a(t-v)\, dt + \int_0^x g_1(v)\, dv$$

$$= v_o \int_0^x dw \int_0^w dv \int_{-\infty}^{\infty} a(t)a(t-v)\, dt + \int_0^x g_1(v)\, dv$$

$$= v_o \int_0^x dw \int_0^w \rho(v)\, dv + \int_0^x g_1(v)\, dv,$$

where

$$g_1(v) = O(1) \int_{-\infty}^{\infty} |a(t)a(t-v)|\, dt = O(1).$$

In a similar way we have

$$R_2 = \int_{-\infty}^{\infty} a(u+\tau)\, du \int_0^x a(w+u+\tau)[F(x-w-u) - F(-u)]\, dw$$

$$= v_o \int_0^x dw \int_0^w \rho(v)\, dv + \int_0^x g_2(v)\, dv$$

where

$$g_2(v) = O(1) \int_{-\infty}^{\infty} |a(t)a(t+v)|\, dt = O(1).$$

Let $g(v) = g_1(v) + g_2(v)$. We then have

$$R = 2v_o \int_0^x dw \int_0^x \rho(v)\, dv + \int_0^x g(v)\, dv.$$

Since $g(x)$ is bounded and hence satisfies (6.5) of Theorem 6.1 and $\rho(v)$ is the mean covariance function which has the form (4.4), R satisfies the condition of Theorem 6.3. Thus the proof is complete.

References

1. A.Blanc-Lapierre and P.Brard: Les fonctions aléatoire et la loi des grand nombres, Bull. Soc. Math. de France, 74(1946) 102-115.
2. M.Kac, R.Salem and A.Zygmund: A gap theorem, Trans. Amer. Math. Soc., 63(1948) 235-243.
3. T.Kawata: Fourier analysis of nonstationary stochastic processes, Trans. Amer. Math. Soc., 118(1965) 276-302.
4. R.Lugannani: Convergence properties of the sample mean and sample correlation for a class of pulse trains, SIAM J. Appl. Math., 21(1971) 1-12.
5. Yu A. Rozanov: Stationary random processes, Eng. Transl., 1967, Holden Day, San Francisco.
6. I.N.Verbitskaya: On conditions for the strong law of large numbers to be applicable to second order stationary processes, Theory of Prob. Appl., 9(1964) 325-331.
7. I.N.Verbitskaya: On conditions for the applicability of the strong law of large numbers to wide sense stationary processes, Theory of Prob. Appl., 11(1966) 632-636.

(Keio University, Yokohama)

A CHARACTERIZATION OF POTENTIAL KERNELS FOR RECURRENT MARKOV CHAINS WITH STRONG FELLER TRANSITION FUNCTION

Ryōji Kondō and Yōichi Ōshima

In [11] the second author of this note has given a necessary and sufficient condition for kernels to be weak potential kernels of recurrent Markov chains in the case of denumerable state space. The purpose of this note is to give the same result for strong Feller recurrent Markov chains with locally compact state space. A similar result has been obtained by Revuz [12] in the case of compact state space.

1. Notations and preliminary results

Let E be a locally compact Hausdorff space with countable base and $\mathcal{E}$ the σ-algebra of all Borel subsets of E. A function K defined on $E \times \mathcal{E}$ is called a kernel on E if, for any $B \in \mathcal{E}$ the function : $x \to K(x, B)$ is $\mathcal{E}$-measurable and if, for any $x \in E$, the function : $B \to K(x, B)$ is a Radon measure on E (restricted to $\mathcal{E}$). As usual, for a function f, a Radon measure μ and kernels K, H, we set

$$Kf(x) = \int_E K(x, dy)f(y) \qquad \text{for } x \in E,$$

$$\mu K(B) = \int_E \mu(dx)K(x, B) \qquad \text{for } B \in \mathcal{E},$$

$$KH(x, B) = \int_E K(x, dy)H(y, B) \qquad \text{for } x \in E,\ B \in \mathcal{E},$$

$$(f \otimes \mu)(x, B) = f(x)\mu(B) \qquad \text{for } x \in E,\ B \in \mathcal{E},$$

$$\langle \mu, f \rangle = \int_E \mu(dx) f(x),$$

so far as they are defined.

In the following, though there is a slight difference between Markov processes with continuous parameter and those with discrete parameter, we shall take mainly notations and terminology from Blumenthal and Getoor [2]. Let P be a strong Feller recurrent Markov kernel on E, that is, a kernel satisfying the following conditions; (P.1) (Markov kernel) $P \geqq 0$ and $P\,1 = 1$, (P.2) (strong Feller property) for any $B \in \mathcal{E}$ the function : $x \to P(x, B)$ is (bounded) continuous and (P.3) (recurrence condition) for any non-empty open set U in E, $\Sigma_{n=1}^{\infty} P^n(x, U) = \infty$ for all $x \in E$. Following [2] we denote by $X = (\Omega, \mathcal{F}^0, \mathcal{F}_n^0, X_n, \theta_n, P^x)$ the canonical Markov process (chain) with transition function P. For any $A \in \mathcal{E}$, we define first hitting times D_A and T_A of A by

$$D_A = \inf\{ n \geqq 0 : X_n \in A\} \qquad \text{and}$$

$$T_A = \inf\{ n \geqq 1 : X_n \in A\} \qquad (= 1 + D_A \circ \theta_1)$$

respectively, where in both cases the infimum of the empty set is understood to be $+\infty$. A Borel set A is said to be <u>polar</u> if $P^x(T_A < \infty) = 0$ for all $x \in E$. By virtue of (P.3), any non-empty open set is non-polar. The proof of the following lemma is similar to those of propositions 4 and 12 of Azema, Duflo and Revuz [1], so will be omitted.

<u>Lemma</u> 1.1. <u>If a set</u> A <u>is non-polar</u>, $P^x(T_A < \infty) = 1$ <u>for all</u> $x \in E$.

As a consequence of this lemma we see that a lower semi-continuous excessive function f, ($f \geq 0$ and $Pf \leq f$) is constant. In particular, if a function in $\underline{B}$ is P-invariant, it is a constant function.

For later use, we shall prove here the following

Theorem 1.1. For any non-polar Borel set A and for any decreasing sequence $(B_n)_{n\geq 1}$ of closed subsets in A^c with intersection $\bigcap_{n\geq 1} B_n = \phi$,

$$\text{(1.1)} \qquad \lim_n \downarrow P^x(D_{B_n} < D_A) = 0,$$

uniformly on each compact set contained in A^c.

Proof. For simplicity we set $h_n(x) = P^x(D_{B_n} < D_A)$ and $h(x) = \lim_n \downarrow h_n(x)$ for all $x \in A^c$. Since $\bigcap_{n\geq 1} B_n = \phi$, for any $x \in A^c$, we can find an n_0 such that $x \notin B_{n_0}$. Since $E^x[h_n(X_1) : 1 < D_A] = h_n(x)$ for all $n \geq n_0$, by the Lebesgue convergence theorem, $E^x[h(X_1) : 1 < D_A] = h(x)$ for all $x \in A^c$. Therefore, for any $k \geq 1$,

$$h(x) = E^x[h(X_k) : k < D_A] \leq P^x(k < D_A)$$

for all $x \in A^c$. Hence, tending k to infinity, we have $h(x) = 0$ for all $x \in A^c$. Since h_n are continuous on $(B_n)^c - A$ and each compact set in A^c is contained in some $(B_n)^c$, by Dini's theorem, the convergence is uniform on each compact set in A^c.

Let A be a non-polar set in E. We define two kernels V_A and G_A on E by

(1.2) $\quad V_A(x, B) = E^x[\sum_{0 \leq n < D_A} I_B(X_n)]$ $\quad$ and

(1.3) $\quad G_A(x, B) = E^x[\sum_{0 \leq n < T_A} I_B(X_n)]$ $\quad$ for $x \in E$, $B \in \mathcal{E}$

respectively, where I_B denotes the indicator of a set B.

Lemma 1.2. If B is contained in a compact set, then the functions : $x \to V_A(x, B)$ and $x \to G_A(x, B)$ are bounded.

Proof. We may assume that B is itself compact. For any $s \in]0, 1[$, we define a function h_s by

$$h_s(x) = E^x[s^{T_A}] \qquad (x \in E).$$

Then by the strong Feller property of P, it is continuous and, of course, $0 < h_s(x) < 1$ for all $x \in E$. Since $\lim_{s \to 1} \uparrow h_s(x) = 1$ for all $x \in E$ and B is compact, for any $c \in]0, 1[$, we can find an $s \in]0, 1[$ such that the set $J_c = \{x : h_s(x) > c\}$ contains B. Since $n < D_A$ implies $D_A = n + T_A \circ \theta_n$,

$$V_A h_s(x) = E^x[\sum_{0 \leq n < D_A} E^{X_n}(s^{T_A})] = E^x[\sum_{0 \leq n < D_A} s^{D_A - n}]$$

$$= E^x[\sum_{1 \leq n \leq D_A} s^n] \leq \sum_{n=1}^{\infty} s^n$$

$$= s/(1 - s)$$

for all $x \in E$. Therefore

$$cV_A(x, B) \leq cV_A(x, J_c) \leq V_A h_s(x) \leq s/(1 - s)$$

for all $x \in E$, which implies that the function $V_A(\ \ , B)$ is bounded. Furthermore, since $G_A(x, B) \leq 1 + \sup_y V_A(y, B)$ for all $x \in E$, $G_A(\ \ , B)$ is also bounded.

Now let F be a non-polar compact subset of E. We introduce kernels H_F and Π_F on E by

$$H_F(x, B) = P^x(X_{D_F} \in B) \qquad \text{and}$$

$$\Pi_F(x, B) = P^x(X_{T_F} \in B) \qquad (= PH_F(x, B)),$$

respectively. Further we denote by P_F the restriction of Π_F to F, which is a strong Feller Markov kernel on F.

In the following, we denote by $\underline{\underline{M}}_F$ the Banach space of all Radon measures on F with the norm (denoted by $\| \ \|$) of total variation.

<u>Lemma</u> 1.3. <u>Let</u> Q_F <u>be a kernel on</u> F <u>defined by</u>

$$(1.4) \qquad Q_F = \sum_{n=1}^{\infty} 2^{-n}(P_F)^n.$$

<u>Then there exists a probability Radon measure</u> ν_F <u>on</u> F <u>such that</u>

$$(1.5) \qquad \sup_x \|(Q_F)^n(x, \ ) - \nu_F\| \leq cr^n \qquad \text{for all } n \geq 0$$

<u>with some constants</u> $c > 0$ <u>and</u> $0 < r < 1$.

Proof. Let B be any Borel subset of F. Since, for any $x \in F$, $Q_F(x, B) > 0$ if and only if $P^x(T_B < \infty) = 1$ and since, for any $x, y \in F$, $P^x(T_B < \infty) = 1$ if and only if $P^y(T_B < \infty) = 1$, $Q_F(x, B) > 0$ is equivalent to $Q_F(y, B) > 0$ for any x and y. Therefore $Q_F(x, \ )$ and $Q_F(y, \ )$ are mutually absolutely continuous for any $x, y \in F$, and hence, so are $(Q_F)^2(x, \ )$ and $(Q_F)^2(y, \ )$. Since Q_F is a strong Feller Markov kernel on F, by Mokobodzki's theorem (see Meyer [7]), $Q_F^2 = (Q_F)^2$ is a strong Feller kernel in the strict sense, that is, the mapping : $x \rightarrow$

$Q_F^2(x,\ \ )$, from E into $\underline{\underline{M}}_F$, is continuous. Therefore

$$\delta(Q_F^2) = \frac{1}{2} \sup_{x,y} \| Q_F^2(x,\ \ ) - Q_F^2(y,\ \ ) \| < 1.$$

Thus, by Ueno's theorem [13], there is a probability measure ν_F on F such that

$$\sup_x \| Q_F^{2n}(x,\ \ ) - \nu_F \| \leqq 2(\delta(Q_F^2))^n \qquad \text{for all } n \geqq 0.$$

For any $x \in F$, since

$$\| Q_F^{2n+1}(x,\ \ ) - \nu_F \| \leqq \sup_x \| Q_F^{2n}(x,\ \ ) - \nu_F \|$$

$$\leqq 2\ (\delta(Q_F^2))^n \qquad \text{for all } n \geqq 0,$$

(1.5) holds with $c = 2\ (\delta(Q_F^2))^{-1/2}$ and $r = (\delta(Q_F^2))^{1/2}$ (when $\delta(Q_F^2) = 0$, since for any $x \in F$ and $n \geqq 2$, $Q_F^n(x,\ \ ) = \nu_F$, (1.5) holds trivially).

Lemma 1.4. The probability measure ν_F introduced above is an invariant probability measure for P_F.

Proof. For any $s \in]0, 1[$, define a kernel G_s^F by

$$G_s^F = \sum_{n=1}^{\infty} s^{n-1} (P_F)^n$$

By definition, it is clear that $\lim_{s\to 0} \sup_x \| G_s^F(x,\ \ ) - P_F(x,\ \ ) \| = 0$. Further, noting that $G_{1/2}^F = 2\ Q_F$, one has the following equations

$$(1.6) \qquad G_s^F - 2\ Q_F - (2s - 1)\ G_s^F\ Q_F = 0$$

$$(1.7) \qquad 2\ Q_F - G_s^F - (1 - 2s)\ Q_F\ G_s^F = 0$$

for all $s \in]0, 1[$. By (1.5), since $\nu_F Q_F = \nu_F$,

$$2\,\nu_F - \nu_F G_s^F - (1 - 2s)\,\nu_F G_s^F = 0$$

for all $s \in]0, 1[$. Therefore, tending s to zero, one has $\nu_F P_F = \nu_F$.

In the following we denote by $\underline{\underline{B}}_F$ the Banach space of all bounded Borel measurable functions defined on F with the supremum norm $\|f\| = \sup_x |f(x)|$ and by $\underline{\underline{N}}_F(\nu_F)$ the space of functions f, in $\underline{\underline{B}}_F$ and $< \nu_F, f > = 0$.

Lemma 1.5. _There exists a bounded kernel_ L_F _on_ F _such that_

(1.8) $\qquad (I - P_F)L_F f = f \qquad$ _for all_ $f \in \underline{\underline{N}}_F(\nu_F)$,

where I _denotes the identity operator._

Proof. We define a kernel L_F by

$$\text{(1.9)} \qquad L_F(x, B) = \sum_{n=0}^{\infty} [Q_F^n(x, B) - \nu_F(B)] + \sum_{n=1}^{\infty} [Q_F^n(x, B) - \nu_F(B)].$$

By (1.5), since $\sup_x \|L_F(x, \ )\| \leq c(1 + r)/(1 - r)$ it is a bounded kernel. If $f \in \underline{\underline{N}}_F(\nu_F)$, since

$$L_F f = \sum_{n=0}^{\infty} Q_F^n(f + Q_F f),$$

it is clear that

$$(I - Q_F)L_F f = f + Q_F f \qquad \text{for all } f \in \underline{\underline{N}}_F(\nu_F).$$

From this and (1.6) it follows that

$$(1 - s)G_s^F L_F f - L_F f + f + sG_s^F f = 0$$

for all $s \in]0, 1[$ and $f \in \underline{\underline{N}}_F(\nu_F)$. Therefore, tending s to zero, one has equation (1.8).

2. Potential kernels

In the following we denote by $\underline{\underline{B}}$ the space of all bounded Borel measurable functions defined on E and by $\underline{\underline{C}}$ bounded continuous functions defined on E. They are Banach spaces with the supremum norm $\|f\| = \sup |f(x)|$. Further we denote by $\underline{\underline{B}}_c$ the space of functions in $\underline{\underline{B}}$ with compact support. When $\underline{\underline{L}}$ is a space of functions, $\underline{\underline{L}}^+$ will denote the cone of non-negative functions in $\underline{\underline{L}}$.

As before let P be a strong Feller recurrent Markov kernel on E and X the canonical Markov chain with transition function P. It is well known that there exists, uniquely except for a constant factor, an invariant measure μ for P (an explicit form of the invariant measure will be given below). A function f in $\underline{\underline{B}}_c$ will be called a null charge with respect to μ (or P) if $< \mu, f> = 0$. We denote by $\underline{\underline{N}}(\mu)$, or $\underline{\underline{N}}$ when there is no confusion, the space of all null charges.

A kernel G on E is called a (weak) potential kernel for P if, for any $f \in \underline{\underline{N}}$, $Gf \in \underline{\underline{B}}$ and

(2.1) $\qquad (I - P)Gf = f \quad$ for all $f \in \underline{\underline{N}}$.

In this section we shall show that there is a potential

kernel for any strong Feller recurrent Markov chain. Let F be a non-polar compact set and ν_F the probability measure on F introduced in lemma 1.3.

The next lemma is known as the Derman-Harris relation, so we shall not prove it.

Lemma 2.1. Let G_F be a kernel defined by (1.3). Then the measure μ defined by

(2.2) $$\mu = \nu_F \; G_F$$

is an invariant measure for P, that is, μ is an everywhere dense positive Radon measure with $\mu P = \mu$.

The next theorem has been proved, by different methods from ours, by Orey [10] for recurrent Markov chains with denumerable state space, by Revuz [12] for those with compact state space and by Duflo [3], Meyer [8] and Neveu [9] for Markov chains which are recurrent in the sense of Harris.

Theorem 2.1. Let P be a strong Feller recurrent Markov kernel with invariant measure μ. Then there exists a potential kernel G_0 for P. Any kernel G on E is a potential kernel for P if and only if it has the form;

$$G = G_0 + h \otimes \mu + 1 \otimes \pi,$$

with a Borel measurable function h and a Radon measure π.

Proof. Take a non-polar compact set F and define the kernel G_0 by

(2.3) $\qquad G_0 = V_F + H_F L_F G_F.$

We shall show that this kernel is the desired one. By lemmas 1.2 and 1.5, G_0 carries functions in $\underline{\underline{B}}_c$ (hence, in $\underline{\underline{N}}$) into $\underline{\underline{B}}$. By the uniqueness of invariant measure we may assume that $\mu = \nu_F G_F$. For any $f \in \underline{\underline{N}}$, since

$$\langle \nu_F, (G_F f)_F \rangle = \langle \nu_F G_F, f \rangle = 0$$

($(G_F f)_F$ denotes the restriction of $G_F f$ to F), $(G_F f)_F$ belongs to $\underline{\underline{N}}_F(\nu_F)$, so that,

$$P_F L_F (Gf)_F = L_F (Gf)_F - (Gf)_F.$$

If $x \in F$,

$$\begin{aligned} PG_0 f(x) &= PV_F f(x) + PH_F L_F (G_F f)_F (x) \\ &= G_F f(x) - f(x) + P_F L_F (G_F f)_F (x) \\ &= (G_F f)_F (x) - f(x) + L_F (G_F f)_F (x) - (G_F f)_F (x) \\ &= L_F (G_F f)_F (x) - f(x) = G_0 f(x) - f(x), \end{aligned}$$

and if $x \notin F$,

$$\begin{aligned} PG_0 f(x) &= V_F f(x) - f(x) + H_F L_F (G_F f)_F (x) \\ &= G_0 f(x) - f(x), \end{aligned}$$

therefore $PG_0 f = G_0 f - f$ on E. Thus G_0 is a potential kernel for P.

The proof of the latter half of the theorem is the same as that of preposition IV-2 of Revuz [12], so will be omitted.

Theorem 2.2. Let G be a potential kernel for P. Then G satisfies the reinforced semi-complete principle of maximum as follows:

(RSCM) For any $f \in \underline{\underline{N}}$ and for any real number m, if $Gf \leq m$ on a set B satisfying $\mu(B) > 0$ and containing the set $\{f > 0\}$, then $Gf - f \leq m$ everywhere.

Proof. Assume that $Gf \leq m$ on a set B satisfying $\mu(B) > 0$ and containing the set $\{f > 0\}$. Since $\mu P(B) = \mu(B) > 0$, $P(y, B) > 0$ for some $y \in E$, which implies that B is non-polar. Therefore, by Dynkin's formula, for any $x \in E$,

$$Gf(x) = E^x[\sum_{0 \leq n < T_B} f(X_n)] + E^x[Gf(X_{T_B})]$$

$$\leq E^x[f(X_0)] + m = f(x) + m,$$

so that, $Gf - f \leq m$ everywhere.

3. A construction of recurrent Markov kernels from potential kernels

We assume that an everywhere dense positive Radon measure μ is given on E. We denote by $\underline{\underline{N}} = \underline{\underline{N}}(\mu)$ the space of all null charges with respect to μ and consider a kernel G which carries $\underline{\underline{N}}$ into $\underline{\underline{B}}$. In the preceding section we have seen that if G is a potential kernel for a strong Feller recurrent Markov kernel P, with invariant measure μ, it satisfies the following conditions;

(G.1) $Gf - f$ is continuous for any $f \in \underline{\underline{N}}$,

(G.2) G satisfies the reinforced semi-complete principle of maximum ((RSCM) in theorem 2.2.).

In the present section we shall consider the converse problem; if (μ, G) satisfies (G.1) and (G.2) does there exist a strong Feller recurrent Markov kernel P which has μ as an invariant measure and G as a potential kernel?

When E is compact, the answer is affirmative, but when E is non-compact, the answer is, in general, negative (for an example, see [4]). However we will give a necessary and sufficient condition for G under which the answer is affirmative for the case where E is non-compact.

Now let μ be an everywhere dense positive Radon measure on E and G a kernel which carries functions in $\underline{\underline{N}}$ into $\underline{\underline{B}}$ and satisfies (G.1) and (G.2). In the rest of this note, we denote by $\mathcal{K}$ the set of all compact sets which are the closures of relatively compact, non-empty open sets in E.

Lemma 3.1. *Let F be a compact set in $\mathcal{K}$ and $g \in \underline{\underline{B}}$. Then there exist a null charge f_F with support in F and a constant $c_F(g)$ such that*

(3.1) $g = Gf_F + c_F(g)$ *on* F.

Such f_F and $c_F(g)$ are unique.

Proof. We may assume that E is compact, F = E and $\mu(F) = 1$, because the restriction (μ_F, G_F) of (μ, G) to F satisfies also (G.1) and (G.2). We shall use the following result (see, for example, T. Watanabe [5]) on linear operators on Banach spaces: Let V be a bounded linear operator from a Banach space $\underline{\underline{L}}$ into itself. If, for any $f \in \underline{\underline{L}}$ and $s \in]0, s_0[$,

(3.2) $\qquad \|sVf\| \leq \|(sV + I)f\|$,

then, $sV + I$ has the inverse $(sV + I)^{-1}$ with domain $\underline{\underline{L}}$ for any $s \in]0, 2s_0[$.

Define a bounded linear operator V from $\underline{\underline{N}}$ into $\underline{\underline{N}}$ by; $Vf = Gf - f - \langle\mu, Gf\rangle$. We will show that (3.2) holds for $s \in]0, 1[$. If $\mu(\{f > 0\}) > 0$, since

$$sGf = sVf + sf + s\langle\mu, Gf\rangle \leq sVf + f + s\langle\mu, Gf\rangle$$
$$\leq \|(sV + I)f\| + s\langle\mu, Gf\rangle \qquad \text{on } \{f > 0\},$$

from (G.2) it follows that

$$sVf = sGf - sf - s\langle\mu, Gf\rangle \leq \|(sV + I)f\|$$

everywhere. If $\mu(\{f > 0\}) = 0$, $f = 0$ μ-almost everywhere and hence

$$sVf = sVf + f \qquad \mu\text{-a.e.}$$
$$\leq \|(sV + I)f\|.$$

Then $sVf \leq \|(sV + I)f\|$ μ-almost everywhere, however, since sVf is continuous, it holds everywhere. In any case, $\sup_x sVf(x) \leq \|(sV + I)f\|$. In the same way, we can prove $\inf_x sVf(x) \geq -\|(sV + I)f\|$, therefore (3.2) holds.
Thus $V + I$ has the inverse with domain $\underline{\underline{N}}$, so that, there exists an $f \in \underline{\underline{N}}$ such that $g - \langle\mu, g\rangle = Gf - \langle\mu, Gf\rangle$, that is,

$g = Gf + c(g)$ with $c(g) = \langle\mu, g\rangle - \langle\mu, Gf\rangle$.

To prove the uniqueness of f and c(g), it is sufficient to show that if Gf = m (constant) on E, then f = 0 and m = 0. Assume that Gf = m on E. Then, by (G.2), $Gf - f \leq m$ everywhere. Hence $f \geq 0$, for, Gf = m on E. In the same way we can show that $f \leq 0$ everywhere, and hence, f = 0 everywhere and m = 0.

Using this lemma, we can prove the following uniqueness theorem.

Theorem 3.1. *Let μ be an everywhere dense positive Radon measure on E and G a kernel which carries functions in* $\underline{N} = \underline{N}(\mu)$ *into* $\underline{B}$ *and satisfies* (G.1) *and* (G.2). *Then, if there is a Markov kernel* P *satisfying*

$$(3.3) \qquad (I - P)Gf = f \qquad \text{for all } f \in \underline{N},$$

it is unique.

Proof. Let P and P′ be Markov kernels satisfying equation (3.3) and, let $(F_n)_{n\geq 1}$ be an increasing sequence of compact subsets in $\mathcal{K}$ with union $\bigcup_{n\geq 1} F_n = E$. For any $g \in \underline{B}$ and for each $n \geq 1$, by lemma 3.1, there is an $f_n \in \underline{N}$ with support in F_n and there is a constant $c_n(g)$ such that

$$g = Gf_n + c_n(g) \qquad \text{on } F_n.$$

Using (G.2), we can show that $\|Gf_n + c_n(g)\| \leq \|g\|$ for each n. Since, $\lim_n (Gf_n + c_n(g)) = g$ on E, by the Lebesgue convergence theorem,

$$Pg(x) = \lim_n P(Gf_n + c_n(g))(x)$$
$$= \lim_n (Gf_n - f_n + c_n(g))(x)$$
$$= \lim_n P'(Gf_n + c_n(g))(x) = P'g(x)$$

for all $x \in E$. Therefore $P = P'$.

Here we state our result in the case that E is compact.

Theorem 3.2. Let E be a compact space. Then a kernel G is a potential kernel for a strong Feller recurrent Markov chain if and only if it satisfies (G.1) and (G.2).

Proof. Since "only if" part has been proved in the preceding section, we have only to prove "if" part. Assume that G satisfies (G.1) and (G.2). For any $g \in \underline{B}$, by lemma 3.1, there is an $f \in \underline{N}$ and there is a constant $c(g)$ such that $g = Gf + c(g)$. We define

(3.4) $\quad Pg = Gf - f + c(g).$

It is clear that P is a bounded linear operator from $\underline{B}$ into $\underline{C}$. If $g = Gf + c(g) \geq 0$, from (G.2), it follows that $Pg = Gf - f + c(g) \geq 0$ and so P is a positive operator. If $1 = Gf + c(1)$, as we have seen in the proof of lemma 3.1. $f = 0$ and $c(1) = 1$. Therefore $P1 = 1$. To show that P is a kernel on E, we take a decreasing sequence $(g_n)_{n\geq 1}$ in $\underline{B}$ with $\lim_n \downarrow g_n = 0$. Let $g_n = Gf_n + c(g_n)$ be the decomposition of g_n in lemma 3.1. Since $Pg_n = g_n - f_n$ and $\|P\| \leq 1$, $\|f_n\| \leq 2\|g_n\| \leq 2\|g_1\|$ for all n. Further, since P is a positive operator, (Pg_n) is a decreasing sequence and $Pg_n \geq 0$ for all n. Therefore (Pg_n) converges and hence, (f_n) converges everywhere. Let $f = \lim_n f_n$. Then, since $0 = \lim_n (Gf_n + c(g_n)) = Gf + \lim_n c(g_n)$, Gf is equal to a constant and so $f = 0$. Thus $\lim_n Pg_n = 0$, which implies P is a kernel on E.

By definition, it is clear that P is strong Feller Markov kernel with μ as an invariant measure and G as a potential kernel. For each $x \in E$, since $P(x, \ )$ is absolutely continuous with respect to μ, so is $P^2(x, \ )$. Then, by Mokobodzki's theorem,

$$r = \frac{1}{2} \sup_x \| P^2(x, \ ) - \mu \| < 1,$$

hence $\sup_x \| P^{2n}(x, \ ) - \mu \| \leq 2r^n$ for all $n \geq 1$. If U is a non-empty open set, then $\mu(U) > 0$ and $P^{2n}(x, U) \geq \mu(U) - 2r^n$ for all $x \in E$ and n. Thus $\Sigma_{n \geq 1} P^n(x, U) = \infty$ for all $x \in E$, which implies that P is recurrent.

From now on we assume that E is not compact and that G is a kernel on E which satisfies (G.1) and (G.2) for μ (μ may or may not be finite). Let F be a compact set in $\mathcal{K}$. For any $g \in \underline{B}$, by lemma 3.1, there are a null charge f_F with support in F and a constant $c_F(g)$ such that $g = Gf_F + c_F(g)$ on F. We set

(3.5) $\quad \bar{H}_F g = Gf_F + c_F(g)$ and

(3.6) $\quad \bar{\Pi}_F g = Gf_F + c_F(g) - f_F = \bar{H}_F g - f_F.$

As was proved in theorem 3.2, $\bar{\Pi}_F$, and hence $\bar{H}_F$, are Markov kernels on E and, for each $x \in E$, the supports of $\bar{\Pi}_F(x, \ )$ and $H_F(x, \ )$ are contained in F.

<u>Lemma</u> 3.2. <u>Let</u> $g \in \underline{\underline{B}}_c$ <u>and let</u> f <u>and</u> h <u>be null charges such that</u> $f = h = g$ <u>on</u> F^c. <u>Then</u>

$$(3.7) \qquad Gf - \bar{H}_F Gf = Gh - \bar{H}_F Gh$$

Proof. Let $Gf = Gf_F + c_F(Gf)$ and $Gh = Gh_F + c_F(Gh)$ be the decompositions of Gf and Gh in lemma 3.1 respectively. Then

$$G((f - f_F) - (h - h_F)) - (c_F(Gf) - c_F(Gh))$$

$$= Gf - \bar{H}_F Gf - Gh + \bar{H}_F Gh = 0 \qquad \text{on } F.$$

Since $(f - f_F) - (h - h_F) = g - g = 0$ on F^c, from (G.2) it follows that $f - f_F = h - h_F$ and $c_F(Gf) = c_F(Gh)$, which proves the lemma.

For any $g \in \underline{\underline{B}}_c$, take a function $f \in \underline{N}$ such that $f = g$ on F^c, and define

$$(3.8) \qquad \bar{V}_F g = Gf - \bar{H}_F Gf.$$

By the preceding lemma, the definition of $\bar{V}_F g$ does not depend on the choice of f. It is clear that, $\bar{V}_F g = 0$ on F and, if $g = 0$ on F^c, $\bar{V}_F g = 0$ everywhere.

We denote by $\underline{B}(F^c)$ the space of all bounded Borel measurable functions vanishing on F and, by $\underline{\underline{B}}_c(F^c)$ functions in $\underline{B}(F^c)$ with compact support.

<u>Lemma</u> 3.3. $\bar{V}_F$ <u>is a non-negative kernel on</u> E <u>which</u>

satisfies the reinforced complete principle of maximum on F^c, that is,

$(RCM)_{F^c}$ For a function $g \in \underline{\underline{B}}_c(F^c)$ and a non-negative constant m, if $\bar{V}_F g \leq m$ on the set $\{g > 0\}$, then $\bar{V}_F g - g \leq m$ everywhere.

Proof. First we show that the linear operator $\bar{V}_F$ from $\underline{\underline{B}}_c(F^c)$ into $\underline{\underline{B}}(F^c)$ satisfies $(RCM)_{F^c}$. Let $g \in \underline{\underline{B}}_c(F^c)$ and $\bar{V}_F g \leq m$ $(m \geq 0)$ on $\{g > 0\}$. Choose $f \in \underline{\underline{N}}$ with $f = g$ on F^c and let $\bar{V}_F g = G(f - f_F) - c_F(Gf)$, where $Gf = Gf_F + c_F(Gf)$ on F. Since $\{f - f_F > 0\}$ is contained in $F \cup \{g > 0\}$ and since $G(f - f_F) - c_F(Gf) \leq m$ on $F \cup \{g > 0\}$, by (G.2), $G(f - f_F) - (f - f_F) - c_F(Gg) \leq m$ everywhere. Thus $\bar{V}_F$ satisfies $(RCM)_{F^c}$. It should be noted that, as a consequence of $(RCM)_{F^c}$, the operator $\bar{V}_F$ is non-negative. To prove that $\bar{V}_F$ is a kernel on E, take a sequence (g_n) in $\underline{\underline{B}}_c^+(F^c)$ with $\lim_n \downarrow g_n = 0$. Let $f_n \in \underline{\underline{N}}$, $f_n = g_n$ on F^c, and let $Gf_n = Gh_n + c_F(Gf_n)$ on F, where $h_n \in \underline{\underline{N}}$, $h_n = 0$ on F^c. Since sequences $(G(f_n - h_n) - c_F(Gf_n))$ and $(G(f_n - h_n) - (f_n - h_n) - c_F(Gf_n))$ of non-negative functions are decreasing, there exists the $\lim_n (f_n - h_n) = k$. Since the support of $f_n - h_n$ is contained in $F \cup \{g_1 > 0\}$ and $\|f_n - h_n\| \leq 2\|\bar{V}_F g_1\|$, $Gk = \lim_n G(f_n - h_n)$. Since the support of k is contained in F and $Gk = \lim_n G(f_n - h_n)$ is equal to a constant on F, $k = 0$ and hence $\lim_n c_F(Gf_n) = 0$. Therefore $\lim_n \downarrow \bar{V}_F g_n = 0$, which implies $\bar{V}_F$ is a kernel.

A non-negative Borel measurable function h vanishing on F

is said to be $\bar{V}_F$ - quasi-excessive if, for any $f \in \underline{B}_c(F^c)$, $\bar{V}_F f \leq h$ on the set $\{f > 0\}$ implies $\bar{V}_F f - f \leq h$ everywhere. Meyer [6] has shown that, for any $\bar{V}_F$ - quasi-excessive function h and for any Borel subset B of F^c, there is the smallest $\bar{V}_F$ - quasi-excessive function that dominates h on B, which we shall call the pseudo-reduite of h to B and denote by ${}^F\bar{H}_B h$. By $(RCM)_{F^c}$, since I_{F^c} is $\bar{V}_F$ - quasi-excessive, we can define ${}^F\bar{H}_B I_{F^c}$, the pseudo-reduite of I_{F^c} to B.

It should be remarked that if G is a potential kernel for a strong Feller recurrent Markov kernel P with invariant measure μ, then $\bar{H}_F$, $\bar{\Pi}_F$ and $\bar{V}_F$ are just equal to H_F, Π_F and V_F in section one, respectively. Furthermore, a non-negative Borel measurable function h, vanishing on F, is $\bar{V}_F$ - quasi-excessive if and only if $E^x[h(X_1) : 1 < D_F] \leq h(x)$ for all $x \in F^c$, and hence, ${}^F\bar{H}_B I_{F^c}(x) = P^x(D_B < D_F)$ for all $x \in F^c$. Therefore, by theorem 1.1, G satisfies:

(G.3) For any compact set F in $\mathcal{K}$ and any decreasing sequence (B_n) of closed subsets in F^c with intersection $\bigcap_{n\geq 1} B_n = \phi$,

$$\lim_n \downarrow {}^F\bar{H}_{B_n} I_{F^c} = 0 \tag{3.9}$$

uniformly on each compact set in F^c.

From now on we assume that a kernel G on E satisfies (G.1), (G.2) and (G.3) for some Radon measure μ, positive on each non-empty open set.

Lemma 3.4. Let F, K and L be compact sets in $\mathcal{K}$ and $F \subset K \subset L$. Then

(3.10) $$\bar{\Pi}_L I_{L-K} \leq {}^F\bar{H}_{L-K} I_{F^c} \quad \text{on } F^c.$$

Proof. Let $I_{L-K} = G(f_L) + c_L(I_{L-K})$ on L, where f_L is the null charge with support in L and $c_L(I_{L-K})$ is a constant Since $\bar{H}_F I_{L-K} = 0$ everywhere, $c_L(I_{L-K}) = -\bar{H}_F G(f_L)$ everywhere, so that,

$$\bar{H}_L I_{L-K} = G(f_L) + c_L(I_{L-K})$$

$$= G(f_L) - \bar{H}_F G(f_L) = \bar{V}_F f_L.$$

Since $0 \leq \bar{\Pi}_L I_{L-K} = \bar{H}_L I_{L-K} - f_L$ and $\bar{H}_L I_{L-K} = 0$ on K, $f_L \leq 0$ on K. On the other hand, since, for each $x \in E$, the measure $\bar{V}_F(x, \ )$ is non-negative and $\bar{V}_F(x, F) = 0$, $\bar{V}_F f_L = \bar{V}_F(I_{F^c} f_L)$. It is clear that $I_{F^c} f_L \in \underline{B}_c(F^c)$ and $\{I_{F^c} f_L > 0\} \subset \{f_L > 0\} \subset L - K$. Now let h be a $\bar{V}_F$ - quasi-excessive function that dominates I_{F^c} on L - K. Since $\bar{V}_F(I_{F^c} f_L) = 1 \leq h$ on $\{I_{F^c} f_L > 0\}$ and since h is $\bar{V}_F$ - quasi-excessive, $\bar{V}_F(I_{F^c} f_L) - I_{F^c} f_L \leq h$ on F^c, so that, $\bar{\Pi}_L I_{L-K} \leq h$ on F^c. Therefore $\bar{\Pi}_L I_{L-K} \leq {}^F\bar{H}_{L-K} I_{F^c}$ on F^c, for, ${}^F\bar{H}_{L-K} I_{F^c}$ is a $\bar{V}_F$ - quasi-excessive function that dominates I_{F^c} on L - K.

Theorem 3.3 Let μ be an everywhere dense positive Radon measure on a locally compact, non-compact, Hausdorff space E with countable base and G a kernel on E. Then G is a (weak) potential kernel for a strong Feller recurrent Markov kernel P with invariant measure μ if and only if (μ, G) satisfies

(G.1), (G.2) and (G.3).

Proof. We have proved already "only if" part, so that, we have only to prove "if" part. Let $g \in \underline{B}$ and let $(F_n)_{n\geq 1}$ be an increasing sequence of compact subsets in $\mathcal{K}$, with $\bigcup_{n\geq 1} F_n = E$. For any compact set K and $\varepsilon > 0$, we choose a compact set F in $\mathcal{K}$ such that $F \cap K = \phi$. By (G.3), we can find an n_0 such that $F_{n_0} \supset F \cup K$ and

$$\sup_{x\in K} {}^{F}\bar{H}_{F^c_{n_0}} I_{F^c}(x) < \varepsilon.$$

Then, for any m, $n \geq n_0$ $(n < m)$ and $x \in K$,

$$|\bar{\Pi}_{F_m} g(x) - \bar{\Pi}_{F_n} g(x)| = |\bar{\Pi}_{F_m} g(x) - \bar{\Pi}_{F_m} \bar{H}_{F_n} g(x)|$$

$$\leq \|g\| \bar{\Pi}_{F_m} I_{F_m - F_n}(x) \leq \|g\| {}^{F}\bar{H}_{F_m - F_n} I_{F^c}(x)$$

$$\leq \|g\| {}^{F}\bar{H}_{F^c_{n_0}} I_{F^c}(x) \leq \varepsilon \|g\|.$$

Hence the sequence $(\bar{\Pi}_{F_n} g)$ converges uniformly on each compact set. We define

$$\text{(3.11)} \qquad Pg = \lim_n \bar{\Pi}_{F_n} g \qquad \text{for } g \in \underline{B}.$$

By definition it is clear that P is a positive linear operator from $\underline{B}$ into $\underline{C}$ with $P1 = 1$. If $(g_n)_{n\geq 1}$ is a decreasing sequence in B_c^+ with $\lim_n \downarrow g_n = 0$, for any $x \in E$ and $\varepsilon > 0$, we can find an m_0 such that $F_{m_0} \supset \{g_1 > 0\} \cup \{x\}$ and $|Pg_n(x) - \bar{\Pi}_{F_{m_0}} g_n(x)| \leq \varepsilon \|g_1\|$ for all n. Therefore

$$\limsup_n Pg_n(x) \leq \varepsilon \|g_1\| + \limsup_n \bar{\Pi}_{F_{m_0}} g_n(x)$$

$$= \varepsilon \|g_1\|.$$

Hence $\lim_n Pg_n(x) = 0$ for all $x \in E$, which implies that P is a kernel on E.

For any $f \in \underline{\underline{N}}$, if F_m contains the support of f, then $\bar{\Pi}_{F_m} Gf = Gf - f$ and hence $PGf = \lim_m \bar{\Pi}_{F_m} Gf = Gf - f$. Therefore P is the unique strong Feller Markov kernel which has G as a potential kernel.

Next we show that μ is an invariant measure for P. Assuming that $g \in \underline{\underline{B}}_c^+$ we choose a compact set F in $\mathcal{K}$ such that $F \cap \{g > 0\} = \phi$. Further let $(F_n)_{n \geq 1}$ be the sequence of compact sets introduced above. We take an n with $F_n \supset F$. Let $g = Gf_{F_n} + c_{F_n}(g)$ (on F_n) be the decomposition of g in lemma 3.1. Then $0 = \bar{H}_F g = \bar{H}_F Gf_{F_n} + c_{F_n}(g)$ and $\bar{\Pi}_{F_n} g = g - f_{F_n}$ on F_n. Since $f_{F_n} \in \underline{\underline{N}}$, $\bar{V}_F f_{F_n} = Gf_{F_n} - \bar{H}_F Gf_{F_n} = Gf_{F_n} + c_{F_n}(g) = g$ on F_n. Hence

$$\begin{aligned}\bar{V}_F(I_{F_n}\bar{\Pi}_{F_n} g) &= \bar{V}_F(I_{F_n} g) - \bar{V}_F(I_{F_n} f_{F_n}) \\ &= \bar{V}_F(I_{F_n} g) - I_{F_n} g \qquad \text{on } F_n.\end{aligned}$$

Since $\lim_n \uparrow [\bar{V}_F(I_{F_n} g) - I_{F_n} g] = \bar{V}_F g - g$, $\bar{V}_F(I_{F_n}\bar{\Pi}_{F_n} g) \leq \bar{V}_F(g) - g$ on F_n. The function $\bar{V}_F g - g$ is $\bar{V}_F$ - quasi-excessive and, for any decreasing sequence $(B_k)_{k \geq 1}$ of closed subsets of F^c with $\bigcap_{k \geq 1} B_k = \phi$,

$$0 \leq \limsup_k {}^F\bar{H}_{B_k}(\bar{V}_F g - g) \leq \| \bar{V}_F g \| \lim_k \downarrow {}^F\bar{H}_{B_k} I_{F^c} = 0,$$

so that, since $\lim_n I_{F_n} \bar{\Pi}_{F_n} g = Pg$, by Meyer's theorem [6],

$$\bar{V}_F P g = \bar{V}_F g - g. \tag{3.12}$$

If $f \in \underline{\underline{N}}$ and $f = g$ on F^c and if $Gf = Gf_F + c_F(Gf)$ on F (the decomposition in Lemma 3.1), then $\bar{V}_F g = G(f - f_F) - c_F(Gf)$ and hence, $P\bar{V}_F g = \bar{V}_F g - (f - f_F)$. Therefore

(3.13) $\langle \mu, I_F P\bar{V}_F Pg\rangle$

$$= \langle \mu, I_F \bar{V}_F g\rangle - \langle \mu, I_F(f - f_F)\rangle - \langle \mu, I_F Pg\rangle$$

$$= - \langle \mu, I_F f\rangle - \langle \mu, I_F Pg\rangle$$

$$= \langle \mu, g\rangle - \langle \mu, I_F Pg\rangle.$$

On the other hand, since $\langle \mu, I_F P\bar{V}_F g\rangle = - \langle \mu, I_F f\rangle = \langle \mu, g\rangle$ for all $g \in \underline{\underline{B}}_c^+(F^c)$,

$$\langle \mu, I_F P\bar{V}_F I_{F_n - F} Pg\rangle = \langle \mu, I_{F_n - F} Pg\rangle$$

for all n with $F_n \supset F$. Thus, tending n to infinity, we have

(3.14) $\langle \mu, I_F P\bar{V}_F Pg\rangle = \langle \mu, I_{F^c} Pg\rangle.$

From (3.13) and (3.14) it follows that $\langle \mu, g\rangle - \langle \mu, I_F Pg\rangle = \langle \mu, I_{F^c} Pg\rangle$, that is, $\langle \mu, g\rangle = \langle \mu, Pg\rangle$, which implies that μ is an invariant measure for P.

Finally we show that P is recurrent. Let X be the canonical Markov chain with transition function P and $H_A(x, B) = P^x(X_{D_A} \in B, D_A < \infty)$, $V_A(x, B) = E^x[\sum_{0 \leqq n < D_A} I_B(X_n)]$. For the proof of recurrence it is sufficient to show that, for any compact set F in $\mathfrak{K}$, $H_F 1(x) = P^x(D_F < \infty) = 1$ for all $x \in E$. Let F be such a compact set. For simplicity, we denote by $\underline{\underline{B}}_0^+$ the set of functions f, in $\underline{\underline{B}}^+$ and $\bar{V}_F f \in \underline{\underline{B}}^+$. For $g \in \underline{\underline{B}}_0^+$, define $P^F g = I_{F^c} P(I_{F^c} g)$, that is, $P^F g(x) = E^x[g(X_1) : 1 < D_F]$. From (3.12) it follows that

$$\bar{V}_F P^F g = \bar{V}_F g - I_{F^c} g$$

so that, for any $n \geq 0$, $(P^F)^n g \in \underline{\underline{B}}_0^+$ ($(P^F)^0 g = I_{F^c} g$) and

$$(3.15) \qquad \bar{V}_F (P^F)^n g \leq \bar{V}_F g - I_{F^c} g,$$

$$(3.16) \qquad \sum_{k=0}^{n} (P^F)^k g = \bar{V}_F g - \bar{V}_F (P^F)^{n+1} g \leq \bar{V}_F g.$$

From (3.16) it follows that $\lim_n (P^F)^n g = 0$. And from (3.15), by Meyer's theorem stated above, it follows that

$$(3.17) \qquad V_F g = \sum_{k=0}^{\infty} (P^F)^k g = \bar{V}_F g.$$

As a consequence ${}^F\bar{H}_B I_{F^c}(x) = P^x(D_B < D_F)$ for any Borel subset B of F^c. Therefore, condition (G.3) implies that $I_{F^c} = V_F g$ with some $g \in \underline{\underline{B}}_0^+$, and hence

$$P^x(D_F = \infty) = \lim_n P^x(n < D_F)$$

$$= \lim_n (P^F)^n V_F g(x) = 0$$

for all $x \in E$. Thus the proof of the theorem is completed.

References

[1] J. Azèma, M. Duflo et D. Revuz: Classes récurrentes d'un processus de Markov, Séminaire de Probabilités II, Springer Verlag, 1968 (Lecture Notes in Math., vol. 51).

[2] R. M. Blumenthal and R. K. Getoor: Markov processes and Potential theory, Academic Press, 1968.

[3] M. Duflo: Opérateurs potentiels des chaines et des processus de Markov irréductibles, Bull. Soc. Math. France, 98 (1970), 127-163.

[4] R. Kondō: A construction of recurrent Markov chains, Osaka J. Math. 6 (1969), 13-28.

[5] R. Kondō, Y. Ōshima and T. Watanabe: Topics in Markov chains II (in Japanese), Seminar on Probability, vol. 32, 1970.

[6] P. A. Meyer: Caractérisation des noyaux potentiels des semi-groupes discrets, Ann. Inst. Fourier, Grenoble 162 (1966), 225-240.

[7] P. A. Meyer: Les résolvantes fortement felleriennes, d'après MOKOBODZKY, Séminaire de Probabilités II, Springer Verlag, 1968 (Lecture Notes in Math., 51).

[8] P. A. Meyer: Solutions de l'equation de Poisson dans le cas récurrent, Seminaire de Probabilités V, Springer Verlag, 1971 (Lecture Notes in Math., 191).

[9] J. Neveu: Potentiel Markovien récurrent des chaines de Harris, Ann. Inst. Fourier, Grenoble 22, 2 (1972), 85-130.

[10] S. Orey: Potential kernels for recurrent Markov chains, J. Math. Anal. and Appl. 8 (1964), 104-132.

[11] Y. Ōshima: A necessary and sufficient condition for a kernel to be a weak potential kernel of recurrent Markov chain, Osaka J. Math. 6 (1969), 29-37.

[12] D. Revuz: Sur la théorie du potentiel pour les processus de Markov récurrents, to appear.

[13] T. Ueno: Some limit theorems for temporally discrete Markov processes, J. Fac. Sci. Univ. Tokyo ser. I. 7 (1957), 449-462.

Department of Mathematics
Faculty of Science
Shizuoka University
Shizuoka, Japan

Department of Mathematics
Faculty of Engineering
Kumamoto University
Kumamoto, Japan

ON A MARKOV PROPERTY FOR STATIONARY GAUSSIAN PROCESSES WITH A MULTIDIMENSIONAL PARAMETER

Shinichi Kotani

1. Introduction

Let $X = \{X(x) : x \in R^d\}$ be a real valued stationary Gaussian process with parameter in the d-dimensional Euclidean space. Throughout this paper we assume that the process X has zero expectation and a spectral density σ. We denote by $Z = L^2(\sigma)$ the complex L^2-space based on $\sigma(\lambda)d\lambda$. For each open set G and closed set F in R^d, we introduce the following subspaces of Z:

(1.1) $Z(G)$ = the closed linear hull of $\{\exp(i \cdot x) : x \in G .\}$.

(1.2) $Z(F) = \bigcap\{Z(G) : G$ is open and $G \supset F .\}$.

In order to prove our theorems it is necessary to set the following hypotheses :

(1.3) σ^{-1} is a locally integrable function.

(1.4) There exists a non-negative and non-decreasing function $T(t)$ $(t \geq c \geq 0)$ such that

$$\int_c^{\infty} \frac{T(t)}{t^2} dt < +\infty \quad \text{and} \quad \sigma^{-1}(x) \leq \exp(T(|x|)),$$

for every sufficiently large $|x|$.

Under these hypotheses, we obtain the following theorems :

<u>Theorem 1.</u> <u>If</u> F <u>is a compact set in</u> R^d, <u>then</u>

$$Z(F) = \{f \in Z : \operatorname{supp} \hat{f} \subset F .\}.$$

<u>Theorem 2.</u> X <u>is Markovian if and only if</u> σ^{-1} <u>agrees a.e. with the restriction to</u> R^d <u>of an entire function of minimal exponential type</u>.

We shall explain a generalized Fourier transformation $\wedge$ in §2 and the Markov property in §3. The idea of proofs consists in discussing local properties in a space of distributions, which is endowed with not only the sheaf property but also a topology. In this connection, we refer the reader to the paper by Y.Okabe [1]. By the Paley-Wiener theorem for Z(see Lemma 4) we see that Theorem 1 is reduced to the theorems of O.I.Presnjakova [2] and O.A.Orebkova [3] when F is a compact convex set. Theorem 2 is a generalization of a result of N.Levinson and H.P.Mckean, Jr. [4] to the multidimensional case. We remark that Theorem 2 has been studied in G.M.Molchan [5] and L.D.Pitt [6] under the conditions that σ^{-1} has a polynomial order. The part of sufficiency of Theorem 2 has been shown by S.Kotani and Y.Okabe [7], but Proposition 1 will be enable us to prove the theorem in the framework of space of ultradistributions.

§2 will be devoted to some preparatory lemmas from the theory of ultradistributions. In §3 we shall prove our theorems. In §4 we shall give examples of a class of minimal exponential type entire functions satisfying the conditions (1.3) and (1.4). It will be noted there that Z(F) has a reproducing kernel for any bounded set F.

The author wishes to express the deepest gratitude to Y.Okabe for his thoughtful guidance and encouragement.

2. Preliminaries

From now on we denote by C_* a constant depending on the index $*$. We define by the following formulas the Fourier transformation $\wedge$ and the inverse Fourier transformation $\vee$ respectively;

$$\hat{\varphi}(\lambda) = \int_{R^d} \exp(-i\,\lambda\cdot x)\varphi(x)\,dx, \quad \check{\varphi}(\lambda) = (1/2\pi)^d \hat{\varphi}(-\lambda).$$

Let $M = \{M_k\}_{k=0}^{\infty}$ be a nomotone increasing sequence of positive numbers satisfying the conditions;

$$(2.1) \qquad M_{k-1}M_{k+1} \leq M_k^2 (k=1,2,\cdots) \quad \text{and} \quad \sum_{k=0}^{\infty} M_k/M_{k+1} < +\infty$$

For a compact set K in R^d and $h > 0$, we introduce a Banach space $\mathcal{E}^{M,h}(K)$ of all infinitely differentiable functions on K such that

(2.2) $$\|f\|_K^{M,h} = \sup_{\alpha, x \in K} |D^\alpha f(x)| / h^{|\alpha|} M_{|\alpha|} < +\infty ,$$

where $D^\alpha = (i\,\partial_{x_1})^{\alpha_1} \ldots (i\,\partial_{x_d})^{\alpha_d}$, for $\alpha = (\alpha_1, \ldots \alpha_d)$. Putting

$$D_K^{M.h} = \{ \varphi \in \mathcal{E}^{M.h}(K) \; : \; \operatorname{supp} \varphi \subset K \},$$

we see that $D_K^{M.h}$ is a Banach space with the norm $\|\cdot\|_K^{M.h}$. The space of ultradifferentiable functions is defined by

(2.3) $$D_M = \varinjlim_{K \subset R^d} \varinjlim_{h \to \infty} D_K^{M.h} .$$

It is known that D_M is a (DFS) - space.

<u>Lemma 1.</u> [C.Roumieu [8]] Suppose that a monotone increasing and positive sequence $\{M_k\}_{k=0}^{\infty}$ satisfies the condition (2.1), then

(i) there exists an approximation of identity

$$\{\rho_\varepsilon\}_{\varepsilon>0} \text{ in } D_M ;$$

(ii) if $f \in \mathcal{E}^{M.h}(K)$ and $\varphi \in D_K^{M.S}$, then

$$\|f\varphi\|_K^{M.h+S} \leq \|f\|_K^{M.h} \; \|\varphi\|_K^{M.S} ;$$

(iii) if φ_0 is a continuous function with support in a compact set K_1 and $\varphi \in D_{K_2}^{M.h}$, then

$$\|\varphi_0 * \varphi\|_{K_1+K_2}^{M.h} \leq C_{K_1,2} \; \|\varphi_0\|_{K_1} \; \|\varphi\|_{K_2}^{M.h}$$

where * is the convolution operator ;

(iv) if φ is a holomorphic function on a complex neighbourhood U of a compact set K in R^d, then

$$\|\varphi\|_K^{M.h} \leq C_U^{M.h} \sup_{z \in U} |\varphi(z)| , \quad \text{for some } h > 0 ;$$

(v) if $\varphi \in D_K^{M.h}$, then

$$|\hat{\varphi}(\lambda)| \leq C_K \|\varphi\|_K^{M.h} \exp\{-M(|\lambda|/dh)\},$$

where $M(t) = \log \sup_{k \geq 0} t^k M_0/M_k$.

We leave the proofs to C.Roumieu [8]. It should be noted, however, that the above lemma implies the existence of a partition of unity by functions in D_M subordinate to any open covering of R^d. Therefore the dual space D_M', which is called the space of ultradistributions, has the sheaf property.

Lemma 2. [C.Roumieu [9]] Suppose that a non-negative and non-decreasing function T fulfils the condition (1.4), then we have a monotone increasing positive sequence $\{M_k\}_{k=0}^{\infty}$ satisfying (2.1) and the condition

$$M(t/h) \geq T(t) + \sqrt{t}, \quad \text{for any } t \geq t_h$$

where t_h is a positive constant depending on h.

Since we assume (1.3) and (1.4), we can construct the space of ultradifferentiable functions D_M by Lemma 2. The following lemma enables us to study the space Z in D_M'.

Lemma 3. If $\varphi \in D_K^{M.h}$, then we have

$$\int_{R^d} |\hat{\varphi}(\lambda)|^p \sigma^{-1}(\lambda)\, d\lambda \leq C_K^{h.p} (\|\varphi\|_K^{M.h})^p, \quad \text{for any } p \geq 1.$$

The proof is an easy consequence of (v) of Lemma 1, Lemma 2 and the condition (1.3).

Now let us put

$$H = \{\varphi : \varphi = \widehat{f\sigma},\ f \in Z\},$$

then H turns out to be a Hilbert space with a reproducing kernel $R = \hat{\sigma}$ and with an inner product

$$(\varphi_1, \varphi_2)_H = (f_1, f_2)_Z$$

where $\varphi_j = \widehat{f_j \sigma}$ $(j = 1,2)$. If $f \in Z$, then we can define f as an element of H' (the dual space of H) in the following way ;

$$\langle \hat{f}, \varphi \rangle = (2\pi)^d \int_{R^d} f(\lambda) g(-\lambda) \sigma(\lambda) d\lambda, \quad (\varphi = \widehat{g\sigma}).$$

Since σ is an even function, it is evident that this $\wedge$ is an extension of the usual Fourier transformation. Further it is easy to see that the inclusion $D_M \subset H$ holds and that the injection is continuous. In fact we have only to put $p = 2$ in Lemma 3. Therefore it is possible to restrict f to D_M, and hence $\hat{f}$ can be regarded as an element of D'_M. Taking this into account, we can extend the Paley-Wiener theorem to the space H' as follows.

<u>Lemma 4.</u> Let K be a compact convex set in R^d and $f \in Z$. Then the following two statements are equivalent.

(i) $\operatorname{supp} \hat{f} \subset K$.

(ii) f coincides a.e. with the restriction to R^d of an entire function F such that

(2.4) $$|F(z)| \leq C_\varepsilon \exp(\varepsilon |z| + H_K(z)), \quad \text{for any } \varepsilon > 0,$$

where $H_K(z) = -\inf_{x \in K} I_m(x \cdot Z)$.

Proof. First we prove that (i) implies (ii). Let $f \in Z$ and $\operatorname{supp} \hat{f} \subset K$, then (ii) and (iv) of Lemma 1 permit us to set

$$F(z) = (1/2\pi)^d \langle \hat{f}, \exp(i \cdot z) \rangle, \quad \text{for } z \in C^d .$$

It is not difficult to see that F is an entire function satisfying (ii). We have to show $F|_{R^d} = f$ a.e.. Let us fix the functions $\{\rho_\varepsilon\}_{\varepsilon > 0}$ of Lemma 1. Set $g_\varepsilon = f \check{\rho}_\varepsilon$, then by Lemma 3

(2.5) $$g_\varepsilon \in L^1(R^d) .$$

Moreover $\widehat{g_\varepsilon} = \hat{f} * \rho_\varepsilon$ holds in D'_M, and hence

(2.6) $\quad \operatorname{supp} \widehat{g_\varepsilon}$ is compacrt .

It follows from (2.5) and (2.6) that $\widehat{g_\varepsilon}$ is a continuous function with compact support. For such a function, however,

(2.7) $\quad g_\varepsilon(x) = (1/2\pi)^d \langle \widehat{g_\varepsilon}, \exp(i\cdot x) \rangle$, a.e. R^d.

Tending ε to zero, we have $f(x) = F(x)$, a.e. R^d. We can prove the converse similarly by using the Paley-Wiener theorem in $L^1(R^d)$. The following two corollaries can be checked easily.

<u>Corollary 1.</u> D_M is embedded into H densely and continuously.

<u>Corollary 2.</u> It is possible to define $\widehat{\sigma^{-1}}$ in the same way as $\hat{f}$ $(f \in Z)$ and the statement of Lemma 4 remains true also for σ^{-1}.

Now we put $\sigma^{-1} = p$, where p is a minimal exponential type entire function. We study a local property of an operator induced by p. We define $p(D)$ by

$$p(D)u = \hat{f}, \quad \text{for} \quad u = \widehat{f\sigma} \in H.$$

$p(D)$ may be regarded as an operator from H to D'_M.

<u>Proposition 1.</u> If $u \in H$ and $u = 0$ on an open set G, then

$$p(D)u = 0 \qquad \text{on } G.$$

Proof. Let $\varphi \in D_M$ and $\operatorname{supp} \varphi \subset G$, then we have

$$\langle p(D)u, \varphi \rangle = \int_{R^d} p(\lambda)\hat{\varphi}(\lambda) f(\lambda)/p(\lambda)\, d\lambda .$$

We set $h = p\hat{\varphi}$, then by Lemma 3

(2.8) $\quad h \in L^1(R^d)$.

On the other hand it follows from Corollary 2 that $\check{h} = \check{p} * \varphi$ in D'_M and that $\operatorname{supp} \check{p} = \{0\}$, and hence

(2.9) $\quad \operatorname{supp} \check{h} = \operatorname{supp} \varphi \subset G.$

(2.8) and (2.9) justify the equalities

$$\langle p(D)u, \varphi \rangle = \int_{R^d} u(\lambda)\check{h}(\lambda)\,d\lambda = 0$$

This completes the proposition.

3. Proofs of theorems

The Markov property defined by H.P.McKean,Jr. [10] is stated in terms of σ-fields. For a Gaussian process, however, we can give the definition in terms of the space H. For each open set G in R^d, we denote by H(G) the closed subspace of H generated by a collection $\{R(\cdot - x) : x \in G\}$. For a closed set F in R^d, H(F) is defined in the same way as in the case of Z. Let T be a bounded open set, then we introduce the following subspaces:

$$H_+(T) = H(T^c),\ H_-(T) = H(\bar{T}) \quad \text{and} \quad \partial H(T) = H(\partial T),$$

where we denote by T^c, $\bar{T}$ and ∂T, the complement, the closure and the boundary of T respectively. We remark that, for Z also, $Z_\pm(T)$ and $\partial Z(T)$ are defined similarly.

Definition 1. [H.P. McKean,Jr. [10] and L.D.Pitt [7]] A stationary Gaussian process X with mean 0 is said to be Markovian if and only if the following equality holds for any bounded open set T in R^d:

$$P_{H_-(T)} H_+(T) = \partial H(T),$$

where $P_{H_-(T)}$ denotes the projection onto $H_-(T)$.

Proof of Theorem 1. Let $f \in Z(F)$. According to the definition of Z(F), for any open set G containing F, we can choose a sequence $\{f_n\}_{n=1}^{\infty}$ converging to f such that each f_n is a finite linear combination of the elements of $\{\exp(i\cdot x) : x \in G\}$. Evidently, $\operatorname{supp} \hat{f}_n \subset G$. Corollary 1, however, says that $\hat{f}_n$ converges to $\hat{f}$ in D'_M, which implies $\operatorname{supp} \hat{f} \subset G$. Hence $\operatorname{supp} \hat{f} \subset F$. Conversely let $f \in Z$ and $\operatorname{supp} \hat{f} \subset F$. In order to show $f \in Z(F)$ it is sufficient to prove that, for an arbitrary fixed open set G including F, $(g, f)_Z = 0$ holds, for any $g \in (Z(G))^{\perp}$. Since the equality $(g, f)_Z = \widehat{g\bar{f}\sigma}(0)$ is valid, we have only to show that $\widehat{g\bar{f}\sigma}$ vanishes in a neighbourhood of the origin of R^d. For $\varphi \in D_M$ with support in $\{x : |x| < \varepsilon\}$, we set $h = f\hat{\bar{\varphi}}$. It follows from Lemma 3 that

$$h \in L^1(R^d). \tag{3.1}$$

Since we assume that $\operatorname{supp} \hat{f} \subset F$, it is obvious that

$$\operatorname{supp} \hat{h} = \operatorname{supp}(\hat{f} * \bar{\varphi}) \subset F_{\varepsilon}, \tag{3.2}$$

where F_{ε} means the ε-neighbourhood of F. (3.1) and (3.2) justify the calculation

$$\langle g\bar{f}\sigma, \varphi \rangle = (1/2\pi)^d \int_{R^d} \widehat{g\sigma}(\lambda) \cdot \overline{\hat{h}(\lambda)}\, d\lambda .$$

On the other hand, $g \in (Z(G))^{\perp}$ implies that

$$\widehat{g\sigma} = 0 \qquad \text{on } G.$$

Therefore by choosing sufficiently small ε, we have $\langle g\bar{f}\sigma, \varphi \rangle = 0$, which completes the proof.

Proof of Theorem 2. Since the part of sufficiency can be shown in the same way as S.Kotani and Y.Okabe [7], we omit the proof. It remains to prove the converse. Let X be Markovian. From Corollary 2 it is possible to regard $\widehat{\sigma}^{-1}$ as an element of D'_M. We have to show $\operatorname{supp} \widehat{\sigma}^{-1} = 0$. Indeed, if $\varphi \in D_M$ and $\operatorname{supp} \varphi \not\ni 0$, then, for sufficiently small $\varepsilon > 0$, $\operatorname{supp} \varphi \cap \operatorname{supp} \rho_{\varepsilon} = \phi$. Then it follows from the Markov property that

$$\langle \widehat{\sigma}^{-1} * \rho_{\varepsilon}, \varphi \rangle = (\rho_{\varepsilon}, \varphi)_H = 0.$$

Tending ε to 0, we have $\operatorname{supp} \widehat{\sigma}^{-1} = 0$, then from Corollary 2 there exists a minimal exponential type entire function P such that $\sigma^{-1} = P$ (a.e. R^d). Thus the theorem is proved.

4. Remarks and examples

1°) It is interesting to rephrase the conditions (1.3) and (1.4) in terms of the process X.

Proposition 2. If the spectral density σ of X satisfies the hypotheses (1.3) and (1.4), then

$$\bigcap_{n=1}^{\infty} \mathcal{B}(G_n) = \text{the trivial field,} \tag{4.1}$$

where $\mathcal{B}(G_n)$ is a σ-field generated by the collection of $\{X(x) : |x| > n\}$.

Proof. Since X is a Gaussian process, (4.1) is equivalent to the following (4.2);

$$\bigcap_{n=1}^{\infty} H(G_n) = \{0\}, \qquad G_n = \{x : |x| > n\}. \tag{4.2}$$

Observing the proof of Theorem 1, we obtain the inclusion $Z(F) \subset \{f \in Z : \text{supp}\,\hat{f} \subset F\}$, for any unbounded closed set F. Therefore if $u \in \bigcap_{n=1}^{\infty} H(G_n)$ and $u = \widehat{f\sigma}$, then $\text{supp}\,\hat{f} \subset \overline{G_n}$, for every n. This means that $\hat{f} = 0$ in D'_M. From corollary 1, we can conclude that $f = 0$ in Z, and hence $u = 0$ in H.

2°) Secondly we consider the prediction theory in the case when the conditions (1.3) and (1.4) are satisfied. To this end the next proposition seems to be useful.

Proposition 3. Let F be a bounded closed set. Then Z(F) has a continuous reproducing kernel.

Proof. Recalling Theorem 1, we can identify any element of Z(F) with an entire function. Put

$$\sigma_F(Z) = \sup\{|f(Z)| : f \in Z(F) \text{ and } \|f\|_Z \leq 1\}.$$

It is well-known that, if $\sigma_F(Z)$ is finite everywhere, then Z(F) has a reproducing kernel. Since the set $\{\hat{f} : f \in Z(F) \text{ and } \|f\|_Z \leq 1\}$ is bounded in D'_M (Corollary 1), we can take C_ε in (ii) of Lemma 4 uniformly for any $\hat{f}$ of the set. Hence, for any $\varepsilon > 0$, $\sigma_F(Z) \leq C_\varepsilon \exp(\varepsilon|Z| + H_F(Z))$. Particularly σ_F is locally bounded. Observing that σ_F is the supremum of a class of holomorphic functions, it is clear that Z(F) has a continuous reproducing kernel.

Now we consider the prediction problem in the Markovian case. Let T be a bounded open set. In view of Proposition 3, $\partial Z(T) = Z(\partial T)$ has a reproducing kernel $J^{\lambda}_{\partial T}(\cdot)$. Using this kernel, we can discuss the prediction problem as a boundary value problem of the following type.

<u>Proposition 4.</u> Let $u \in H_+(T)$ and $v = P_{H-(T)}u$, then v is characterized by the operator as follows,

$$v \in H, \quad P(D)v = 0 \quad \text{on} \quad (\bar{T})^c \quad \text{and} \quad \langle \widehat{J^{\lambda}_{\partial T}}, u - v \rangle = 0,$$

for any $\lambda \in R^d$.

Therefore the study of the kernel seems to be important. When P is a rotation invariant polynomial and ∂T is a sphere, we can calculate $J_{\partial T}$ by the Fourier analysis on the sphere.

3) We finish this section by giving examples of a class of minimal exponential type entire functions satisfying (1.3) and (1.4).

<u>Example 1.</u> (the case d = 1). Let P be a minimal exponential type entire function such that

$$\sum_{n=1}^{\infty} 1/|Z_n| < +\infty,$$

where $\{Z_n\}_{n=1}^{\infty}$ is the set of all zero points of P, then $|P|^{-1}$ satisfies (1.3) and (1.4).

<u>Example 2.</u> (the case $d \geq 1$). Let $\{d_k\}_{k=1}^{m}$ be positive integers such that the total sum equals d and let $\{P_k\}_{k=1}^{m}$ be minimal exponential type entire

functions with d_k variables respectively satisfying (1.3) and (1.4), then

$$P = P_1 \cdots P_m$$

also satisfies (1.3) and (1.4).

References

[1] Y. Okabe : On stationary Gaussian processes with Markovian property and M. Sato's hyperfunctions, to appear in J. Math. Soc. Japan.

[2] O. I. Presnjakova : On the analytic structure of subspaces generated by random homogeneous fields, Soviet Math. Dokl. 11 (1970) 622-625.

[3] O. A. Orebkova : Some problems for extrapolation of random fields, Dokl. Akad. Nauk SSSR. 196 (1971) 776-778.

[4] N. Levinson and H. P. McKean, Jr. : Weighted trigonometrical approximation on R^1 with application to the germ field of a stationary Gaussian noise, Acta Math. 112 (1964), 99-143.

[5] G. M. Molchan : On the characterization of a Gaussian field with a Markov Property, Dokl. Akad. Nauk SSSR. 197 (1971), 784-787.

[6] L. D. Pitt : A Markov property for Gaussian processes with a multidimensional parameter, Arch. Rational Mech. Anal. 43 (1971), 367-391.

[7] S. Kotani and Y. Okabe : On a Markovian property of stationary Gaussian processes with a multidimensional parameter, to appear in Lecture Notes in Math. Proc. Katata Symp. Springer, Berlin.

[8] C. Roumieu : Ultra-distributions definies sur R^n et sur certains classes de variétés différentiables, J. Analyse Math. 10 (1962/3), 153-192.

[9] C. Roumieu : Sur quelque extensions de la notion de distribution, Ann. Sci. Ecole Norm. Sup. 3 ser. 77 (1960), 41-121.

[10] H. P. McKean, Jr. : Brownian motion with a several dimensional time, Theory of Prob. Appl. 8 (1963), 335-354.

Department of Mathematics
Kyoto University.

CONVERGENCE RATE IN THE ERGODIC THEOREM FOR AN ANALYTIC FLOW ON THE TORUS

Masasi Kowada

For some class of analytic flows on the 2-dimensional torus, we shall show that the time mean of a quantity $f \in C^{(k)}$ converges to its phase mean $\mathbb{E}(f)$ with the speed $1/n$ in the sense of both G.Birkoff and J.von Neumann.

We consider a differential equation on the 2-dimensional torus M_2: $dx/dt = F(x,y)$ $dy/dt = G(x,y)$ (mod 1), where $F(x,y)$ and $G(x,y)$ are analytic functions on M_2 such that

$$F^2(x,y) + G^2(x,y) \neq 0 .$$

Let T_t be the transformation on M_2 defined by

$$T_t(x(0),y(0)) = (x(t),y(t)),$$

where $(x(t),y(t))$ is a solution of the above equation with an initial condition $(x(0),y(0))$, and let $\mathcal{B}$ be the topological Borel field in M_2. We assume there exists an invariant probability measure P with an analytic density $p(x,y)$ with respect to Lebesgue measure $dxdy$. Then $\mathcal{T} = (M_2,\mathcal{B},P,T_t)$ is an analytic flow. Denote the rotation number of $\mathcal{T}$ by γ : $\gamma = \int_{M_2} F(\omega)dP(\omega) / \int_{M_2} G(\omega)dP(\omega)$. We put $\mathbb{E}(f) = \int_{M_2} f(\omega)dP(\omega)$.

We shall prove the following theorem:

Theorem. Let $\mathcal{T} = (M_2,\mathcal{B},P,T_t)$ be the flow defined above, and assume that the rotation number γ of $\mathcal{T}$ is irrational and moreover there exist positive numbers L and H such that

$$|m + n\gamma| \geq L/|n|^H \quad \text{for any integers } m \text{ and } n.$$

Then, if $k-H > 1$,

$$\left| \frac{1}{S}\int_0^S f(T_t(x,y))dt - E(f) \right| = O(\tfrac{1}{S}) \quad \text{a.e. } (x,y) \in M_2$$

holds for any $C^{(k)}$-function f, and moreover if $k-H > 1/2$

$$\left\| \frac{1}{S}\int_0^S f(T_t(x,y))dt - E(f) \right\| = O(\tfrac{1}{S}),$$

holds for any $C^{(k)}$-function f, where $\| \ \|$ means the L^2-norm.

Proof. Let $\mathcal{S} = (M_2, \mathcal{B}, dxdy, S_s)$ be a flow defined by

$$dx/dt = 1 \qquad dy/dt = \gamma.$$

It is known [1] and [2] that the flow $\mathcal{T}$ is isomorphic to the flow $\mathcal{S}$ by an analytic transformation ρ on M_2:

$$\rho T_t \rho^{-1}(x,y) = S_t(x,y)$$
$$dP(\rho(x,y)) = dxdy.$$

We consider first the flow $\mathcal{S}$. The flow $\mathcal{S}$ has the discrete spectrum $\mu_{n,m} = n + \gamma m$ (n and m run over all integers) and eigenfunctions $\chi_{n,m}(x,y) = \exp i(nx+my)$:

$$\chi_{n,m}(S_t(x,y)) = \chi_{n,m}(x+t, y+t\gamma) = e^{it\mu_{n,m}}\chi_{n,m}(x,y).$$

Since γ is irrational, $\mathcal{S}$ is ergodic and hence there exists an ergodic automorphism S_{t_0}. Putting $T = S_{t_0}$, and $F(\omega) = \int_0^{t_0} f(S_t\omega)dt$, we get

$$E(F) = \int_{M_2}\int_0^{t_0} f(S_t\omega)dtdP(\omega) = \int_0^{t_0}\int_{M_2} f(S_t\omega)dP(\omega)dt = t_0 E(f)$$

and

$$\frac{1}{S}\int_0^S f(S_t\omega)dt = \frac{1}{S}\sum_0^{\ell-1} F(T^j\omega) + \frac{1}{S}\int_{\ell t_0}^S f(S_t\omega)dt,$$

where ℓ is the integral part of S/t_0. We shall show the last term of the right side of the above equation is bounded in S. If it is not bounded, we can find sequences $A_n \uparrow \infty$ and $S_n \uparrow \infty$ such that

$$A_n < \left| \int_{\ell_n t_0}^{S_n} f(S_t\omega)dt \right| = \left| \int_0^{S_n - \ell_n t_0} f(S_t\omega)dt \right|$$

and hence

$$A_n < \int_{M_2} \left| \int_0^{S_n - \ell_n t_0} f(T_u T_{[S_n]}\omega)\,du \right| dP$$

$$\leq \int_0^{S_n - \ell_n t_0} \int_{M_2} \left| f(T_u T_{[S_n]}\omega) \right| dP\,du = \mathbb{E}(|f|)\,(S_n - \ell_n t_0) \leq \mathbb{E}(|f|)$$

for any n. This is a contradiction and so there exists $K_1 > 0$ such that $\left| \int_{\ell \cdot t_0}^{S} f(S_t\omega)\,dt \right| < K_1$ for any S.

Let $F = \sum F_{n,m}\chi_{n,m}$ be a Fourier expansion of F. Since F is $C^{(k)}$-function, we obtain the following estimation of $F_{n,m}$:

$$|F_{n,m}| \leq M/|m|^k .$$

The spectrum of T consists of $\lambda_{n,m} = \exp i\mu n,m$, and since T is ergodic, $\lambda_{n,m} \neq 1$ if $(n,m) \neq (0,0)$. We can find a constant $K > 0$ and an integer h such that $|\lambda_{n,m} - 1| > K|m\gamma - h|$. Hence we get

$$|F_{n,m}|/|\lambda_{n,m} - 1| \leq M/KL\,|m|^{k-H}$$

for any integers n,m. Put $G(\omega) = \sum_{(n,m)\neq(0,0)} \frac{F_{n,m}}{\lambda_{n,m} - 1}\chi_{n,m}(\omega)$. Then, if $k-H > 1/2$, $G(\omega)$ belongs to $L^2(M_2)$, and if $k-H > 1$, $G(\omega)$ is continuous. Suppose at first $k-H > 1$. Then it follows

$$G(T\omega) - G(\omega) = \sum \frac{F_{n,m}}{\lambda_{n,m} - 1}\chi_{n,m}(T\omega) - \sum \frac{F_{n,m}}{\lambda_{n,m} - 1}\chi_{n,m}(\omega)$$

$$= \sum_{(n,m)\neq(0,0)} F_{n,m}\chi_{n,m}(\omega) = F(\omega) - \mathbb{E}(F)$$

and then $\left| \sum_0^{\ell-1} F(T^i\omega) - \mathbb{E}(F) \right| = |G(T^\ell\omega) - G(\omega)| \leq 2\mathrm{Max}\,|G(\omega)| = K_2$.

Now we get

$$\left| \int_0^S f\circ\rho(T_t\rho^{-1}\omega)\,dt - \mathbb{E}(f\circ\rho) \right| = \left| \int_0^S f(S_t\omega)\,dt - \mathbb{E}(f) \right|$$

$$\leq \left| \sum_0^{\ell-1} F(T^j\omega) - \tfrac{1}{t_0}\mathbb{E}(F) \right| + \left| \int_{\ell t_0}^S f(S_t\omega)\,dt \right|$$

$$< K_1 + K_2 .$$

Since the isomorphism ρ mentioned before is analytic, $f\circ\rho$ is $C^{(k)}$-function if and only if f is so. This provides us the first assertion in the theorem. In the case where $k-H > 1/2$, we can do similarly and we omit the proof.

References

[1] V.I.Arnold: Small denominators. I. Mappings of the circumference onto itself, Izv.Akad.Nauk SSSR 26(1961),21-86

[2] A.N.Kolmogorov: On dynamical systems with an integral invariant on the torus,D.A.N.SSSR 93(1953),763-766

[3] M.Kowada: Convergence rate in ergodic theorem J. of Tsuda College 3(1971) (Japanese), 1-4

Tsuda College

Present Address
Tokyo University of Education

CONVERGENCE OF NUMERIC CHARACTERISTICS OF SUMS OF INDEPENDENT RANDOM VARIABLES AND GLOBAL THEOREMS

V.M.Kruglov

1. One of the fundamental problems in the probabilitic limit theorem theory is to discover criteria for the convergence of the distributions of sums of independent random variables to a given distribution. This problem has been solved in a general way by V.M.Zolotarev. It is of interest to generalize the above problem as follows. If the distributions of sums of independent random variables converge to a limit what conditions guarantee the convergence of some numeric characteristics of these distributions to corresponding characteristics of the limit distribution? The numeric characteristics in question may be chosen to be some moments or mean values with respect to the distribution considered of a fixed function from a particular class as is the case in this paper (see Section 3). The first step in answering this question seems to have been done by S.N.Bernstein in [1] , p. 358, where he gave a sufficient condition for the normal weak convergence of normed centered sums of independent random variables combined with the convergence of moments of sums up to a given order to those of the normal distribution. Sufficient S.N.Bernstein's condition becomes necessary if we assume that random variables are infinitesimal. Note that the sufficient condition above involves uniform smallness of random variables. S.N.Bernstein's theorem is generalized to the case of differently normed random variables in [7] . The most general case was considered in [8] , [9] . In these papers, necessary and sufficient conditions for the convergence of moments of infinitesimal independent random variables to those of an infinitely divisible distribution are given.

Assuming the existence of moments for sums of independent random variables and their convergence to the moments of the limiting distribution, one can prove limit theorems in L_p-metric. We shall call them global theorems. A global version of the central limit theorem for identicaly distributed random variables with finite second momets was proved in [10] . Note that [10] only contains sufficient conditions. This result was generalized to the case of differently distributed random variables in [3] (p.156). In [4] (p.172), a global theorem is proved which asserts that convergence of distributions of normalized sums of identicaly distributed random variables to a stable distribution is equivalent to convergence of these distributions in L_p-metric for some $p>0$.

In our paper, necessary and sufficient conditions are given for the convergence of the expectations of functions from a particular class with sums of independent random variables as argument to the corresponding expectation with respect to the limit distribution. It is shown in particular that theorems on convergence of moments are contained in our theorems.

Theorems on convergence of moments of sums of independent random variables are then used to establish necessary and sufficient conditions in global theorems, which contain the theorems proved in [3] , [10] as particular cases. In the last section a generalization of I.A.Ibragimov's theorem ([4] , p. 172) is formulated.

The author is grateful to V.M.Zolotarev for fruitful discussions.

2. In this section we introduce some notation which will be later used without further explanation

Let $\xi_{n1}, \xi_{n2}, \ldots$ be a double sequence of random variables which are row-wise indepcudent such that, for any $n = 1,2,\ldots,$ the series $\xi_n = \sum_j \xi_{nj}$ converges with probability 1. We introduce the following notation: F_n, F_{n_j} are cumulative distribution functions (d.f.) of random variables ξ_n, ξ_{n_j} respectively; $\xi^{(s)} = \xi - \xi'$, where ξ and ξ' are independent identicaly distributed random variables; $E(a)$ is the degenerate d.f. of the random variable $\xi = a$ with probability 1; $\lambda(\mathcal{D})$ is the characteristic function of set $\mathcal{D}$; L is the Lévy-metric; $F = F[\alpha, \sigma, \mu]$ is the infinitely divisible (i.d.) d.f. defined by the characteristic function (ch.f.) of the form (introduced by P.Lévy ([4] , p.32))

$$f(t) = \exp\left\{it\alpha - \frac{\sigma^2}{2}t^2 + \int' K(t,u)\,\mu(du)\right\},$$

$$K(t,u) = \exp(itu) - 1 - \frac{itu}{1+u^2},$$

where α, $\sigma \geqslant 0$ are arbitrary number, μ is a spectral function such that

a) μ is non-decreasing over the half-lines $(-\infty, 0)$, $(0, +\infty)$ and $\mu(+\infty) = \mu(-\infty) = 0$,

b) $\int'_{|u|<1} u^2 \mu(du) < \infty$.

The symbol $\int'$ means that integration is spread over the whole line except the point zero. It is assumed further that the integration set is the whole line unless otherwise stated. We shall denote the d.f. with a capital letter and its ch.f. with the corresponding small letter.

3. We shall say that a continuous function φ defined on the real line is in the class $\mathcal{B}$ if $\inf\limits_u \varphi(u) > 0$ and, for arbitrary numbers x and y,

$$\varphi(x+y) \leq \varphi(x)\,\varphi(y).$$

Note that a function $\varphi \in \mathcal{B}$ cannot increase faster than an exponent. Let $x > 0$ be an arbitrary number and n be the positive integer such that $n \le x < n+1$.

From the definition of class $\mathcal{B}$ it follows that

$$\varphi(x) \le \varphi(x-n)\,\varphi(n) \le \varphi(x-n)\left[\varphi(1)\right]^n \le$$

$$\le \max_{|u| \le 1} \varphi(u) \exp\left(x \,|\ln \varphi(1)|\right).$$

Remark 1. Let $\Psi > 0$ be defined on the real line and continuous.

a) If, for arbitrary numbers x and y and some $B \ge 1$

$$\Psi(x+y) \le B\,\Psi(x)\cdot\Psi(y),$$

then $\varphi = B\Psi \in \mathcal{B}$.

b) If, for arbitrary numbers x and y and some $B \ge 1$,

$$\Psi(x+y) \le B\left[\Psi(x) + \Psi(y)\right],$$

then $\varphi = B(\Psi + 2) \in \mathcal{B}$.

The following functions are in $\mathcal{B}$:

$$\varphi(u) = C\exp\left(\alpha |u|^{\gamma}\right), \quad C \ge 1, \ \alpha > 0, \ 0 < \alpha \le 1,$$

$$\varphi(u) = \left(\frac{4}{\delta}\right)^p \cdot |\delta|^p, \quad |u| \le \delta, \quad \varphi(u) = \left(\frac{4}{\delta}\right)^p |u|^p, \quad |u| > \delta,$$

$$\varphi(u) = 2^p\left(2 + |u|^p\right),$$

for any fixed numers $\delta > 0$, $p > 0$.

We shall also use the notation

$$A = A(h, \varphi) = \left(\max_{|u| \le h} \varphi(u)\right)^{-1}, \quad \varphi \in \mathcal{B}.$$

4. In this section we shall formulate the main result on the convergence of numeric characteristics of sums of random variables.

Let ξ be a random with d.f. F, $\xi_{n1}, \ldots$ be a double sequence of row-wise independent random variables (as deseribed in section 2) and a function $\varphi \in \mathcal{B}$. Assume that the expectations $M\varphi(\xi)$, $M\varphi(\xi_n)$, $M\varphi(\xi_{nj})$ $n = 1,2,\ldots$ are finite. Note that when m_n (the number of random variables in the n-th row is finite, the finiteness of $M\varphi(\xi_n)$ follows from that of $M\varphi(\xi_{nj})$, $1 \le j \le m_n$.

Denote

$$m_{nj} = \operatorname{med} F_{nj}, \quad \bar{F}_{nj}(u) = F_{nj}(u + m_{nj}), \quad n, j = 1, 2, \ldots$$

Theorem 1. Let $L(F_n, F) \to 0$ $n \to \infty$, $\varphi \in \mathcal{B}$. φ is a fixed function. In order that

$$\lim_{n \to \infty} M\varphi(\xi_n) = M\varphi(\xi)$$

it is necessary and sufficient that

$$\lim_{R \to \infty} \sup_n \sum_j \int_{|u| > R} \varphi(u) \bar{F}_{nj}(du) = 0. \tag{1}$$

The following theorem gives necessary and sufficient conditions for convergence to an i.d.d.f. $F = F[\alpha, \sigma, \mu]$. Let $\xi_{n1}, \xi_{n2}, \ldots, \xi_{nm_n}$ be an infinitesimal system of row-wise independent random variables. Fix a number $\tau > 0$ and denote

$$a_{nj} = \int_{|u| < \tau} u F_{nj}(du), \quad G_{nj}(u) = F_{nj}(u + a_{nj}),$$

$$\alpha_n = \sum_{j=1}^{m_n} \left\{ a_{nj} + \int \frac{u}{1+u^2} G_{nj}(du) \right\},$$

$$\Psi_n(u)=\sum_{j=1}^{m_n}\int_{-\infty}^{u}\frac{y^2}{1+y^2}\,G_{n_j}(dy),$$

$$\Psi(u)=\begin{cases}\displaystyle\int_{-\infty}^{u}\frac{y^2}{1+y^2}\,\mathcal{M}(dy), & u<0\\[2ex] \displaystyle 6^2+\int_{-\infty}^{0}\frac{y^2}{1+y^2}\,\mathcal{M}(dy)+\int_{+0}^{u}\frac{y^2}{1+y^2}\,\mathcal{M}(dy), & u>0\end{cases}$$

Theorem 2. Let ξ_{n_j}, $1\le j\le m_n$, be an infinitesimal system row-wise independent random variables and let φ be a fixed function from $\mathcal{B}$. In order that

$$L(F_n,F)\to 0,\quad M\varphi(\xi_n)\to M\varphi(\xi)\qquad n\to\infty$$

it is necessary and sufficient that, as $n\to\infty$,

a) $\alpha_n\to\alpha$,

b) $\Psi_n\to\Psi$ completely

c) $\int\varphi(u)\,\Psi_n(du)\longrightarrow\int\varphi(u)\,\Psi(du)$.

5. The case when the limit distribution is normal, permits an easier and more elegant treatment. Let a double sequence of row-wise independent random variables (as described in Section 2) be given. Assume that the random variables ξ_{n_j}, n,j = 1,2,... satisfy the conditions:

$$M\xi_{n_j}=0,\quad var\,\xi_{n_j}=6^2_{n_j}<\infty,\quad M\varphi(\xi_n)<\infty,\quad M\varphi(\xi_{n_j})<\infty,$$

(2) $$\sum_j 6^2_{n_j}=1.$$

Denote by Φ (resp. Φ_{nj}) the normal (0,1), (resp. $(0, \sigma_{nj}^2$) d.f., and by a random variable distributed according to Φ .

Theorem 3. Let condition (2) hold and φ be a fixed function from $\mathcal{B}$. In order that

$$L(F_n, \Phi) \to 0, \quad M\varphi(\xi_n) \to M\varphi(\xi) \qquad n \to \infty,$$

it is necessary and sufficient that

a) $\lim\limits_{n\to\infty} \beta_n = 0, \qquad \beta_n = \sup\limits_j L(F_{nj}, \Phi_{nj}),$

b) for any $\delta > 0$,

$$\lim_{n\to\infty} \sum_j \int_{|u|>\delta} \varphi(u)\left[F_{nj} - \Phi_{nj}\right](du) = 0,$$

as $n \to \infty$.

6. In this section, we state global theorems. We assume that a random variable ξ with d.f. F and random variables ξ_{nj}, $n, j = 1,2,\ldots$ (as described in Section 2) possess finite absolute moments of order $\alpha > 0$.

Theorem 4. Let $p > \alpha^{-1}$ be any fixed number. In order that

$$\int |F_n(u) - F(u)|^p du \to 0, \qquad M|\xi_n|^\alpha \to M|\xi|^\alpha,$$

as $n \to \infty$ it is necessary and sufficient that

a) $\lim\limits_{n\to\infty} L(F_n, F) = 0,$

b) $\lim\limits_{n\to\infty} \sup\limits_n \sum\limits_j \int_{|u|>R} |u|^\alpha \bar{F}_{nj}(du) = 0.$

Here is an important particular case. Let us assume that the random variables ξ_n, ξ_{nj}, $n, j = 1,2,\ldots$ have finite means.

Theorem 5. In order that

$$\lim_{n\to\infty} \int |F_n(u) - F(u)|\, du = 0$$

it is necessary and sufficient that

a) $\lim_{n\to\infty} L(F_n, F) = 0,$

b) $\lim_{R\to\infty} \sup_n \sum_j \int_{|u|>R} |u|\, \bar{F}_{n_j}(du) = 0.$

7. In this section, several auxiliary assertions will be stated as lemmas. We shall need the following simple inequalities. Let ξ, η be independent random variables, $\varphi \in \mathcal{B}$ then, for any $0 < h < R$

$$M[\varphi(\xi+\eta)\lambda(|\xi+\eta|>R)] \leq M[\varphi(\xi)\lambda(|\xi| > \tfrac{1}{2}R)]\cdot M\varphi(\eta) + \tag{3}$$
$$+ M[\varphi(\eta)\lambda(|\eta| > \tfrac{1}{2}R)]\cdot M\varphi(\xi),$$

$$M[\varphi(\xi+\eta)\lambda(|\xi+\eta|>R)] \geq AP(|\eta| \leq h)M[\varphi(\xi)\lambda(|\xi|>2R)], \tag{4}$$

$$M\varphi(\xi+\eta) \geq M\varphi(\xi)\cdot M\varphi^{-1}(-\eta) \geq AP(|\eta|\leq h)M\varphi(\xi). \tag{5}$$

(For the definition of $A = A(h, \varphi)$ cf. Section 3.)

Lemma 1. If $\xi_1, \xi_2, \ldots$ are independent symmetric random variables such that the series $\sum_j \xi_j$ converges with probability 1, then, for any $h>0$ and any subset $\nu \subset \{1,2,3,\ldots\}$,

$$P\left(\left|\sum_{j\in\nu} \xi_j\right| > h\right) \leq 2P\left(\left|\sum_j \xi_j\right| > h\right).$$

The proof of this lemma follows from P.Lévy's inequality ([5], p. 261), if we take into account that the sum of the series $\sum_j \xi_j$ does not depend on any transposition of random variables.

Let ξ_{nj} , n,j = 1,2,..., be a double sequence of row-wise independent random variables (as described in Section 2). For each random variable ξ_{nj}, n,j = 1,2,..., we consider i.d.d.f. $\hat{F}_{nj}$, defined by ch.f.

$$\hat{f}_{n_j}(t) = exp\left(\ln 2 \int (e^{itu} - 1)\, F_{n_j}^{(s)}(du)\right).$$

The convergence with probability 1 of the series $\sum_j \xi_{nj}$ permits by the three series theorem ([5], p.251) to conclude that the following series converge:

$$\sum_j F_{n_j}^{(s)}(u), \quad u<0, \quad \sum_j \left[F_{n_j}^{(s)}(u) - 1\right], \quad u>0,$$

$$\sum_j \int_{|u|<1} u^2 F_{n_j}^{(s)}(du), \quad n = 1, 2, \dots$$

Convergence of the convolution $Q_n = \prod_j^* F_{n_j}^{(s)}$ follows from the convergence of these series. We have from the formula for ch. f. $\hat{f}_{n_j}$:

$$\hat{F}_{n_j} = \frac{1}{2}\left[E(0) + \sum_{m=1}^{\infty} \frac{(\ln 2)^m}{m!} F_{n_j}^{(s)*m}\right].$$

Denote

$$G_{n_j} = \sum_{m=1}^{\infty} \frac{(\ln 2)^m}{m!} F_{n_j}^{(s)*(m-1)}.$$

We have

$$\hat{F}_{n_j} = \frac{1}{2}\left[E(0) + F_{n_j}^{(s)} * G_{n_j}\right].$$

Let $\mathcal{J} \subset \{1,2,\ldots\}$ be an arbitrary set. Define for any number $h > 0$ two other numbers

$$c_n(h) = \inf_{\mathcal{J}} \left(\prod_{j\in\mathcal{J}}^{*} G_{n_j}\right)\left(\{u : |u| \le h\}\right),$$

$$c_n'(h) = \inf_{\mathcal{J}} \left(\prod_{j\in\mathcal{J}}^{*} F_{n_j}^{(s)}\right)\left(\{u : |u| \le h\}\right).$$

Lemma 2. If the sequence of d.f. $\{F_n^{(s)}\}$ is weakly compact then there exists a number ℓ such that, for $h \ge \ell$

$$\inf_n c_n(h) = c(h) > 0, \qquad \inf_n c_n'(h) = c'(h) > 0.$$

Proof. Due to Lemma 1 the second assertion is obvious. Let us prove the first assertion. Fix a set $\mathcal{J} \subset \{1,2,\ldots\}$ and assume that the number of elements $d(\mathcal{J})$ in $\mathcal{J}$ is finite. Denote by $\mathcal{V} \subset \mathcal{J}$ an arbitrary set, and let $d(\mathcal{V}) = d$ be the number of elements in $\mathcal{V}$, $\hat{F}_{n,\mathcal{J}}$ be the convolution of d.f. $\hat{F}_{n_j}$, $j \in \mathcal{J}$. We have

$$\hat{F}_{n,\mathcal{J}} = \prod_{j\in\mathcal{J}}^{*} \frac{1}{2}\left[E(0) + F_{n_j}^{(s)} * G_{n_j}\right] =$$

(6)

$$= \frac{1}{2^{d(\mathcal{J})}} \sum_{d=0}^{d(\mathcal{J})} \sum_{\mathcal{V}: d(\mathcal{V}) = d} \left[\left(\prod^{*} G_{n_j}\right) * H_{n,\mathcal{V}}\right].$$

D.f. $H_{n,\mathcal{V}}$ is the convolution of $F_{n_j}^{(s)}$, where j runs through some subset of the set $\mathcal{J}\setminus\mathcal{V}$. Let $\eta_{n_j}, \rho_{n,\mathcal{V}}, j \in \mathcal{V}$ be independent random variables with d.f. G_{nj}, $H_{n,\nu}$ respectively. Put $\eta_{n,\nu} = \sum_{j\in\nu} \eta_{nj}$. We have, for an arbitrary $h>0$,

$$P(|\eta_{n,\nu}+\rho_{n,\nu}|>h) \geqslant P(|\eta_{n,\nu}|>2h)\cdot P(|\rho_{n,\nu}|\leq h).$$

D.f. $H_{n,\nu}$ is a symmetric component of d.f. $F_n^{(s)}$. On account of Lemma 1 the weak compactness of $\{F_n^{(s)}\}$ gives that $C'(h)\to 1$ as $h\to\infty$. Put $\bar{\nu} = \mathcal{I}\setminus\nu$. It follows from (6) that

$$\hat{F}_{n,\mathcal{I}}(\{u:|u|>h\}) \geqslant \frac{c'(h)}{2^{d(\mathcal{I})}} \sum_{d=0}^{d(\mathcal{I})} \sum_{\nu:d(\nu)=d} P(|\eta_{n,\nu}|>2h) =$$

$$= \frac{c'(h)}{2^{d(\mathcal{I})+1}} \sum_{d=0}^{d(\mathcal{I})} \sum_{\nu:d(\nu)=d} [P(|\eta_{n,\nu}|>2h)+P(|\eta_{n,\bar{\nu}}|>2h)] \geqslant$$

$$\geqslant \frac{c'(h)}{2} P(|\sum_{j\in\mathcal{I}} \eta_{nj}|>4h).$$

If the number of elements in $\mathcal{I}$ is infinite, then the inequality above is valid for any finite part of $\mathcal{I}$. The passage to the limit gives us the general assertion: for any $\mathcal{I}\subset\{1,2,\dots\}$

$$\hat{F}_{n,\mathcal{I}}(\{u:|u|>h\}) \geqslant \frac{c'(h)}{2} P(|\sum_{j\in\mathcal{I}} \eta_{nj}|>4h).$$

Our lemma will be proved if we show that

$$\lim_{h\to\infty} \sup_{n} \sup_{\mathcal{I}} \hat{F}_{n,\mathcal{I}}(\{u:|u|>h\}) = 0.$$

Note that d.f. $\hat{F}_{n,\mathcal{I}}$ is the symmetric component of d.f. Q_n. On account of Lemma 1 we need to prove that the sequence of d.f. $\{Q_n\}$

is weakly compact. Now the inequality

$$\hat{f}_{n_j}(t) \geq |f_{n_j}(t)|^2$$

gives

$$(7) \qquad 1 - \prod_{j=1}^{\infty} \hat{f}_{n_j}(t) \leq 1 - \prod_{j=1}^{\infty} |f_{n_j}(t)|^2 = 1 - |f_n(t)|^2.$$

The compactness of $\{F_n^{(s)}\}$ implies that the set of ch.f. $\{\prod_{j=1}^{\infty} \hat{f}_{n_j}\}$ is uniformly equicontinuous and, therefore $\{Q_n\}$ is a weakly compact sequence of d.f. ([5], p. 206). The lemma is proved.

Lemma 3. Let $\{F_n\}$ be a sequence of i.d.d.f. defined by i.d. ch.f.

$$exp(\int(e^{itu} - 1)\mathcal{M}_n(du)), \quad n = 1, 2, \ldots$$

such that

$$\sup_n [\mathcal{M}_n(-0) - \mathcal{M}_n(+0)] < \infty.$$

If some function $\varphi \in \mathcal{B}$ satisfies the condition

$$(8) \qquad \lim_{R\to\infty} \sup_n \int_{|u|>R} \varphi(u)\mathcal{M}_n(du) = 0.$$

then (8) holds with the spectral functions $\mathcal{M}_n$ replaced by d.f. F_n, $n = 1,2,\ldots$

Proof. Denote $r_n = \mathcal{M}_n(-0) - \mathcal{M}_n(+0)$, $n = 1,2,\ldots$. Without loss of generality we can assume that $r_n > 0$, $n = 1,2,\ldots$. Put

$$K_n(u) = \begin{cases} r_n^{-1} \mathcal{M}_n(u), & u < 0, \\ r_n^{-1}[r_n + \mathcal{M}_n(u)], & u > 0, \end{cases}$$

$$K_n(-0) = r_n^{-1}(-0), \quad K_n(+0) = r_n^{-1}[r_n + \mathcal{M}_n(+0)].$$

We have the following representation for F_n , $n = 1,2,\dots$:

$$F_n = e^{-r_n}\Big[E(0) + \sum_{m=1}^{\infty} \frac{(r_n)^m}{m!} K_n^{*m}\Big].$$

It is quite easy to verify that

$$\int \varphi(u)\, K_n^{*m}(du) \le \Big[\int \varphi(u)\, K_n(du)\Big]^m, \qquad n, m = 1,2,\dots$$

It follows from the inequality (3) and the assertion of the lemma that, for any fixed $m = 1,2,\dots$,

$$\lim_{R\to\infty} \sup_n r_n^m \int_{|u|>R} \varphi(u)\, K_n^{*m}(du) = 0.$$

The terms of the series

$$(10) \qquad \int_{|u|>R} \varphi(u)\, F_n(du) = e^{-r_n}\Big[\sum_{m=1}^{\infty} \frac{r_n^m}{m!} \int_{|u|>R} \varphi(u)\, K_n^{*m}(du)\Big]$$

are not greater than the corresponding terms of the series

$$e^{-r_n}\Big[\sum_{m=1}^{\infty} \frac{[r_n \int \varphi(u)\, K_n(du)]^m}{m!}\Big], \quad \sup_n r_n \int \varphi(u)\, K_n(du) < \infty.$$

The passage to the limit in the summands of (10) as $R \to \infty$ completes the proof of our lemma.

Lemma 4. Let $F_n = F_n[0,0,M_n]$ be a sequence of i.d. d.f. such that

a) spectral functions M_n are constant outside $(-A, A)$ for some fixed $A > 0$.

b) $\sup_n \int_{|u|\le A} u^2 M_n(du) < \infty$,

then there exists a number $\gamma > 0$ such that

$$\sup_n \int \exp(\gamma |u| \ln(|u|+1))\, F_n(du) < \infty .$$

Proof. The proof of this lemma is based on the consideration of the decrease rate of functions $F_n(-u)$ and $1 - F_n(u)$ as $u \to \infty$. Note that

$$F_n(-u) = 1 - \tilde{F}_n(u), \quad u > 0,$$

$$\tilde{F}_n = \tilde{F}[0, 0, \tilde{\mu}_n], \quad \tilde{\mu}_n(u) = -\mu_n(-u).$$

Therefore, it is enough to consider the behaviour of $1 - F_n(u)$, as $u \to \infty$. Represent d.f. F_n in the form

$$F_n = E(a_n) * \check{F}_n, \qquad a_n = \int_{-A}^{A} \left(u - \frac{u}{1+u^2}\right) \mu_n(du),$$

where i.d. d.f. $\check{F}_n$ is defined by ch.f.

$$\exp\left(\int_{-A}^{A} (e^{itu} - 1 - itu)\, \mu_n(du)\right).$$

It follows from condition b) that

$$\sup_n |a_n| < \infty .$$

Therefore, it suffices to prove our lemma for the sequence of d.f. $\{\check{F}_n\}$. For each d.f. $\check{F}_n$ we consider the fuction

$$\Psi_n(z) = \int_{-A}^{A} (e^{zu} - 1 - zu)\, \mu_n(du), \quad z \in (-\infty, +\infty).$$

Since $\Psi_n''(z) \geq 0$, the function $\Psi_n'(z)$ is non-decreasing. There exists a function $\vartheta_n(z)$, defined by

(11) $$\Psi_n'(\vartheta_n(u)) = u .$$

In [6], it is proved that, for $u>0$,

$$1-\check{F}_n(u)\le \exp\Big(-\int_0^u \nu_n(z)\,dz\Big), \tag{12}$$

$$\nu_n(u)>0, \quad \nu_n(u)\uparrow\infty, \quad u\uparrow\infty.$$

It follows from (11) that, for $u\geqslant 1$,

$$u=\int_{-A}^{A}\big(e^{\nu_n(u)z}-1\big)z\,\mu_n(du)\le \nu_n(u)e^{\nu_n(u)A}\cdot\int_{-A}^{A} z^2\mu_n(dz),$$

$$\ln u\le A\nu_n(u)+\ln\nu_n(u)+\ln\Big(\sup_n\int_{-A}^{A} z^2\mu_n(dz)+1\Big).$$

Fix $u_0\geqslant 1$ such that, for $u\geqslant u_0$,

$$2\ln\Big(\sup_n\int_{-A}^{A} z^2\mu_n(dz)+1\Big)\le \ln u.$$

We have for any $u\geqslant u_0$ and all $n=1,2,\ldots$

$$\frac{\ln\nu_n(u)}{\nu_n(u)}\le 1.$$

Therefore, for any $u\geqslant u_0$ and all $n=1,2,\ldots$, we have

$$\frac{2}{A+1}\ln u\le \nu_n(u).$$

Quite easy calculations complete the proof.

Lemma 5. For arbitrary d.f.

$$L\Big(\prod_{j=1}^{n}{}^{*}F_j,\ \prod_{j=1}^{n}{}^{*}G_j\Big)\le \sum_{j=1}^{n}L(F_j,G_j).$$

For the proof see [2].

8. Proof of Theorem 1. Necessity. Denote

$$B_n=\{|\xi_n^{(s)}|>R\}, \quad B_{n_j}=\{|\xi_{n_j}^{(s)}|>2R\},$$

$$D_{n_j}=\{|\xi_n^{(j)}|<R\}\cap\bigcap_{i\neq j}\{|\xi_{n_i}^{(s)}|\leq 2R\}, \quad \xi_n^{(j)}=\xi_n^{(s)}-\xi_{n_j}^{(s)}, \quad n,j=1,2,\ldots$$

Note that the sets $B_{n_j}\cap D_{n_j}$ are disjoint for each $n=1,2,\ldots$ and

$$B_n\supseteq\bigcup_j(B_{n_j}\cap D_{n_j}).$$

It follows from the independence of random variables and the property of the function $\varphi\in\mathcal{B}$ that

$$M[\varphi(\xi_n^{(s)})\lambda(B_n)]\geq\sum_j M[\varphi(\xi_n^{(s)})\lambda(B_{n_j}\cap D_{n_j})]\geq$$

$$\geq\sum_j M[\varphi(\xi_{n_j}^{(s)})\lambda(B_{n_j})]\,M[\varphi^{-1}(-\xi_n^{(j)})\lambda(D_{n_j})]. \tag{13}$$

Let us prove that

$$\lim_{R\to\infty}\inf_n\inf_j P(D_{n_j})=1. \tag{14}$$

To establish (14) it suffices to prove the following two assertions: as $R\to\infty$,

$$\sup_n\sup_j P(|\xi_n^{(j)}|>R)\longrightarrow 0, \tag{15}$$

$$\sup_n P(\bigcup_j\{|\xi_{n_j}^{(s)}|>R)\longrightarrow 0. \tag{16}$$

The first of them follows from the condition $L(F_n,F)\longrightarrow 0$ and Lemma 1. The second is a consequence of the fact that $L(F_n,F)\to 0$ and P.Lévy's inequality ([5], p.261)

$$P\Big(\bigcup_j\{|\xi_{n_j}^{(s)}|>R\}\Big)=P\{\sup_j|\xi_{n_j}^{(s)}|>R\}\leq 2P(\{|\xi_n^{(s)}|>R\}).$$

It follows from (14) and (15) that there exists a number $h>0$ such that, for all large enough R,

$$M[\varphi^{-1}(-\xi_n^{(j)})\lambda(D_{n_j})]\geqslant\tfrac{1}{2}A>0.$$

(The definition of $A=A(h,\varphi)$ is given in Section 3.) By the assumption of the theorem, we have

$$\lim_{R\to\infty}\sup_n M[\varphi(\xi_n^{(s)})\lambda(B_n)]=0.$$

In view of (13)

(17) $$\lim_{R\to\infty}\sum_j M[\varphi(\xi_{n_j}^{(s)})\lambda(|\xi_{n_j}^{(s)}|>2R)]=0.$$

It is well-known ([5], p. 261) that

(18) $$\tfrac{1}{2}P(|\xi-m|>R)\leq P(|\xi^{(s)}|>R)\leq 2P(|\xi-m|>\tfrac{1}{2}R),$$

where m is the median of ξ. We have from (16):

$$\lim_{R\to\infty}\inf_n M[\varphi^{-1}(\xi'_{n_j}-m_{n_j})\lambda(|\xi'_{n_j}-m_{n_j}|\leq R)]>0,$$

where ξ'_{n_j} is independent of ξ_{n_j} and has the same distribution. Relations (13), (17) and the inequality

$$M[\varphi(\xi_{n_j}^{(s)})\lambda(|\xi_{n_j}^{(s)}|>2R)]\geqslant$$

$$\geqslant M[\varphi(\xi_{n_j}-m_{n_j})\lambda(|\xi_{n_j}-m_{n_j}|>3R)]\cdot M[\varphi^{-1}(\xi'_{n_j}-m_{n_j})\lambda(|\xi'_{n_j}-m_{n_j}|\leq R)]$$

imply the necessity of the condition of the theorem.

Proof of sufficiences. Consider the sequence of i.d. d.f. $\{Q_n\}$ defined by ch.f.

$$\exp\Big(\sum_j \ln 2 \int (e^{itu}-1) F_{n_j}^{(s)}(du)\Big), \quad n=1,2,\ldots$$

As in Lemma 2 (see (7)), we shall prove that the sequence of d.f. $\{Q_n\}$ is weakly compact. Fix a number $\nu>0$ such that points $\pm\nu$ are points of continuity of the spectral functions

$$\mathcal{M}_n(u)=\begin{cases}\ln 2 \sum_j F_{n_j}^{(s)}(u), & u<0,\\ \ln 2 \sum_j \big[F_{n_j}^{(s)}(u)-1\big], & u>0,\end{cases}$$

$n = 1,2,\ldots$. Represent d.f. Q_n in the form $Q_n = Q_{n_1} * Q_{n_2}$, where Q_{n_1} and Q_{n_2} are d.f. defined by ch.f.

$$q_{n_1}(t) = \exp\Big(\int_{|u|\le\nu} (e^{itu}-1)\mathcal{M}_n(du)\Big),$$

$$q_{n_2}(t) = \exp\Big(\int_{|u|>\nu} (e^{itu}-1)\mathcal{M}_n(du)\Big).$$

Because of the weak compactness of $\{Q_n\}$ Lemma 4 holds true for the sequence $\{Q_{n_1}\}$. We have by Lemma 4:

$$\lim_{R\to\infty}\ \sup_n \int_{|u|>R} \varphi(u)\, Q_{n_1}(du) = 0. \tag{19}$$

Here we take into account that the function φ increases not faster than an exponent (see Section 3). We should keep in mind that $\xi_{n_j}^{(s)} = \xi_{n_j} - \xi'_{n_j}$ (see Section 2). In view of (3) for $\xi_{n_j} - m_{n_j}$ and $m_{n_j} - \xi'_{n_j}$ and the conditions of the theorem, we have (17). All the conditions of Lemma 3 for the sequence of d.f. $\{Q_{n_2}\}$ are satisfied. Therefore (19) is true if we replace Q_{n_1} by Q_{n_2}, $n = 1,2,\ldots$. From (19) (for Q_{n_1} and Q_{n_2}) and the inequality (3) it follows that

(20) $$\lim_{R\to\infty} \sup_n \int_{|u|>R} \varphi(u)\, Q_n(du) = 0.$$

Put $\mathcal{I} = \mathcal{I}_m = \{1, 2, \dots m\}$, and let Q_{nm} be an i.d. d.f., defined by ch.f.

$$\exp\Big(\sum_{j=1}^{m} \ln 2 \int (e^{itu} - 1)\, F_{nj}^{(s)}(du)\Big), \quad n, m = 1, 2, \dots$$

Keeping the notation of Section 2, we obtain as in (6):

(21) $$Q_{nm} = \frac{1}{2^m} \sum_{d=0}^{m} \sum_{\nu: d(\nu)=d} \Big[\Big(\prod_{j\in\nu}{}^{*} F^{(s)} * H_{n,\nu}\Big)\Big].$$

D.f. $H_{n,\nu}$ is the convolution of d.f. G_{nj}, where j runs through some subset of the set $\mathcal{I}\setminus\nu$. Each term of (21) may be considered as the d.f. of the sum of independent random variables

$$\sum_{j\in\nu} \xi_{nj}^{(s)} + \rho_{n,\nu}.$$

Denoting $\zeta_{n,\nu} = \sum_{j\in\nu} \xi_{nj}^{(s)}$, we have for any $R>0$:

$$\int_{|u|>R} \varphi(u)\, Q_{nm}(du) = \frac{1}{2^m} \sum_{d=0}^{m} \sum_{\nu: d(\nu)=d} M[\varphi(\zeta_{n,\nu}+\rho_{n,\nu})\lambda(|\zeta_{n,\nu}+\rho_{n,\nu}|>R)].$$

Since $L(F_n, F) \to 0$, and by Lemma 2, there exists a number $h>0$ such that $C(h)>0$, $C'(h)>0$. By the inequality (4), we have

$$M[\varphi(\zeta_{n,\nu}+\rho_{n,\nu})\lambda(|\zeta_{n,\nu}+\rho_{n,\nu}|>R)] \geq A c(h) M[\varphi(\zeta_{n,\nu})\lambda(|\zeta_{n,\nu}|>2R)].$$

Denoting $\bar{\nu} = \mathcal{I}\setminus\nu$, we have

$$\int_{|u|>R} \varphi(u)\, Q_{nm}(du) \geq \frac{A c(h)}{2^{m+1}} \sum_{d=0}^{m} \sum_{\nu: d(\nu)=d} \Delta(\nu),$$

$$\Delta(\nu)=M[\varphi(\zeta_{n,\nu})\lambda(|\zeta_{n,\nu}|>2R)]+M[\varphi(\zeta_{n,\bar\nu})\lambda(|\zeta_{n,\bar\nu}|>2R)].$$

Put

$$l_{nm}=\max_{\nu\subset\mathcal{I}} M\varphi(\zeta_{n,\nu}),\qquad \hat\zeta^{(s)}_{nm}=\sum_{j\in\mathcal{I}}\xi^{(s)}_{nj}.$$

Using independence of random variables $\zeta_{n,\nu}$ and $\zeta_{n,\bar\nu}$, we have for any subset $\nu\subseteq\mathcal{I}$

$$M[\varphi(\zeta_{n,\nu})\lambda(|\zeta_{n,\nu}|>2R)]\geqslant l^{-1}_{nm}M[\varphi(\zeta_{n,\nu})\varphi(\zeta_{n,\bar\nu})\lambda(|\zeta_{n,\nu}|>2R)]$$

Recall that

$$\zeta_{n,\nu}+\zeta_{n,\bar\nu}=\hat\zeta^{(s)}_{nm},\qquad \varphi(\zeta_{n,\nu})\varphi(\zeta_{n,\bar\nu})\geqslant\varphi(\hat\zeta^{(s)}_{nm}).$$

Therefore

$$\Delta(\nu)\geqslant l^{-1}_{nm}M[\varphi(\hat\zeta^{(s)}_{nm})\lambda(|\hat\zeta^{(s)}_{nm}|>4R),$$

and

$$(22)\qquad \int_{|u|>R}\varphi(u)Q_{nm}\geqslant\frac{Ac(h)}{2l_{nm}}M[\varphi(\hat\zeta^{(s)}_{nm})\lambda(|\hat\zeta^{(s)}_{nm}|>4R)].$$

Let us prove that, for each $n=1,2,\ldots$,

$$(23)\qquad l_n=\overline{\lim_{m\to\infty}}\,l_{nm}\leqslant\frac{1}{A^3c'(h)[c(h)]^2}\left[\int\varphi(u)Q_n(du)\right]^2$$

By inequality (5) and (21)

$$\int\varphi(u)Q_{nm}(du)\geqslant\frac{Ac(h)}{2^{m+1}}\sum_{d=0}^{m}\sum_{\nu:d(\nu)=d}M[\varphi(\zeta_{n,\nu})+\varphi(\zeta_{n,\bar\nu})]\geqslant$$

$$\geq \frac{Ac(h)}{2^m}\sum_{d=0}^{m}\sum_{\nu:d(\nu)=d}\sqrt{M\varphi(\zeta_{n,\nu})M\varphi(\zeta_{n,\bar\nu})}\geq Ac(h)\sqrt{M\varphi(\hat\zeta^{(s)}_{nm})}$$

From (5) and the inequality just obtained it follows:

$$\ell_{nm}\leq\left[Ac'(h)\right]^{-1}M\varphi(\hat\xi^{(s)}_{nm})\leq$$

$$\leq\left[Ac'(h)\right]^{-1}\left[(Ac(h))^{-1}\int\varphi(u)Q_{nm}(du)\right]^2.$$

This inequality will imply (23) if we show that

(24) $$\lim_{m\to\infty}\int\varphi(u)\,Q_{nm}(du)=\int\varphi(u)\,Q_n(du).$$

To prove (24) we note that the argument which lead us to (20) show also that

$$\lim_{R\to\infty}\ \sup_{m}\int_{|u|>R}\varphi(u)\,Q_{nm}(du)=0.$$

In order to complete the proof of (24) we recall that $L(Q_{nm},Q_n)\to 0$ as $n\to\infty$. Keeping in mind that (see Section 2) $\xi^{(s)}_n=\xi_n-\xi'_n$ we have by the inequality (4)

$$M[\varphi(\xi^{(s)}_n)\lambda(|\xi^{(s)}_n|>R)]\geq Ac'(h)M[\varphi(\xi_n)\lambda(|\xi_n|>2R)].$$

From this inequality, Lemma 2, (20), (22) and (23) we have

$$\lim_{R\to\infty}\ \sup_{n}\int_{|u|>R}\varphi(u)\,F_n(du)=0$$

Because of the condition $L(F_n,F)\longrightarrow 0$, the sequence $\{M\varphi(\xi_n)\}$ converges to $M\varphi(\xi)$. The proof is completed.

Proof of Theorem 2. It is well-known ([5], p.323) that conditions a) and b) are equivalent to weak convergence of d.f. F_n to d.f. F. Therefore, we need to prove that, if $L(F_n,F)\to 0$, then the convergence of $M\varphi(\xi_n)$ to $M\varphi(\xi)$ is equivalent to the condition c). By Theorem 1, it suffices to prove that c) is equivalent to

$$(25) \quad \lim_{R\to\infty} \sup_n \sum_{j=1}^{m_n} M[\varphi(\xi_{nj}-m_{nj})\lambda(|\xi_{nj}-m_{nj}|>R)]=0,$$

where m_{nj} is the median of ξ_{nj}. Note that condition c) is equivalent to

$$(26) \quad \lim_{R\to\infty} \sup_n \sum_{j=1}^{m_n} M[\varphi(\xi_{nj}-a_{nj})\lambda(|\xi_{nj}-a_{nj}|>R)]=0.$$

From infinite smallness of the random variables ξ_{nj}, $1\le j\le m_n$, it follows that there exists a number $c>0$ such that

$$\sup_n \max_{1\le j\le m_n} |a_{nj}-m_{nj}|\le c.$$

The equivalence of (25) and (26) follows from the inequality

$$M[\varphi(\xi_{nj}-m_{nj})\lambda(|\xi_{nj}-m_{nj}|>R)=$$

$$=M[\varphi(\xi_{nj}-a_{nj}+a_{nj}-m_{nj})\lambda(|\xi_{nj}-a_{nj}+a_{nj}-m_{nj}|>R)]\le$$

$$\le \varphi(a_{nj}-m_{nj})M[\varphi(\xi_{nj}-a_{nj})\lambda(|\xi_{nj}-a_{nj}|>R-c)].$$

Theorem 2 is proved.

9. Proof of Theorem 3. Necessity. In [2], it is demonstrated that, if condition (2) holds, then the weak convergence of d.f. F_n to d.f. Φ is equivalent to conditions a) and b) with $\varphi(u)=u^2$. Therefore, we need only to prove the necessity of condition b). We shall first prove that

$$(27) \quad \lim_{R\to\infty} \sup_n \sum_j \int_{|u|>R} \varphi(u)\, F_{nj}(du)=0.$$

(28) $$\lim_{R\to\infty} \sup_{n} \sum_{j} \int_{|u|>R} \varphi(u)\, \Phi_{n_j}(du) = 0.$$

From Theorem 1, (27) and (28) follow with d.f. F_{n_j} and Φ_{n_j} replaced by

$$\bar{F}_{n_j}(u) = F_{n_j}(u + m_{n_j}) \qquad \text{and} \qquad \bar{\Phi}_{n_j}(u) = \Phi_{n_j}(u + \mu_{n_j}),$$

where m_{n_j} and μ_{n_j} are medians of F_{n_j} and Φ_{n_j}. By (2), $|m_{n_j}| \le 1$, $|\mu_{n_j}| \le 1$ for any $n = 1,2,\ldots$ ([5], p. 256). The relations (27) and (28) follow from relations (27) and (28) for $\bar{F}_{n_j}$ and Φ_{n_j}, if we take into account the inequality (see inequality in the end of the proof of Theorem 2)

$$M[\varphi(\xi_{n_j})\lambda(|\xi_{n_j}|>R)] \le \varphi(m_{n_j}) M\, \varphi(\xi_{n_j} - m_{n_j})\lambda(|\xi_{n_j} - m_{n_j}| > R-1)]$$

and a similar inequality for d.f. Φ_{n_j}, $n, j = 1,2,\ldots$.

It follows from a), (27), (28) that, for any number $\delta > 0$,

(29) $$\lim_{n\to\infty} \Delta_n = 0$$

$$\Delta_n = \Delta_n(\delta) = \sup_j \Big| \int_{|u|>\delta} \varphi(u)[F_{n_j} - \Phi_{n_j}](du) \Big|.$$

Fix a number $\delta > 0$ and define the sets

$$L_n = \{ j : \sigma^2_{n_j} < \varepsilon_n \}, \quad \varepsilon_n = \max\{\sqrt{\Delta_n}, \sqrt{\beta_n}\}, \quad L'_n = \{1, 2, \ldots\} \setminus L_n.$$

In view of (2), the number of elements in L'_n does not exceed ε_n^{-1}. Because of (29),

(30) $$\sum_{j\in L'_n} \int_{|u|>\delta} \varphi(u)[F_{n_j} - \Phi_{n_j}](du)] \le \sqrt{\Delta_n} \longrightarrow 0$$

as $n \to \infty$.

Lemma 5 and the condition a) have as consequence the relation

$$(31) \qquad L\left(\prod_{j\in L_n'}^{*} F_{nj},\ \prod_{j\in L_n'}^{*} \Phi_{nj}\right) \le \sqrt{\beta_n} \to 0,$$

as $n \to \infty$.

Let us prove that

$$(32) \qquad \lim_{n\to\infty} \gamma_n = 0,$$

where

$$\gamma_n = \gamma_n(\delta) = \left| \sum_{j\in L_n} \int_{|u|>\delta} \varphi(u)[F_{n_j} - \Phi_{n_j}](du) \right|.$$

Assume that (32) is not true. Then, for some sequence $\{n'\}$ and some $\alpha > 0$, $\gamma_{n'} \geq \alpha > 0$. On account of (2) the sequence of d.f.

$$G_{n_1} = \prod_{j\in L_n}^{*} F_{nj}, \qquad G_{n_2} = \prod_{j\in L_n}^{*} \Phi_{nj}, \qquad n = 1, 2, \ldots$$

is weakly compact ([5], p. 197). Without loss of generality we may assume that d.f. $G_{n'_1}$ and $G_{n'_2}$ convege to d.f. G_1 and G_2 respectively and

$$\lim_{n'\to\infty} \sum_{j\in L_{n'}} \sigma^2_{n'_j} = \sigma^2.$$

Since $L(F_n, F) \to 0$, from (31) it follows that

$$\lim_{n\to\infty} L(G_{n_1}, G_{n_2}) = 0.$$

Therefore $G_1 = G_2$ is a normal d.f. Using the classical theorem ([5], p. 329) on convergence of sequences of d.f. $\{G_{n'_1}\}$ and $\{G_{n'_2}\}$ to the normal d.f., we obtain

$$(33) \qquad \lim_{n'\to\infty} \sum_{j\in L_{n'}} \int_{|u|>\delta} [F_{n'_j} + \Phi_{n'_j}](du) = 0.$$

Note that the classical result above refers to finite sums of random variables. In our case the number of elements in L_n may be infinite. But all the argumentations of the classical proof may be used in our case too (see also [2]). Because of (27), (28), (33), we conclude that $\gamma_{n'} \to 0$ as $n' \to \infty$. This contradiction proves (32). Necessity of the condition b) follows from (30) and (32).

Proof of sufficiency. Applying. Theorem 1 for normal d.f. Φ_{n_j}, $n, j = 1,2,\ldots$, we obtain (28) again. Let us prove that (27) is also true. Assume that (27) does not hold. Then we may choose sequences $R_m \uparrow \infty$, $n_m \uparrow \infty$ $(m \to \infty)$ and a number $c > 0$ such that

$$(34) \qquad \lim_{R_m \to \infty} \sum_j \int_{|u|>R_m} \varphi(u) F_{n_{m}j}(du) = c > 0.$$

Fix a number $R > 0$ such that

$$\sup_n \sum_j \int_{|u|>R} \varphi(u)\Phi_{n_j}(du) < \frac{1}{8}c.$$

Under the condition b), for $R_m > R$ and for all large enough n_m,

$$\frac{c}{2} \le \sum_j \int_{|u|>R} \varphi(u) F_{n_m j}(du) =$$

$$= \sum_j \int_{|u|>R} \varphi(u)[F_{n_m j} - \Phi_{n_{m_j}}](du) + \sum_j \int_{|u|>R} \varphi(u)\Phi_{n_m j}(du) < \frac{1}{4}c.$$

This contradiction proves (27), Repeating the argument following (28), we pass from (27) to

$$\lim_{R \to \infty} \sup_n \sum_j \int_{|u|>R} \varphi(u) \bar{F}_{n_j}(du) = 0.$$

The latter is the condition of Theorem 1.

To complete the proof of our theorem, we need, on account of

Theorem 1, to prove that conditions a) and b) of our theorem quarantee the weak convergence of d.f. F_n to d.f. Φ. To prove it we note that weak compactness of the sequence of d.f. $\{F_n\}$ follows from (2) ([5], p. 197). It suffices therefore to show that an arbitrary weakly convergent sequence $\{F_{n'}\}$ converges to Φ. Note that (29), (30), (31) follow from condition a), (27), (28). By the condition b) and (30)

$$\lim_{R\to\infty} \sum_{j\in L_n} \int_{|u|>R} \varphi(u)[F_{nj}-\Phi_{nj}](du)=0 \tag{35}$$

Put

$$K_{n1}=\prod_{j\in L_n}^{*} F_{nj},\quad K_{n2}=\prod_{j\in L_n'}^{*} F_{nj},\quad K_{n3}=\prod_{j\in L_n}^{*}\Phi_{nj},\quad n=1,2,\ldots$$

Weak compactness of sequences d.f. $\{K_{n1}\}$, $\{K_{n2}\}$ and $\{K_{n3}\}$ follows from (2) ([5], p. 197). Without loss of generality we may consider sequences d.f. $\{K_{n1}\}$, $\{K_{n2}\}$ and $\{K_{n3}\}$ to be weakly convergent to K_1, K_2 and K_3 respectively and the condition

$$\lim_{n'\to\infty} \sum_{j\in L_{n'}} \sigma_{n'j}^2 = \sigma^2$$

to be satisfied. We have the central limit theorem already mentioned:

$$\lim_{n'\to\infty} \sum_{j\in L_{n'}} \int_{|u|>\delta} \Phi_{n'j}(du)=0.$$

By the criterion of weak convergence of infinitesimal random variables ([5], p. 325), we conclude that d.f. K_1 is i.d. d.f. Note again that, in our case, this criterion may be used even if L_n consists of an infinite number of elements. Thus the weak limit $K_1 * K_2$ for $\{F_{n'}\}$ is i.d. d.f., whose spectral function is concentrated in the interval $(-\delta,\delta)$. The number $\delta>0$ was arbitrary, therefore $K_1 * K_2$ is the normal d.f. From (2) it follows that $K_1 * K_2 = \Phi$.

The theorem is proved.

10. Proof of Theorem 4. Necessity. In order to prove condition a) we repeat the arguments from [4], p. 173. Assume that a) is not true, Therefore there exist a number x_0 which is a point of continuity for function F and such a sequence $\{n_j\}$ that $|F_{n_j}(x_0)-F(x_0)|\geqslant \geqslant \delta>0$ for all n_j. Assume that

$$F_{n_j}(x_0)-F(x_0)\geqslant\delta>0.$$

Select a number $\varepsilon>0$ such that

$$F(x_0+\varepsilon)-F(x_0)<\tfrac{1}{2}\delta.$$

If $x_0\leqslant y\leqslant x_0+\varepsilon$, then

$$F_{n_j}(y)-F(y)\geqslant F_{n_j}(x_0)-F(x_0)-[F(x_0+\varepsilon)-F(x_0)]\geqslant\tfrac{1}{2}\delta;$$

hence, for all n_j,

$$\int|F_{n_j}-F|^p du\geqslant\int_{x_0}^{\varepsilon+x_0}|F_{n_j}-F|^p du\geqslant\varepsilon\left(\frac{\delta}{2}\right)^p.$$

This contraction proves the condition a). On account of Theorem 1 the condition a) and the convergence of moments imply the condition b). Necessity of conditions a) and b) is proved.

Proof of sufficiency. The conditions a) and b) imply convergence of moments. We have as a consequence:

$$\lim_{R\to\infty}\ \sup_{n}\int_{|u|>R}|u|^{\alpha}F_n(du)=0.$$

It follows that

$$\lim_{R\to\infty}\ \sup_{n}\int_{|u|>R}|F_n-F|^p du=0.$$

Fix a number $\varepsilon>0$ and select a number $R>0$ such that

$$\sup_{n} \int_{|u|>R} |F_n - F|^p du < \varepsilon.$$

we have

$$\int |F_n - F|^p du = \int_{-R}^{R} |F_n - F|^p du + \int_{|u|>R} |F_n - F|^p du.$$

By the Lebesque dominated convergence theorem

$$\lim_{n\to\infty} \int_{-R}^{R} |F_n - F|^p du = 0.$$

Thus

$$\overline{\lim_{n\to\infty}} \int |F_n - F|^p du \leq \varepsilon.$$

Now $\varepsilon > 0$ being arbitrary this limit is zero. The theorem is proved.

Proof of Theorem 5. By Theorem 4 we only need to prove the necessity. We shall reduce the proof of necessity to the last theorem. With this in mind we prove that

$$\lim_{n\to\infty} M|\xi_n| = M|\xi|$$

The latter follows from inequalities

$$\left| \int_0^\infty u F_n(du) - \int_0^\infty u F(du) \right| \leq \int_0^\infty |F_n - F| du,$$

$$\left| \int_{-\infty}^0 u F_n(du) - \int_{-\infty}^0 u F(du) \right| \leq \int_{-\infty}^0 |F_n - F| du,$$

which are obtained integrating by parts. The theorem is proved.

11. This section deals with a generalization of I.A.Ibragimov's theorem ([4] , p. 172). To state our theorem we need some additional definitions and assertions.

Let $\{\xi_n\}$ be a sequence of independent identically distributed random variables with d.f. F. Consider a new sequence of random variables

$$\eta_n = \frac{\xi_1 + \xi_2 + \ldots + \xi_{m_n}}{b_n} + a_n , \tag{36}$$

where $b_n > 0$, a_n are some numbers, and a sequence $\{m_n\}$ satisfying the condition

$$m_n < m_{n+1}, \qquad n = 1, 2, \ldots, \qquad \lim_{n\to\infty} \frac{m_{n+1}}{m_n} = r < \infty . \tag{37}$$

Denote by $\mathfrak{A}$ the set of limit d.f. for d.f. of random variables η_n, $n = 1,2,\ldots$. The class $\mathfrak{A}$ is described as follows.

Theorem 6. In order that d.f. $G \in \mathfrak{A}$, it is necessary and sufficient that d.f. G be normal, or ch.f. of d.f. G be of the form

$$g(t) = \exp(i\gamma t + \int' K(t,u)\mu(du)),$$

with the spectral function μ of the form

$$\mu(-u) = u^{-\alpha}\theta_1(\ln u), \qquad u > 0,$$

$$\mu(u) = -u^{-\alpha}\theta_2(\ln u), \qquad u > 0,$$

where $0 < \alpha < 2$, θ_i are periodical functions with a common period such that, for all x and for all $h > 0$,

$$e^{\alpha h}\theta_i(x-h) - e^{-\alpha h}\theta_i(x+h) \geqslant 0,$$

$$d_i \geqslant \theta_i(u) \geqslant c_i \geqslant 0, \qquad c_1 + c_2 > 0,$$

d_i, c_i are some numbers.

Note that numbers α and r satisfy the relation $ra^{\alpha} = 1$,

$$a = \lim_{n \to \infty} \frac{b_n}{b_{n+1}}, \qquad 0 < a < 1.$$

For proof of this theorem see in [11] .

Let $G(x;\alpha)$ be d.f. of the class $\mathfrak{A}$ different from the normal d.f. As in the classical case, we say that d.f. F is in the domain of attraction of d.f. $G(x;\alpha)$ if the distributions of random variables η_n converge to $G(x;\alpha)$ and, a sequence of indices $\{m_n\}$ satisfies the condition (37).

Theorem 7. If d.f. F is in the domain of attraction of d.f. $G(x;\alpha) \in \mathfrak{A}$, then the d.f. F possesses all moments of order δ, $0 \le \delta < \alpha$.

Proof. This theorem is a particular case of the theorem in [12].

The following assertion is an analogue of a lemma from [4] (p. 174). Denote by F_n d.f. of the random variable η_n, by f_n ch.f. of d.f. F_n. If d.f. F is in the domain of attraction of d.f. $G(x;\alpha) \in \mathfrak{A}$, $\alpha > 1$, then we assume that the expectations of random variables η_n are zero.

Lemma 6. If d.f. F is in the domain of attraction of d.f. $G(x;\alpha)$, then there exists a neighbourhood of zero where

$$|1 - f_n(t)| \le C(\delta)|t|^{\delta}, \qquad 0 < \delta < \alpha,$$

($C(\delta) > 0$ does not depend on n).

For the proof of this lemma see [11]).

Repeating the proof of Lemma 5.2.2 [4], p. 174, we obtain the following assertion.

Lemma 7. If d.f. F is in the domain of attraction of d.f. $G(x;\alpha) \in \mathfrak{A}$, then, for any fixed number δ, $0 < \delta < \alpha$, the moments:

$$\int |u|^{\alpha} F_n(du)$$

are uniformly bounded.

We are able now, repeating I.A.Ibragimov's proof, to get a generalization of his theorem.

Theorem 8. If $G(x;\alpha)$ is the limit in L_p-metric, $p>0$ d.f. of d.f. of sums (36) satisfying the condition (37), then it is in the class $\mathfrak{A}$. Conversely, if d.f. F_n of sums (36) converge to d.f. $G(x;\alpha)\in\mathfrak{A}$, then

$$\lim_{n\to\infty} \int \left|F_n(u)-G(u;\alpha)\right|^p du = 0$$

for any $p>\alpha^{-1}$.

REFERENCES

[1] S.N.Bernstein, Collected works, 4, M., "Nauka", 1964.(Russion)

[2] V.M.Zolotarev, A generalization of the Lindeberg-Feller theorem, Teorija veroyatnostei i ee primen. 12, 4 (1967), 666-677 (Russion)

[3] V.V.Petrov, Sums of independent random variables, Moscow, "Nauka", 1972 (Russion)

[4] I.A.Ibragimov,U.V.Linnik, Independent and Stationarily connected variables, Moscow, "Nauka", 1965 (Russion)

[5] M.Loève, Probability theory, Publishing House of Foreign literature, Moscow, 1962 (Russion edition)

[6] V.M.Zolotarev, Asymptotic behaviour of the distributions of processes with independent increments. Teoriya veroyatnostei i ee primen. 10, 1 (1965), 30-50 (Russion)

[7] A.K.Zaremba, Note on the central limit theorem, Math. Z., 69, 3 (1958), 295-298

[8] B.M.Brown, A note on convergence of moments, Ann. Math. Statist., 42, 2 (1971), 777-779

[9] B.M.Brown, G.K.Eaglson, Behaviour of moments of row sums of ele-

mentary systems, Ann. Math. Statist., 41, 6, (1970), 1853-1860

[10] G.Agnew, Global versions of the central limit theorem, Proc. Nat. Sci. (USA), 40 (1954), 800-804

[11] V.M.Kruglov, On an extension of the class of stable distributions, Teoriya veroyatnostei i ee primen. 17, 4 (1972), 723-732 (Russion)

[12] V.M.Kruglov, A note to the theory of infinitely divisible laws, Teoriya veroyatnostei i ee primen. 15, 2 (1970), 330-336 (Russion)

Moscow State University, Moscow

ERGODICITY OF THE DYNAMICAL SYSTEM OF A PARTICLE ON A DOMAIN WITH IRREGULAR WALLS

Izumi Kubo

§1. Introduction

In this report, we consider a dynamical system $\{S_t\}$ resulting from a motion of a material point in a connected domain with piecewise convex and smooth boundary. This system is a generalization of Sinai's "Billiard Problem" [3].

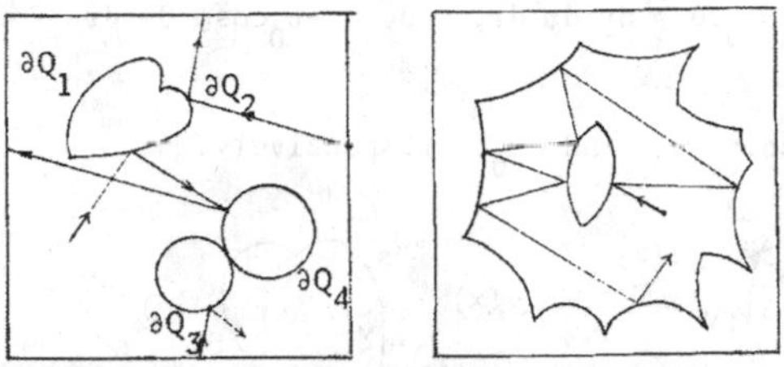

Fig. 1.

Now, we describe our system exactly. Let $\mathbf{T}$ be a two-dimensional torus and Q be a connected domain on $\mathbf{T}$ with piecewise convex and smooth boundary ∂Q, that is, ∂Q consists of walls ∂Q_s ; $s = 1, \cdots, s_0$, which are smooth curves with curvatures $k(q) \geq k_{min} (> 0)$, $q \in \partial Q_s$. We say that $q \in \partial Q$ is an irregular point if ∂Q is not smooth at q.

We denote by $\{S_t\}$ the dynamical system generated by the motion of a material point of mass 1 with unit velocity. The law of the motion is as follows : The point moves along straight lines inside Q and is reflected at ∂Q by the law "the angle of incidence is equal to the angle of reflection". $\{S_t\}$ acts on the unit tangent vector bundle M of Q. Let $\pi(x)$ be the projection of $M = \{x = (q, p) ; q \in Q, |p| = 1\}$ onto Q defined by $\pi(q, p) = q$. Let M_1 be all incidence vectors at ∂Q. M_1 has a natural coordinate system (s, r, ψ) ; r is the arc length in ∂Q_s, ψ is the angle between the incidence vector and the inward normal at $\pi(y)$ relative to Q. Put

$$v(x) \equiv \inf\{t \geq 0 \ ; \ \pi(S_t x) \in \partial Q\}, \quad \tau(x) \equiv \sup\{t < 0 \ ; \ \pi(S_t x) \in \partial Q\}.$$

Almost every element of M can be represented by (s, r, ψ, v). In this representation, the action of $\{S_t\}$ is described by the shift of v-axis and by an automorphism T of M_1 ; $Ty \equiv S_{\tau(y)-0}y$. On the other hand, invariant measures of $\{S_t\}$ and of T are given by

$$d\mu = -\mu_0 \cos\psi \, dv \, d\psi \, dr, \quad d\nu = -\nu_0 \cos\psi \, d\psi \, dr$$

with normalized constants μ_0 and ν_0, respectively.

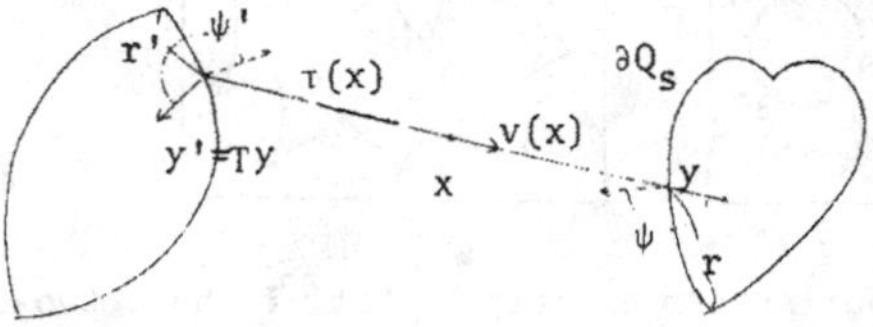

Fig. 2.

It is easily seen that $\{S_t\}$ is ergodic if and only if T is ergodic. The ergodicity of T can be proved in a similar story of "Billiard Problem". In the following sections, we shall explain two important points of the proof. Firstly, we give a simple construction of transversal fibres. Furthermore, we give a partition α of finite entropy which generates $\zeta^{(c)} = \bigvee_{k=0}^{\infty} T^k\alpha$ and $\zeta^{(e)} = \bigvee_{k=-\infty}^{-1} T^k\alpha$, where the elements of $\zeta^{(c)}$ (resp. $\zeta^{(e)}$) are local contracting (expanding) fibres. Secondly, we assert a main lemma which assures the absolute continuity of the canonical maps, under some assumptions for irregular points of walls. Using the lemma repeatedly, we can show that $\bigwedge T^k\zeta^{(c)} = \bigwedge T^k\zeta^{(e)}$ is the trivial partition.

§2. Construction of Transversal Fibres

In the arguments of this section, the numbers of boundaries "s" do not play any important role, and we shall omit it. Put $S = \{(r, \psi)\ ;\ r \in \partial Q, \psi = \frac{\pi}{2}$ or $\frac{3\pi}{2}\} \cup \{(r, \psi)\ ;\ r$ runs over irregular points, $\frac{\pi}{2} \le \psi \le \frac{3\pi}{2}\}$. Then $S \cup TS$ (resp. $S \cup T^{-1}S$) separates M_1 into at most countable connected components $\{X_j^{(c)}\}$ (resp. $\{X_j^{(e)}\}$). The boundary of each $X_j^{(c)}$ (resp. $X_j^{(e)}$) consists of piecewise smooth decreasing (resp. increasing) curves in M_1. Moreover, T^{-1} maps $X_j^{(c)}$ onto some $X_{j'}^{(e)}$, smoothly. We denote the partition $\{X_j^{(c)}\}$ by α. Then $T^{-1}\alpha = \{X_j^{(e)}\}$. The following lemma is elementary but fundamental.

Lemma 1. Let γ be a smooth curve in $M_1 = \{(r, \psi)\ ;\ r \in \partial Q, \frac{\pi}{2} \le \psi \le \frac{3\pi}{2}\}$ and assume that T^{-1} is smooth on γ. Put $\gamma_1 \equiv T^{-1}\gamma$ and $(r_1, \psi_1) \equiv T^{-1}(r, \psi)$ for $(r, \psi) \in \gamma$. Then, we have

$$\frac{d\psi}{dr} = -k(r) - \cfrac{\cos\psi}{\tau(r_1, \psi_1) - \cfrac{\cos\psi_1}{\frac{d\psi_1}{dr_1} - k(r_1)}}$$

$$\frac{dr}{dr_1} = -\frac{\cos\psi_1}{\cos\psi}\left[1 - \frac{\tau(r_1, \psi_1)}{\cos\psi}\left(\frac{d\psi_1}{dr_1} - k(r_1)\right)\right]$$

$$\frac{d\psi}{d\psi_1} = \left[\frac{k(r)}{\cos\psi}(\cos\psi_1 + \tau(r_1, \psi_1)k(r_1)) + k(r_1)\right]\frac{dr_1}{d\psi_1} - \left[1 + \frac{\tau(r_1, \psi_1)k(r)}{\cos\psi}\right]$$

$$\frac{d\psi_1}{dr_1} = k(r_1) + \cfrac{\cos\psi_1}{\tau(r_1, \psi_1) + \cfrac{\cos\psi}{\frac{d\psi}{dr} + k(r)}}$$

$$\frac{dr_1}{dr} = -\frac{\cos\psi}{\cos\psi_1}\left[1 + \frac{\tau(r_1, \psi_1)}{\cos\psi}\left(\frac{d\psi}{dr} + k(r)\right)\right]$$

$$\frac{d\psi_1}{d\psi} = -\left[\frac{k(r_1)}{\cos\psi_1}(\cos\psi + \tau(r_1, \psi_1)k(r)) + k(r)\right]\frac{dr}{d\psi} - \left[1 + \frac{\tau(r_1, \psi_1)k(r_1)}{\cos\psi_1}\right].$$

We say that a monotone curve in M_1 is admissible if the function $\psi = u(r)$ defining γ satisfies $0 \le |\frac{du}{dr}| \le \frac{1}{k_{min}}$. An increasing (resp. decreasing) curve is K-increasing (resp. K-decreasing), if $\frac{1}{K} \le |\frac{du}{dr}| \le \frac{1}{k_{min}}$. An increasing curve belongs to class H(L), if

$$\left|\cos\psi \frac{d}{d\psi}\left(\frac{d\psi}{du} + k(u)\right)^{-1}\right| \le L.$$

We denote by $\theta(\gamma)$ (resp. $\rho(\gamma)$) the total variation of γ with respect to ψ-axis (resp. r-axis). We denote by $\sigma = \sigma_\gamma$ the measure of γ induced by θ. Put

$$d(x) \equiv dis(x, S \cup T^{-1}S),$$

where $dis(x, y)$ means the Euclidean distance between x and $y \in M_1$. The following four facts are easily seen.

(1°) The function $\log d(x)$ is integrable.

(2°) There exists a constant $c_1 > 0$ such that $-\tau(x)\, k_{min} \ge 4c_1 d^2(x)$.

(3°) $D(x) \equiv \sum_{i=1}^{\infty} \frac{2c_2}{d(x_i)} \prod_{j=1}^{i} \frac{1}{(1 + c_1 d^2(x_j))} < \infty$ a.e. x,

where $x_j \equiv T^{-j}x$ and $c_2 = (1 + k_{min}^{-2})^{\frac{1}{2}}$.

(4°) If a connected admissible decreasing curve $\tilde{\gamma}_i \ni x_i$ satisfies $\theta(\tilde{\gamma}_i) \le \frac{1}{2c_2} d(x_i)$, then

$$d(y) \le \frac{1}{2} d(x_i), \quad -\tau(y)k_{min} \ge c_1 d^2(x_i) \quad \text{for } y \in \tilde{\gamma}_i.$$

And hence, T is smooth on $\tilde{\gamma}_i$ and

$$\theta(T\tilde{\gamma}_i) \geqq (1 + c_1 d^2(x_i))\theta(\tilde{\gamma}_i).$$

We now construct a local contracting fibre of x, $D(x) < \infty$. Let $\tilde{\gamma}_n^{(n)}$ be a fixed admissible decreasing curve passing through $x_n \equiv T^{-n}x$. We can take a sequence of connected subcurves $\tilde{\gamma}_i^{(n)} \subset T^{n-i}\gamma_n^{(n)}$ inductively in the following manner,

$$\theta(\tilde{\gamma}_n^{(n)}) = \frac{1}{2c_2} d(x_n), \quad x_n \in \tilde{\gamma}_n^{(n)},$$

$$\theta(\tilde{\gamma}_{i-1}^{(n)}) = \min\left(\frac{1}{2c_2} d(x_{i-1}), \quad (1 + c_1 d^2(x_i))\theta(\tilde{\gamma}_i^{(n)})\right),$$

$$x_{i-1} \in \tilde{\gamma}_{i-1}^{(n)} \subset T\tilde{\gamma}_i^{(n)} \qquad 1 \leq i \leq n.$$

We can easily see inequalities

$$\theta(\tilde{\gamma}_i^{(n)}) \geqq \frac{1}{D(x)} \prod_{j=1}^{i} \frac{1}{(1 + c_1 d^2(x_j))}, \quad 0 \leq i \leq n,$$

by (4°), inductively. Especially, we obtain an estimate

$$\theta(\tilde{\gamma}_0^{(n)}) \geqq \frac{1}{D(x)}.$$

Obviously, T^n is smooth on $\tilde{\gamma}_0^{(n)}$. We now prove that $\tilde{\gamma}_0^{(n)}$ converges uniformly to some connected decreasing curve $\tilde{\gamma}_0(x)$ on the interval $[\psi - \frac{1}{D(x)}, \psi + \frac{1}{D(x)}]$ as $n \to \infty$. Let $\hat{\gamma}^{(n)}$ be a segment of horizontal line with a level $\hat{\psi}$ such that $\hat{\gamma}^{(n)}$ joins $\tilde{\gamma}_0^{(n)}$ and $\tilde{\gamma}_0^{(n+1)}$. Then we can prove that T^{-n} is smooth on $\hat{\gamma}^{(n)}$ and that $d(y) \geq \frac{1}{2} d(x_i)$, for $y \in T^{n-i}\hat{\gamma}^{(n)}$. By Lemma 1, we obtain

$$\rho(\hat{\gamma}^{(n)}) \leqq \frac{1}{|\cos\hat{\psi}|} \prod_{j=1}^{n} [1 + c_1 d^2(x_j)]^{-1} \rho(T^{-n}\hat{\gamma}^{(n)}),$$

$$\sum_n \rho(\hat{\gamma}^{(n)}) \le \frac{\pi^2 D(x)}{|\cos\hat{\psi}|} < \infty.$$

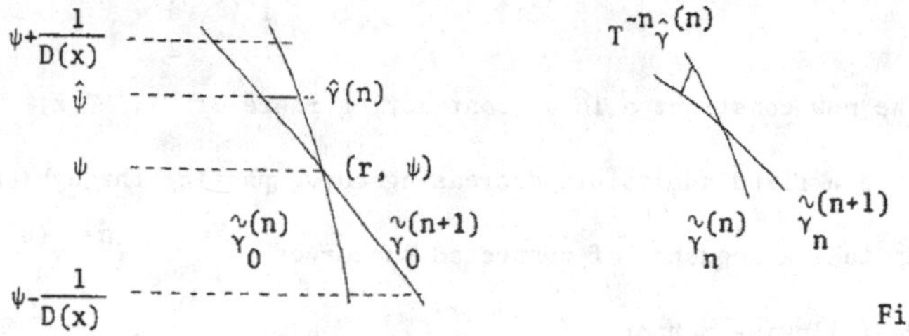

Fig. 3.

The limiting curve $\tilde{\gamma}_0(x)$ is smooth. Put $\Gamma^{(c)}(x) \equiv \bigcup_{n=o}^{\infty} T^n\tilde{\gamma}_0(T^{-n}x)$. Then $\Gamma^{(c)}(x)$ consists of countable smooth decreasing curves. Let $\gamma^{(c)}(x)$ be the connected component of x in $\Gamma^{(c)}(x)$. It can be remarked that $\gamma^{(c)}(x)$ is a solution curve of the differential equation

$$\frac{d\psi}{dr} = -\kappa^{(c)}(r, \psi)\cos\psi + k(r),$$

$$\kappa^{(c)}(r, \psi) \equiv \frac{2k(r)}{\cos\psi} + \cfrac{1}{\tau(r_1, \psi_1)} + \cdots + \cfrac{1}{\cfrac{2k(r_n)}{\cos\psi_n}} + \cfrac{1}{\tau(r_{n+1}, \psi_{n+1})} + \cdots$$

We denote by $\zeta^{(c)}$ the partition of M_1 into $\{\gamma^{(c)}\}$. Noting that T^{-n} is smooth on each element of $\bigvee_{j=0}^{n-1} T^j\alpha$, we can show that

$$\bigvee_{n=0}^{\infty} T^n\alpha = \zeta^{(c)}.$$

Similarly, we can construct $\Gamma^{(e)}$, $\gamma^{(e)}$ and $\zeta^{(e)}$. Combining these facts with a theorem of Rohlin = Sinai [2], we have the following theorem immediately :

Theorem 1.

(i) α is a generator with finite entropy.

(ii) $\bigvee_{k=0}^{\infty} T^k\alpha = \zeta^{(c)}$, $\bigvee_{k=-\infty}^{-1} T^i\alpha = \zeta^{(e)}$, where elements of $\zeta^{(c)}$ (resp. $\zeta^{(e)}$) are locally contracting (resp. expanding) fibres,

(iii) $T^{-1}\zeta^{(c)} > \zeta^{(c)}, \quad T\zeta^{(e)} > \zeta^{(e)}$

(iv) $\bigvee_k T^k\zeta^{(c)} = \bigvee_k T^k\zeta^{(e)} = \varepsilon,$

(v) $\bigwedge_k T^k\zeta^{(c)} = \bigwedge_k T^k\zeta^{(e)} = \pi(T)$

is the measurable covering of $\{\Gamma^{(c)}\}$ and $\{\Gamma^{(e)}\}$.

(vi) $h(T) = h(T^{-1}\zeta^{(c)}|\zeta^{(c)}) = h(T\zeta^{(e)}|\zeta^{(e)})$.

§3. Ergodicity.

By Theorem 1, if $\pi(T)$ is the trivial partition, then $\zeta^{(c)}$ and $\zeta^{(e)}$ are K-partitions. The following Main Lemma plays an essential role in showing that $\pi(T) = \bigwedge_k T^k\zeta^{(c)} \wedge \bigwedge_k T^k\zeta^{(e)}$ is trivial. The lemma is similar to Theorem 6.1 of Sinai [3]. But, it contains more general and more delicate assertions than Sinai's. We now assume some assumptions for irregular points of ∂Q.

Assumption A.

(i) Walls ∂Q_s, $s = 1,\cdots, s_0$, do not tangentially touch each other.

(ii) There are no triple of irregular points which lie on the same straight line inside Q.

(iii) For any irregular point $r \in \partial Q$, $|\tau(r, \psi)| \geq \frac{2}{k_{min}}$.

Under the above assumption, there exist a natural number ℓ_1 and a positive number λ, $0 < \lambda < 1$, such that

$$\prod_{j=1}^{n} [1 - \tau(r_j, \psi_j)k_{min}]^{-1} \leq \lambda_1^n \quad \text{for } n \geq \ell_1.$$

Moreover, there exists a positive number L_0 which assures that $\gamma^{(e)} \in H(L_0)$ and that $T^{-n}\gamma \in H(L_0)$ for increasing curve $\gamma \in H(L_0)$, $n \geq \ell_1$.

A (K-) <u>quadrilateral</u> G is a domain whose boundary consists of four piecewise monotone curves, one pair of opposite curves being (K-) increasing and the other being (K-) decreasing.

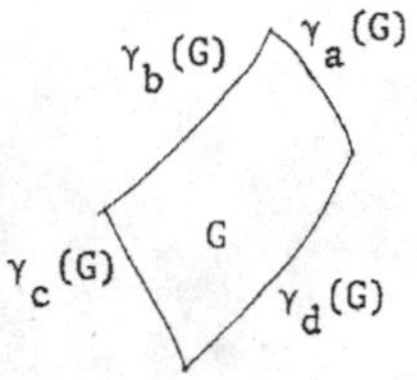

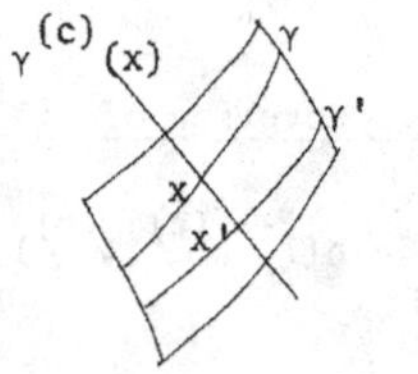

Fig. 4.

<u>Lemma 2</u>. For any $\delta(0 < \delta < 1)$, $\omega(0 < \omega)$ and $\Omega(1 \leq \Omega)$, there exists a natural number $\ell_0 = \ell_0(\delta, \omega, \Omega)$ which has the following property with $K(\omega) = (1 + \frac{1}{4} c_1^{-1} \omega^{-2}) \max k(r)$:

Let G be a $K(\omega)$-quadrilateral for which $\gamma_b(G)$, $\gamma_d(G) \in H(L_0)$,

$$\frac{1}{\Omega} \leq \frac{\theta(\gamma_a(G))}{\theta(\gamma_b(G))} \leq \Omega, \quad \text{dis } (G, S) \geq \omega$$

$$\text{dis } (T^{-m}G, S) \geq \lambda^m \theta(\gamma_b(G)) \qquad m = 0, 1, \cdots, \ell_0$$

and $T^{-\ell_0}G$ is a quadrilateral. Then there exists a subset $G^{(\delta)}$ of G such that

(a) For any $x \in G^{(\delta)}$, $\gamma^{(c)}(x)$ joins $\gamma_b(G)$ and $\gamma_d(G)$, and $\gamma^{(c)}(x) \cap G = \gamma^{(c)}(x) \cap G^{(\delta)}$ holds.

(b) $\nu(G^{(\delta)}) \geq (1 - \delta)\nu(G)$.

(c) Let γ, γ' $(\subset G)$ be any pair of $K(\omega)$-increasing curves joining $\gamma_a(G)$ and $\gamma_c(G)$. Then the canonical map

$$
\begin{array}{rcl}
\Psi : \gamma \cap G^{(\delta)} & \to & \gamma' \cap G^{(\delta)} \\
\cup\!\!\!\shortmid & & \cup\!\!\!\shortmid \\
x & \to & \gamma' \cap \gamma^{(c)}(x)
\end{array}
$$

is absolutely continuous and satisfies the inequality

$$\frac{1}{\beta} \le \frac{d\Psi\sigma_{\gamma'}}{d\sigma_{\gamma}} \le \beta$$

with some $\beta = \beta(\omega, \Omega)$.

Applying Lemma 2 repeatedly, we can prove that $\pi(T)$ is the trivial partition. The complete proof will be presented separately (together with the proof of Lemma 2).

Theorem 2. Under Assumption (A), T and $\{S_t\}$ are K-systems.

REFERENCES

[1] Kubo, I ; Review on ergodic theory, Research Inst. Math. Sci., Seminar Report 56 (1968), 6-29. (in Japanese)

[2] Rohlin, V. A. and Sinai, Ya. G. ; Construction and properties of invariant measurable partitions, Soviet Math. Dokl. Vol. 2 No. 6 (1961), 1611-1614.

[3] Sinai, Ya. G. ; Dynamical systems with elastic reflections, Russian Math. Surveys Vol. 25, No. 2 (1970), 137-189.

Mathematical Institute,
Faculty of Science,
Nagoya University.

ON "ATTRACTION DOMAINS" IN THE THEORY OF SEQUENTIAL ESTIMATION

Yu.V.Linnik, L.B.Klebanov, A.L.Rukhin

In the theory of sequential estimation the exact determination of optimal in some sense (if any) procedure is in general a very difficult problem. It seems natural therefore to investigate asymptotically optimal procedures when the cost of observation tends to zero. Here we observe some phenomena resembling the situation in the asymptotical theory of summation of independent random variables. More precisely, it is convenient to speak about "attraction domains" of asymptotically optimal procedures to different types of sequential decisions. In this paper, we shall give two such "attraction domains" which are obtained under various assumptions about the mean observation time.

Let $(\mathcal{X}, \mathfrak{A}, P_\theta)$ be a probability space with a family of measures $\{P_\theta\}$ depending on a real parameter $\theta \in \Theta \subset R^1$ which we have to estimate from a given sequence of independent variables $x_1, x_2, \dots$ distributed according to one of P_θ's with limitations of two types on the mean observation time: $E_\theta \tau \le n$ for all $\theta \in \Theta$ or $\int E_\theta \tau \, d\mu(\theta) \le n$ (Here τ denotes a Markovian stopping rule).

Let us fix differentiable nonnegative even and convex functions $W_1, \dots, W_K$, $W_j(0)=0$, $j=1,\dots,K$, which will serve as loss functions, and convex functions $\Psi_1,\dots,\Psi_K$. The quality of sequential estimator $(\tau, T_\tau(x_1,\dots,x_\tau)) = (\tau, T_\tau(z_\tau))$ where T_τ is the terminal decision will be characterized by the quantity

(1) $\mathcal{F}\left(\int \Psi_1(E_\theta W_1(T_\tau(z_\tau)-\theta))\,d\mu_1(\theta),\ldots,\int \Psi_K(E_\theta W_K(T_\tau(z_\tau)-\theta))\,d\mu_K(\theta)\right)$

where $\mathcal{F}$ is a continuous function monotonically increasing in each of its arguments, $\mathcal{F}(0,\ldots,0)=0$ and $\mu_1,\ldots,\mu_K$ are absolutely continuous mutually equivalent probability measures on Θ. We were led to consideration of quantities of type (1) by some problems of unbiased sequential estimation of the parameter θ which we shall study separately.

Theorem 1. Let the following conditions be satisfied:

(2) $$0<\gamma_j=\lim_{t\to 0}\frac{W_j'(t)}{t^{\sigma_j}}<\infty,$$

$$\overline{\lim_{t\to\infty}}\frac{W_j'(t)}{t^{\tau}}<\infty,\quad j=1,\ldots,K,$$

(3) $$0<\delta_j=\lim_{t\to 0}\frac{\Psi_j(t)}{t}<\infty,$$

(4) $$0<\lim_{\substack{t\to 0\\ t_i=t_j=t}}\frac{\mathcal{F}(0,\ldots,t_i,\ldots,0)}{\mathcal{F}(0,\ldots,t_j,\ldots,0)}=\rho_{ij},\quad i,j=1,\ldots,K,$$

(5) $$\int |t|^{\tau}\,d\mu_j(t)<+\infty,\quad j=1,\ldots,K,$$

where τ, σ_j, $j=1,\ldots,n$, are constants. Assume that $p(x,\theta)$ – the density of P_θ – is positive, twice continuously differentiable in θ,

$$E_\theta \sup_{|\theta-\theta_1|>\varepsilon} \log\frac{p(x,\theta)}{p(x,\theta_1)}<0$$

and

$$E_\theta \sup_{|\theta-\theta_1|<\varepsilon} \frac{\partial^2 \log p(x,\theta)}{\partial\theta^2}<\infty.$$

Then

(6) $$\lim_{n\to\infty}\frac{\inf\mathcal{F}\left(\int\Psi_1(E_\theta W_1(T_\tau(z_\tau)-\theta))\,d\mu_1(\theta),\ldots,\int\Psi_K(E_\theta W_K(T_\tau(z_\tau)-\theta))\,d\mu_K(\theta)\right)}{\mathcal{F}\left(\int\Psi_1(E_\theta W_1(\tilde{T}_{[n]}(z_{[n]})-\theta))\,d\mu_1(\theta),\ldots,\int\Psi_K(E_\theta W_K(\tilde{T}_{[n]}(z_{[n]})-\theta))\,d\mu_K(\theta)\right)}=1$$

where the infimum is taken over all procedures (τ, T_τ) with the condition

$$(7) \qquad E_\theta \tau \leq n$$

and $\widetilde{T}_m$ is the maximum likelihood estimator of θ based on the sample of size m, or the Bayes estimator of θ, the prior distribution being one of $\mu_j,\ j = 1, \dots, K$.

Theorem 2. Let the conditions of Theorem 1 be satisfied and

$$(8) \qquad \frac{\partial}{\partial t_i} \mathcal{F} \Big|_{t_1 = 0, \dots, t_K = 0} \neq 0, \quad i = 1, \dots, K.$$

Let

$$(9) \qquad E\tau = \int E_\theta \tau \, d\mu(\theta) \leq n$$

where μ is a measure equivalent to all the measures μ_j, $j = 1, \dots, K$. Then

$$(10) \qquad \lim_{n \to \infty} \frac{\inf \mathcal{F}\left(\int \Psi_1(E_\theta W_1(T_\tau(z_\tau) - \theta))\, d\mu_1(\theta), \dots, \int \Psi_K(E_\theta W_K(T_\tau(z_\tau) - \theta))\, d\mu_K(\theta)\right)}{\mathcal{F}\left(\int \Psi_1(E_\theta W_1(\widetilde{T}_{\tau_0}(z_{\tau_0}) - \theta)\, d\mu_1(\theta), \dots, \int \Psi_K(E_\theta W_K(\widetilde{T}_{\tau_0}(z_{\tau_0}) - \theta))\, d\mu_K(\theta)\right)} = 1$$

where the infimum is taken over all estimators T_τ which satisfy the condition (9) and $(\tau_0, \widetilde{T}_{\tau_0})$ is a procedure, τ_0 being defined in (25) and $\widetilde{T}_m$ being the same as in Theorem 1.

Note that, considering the quantity (1), we are beyond the bounds of classical theory of Wald.

In the particular case when $K = 1$, Ψ and $\mathcal{F}$ are linear functions, we get results close to those of [1,2].

Proof of Theorem 1. We shall show first that, for every procedure (τ, T_τ), satisfying the condition $E_\theta \tau \leq n$ the following inequality holds true:

$$(11) \qquad \lim_{n \to \infty} \frac{\int_\Theta E_\theta W_j(T_\tau(z_\tau) - \theta)\, d\mu_j(\theta)}{\int_\Theta E_\theta W_j(\widetilde{T}_{[n]}(z_{[n]}) - \theta)\, d\mu_j(\theta)} \geq 1, \quad j = 1, \dots, K.$$

If for procedure (τ, T_τ)

$$\varlimsup_{n\to\infty} \frac{\int_\Theta E_\theta W_j(T_\tau(z_\tau)-\theta)\, d\mu_j(\theta)}{\int_\Theta E_\theta W_j(\tilde{T}_{[n]}(z_{[n]})-\theta)\, d\mu_j(\theta)} \le 1$$

for some j , then

$$\int_\Theta P_\theta\{\tau \le \sqrt{n}\}\, d\mu_j(\theta) = o(1), \quad n\to\infty. \tag{12}$$

In fact, we can assume that, under the condition $\tau=[n]$. T_τ is a Bayes estimator based on a sample of size n , and hence

$$E_\theta\{W_j(T_\tau(z_\tau)-\theta)/\tau=m\} \ge E_\theta\{W_j(T_\tau(z_\tau)-\theta)/\tau=s\} \tag{13}$$

for $s \ge m$. From (13), we deduce:

$$\frac{\int_\Theta E_\theta W_j(T_\tau(z_\tau)-\theta)\, d\mu_j(\theta)}{\int_\Theta E_\theta W_j(\tilde{T}_{[n]}(z_{[n]})-\theta)\, d\mu(\theta)} = \frac{\int_\Theta \sum_i E_\theta\{W_j(T_\tau(z_\tau)-\theta)/\tau=i\}\, P_\theta\{\tau=i\}\, d\mu_j(\theta)}{\int_\Theta E_\theta W_j(\tilde{T}_{[n]}(z_{[n]})-\theta)\, d\mu_j(\theta)} \ge \tag{14}$$

$$\ge \frac{\int_\Theta E_\theta\{W_j(T_\tau(z_\tau)-\theta)/\tau=\sqrt{n}\}\, P_\theta\{\tau\le n\}\, d\mu_j(\theta)}{\int_\Theta E_\theta W_j(\tilde{T}_{[n]}(z_{[n]})-\theta)\, d\mu_j(\theta)}.$$

In [1] , it was shown that, if

$$y_n = \min_{d\in R^1}\Big\{\int_\Theta W_j(\theta-d)\prod_1^n f(z_i,\theta)\Big[\int_\Theta \prod_1^n f(z_j,\eta)\, d\mu_j(\eta)\Big]^{-1} d\mu_j(\theta)\Big\},$$

then

$$n^{\frac{\sigma_j+1}{2}}\, y_n \longrightarrow \frac{\alpha_{\sigma_j+1}}{[I(\theta)]^{\sigma_j+1}}. \tag{15}$$

Here α_{σ_j+1} is the (σ_j+1) -th moment of the standard normal law and $I(\theta)$ the Fisher information. An analogous result holds for maximum likelihood estimators.

These results imply:

$$\int_\Theta E_\theta\{W_j(T_\tau(z_\tau)-\theta)/\tau\le\sqrt{n}\}\, P_\theta\{\tau\le n\}\, d\mu_j(\theta) \sim \tag{16}$$

$$\sim \frac{\int\limits_{\Theta} \frac{\alpha_{\sigma_j+1}}{[I(\theta)]^{\sigma_j+1}} P_\theta(\tau \le n)\, d\mu_j(\theta)}{n^{\frac{\sigma_j+1}{4}}},$$

so that

$$(17) \quad \int\limits_{\Theta} E_\theta W_j(\tilde{T}_{[n]}(Z_{[n]})-\theta)\, d\mu_j(\theta) \sim \frac{\int\limits_{\Theta} \frac{\alpha_{\sigma_j+1}}{[I(\theta)]^{\sigma_j+1}}\, d\mu_j(\theta)}{n^{\frac{\sigma_j+1}{2}}}.$$

Since the Fisher information quantity is bounded and separated from zero, with the help of (14) from (16) and (17), we get formula (12).

Now we can deduce from (15) that

$$\varliminf_{n\to\infty} \frac{\int\limits_{\Theta} E_\theta W_j(T_\tau(Z_\tau)-\theta)\, d\mu_j(\theta)}{\int\limits_{\Theta} E_\theta W_j(\tilde{T}_{[n]}(Z_{[n]})-\theta)\, d\mu_j(\theta)} \geqslant$$

$$\geqslant \varliminf_{n\to\infty} \frac{\sum\limits_{i=[\sqrt{n}]}^{\infty} \int\limits_{\Theta} E_\theta\{W_j(T_\tau(Z_\tau)-\theta)/\tau=i\} P_\theta\{\tau=i\}\, d\mu_j(\theta)}{\int\limits_{\Theta} E_\theta W_j(\tilde{T}_{[n]}(Z_{[n]})-\theta)\, d\mu_j(\theta)} =$$

$$= \varliminf_{n\to\infty} \frac{\sum\limits_{i=[\sqrt{n}]}^{\infty} \int\limits_{\Theta} \frac{\alpha_{\sigma_j+1}}{[I(\theta)]^{\sigma_j+1}} P_\theta\{\tau=i\}\, d\mu_j(\theta) \frac{1}{i^{\frac{\sigma_j+1}{2}}}}{\int\limits_{\Theta} \frac{\alpha_{\sigma_j+1}}{[I(\theta)]^{\sigma_j+1}}\, d\mu_j(\theta) \frac{1}{n^{\frac{\sigma_j+1}{2}}}}$$

and the condition $\sum\limits_{i=1}^{\infty} i\, P_\theta(\tau=i) \le n$ must hold.

Let us find the minimal value of

$$(18) \quad \sum_{i=[\sqrt{n}]}^{\infty} \int\limits_{\Theta} \frac{P_\theta\{\tau=i\}}{[I(\theta)]^{\sigma_j+1}}\, d\mu_j(\theta) \frac{1}{i^{\frac{\sigma_j+1}{2}}}$$

subject to the conditions

$$(19) \quad \sum_{i=1}^{\infty} i P_\theta\{\tau=i\} \le n, \quad \sum_{i=1}^{\infty} P_\theta\{\tau=i\}=1, \quad P_\theta\{\tau=i\} \ge 0.$$

We have thus a linear optimization problem to minimize (18) subject to conditions (19). It is clear from geometric reasons that the minimum in this problem is obtained if

$$(20) \quad P_\theta\{\tau \in [[n],[n]+1]\}=1, \quad \theta \in \Theta.$$

Let now (τ, T_τ) be any procedure with $E_\theta \tau \leq n$. From the previous results, we obtain

(21) $$\int_\Theta E_\theta w_j(T_\tau(z_\tau)-\theta)\,d\mu_j(\theta) \geq \int_\Theta E_\theta w_j(\tilde{T}_{[n]}(z_{[n]})-\theta)\,d\mu_j(\theta)[1+o(1)].$$

From the monotonicity property of $\mathcal{F}$ and by Jensen's inequality, we get

(22) $$\mathcal{F}\Big(\int_\Theta \Psi(E_\theta w_1(T_\tau(z_\tau)-\theta))\,d\mu_1(\theta),\ldots,\int_\Theta \Psi_K(E_\theta w_K(T_\tau(z_\tau)-\theta))\,d\mu_K(\theta)\Big) \geq$$

$$\geq \mathcal{F}\Big(\Psi_1\Big(\int_\Theta E_\theta w_1(\tilde{T}_{[n]}(z_{[n]}))-\theta)\,d\mu_1(\theta)[1+o(1)]\Big),\ldots,\Psi_K\Big(\int_\Theta E_\theta w_K(\tilde{T}_{[n]}(z_{[n]})-\theta)\,d\mu_K(\theta)[1+o(1)]\Big)\Big)\sim$$

$$\sim \mathcal{F}\left(\delta_1 \frac{\int_\Theta \frac{\alpha_{\sigma_1+1}}{[I(\theta)]^{\sigma_1+1}}\,d\mu_1(\theta)}{n^{\frac{\sigma_1+1}{2}}},\ldots,\delta_K \frac{\int_\Theta \frac{\alpha_{\sigma_K+1}}{[I(\theta)]^{\sigma_K+1}}\,d\mu_K(\theta)}{n^{\frac{\sigma_K+1}{2}}}\right).$$

On the other hand,

(23) $$\mathcal{F}\Big(\int_\Theta \Psi_1(E_\theta w_1(\tilde{T}_{[n]}(z_{[n]})-\theta))\,d\mu_1(\theta),\ldots,\int_\Theta \Psi_K(E_\theta w_K(\tilde{T}_{[n]}(z_{[n]})-\theta))\,d\mu_K(\theta)\Big)\sim$$

$$\sim \mathcal{F}\left(\delta_1 \frac{\int_\Theta \frac{\alpha_{\sigma_1+1}}{[I(\theta)]^{\sigma_1+1}}\,d\mu_1(\theta)}{n^{\frac{\sigma_1+1}{2}}},\ldots,\delta_K \frac{\int_\Theta \frac{\alpha_{\sigma_K+1}}{[I(\theta)]^{\sigma_K+1}}\,d\mu_K(\theta)}{n^{\frac{\sigma_K+1}{2}}}\right).$$

Comparison of (22) and (23) completes the proof of Theorem 1.

We shall give now the sketch of the proof of Theorem 2. Consider the problem of minimization of

(24) $$\mathcal{F}\Big(\int_\Theta \Psi_1(E_\theta W_1(T_\tau(z_\tau)-\theta))\,d\mu_1(\theta),\ldots,\int_\Theta \Psi_K(E_\theta W_K(T_\tau(z_\tau)-\theta))\,d\mu_K(\theta)\Big)+cE\tau \quad (c>0).$$

Let $(\tau(c),\ T^*_{\tau(c)}(z_{\tau(c)}))$ be a solution of this problem. It will be clear from the following that $E\tau(c)$ is a continuous function of c and $E\tau(c)\to\infty$, $c\to 0$.

Hence by an appropriate choice of c we can achieve the equality $E\tau(c)=n$ at least for sufficiently large values of n. For this value of c, the procedure $(\tau(c),\ T_{\tau(c)}(z_{\tau(c)}))$ will be a solution of the initial problem as well. Thus we can restrict ourselves to considering the quantity (24).

It may be shown as in the proof of Theorem 1 and Theorem 2.1 of [1] that the problem of minimization of (24) can be solved by determining the minimal value of

$$\mathcal{F}\Big(\delta_1 \int_{\Theta} \frac{\alpha_{\sigma_1+1}}{[I(\theta)]^{\sigma_1+1} m^{\frac{\sigma_1+1}{2}}}[1+o(1)]\,d\mu_1(\theta),\dots$$

$$\dots,\delta_K \int_{\Theta} \frac{\alpha_{\sigma_K+1}}{[I(\theta)]^{\sigma_K+1} m^{\frac{\sigma_K+1}{2}}}[1+o(1)]\,d\mu_K(\theta)\Big) + c\int_{\Theta} m\,d\mu(\theta).$$

From the assumptions of Theorem 2, we see that, for all $\theta \in \Theta$, it is necessary to minimize

$$\sum_{j=1}^{K} \delta_j \frac{\alpha_{\sigma_j+1}}{[I(\theta)]^{\sigma_j+1} m^{\frac{\sigma_j+1}{2}}} \mu_j'(\theta)[1+o(1)] + cm\mu'(\theta)$$

(here μ_j', μ' are the densities of measures μ_j and μ, m depends on θ).

As in Theorem 2.1 of [1] , we deduce that the following stopping rule

$$(25)\qquad \tau_0 = \min\Big\{ \ell : c\mu'(\hat\theta_\ell(z_\ell)) - \sum_{j=1}^{K} \delta_j \frac{\sigma_j+1}{2} \frac{\alpha_{\sigma_j+1}}{[I(\hat\theta_\ell(z_\ell))]^{\sigma_j+1}} \times$$

$$\times \frac{\mu_j'(\hat\theta_\ell(z_\ell))}{\ell^{\frac{\sigma_j+3}{2}}} \geqslant 0 \Big\}$$

has the property stated in Theorem 2. In (25) $\hat\theta_\ell(z_\ell)$ is a maximum likelihood estimator or Bayes estimator based on the sample of size ℓ . The value of c is chosen so that $E\tau_0 = n$.

References

[1] P.J. Bickel, J.A. Yahav, Some contributions to the asymptotic theory of Bayes solutions, Z. Wahrscheinlichkeitstheorie verw. Geb. 11, 4(1969), 258-276.

[2] P.J. Bickel, J.A. Yahav, Asymptotically pointwise optimal procedures in sequential analysis, Proc. 5-th Berkeley Sympos. math. Statist. Probability, 1965.

[3] Yu.V. Linnik, I.V. Romanovskii, On the theory of sequential estimation, Soviet Math. Dokl., 5(1970), 1196-1198.

Steklov Mathematical Institute
of the Academy of Sciences of the USSR
Leningrad

APPLICATIONS OF ORNSTEIN'S THEORY TO STATIONARY PROCESSES

G. Maruyama

Ornstein's theory developed in a series of recent papers has attained a great advance in ergodic theory, providing directly or indirectly solutions to long-standing unsolved problems. It would be natural that one can draw from his theory conclusions that throw light on classical problems for stationary processes.

Among Gaussian stationary processes, characteristics used for distinguishing metrical types of generated automorphisms were spectral measures ; if the spectral measures of the given two Gaussian stationary processes with zero means are absolutely continuous each other, then they generate isomorphic automorphisms. However, many essential cases escape from this criterion, which in its essence rests on linear theory of stationary processes.

For instance, suppose we are given two discrete-time Gaussian stationary processes with spectral measures having densities relative to Lebesgue measure and suppose further that they have separate supports. What can be said about isomorphism for generated automorphisms ? Although, the above criterion based on spectral measures can never answer to the question, Ornstein's theory provides a satisfactory answer that they are Bernoulli automorphisms as factors of Bernoulli automorphisms having infinite entropy and therefore isomorphic.

The notion of weak Bernoulli automorphism that plays a fundamental role in his theory is concerned with a sort of mixing conditions expressed in terms of discrepancy between σ-algebras, as a measure of independency. Let $(\Omega, \mathcal{B}, m)$ be the basic probability measure space isomorphic with [0, 1] endowed with ordinary Lebesgue measure, and $\mathcal{P}$, $\mathcal{Q}$ be measurable partitions (finite or infinite) of Ω. Use has been made of the quantity

$$V(\mathcal{P},\mathcal{Q}) = \int_{\mathcal{P}} m(dp) \int_{\mathcal{Q}} | m(dq \mid p) - m(dq) |$$

to measure dependency between $\mathcal{P}$, $\mathcal{Q}$, where $m(dq|\,p)$ signifies the measure conditional under $\mathcal{P}$, so that if $\mathcal{P}$, $\mathcal{Q}$ are finite partitions the right-hand member is reduced to the form

$$\sum_{p \in \mathcal{P},\ q \in \mathcal{Q}} | m(p \cap q) - m(p)m(q) | ,$$

where p, q are respectively $\mathcal{P}$-, $\mathcal{Q}$-atoms. $V(\mathcal{P},\mathcal{Q})$ is then monotone increasing in $(\mathcal{P},\mathcal{Q})$, i.e.

(1) $\quad V(\mathcal{P},\mathcal{Q}) \leq V(\mathcal{P}',\mathcal{Q}')$

if $\mathcal{P} \subset \mathcal{P}'$, $\mathcal{Q} \subset \mathcal{Q}'$, where $\mathcal{P} \subset \mathcal{P}'$ means that $\mathcal{P}'$ refines $\mathcal{P}$, and further $V(\mathcal{P},\mathcal{Q})$ satisfies that

(2) $\quad \lim_{m,n\to\infty} V(\mathcal{P}_m, \mathcal{Q}_n) = V(\mathcal{P},\mathcal{Q})$,

if $\mathcal{P}_n \uparrow \mathcal{P}$, $\mathcal{Q}_n \uparrow \mathcal{Q}$.

Consider an automorphism T acting on the above probability space. $(T,\mathcal{P})$ is called a weak Bernoulli pair if

$$\Delta_k = V(\mathcal{P}_k^{\infty}, \mathcal{P}_{-\infty}^{0}) \to 0 , \quad \text{as} \quad k \to \infty ,$$

where

$$\mathcal{P}_a^b = \bigvee_{i=a}^{b} T^i \mathcal{P} \quad (a \leq b) .$$

In this connection, recall mixing conditions proposed by several authors before Ornstein in the study of stationary processes. Suppose we are given a real stationary process $X = (x_n, -\infty < n < \infty)$ on a probability space $(\Omega , \mathcal{A}, P)$, and consider an automorphism T acting on the σ-algebra

$$\mathcal{A}_{-\infty}^{\infty} = \lim_{\substack{a\to -\infty \\ b\to\infty}} \mathcal{A}_a^b ,$$

where $\mathcal{A}_a^b$ $(a \leq b)$ is the σ-algebra generated by $\{x_n , a \leq n \leq b\}$.

Clearly every $\mathcal{A}_n^n$ is its generator.

Among mixing conditions under consideration, the weakest is

$$(R_e) \qquad \bigcap_n \mathcal{A}_{-\infty}^n = \text{trivial } \sigma\text{-algebra},$$

X subject to which being called a regular process.

Now we may ask under what conditions would it be possible to write X in the form

$$(3) \qquad x_n = f(\cdots, \theta^n \xi_{-1}, \theta^n \xi_0, \theta^n \xi_1, \cdots) ,$$

a function of independent identically distributed ξ_n , associated with shifts $\theta^n \xi_k = \xi_{n+k}$? First of all, an easy application of [4], [5] leads us to the

Proposition 1. X is capable of the representation (3) if and only if the generated automorphism T is weak Bernoulli.

Let $\mathcal{P}$ be a measurable partition of Ω generated by $\mathcal{B}$-measurable sets A_n , $1 \leqq n < \infty$, i.e. a partition whose atom is defined by

$$p = \bigcap_{n=1}^{\infty} A_n^{\varepsilon_n} , \qquad \varepsilon_n = 1 \text{ or } 0 ,$$

and define a stationary process $X = \{x_n(\omega) , -\infty < n < \infty\}$, where $x_n(\omega) = f(T^{-n}\omega)$, and

$$f(\omega) = \sum_{n=1}^{\infty} \frac{A_n(\omega)}{3^n} ,$$

$A_n(\omega)$ the indicator function of A_n . Then the σ-algebra generated by $x_m , x_{m+1} , \cdots , x_n$ $(m \leqq n)$ is $\mathcal{P}_m^n$. If in particular $\mathcal{P}$ is a generator of T, then T is isomorphic with the automorphism generated by X . Also , if T is a Kolmogorov automorphism with defining increasing partition $\mathcal{P}$, i.e.

$$\mathcal{P} \subset T\mathcal{P} , \qquad \bigwedge_n T^n \mathcal{P} = \text{trivial partition},$$

$$\bigvee_n T^n \mathcal{P} = \text{partition into individual points} ,$$

then X defined as above satisfies (R_e) . Thus , employing Ornstein's result [6] that there exists a Kolmogorov automorphism which is not weak

Bernoulli , we are led to the

Proposition 2. There is a regular stationary process which does not admit the representation (3)

The condition

$$(R_0) \qquad \lim_{k\to\infty} \sup_{A\in\mathcal{A}_k^\infty,\ B\in\mathcal{A}_{-\infty}^0} |P(A\cap B) - P(A)P(B)| = 0$$

proposed by Rosenblatt as a stronger mixing condition than (R_e) , has proved useful for obtaining central limit theorems of stationary processes , but its significance in our problem has never been clarified. A more stronger condition was introduced by Ibragimov for the same purpose as (R_0) :

$$(I) \qquad \varphi(k) = \sup_{A\in\mathcal{A}_k^\infty} \text{ess. sup} \, |P(A \mid \mathcal{A}_{-\infty}^0) - P(A)| \to 0 , \quad \text{as } k\to\infty .$$

Rewriting the weak Bernoulli conditon with $\mathcal{A}_0^0$ as generator for the generated automorphism T by X we have a sufficient condition [8] for the representation (3) . Δ_k is expressed in the form

$$\Delta_k = \int_\Omega 2\alpha_{-\infty}^{-k} P(d\omega) ,$$

where

$$2\alpha_{-\infty}^{-k} = 2 \sup_{A\in\mathcal{A}_0^\infty} |\mathcal{Q}(A)| = \text{total variation of } \mathcal{Q}(A) \text{ over } \mathcal{A}_0^\infty ,$$

$$\mathcal{Q}(A) = P(A \mid \mathcal{A}_{-\infty}^{-k}) - P(A).$$

Then the weak Bernoulli condition is

$$(C_1) \qquad \lim_{k\to\infty} \Delta_k = 0 , \quad \text{or equivalently} \quad P(\lim_{k\to\infty} \alpha_{-\infty}^{-k} = 0) = 1 .$$

Since

$$\Delta_k \leqq \varphi(k) ,$$

(I) is stronger than (C_1) .

As was observed in [8] , when applied to a stationary Markov chain $\{x_n , -\infty < n < \infty\}$ with state space S , k-step transition probability $p^{(k)}(x,dy)$

and stationary probability measure $p(dx)$, (C_1) leads us to the condition

$$(C_2) \quad \lim_{k\to\infty} \int_S p(dx) \int_S | p^{(k)}(x, dy) - p(dy) | = 0 ,$$

which is also sufficient for (3) .

On the other hand it is known [1] that (I) is equivalent to the condition that there exists $0 < \rho < 1$ such that

$$(C_3) \quad \sup_x \int_S | p^{(k)} (x, dy) - p(dy) | = O(\rho^k) .$$

McCabe and Shields gave a sufficient condition (C_4) for the automorphism generated by the Markov chain in the above to be weak Bernoulli . The kernel $p(x, dy)$ difines a positive bounded operator P in $L^1(p)$, $u \in L^1(p) \longrightarrow uP \in L^1(p)$, and its dual P^* in $L^\infty_{(p)}$, $f \in L^\infty_{(p)} \to P^*f$ such that

$$P^*f(x) = \int_S p(x, dy) f(y) ,$$

$$\int_S uP(y) f(y) p(dy) = \int_S u(x) P^*f(x) p(dx).$$

The bounded positive operator E in $L^1(p)$, $u \in L^1(p) \longrightarrow uE = \int_S u(x)p(dx)$ corresponds to the kernel

$$Q(x, dy) = p(y) .$$

The condition by McCabe-Shield is that

$$(C_4) \quad \rho = \| P - E \|_1 < 1 .$$

Taking into account of the fact that $f \in L^\infty(p) \longrightarrow E^*f(x) = \int_S f(y)p(dy)$, easy calculations show that

$$(E^*)^n = E^* , \quad E^*P^* = P^*E^* = E^* ,$$

so that

$$\{(P - E)^*\}^n = P^{*n} - E^* ,$$

whereas

$$\rho^n \geqq \|(P - E)^n\|_1 = \|P^{*n} - E^*\|_\infty$$

$$= \operatorname{ess.sup}_x \sup_{\|f\|_\infty \leqq 1} \left| \int_S p^{(n)}(x, dy)f(y) - \int_S p(dy)f(y) \right|$$

$$= \operatorname{ess.sup}_x \{\text{total var. of } (p^{(n)}(x, \cdot) - p(\cdot))\}$$

Therefore we are led to the following conclusion.

Proposition 3. (C_4) implies (I).

We may generalize (C_4) to the condition that there exists k_0 such that

$$(C_4') \qquad \rho = \|(P - E)^{k_0}\|_1 < 1 .$$

This is an operator-theoretical condition equivalent to (I).

References

[1] Ибрагимов, И. А. и Ю. В. Линник, Независимые и стачионарно связанные величины, Изд. Наука, Москва, 1965.

[2] Ornstein, D., Bernoulli shifts with the same entropy are isomorphic, Advances in Math. 4(1970), 337-352.

[3] ________, Two Bernoulli shifts with infinite entropy are isomorphic, Advances in Math. 5(1971), 339-348.

[4] ________, Factors of Bernoulli shifts are Bernoulli shifts, Advances in Math. 5(1971), 349-364.

[5] Friedman N. and D. Ornstein, On isomorphism of weak Bernoulli transformations, Advances in Math. 5(1971), 365-394.

[6] Ornstein, D., A Kolmogorov automorphism that is not a Bernoulli shift.

[7] McCabe R. and P. Shields, A class of Markov shifts which are Bernoulli shifts, Advances in Math. 6(1971), 323-328.

[8] Maruyama G., Some aspects of Orntein's theory of isomorphism problems in ergodic theory, Pub. Res. Inst. Math. Sci., 7(1972), 511-539.

Tokyo University of Educatior

COMPARISON THEOREMS FOR SOLUTIONS OF ONE-DIMENSIONAL STOCHASTIC DIFFERENTIAL EQUATIONS

Shintaro Nakao

§ 0. Introduction

The present paper is a continuation of [1] in which we have investigated the pathwise uniqueness of stochastic differential equations. Here, we shall discuss problems of the comparison and the pathwise uniqueness for solutions of one-dimensional stochastic differential equations allowing reflection. Let $a(x)$ and $b(x)$ be bounded Borel measurable functions. We shall consider the following one-dimensional Itô's stochastic differential equation:

$$dx_t = a(x_t)dB_t + b(x_t)dt \tag{0.1}$$

and the following Skorokhod's stochastic differential equation with reflection:

$$dx_t = a(x_t)dB_t + b(x_t)dt + d\phi_t. \tag{0.2}$$

A comparison theorem for solutions of the equation (0.1) was established by A.V. Skorokhod (cf. [3]) and recently T. Yamada [4] improved it by a simpler method. On the other hand, for the equation (0.2), if $a(x)$ and $b(x)$ are Lipschitz continuous, the existence and the uniqueness of solutions was proved nicely by A.V. Skorokhod [2].

In this paper, we shall obtain comparison theorems for solutions of (0.1) and (0.2) under the assumption that $a(x)$ is uniformly positive and of bounded variation on any compact interval. The pathwise uniqueness of solutions follows immediately from these theorems.

Finally the author wishes to express his hearty thanks to Professor S. Watanabe for his valuable suggestions.

§ 1. Preliminaries

We refer to [1] for a precise definition of solutions of the equation (0.1). Here, we give a definition of solutions of the equation (0.2). Let $(\Omega, \mathcal{F}, P : \mathcal{F}_t)$ stand for a probability space $(\Omega, \mathcal{F}, P)$ with an increasing family $\{\mathcal{F}_t\}_{t\in[0,\infty)}$ of sub-σ-algebras of $\mathcal{F}$.

Definition 1.1. By a solution of (0.2), we mean a quadruplet $(\Omega, \mathcal{F}, P : \mathcal{F}_t)$ and a stochastic process $X_t = (x_t, B_t, \phi_t)$ defined on it satisfying the following conditions:

(i) with probability one, X_t is continuous in t, $B_0 = 0$, $\phi_0 = 0$ and x_t is non-negative,

(ii) X_t is an $\mathcal{F}_t$-adapted process and B_t is an $\mathcal{F}_t$-Brownian motion,

(iii) with probability one, ϕ_t is nondecreasing and

$$\int_0^t I_{\{0\}}(x_s)\,d\phi_s = \phi_t,$$

where $I_{\{0\}}$ is the indicator function of the set $\{0\}$,

(iv) X_t satisfies

$$x_t = x_0 + \int_0^t a(x_s)\,dB_s + \int_0^t b(x_s)\,ds + \phi_t \quad \text{a.s.},$$

where the integral by dB_s is understood in the sense of the stochastic integral of Itô.

The following lemma is an extension of Lemma in [1] and plays a fundamental role in this paper.

Lemma 1.1. Let S_t, ψ_t and V_t be continuous real processes defined on a quadruplet $(\Omega, \mathcal{F}, P : \mathcal{F}_t)$. Suppose that the total variation of V_t on $[0, T]$ has finite expectation. Further, suppose that ψ_t is $\mathcal{F}_t$-adapted and S_t is an $\mathcal{F}_t$-supermartingale satisfying the following conditions:

(i) $S_0 + \psi_0 \leqq 0$ a.s.,

(ii) ψ_t is flat off $\{t : S_t + \psi_t \leqq 0\}$ a.s.,

(iii) there exist positive constants m_1 and m_2 such that $m_1(S_t + \psi_t) \leqq V_t \leqq m_2(S_t + \psi_t)$ a.s.

for $(t, \omega) \in \{(t, \omega) : t \in [0, T]$ and $S_t(\omega) + \psi_t(\omega) > 0\}$.

Then, $S_t + \psi_t \leqq 0$ a.s. for $0 \leqq t \leqq T$.

The proof proceeds in a same way as in [1].

§ 2. A comparison theorem for the equation (0.1)

Theorem 2.1. (Comparison theorem)

Suppose that $a(x)$, $b_1(x)$ and $b_2(x)$ satisfy the following conditions:

(i) $a(x), b_1(x)$ and $b_2(x)$ are bounded Borel measurable on R,

(ii) $a(x)$ is of bounded variation on any compact interval,

(iii) there exists a constant $c > 0$ such that

$$a(x) \geqq c \qquad \text{for} \quad x \in R.$$

Further, suppose that $X_1(t) = (x_1(t), B(t))$ and $X_2(t) = (x_2(t),$

B(t)) are solutions of (0.1) for (a, b_1) and (a, b_2), respectively, defined on a same quadruplet such that $x_1(0) \leqq x_2(0)$ a.s.. Under these conditions, if $b_1(x) \leqq b_2(x)$ almost everywhere, then, with probability one, $x_1(t) \leqq x_2(t)$ for $t \geqq 0$.

Proof. It is not difficult to see by a change of scale that we may prove this theorem in case that $b_1(x) = 0$. Let

$$h(x) = \int_0^x \frac{1}{a(y)}\, dy \qquad \text{for } x \in R,$$

$$\tau_N^i = \inf\{t \geqq 0 : |x_i(t)| = N\} \qquad i = 1,2,$$

$$\gamma_N = \tau_N^1 \wedge \tau_N^2.$$

Using the result in [1], we see that the total variation of $h(x_1(t\wedge \gamma_N)) - h(x_2(t\wedge\gamma_N))$ has finite expectation. Applying Lemma 1.1 by setting

$$S_t = x_1(0) - x_2(0) + \int_0^{t\wedge\gamma_N} \{a(x_1(s))-a(x_2(s))\}\, dB_s - \int_0^{t\wedge\gamma_N} b_2(x_2(s))\,ds,$$

$\psi_t = 0$ and $V_t = h(x_1(t\wedge\gamma_N)) - h(x_2(t\wedge\gamma_N))$, we have $V_t \leqq 0$, i.e., $x_1(t\wedge\gamma_N) \leqq x_2(t\wedge\gamma_N)$. By letting $N \longrightarrow \infty$, we have $x_1(t) \leqq x_2(t)$.

Q.E.D.

§ 3. A comparison theorem and the pathwise uniqueness for the equation (0.2)

Theorem 3.1. (Comparison theorem)

Suppose that $a(x)$, $b_1(x)$ and $b_2(x)$ satisfy the following conditions:

(i) $a(x)$, $b_1(x)$ and $b_2(x)$ are bounded Borel measurable on $[0, \infty)$,

(ii) $a(x)$ is of bounded variation on any compact interval,

(iii) <u>there exists a constant</u> $c > 0$ <u>such that</u>

$$a(x) \geq c \qquad \text{for } x \in [0, \infty).$$

<u>Further, suppose that</u> $X_1(t) = (x_1(t), B(t), \phi_1(t))$ <u>and</u> $X_2(t) = (x_2(t), B(t), \phi_2(t))$ <u>are solutions of</u> (0.2) <u>for</u> (a, b_1) <u>and</u> (a, b_2), <u>respectively, defined on a same quadruplet such that</u> $x_1(0) \leq x_2(0)$ <u>a.s.. Under these conditions, if</u> $b_1(x) \leq b_2(x)$ <u>almost everywhere, then, with probability one,</u> $x_1(t) \leq x_2(t)$ <u>for</u> $t \geq 0$.

<u>Proof</u>. We can assume, as before, $b_1(x) = 0$. Let

$$h(x) = \int_0^x \frac{1}{a(y)}\, dy \qquad \text{for } x \in [0, \infty),$$

$$\tau_N^i = \inf\{t : x_i(t) = N\} \qquad i = 1,2,$$

$$\gamma_N = \tau_N^1 \wedge \tau_N^2.$$

In a similar way as in the proof of Theorem 2.1, we see that the total variation of $h(x_1(t \wedge \gamma_N)) - h(x_2(t \wedge \gamma_N))$ has finite expectation. Since $\int_0^t I_{\{0\}}(x_1(s))d\phi_1(s) = \phi_1(t)$ and $x_2(t) \geq 0$, $S_t = x_1(0) - x_2(0) + \int_0^{t\wedge\gamma_N}\{a(x_1(s)) - a(x_2(s))\}dB_s - \int_0^{t\wedge\gamma_N} b_2(x_2(s))ds - \phi_2(t \wedge \gamma_N)$, $\psi_t = \phi_1(t \wedge \gamma_N)$ and $V_t = h(x_1(t \wedge \gamma_N)) - h(x_2(t \wedge \gamma_N))$ satisfy the conditions of Lemma 1.1. Then, by Lemma 1.1, we have $V_t \leq 0$ implying $x_1(t \wedge \gamma_N) \leq x_2(t \wedge \gamma_N)$. Letting $N \longrightarrow \infty$, we have $x_1(t) \leq x_2(t)$. Q.E.D.

The following theorem is a immediate consequence of Theorem 3.1.

<u>Theorem 3.2</u>. (<u>Pathwise uniqueness</u>)

<u>Let</u> $a(x)$ <u>and</u> $b(x)$ <u>be bounded Borel measurable functions defined on</u> $[0, \infty)$. <u>Suppose that</u> $a(x)$ <u>is of bounded variation on</u>

any compact interval. Further, suppose that there exists a constant $c > 0$ *such that*

$$a(x) \geq c \qquad \textit{for} \quad x \in [0, \infty).$$

Under these conditions, if $X_1(t) = (x_1(t), B(t), \phi_1(t))$ *and* $X_2(t) = (x_2(t), B(t), \phi_2(t))$ *are solutions of* (0.2) *for* (a, b) *defined on a same quadruplet such that* $x_1(0) = x_2(0)$ *a.s., then, with probability one,* $x_1(t) = x_2(t)$ *for* $t \geq 0$.

A consequence of Theorem 3.2 is that, for any solutions of (0.2), $x(t)$ is a function of the Brownian path $\{B(t)\}$.

References

[1] S. Nakao: On the pathwise uniqueness of solutions of one-dimensional stochastic differential equations, (to appear).

[2] A.V. Skorokhod: Stochastic equations for diffusion processes in a bounded region, Theory of Prob. and its Appl. 6 (1961), 264-274.

[3] A.V. Skorokhod: Studies in the theory of random processes, Kiev Univ., Kiev, 1961.

[4] T. Yamada: On a comparison theorem for solutions of stochastic differential equations and its applications, (to appear).

Department of Mathematics
Osaka University

REMARKS ON PROBABILISTIC SOLUTIONS OF CERTAIN QUASILINEAR PARABOLIC EQUATIONS

Makiko Nisio

1. Introduction. Let us consider the Cauchy problem for quasilinear parabolic system

$$(1)\quad \begin{cases} \frac{\partial u^{\ell}}{\partial s} + L_1^{\ell} u^{\ell} + L_2^{\ell} u^{\ell} = f^{\ell}(x, \bar{u}) & \text{on } [0\ T) \times R^n \\ u^{\ell}(T, x) = h^{\ell}(x), & \ell = 1, \cdots m, \end{cases}$$

where $L_1 = \frac{1}{2}\sum_{i,j} a^{\ell}_{ij}(x)\frac{\partial^2}{\partial x_i \partial x_j} + \sum_i b_i(x)\frac{\partial}{\partial x_i}$, $L_2 = \sum_i \gamma_i^{\ell}(x, \bar{u})\frac{\partial}{\partial x_i} - c^{\ell}(x, \bar{u})$

and $\bar{u}(s, x) = (u^1(s, x), \cdots u^m(s, x))$.

We shall assume that the coefficient matrix $(a^{\ell}_{ij}) = (\alpha^{\ell}_{ij})^2$ with a symmetric and non-negative definite $n \times n$ matrix α^{ℓ}, and the linear algebraic system $\alpha^{\ell}(x)\delta^{\ell}(x, v) = \gamma^{\ell}(x, v)$ has a bounded solution $\delta^{\ell} = (\delta^{\ell}_1, \cdots \delta^{\ell}_n)$. Moreover let us assume that $\alpha^{\ell}(x)$ and $b^{\ell}(x)$ are Lipschitz continuous.

According to [1], we have the following probabilistie version of (1). By the assumption of α^{ℓ} and b^{ℓ}, the stochastic integral equation (2),

$$(2)\qquad X(t) \equiv X(t; s, x) = x + \int_s^t \alpha^{\ell}(X(\tau))dB(\tau) + \int_s^t b^{\ell}(X(\tau))d\tau$$

has a unique solution X^{ℓ}. So, the equation (1) turns out the following

$$(3)\quad u^{\ell}(s, x) = -E_{(sx)}\int_s^T f^{\ell}(X^{\ell}(t), \bar{u}(t, X^{\ell}(t))J(\ell, t, s, \bar{u})dt$$

$$+ E_{(sx)}h^{\ell}(X^{\ell}(T))J(\ell, T, s, \bar{u})$$

where $\phi(\ell, t, s, \bar{u}) = -\int_s^t c^{\ell}(X^{\ell}(\tau), \bar{u}(\tau, X^{\ell}(\tau))d\tau$

$$+ \int_s^t \delta^\ell(X^\ell(\tau), \overline{u}(\tau, X^\ell(\tau))dB(\tau) - \frac{1}{2}\int_s^t \| \delta^\ell(X^\ell(\tau), \overline{u}(\tau, X^\ell(\tau))\|^2 d\tau$$ (1)

and $J(\ell, t, s, \overline{u}) = e^{\phi(\ell, t, s, \overline{u})}$.

It is well-known that a smooth solution of (1) satisfies (3) and conversely a smooth solution of (3) satisfies (1), under the suitable conditions for coefficients.

In this note we shall remark on the solvability of (3) and the limit behavior of a solution, as $T \to \infty$.

Put $\overline{\theta}(x, v) = (\theta^1(x, v), \cdots \theta^m(x, v))$ and let us introduce the following five conditions,

(C 1). $\overline{c}$, $\overline{f}$, $\overline{\delta}$ and $\overline{h}$ are bounded and, for any x, continuous in v.

(C 2). $\overline{c}$, $\overline{f}$, $\overline{\delta}$ and $\overline{h}$ are bounded and Lipschitz continuous in v, i.e

$$|\theta(x, v) - \theta(x, v')| \leq K\|v - v'\|, \qquad \theta = c^\ell, f^\ell, \delta^\ell,$$

with a constant K.

(C 3). $c^\ell(x, v) \geq A, \qquad \ell = 1, \cdots m,$

with a positive constant A.

(C 4). the constant A of (C 3) is sufficiently large.

(C 5). the transition probability $P(X^\ell(t; s, x) \in dy)$ has the density $p^\ell(t-s, x, y)$. Moreover, for and $\tau > 0$ and any $x \in R^n$,

(4) $$\int_{R^n} |p^\ell(\tau, x, y) - p^\ell(\tau, x', y)|dy \to 0 \qquad \text{as } x \to x'.$$

(C 5) is rather restrictive. But if α^ℓ is uniformly elliptic and both coefficients, α^ℓ and b^ℓ, are smooth, then it holds. In the case where α^ℓ is degenerate, Sonin [4] treated an iteresting class which satisfies (C 5).

---

(1) $\|\xi\|^2 = \sum_{i=1} \xi_i^2$ for $\xi = (\xi_1 \cdots \xi_n)$.

Theorem 1. If (C 1) and (C 5) hold, then we have a bounded solution $\bar{u}^T$ of (3), which is continuous on $[0\ T) \times R^n$. Moreover if $\bar{h}$ is continuous at x_0, then $\bar{u}^T$ is continuous at (T, x_0).

Theorem 2. If (C 1)(C 3) and (C 5) hold, then any family of solutions $\{\bar{u}^T, T > 0\}$ has a sequence $\bar{u}^{T_k}$, $(T_k \to \infty)$, which converges uniformly on any compact set of $[0\ \infty) \times R^n$. Furthermore the limit function $\bar{u}$ is continuous and satisfies the following equation,

$$(5) \qquad u^\ell(s, x) = -E_{(sx)}\int_s^\infty f^\ell(X^\ell(t), \bar{u}(t, X^\ell(t))J(\ell, t, s, \bar{u})dt.$$

Theorem 3. Under the same assumptions of Theorem 2, there exists a continuous stationary solution $\bar{w}$ of (5), i.e. $\bar{w}(s, x)$ is independent of s. Furthermore if $\bar{u}^T$ is the unique solution of (3) and converges to $\bar{u}$ as $T \to \infty$, then $\bar{u}$ is a stationary solution of (5).

Theorem 4. [1]. If (C 2) holds, then there exists a bounded solution $\bar{u}^T$ of (3) uniquely. Moreover if $\bar{\alpha}$, $\bar{b}$, $\bar{c}$, $\bar{\delta}$ and $\bar{h}$ have bounded and continuous derivatives with respect to x and v, then $\bar{u}^T$ is Lipschitz continuous in x.

Theorem 5. If (C 2) and (C 4) hold, then there exists a bounded solution $\bar{u}$ of (5) uniquely. $\bar{u}$ is stationary and for any lateral data $\bar{h}$, the unique solution $\bar{u}^T$ tends $\bar{u}$ uniformly on $[0, s_0]\times R^n$, as $T \to \infty$. Moreover if $\bar{\alpha}$, $\bar{b}$, $\bar{c}$, $\bar{\delta}$ and $\bar{h}$ have bounded and continuous derivatives with respect to x and v, then $\bar{u}$ is Lipschitz continuous.

We shall prove above theorems in the following sections.

2. Proof of Theorem 1. Let C be the space of all continuous R^m - valued functions defined on $[0\ T) \times R^n$. C is a separable complete metric space with the metric ρ,

$$\rho(f, g) = \sum_{p=1}^{\infty} 2^{-p}\frac{\|f - g\|_p}{1 + \|f - g\|_p}$$

where $\|f\|_p = \max_{\|x\|\leq p,\ 0\leq t\leq T-\frac{1}{p}} \|f(t, x)\|$.

Define $F\bar{u} = ((F\bar{u})^1, \cdots (F\bar{u})^m)$ by the right side of (3) of $\bar{u} \in C$. Then we have

Lemma. $\{F\bar{u};\ \bar{u} \in C\}$ is totally bounded in C.

Proof. By (C 1), $F\bar{u}$ is bounded uniformly in $\bar{u}$. We shall evaluate the moduli of continuity of $F\bar{u}$. Fix s_0 and take a small positive $\tau(<1)$ so that $s_0 + \tau < T$. For simplicity we drop the suffix ℓ in (3) and put $v = (F\bar{u})^{\ell}$. For $s \leq s_0$, we have

$$v(s, x) - v(s, y) = -I_1 + I_1' - I_2 + I_2' - I_3 + I_4 - I_4' + I_5,$$

where

$$I_1 = E_{(sx)}\int_s^{s+\tau} f(X(t), \bar{u}(t, X(t))J(\ell, t, s, \bar{u})dt$$

$$I_2 = E_{(sx)}\int_{s+\tau}^{T} f\cdot(J(\ell, t, s, \bar{u}) - J(\ell, t, s+\tau, \bar{u}))dt$$

$$I_3 = E_{(sx)}\int_{s+\tau}^{T} f\cdot J(\ell, t, s+\tau, \bar{u})dt - E_{(sy)}\int_{s+\tau}^{T} f\cdot J(\ell, t, s+\tau, \bar{u})dt$$

$$I_4 = E_{(sx)}h(X(T))(J(\ell, T, s, \bar{u}) - J(\ell, T, s+\tau, \bar{u}))$$

$$I_5 = E_{(sx)}h(X(T))J(\ell, T, s+\tau, \bar{u}) - E_{(sy)}h(X(T))J(\ell, T, s+\tau, \bar{u})$$

and I_i' is defined by switching x and y in I_i. Easily verify the following estimates, where K_i stands for a constant independent of x, y, s and $\bar{u}$;

$$|I_1| \leq K_1\tau,$$

$$|I_2| \leq E_{(sx)}|J(\ell, s+\tau, s, \bar{u}) - 1|E_{(s+\tau,\ X(s+\tau))}\left|\int_{s+\tau}^{T} f\cdot J(\ell,t,s+\tau,\bar{u})dt\right|$$

$$\leq K_2E_{(sx)}|J(\ell, s+\tau, s, \bar{u}) -1| \leq K_3\sqrt{\tau},$$

$$|I_4| \leq K_4\sqrt{\tau}$$

and $|I_i'|$ has the same estimate as $|I_i|$. Moreover

$$-I_3 + I_5 = E_{(sx)}v(s+\tau, X(s+\tau)) - E_{(sy)}v(s+\tau, X(s+\tau)).$$

Therefore, with the use of boundedness of v, we have

$$|-I_3 + I_5| \leq K_5 \int |p(\tau, x, z) - p(\tau, y, z)| dz.$$

Hence, recalling (C 5), we can choose, for any $\varepsilon > 0$ and a compact set σ, a positive $\theta = \theta(\tau, \sigma, \varepsilon)$ so that

$$|-I_3 + I_5| < \varepsilon \qquad \text{for } x \in \sigma \text{ and } \|x - y\| < \theta.$$

Consequently, taking τ small enough, we can choose a positive $\Delta = \Delta(\varepsilon, \sigma, s_0)$ so that

$$|v(s, x) - v(s, y)| < \varepsilon \qquad \text{for } s \leq s_0,\ x \in \sigma \text{ and } \|x - y\| < \Delta.$$

Repeating similar calculations, we can take a positive $\Delta' = \Delta'(\varepsilon, \sigma, s_0)$ such that

$$|v(s, x) - v(t, y)| < \varepsilon \qquad \text{for } s, t \leq s_0,\ |s - t| < \Delta' \text{ and } x \in \sigma.$$

This completes the proof of Lemma.

Suppose that $\bar{u}_k$ converges to $\bar{u}$ in C, as $k \to \infty$. Then by (C 1), $\phi(\ell, t, s, \bar{u}_k)$ tends to $\phi(\ell, t, s, \bar{u})$ in probability. Because of the boundedness of δ^ℓ and c^ℓ, $J(\ell, t, s, \bar{u}_k)$ is uniformly integrable. Hence $(F\bar{u}_k)(s, x)$ converges to $(F\bar{u})(s, x)$ at each (s, x). Therefore Lemma tells us that $F\bar{u}_k$ converges to $F\bar{u}$ in C, namely F is continuous. Consequently the fixed point theorem guarantees the existence of a solution of (3).

Let $\bar{h}$ be continuous at x_0. We use the following estimate,

$$|v(s, x) - h(x_0)| \leq E_{(sx)} \int_s^T |f| J(\ell, t, s, \bar{u}) dt$$

$$+ E_{(sx)} |h(X(T)) - h(x)| J(\ell, T, s, \bar{u})$$

$$+ h(x) E_{(sx)} |J(\ell, T, s, \bar{u}) - 1| + |h(x) - h(x_0)|.$$

In the same way, we can show the continuity at (T, x_0).

This completes the proof of Theorem 1.

Example.

$$(6)\begin{cases} \dfrac{\partial u^\ell}{\partial s} + \dfrac{1}{2}\sum_{i=1}^{n} \dfrac{\partial^2 u^\ell}{(\partial x_i)^2} + \sum_{i=1}^{n} x_i \dfrac{\partial u^\ell}{\partial y_i} - c^\ell(\bar{u})u^\ell = f^\ell(\bar{u}) \quad \text{on} \quad [0\ T) \times R^{2n}. \\ u^\ell(T, x, y) = h^\ell(x, y). \qquad \ell = 1,\cdots m. \end{cases}$$

Suppose that h^ℓ, c^ℓ and f^ℓ are bounded and Hölder continuous in v, i.e

$$|\theta(v) - \theta(v')| \leq K\|v - v'\|^r, \qquad \theta = c^\ell, f^\ell \quad \text{with} \quad r > \frac{1}{2}.$$

In this case, we have

$$X^\ell(t; s, x, y) = x + B(t) - B(s), \quad Y^\ell(t; s, x, y) = y + \int_s^t X^\ell(\tau)d\tau.$$

Hence, for each ℓ, $Z_i^\ell(t) \equiv (X_i^\ell(t), Y_i^\ell(t))$, $i = 1\cdots n$, is a system of mutually independent gaussian processes with probability density $p^\ell(t - s, (x, y), (\xi, \eta))$,

$$p^\ell(\tau, (x\ y), (\xi, \eta)) = \prod_{i=1}^{n} \frac{\sqrt{3}}{\pi\tau^2} \exp(-2\frac{\mu_i^2}{\tau} + 6\frac{\mu_i\nu_i}{\tau^2} - 6\frac{\nu_i^2}{\tau^3})$$

where $\mu_i = \xi_i - x_i$ and $\nu_i = \eta_i - y_i - x_i\tau$. By the Markov property of Z^ℓ, (3) turns out the following form,

$$u^\ell(s, z) = -\int_s^T\int_{R^{2n}} g(t, \zeta)p\ (t-s, z, \zeta)d\zeta + \int_{R^{2n}} h^\ell(\zeta)p^\ell(T-s, z, \zeta)d\zeta$$

where $g(t, \zeta) = f^\ell(\bar{u}(t, \zeta)) - c^\ell(\bar{u}(t, \zeta))u^\ell(t, \zeta)$.
On account of Hölder continuity, we can see that $\dfrac{\partial u^\ell}{\partial y_i}$ and $\dfrac{\partial^2 u^\ell}{\partial x_i \partial x_j}$ are continuous in x. Hence a formula of stochastic differentials [2] implies that $\bar{u}$ is a solution of parabolic equation (6).

3. Proof of Theorems 2 and 3. In this section, we denote by C the space of all continuous R^m-valued functions defined on $[0\ \infty) \times R^n$. C is a separable complete metric space with the metric ρ,

$$\rho(f, g) = \sum_{p=1}^{\infty} 2^{-p}\frac{\|f-g\|_p}{1+\|f-g\|_p}$$

where $\|f\|_p = \max_{\|x\| \leq p,\ t \leq p} \|f(t, x)\|$.

Let $\bar{u}^T$ be a solution of (3). Then, according to evaluations of Section 2, (C 3) guarantees that $\sup_{T,s,x} \|\bar{u}^T(s, x)\|$ is finite and each K_i is independent of T. Hence $\{\bar{u}^T;\ T > 0\}$ may be regarded as a totally bounded subset of C. Therefore we have a convergent sequence $\bar{u}^{T_k}$, as $T_k \to \infty$, and its limit function $\bar{u}$ is a solution of (5) by (C 1).

For the proof of Theorem 3, we shall introduce the following auxiliary equation,

(7) $$w^\ell(x) = -E_{(0x)}\int_0^\infty f^\ell(X^\ell(t)\ \bar{w}(X^\ell(t))J(\ell, t, 0, \bar{w})dt.$$

Applying the same calculation as in Section 2, we can obtain a bounded and continuous solution $\bar{w}$ of (7). On the other hand, $\{X^\ell(t+s, s, x), dB(t+s), t \geq 0\}$ has the same law as $\{X^\ell(t; 0, x), dB(t), t \geq 0\}$, since α^ℓ and b^ℓ are independent of time variable. Hence $E_{(sx)}\int_s^\infty f^\ell(X^\ell(t), \bar{w}(X^\ell(t))J(\ell, t, s, \bar{w})dt$ is independent of s. Therefore putting $\bar{w}(s, x) = \bar{w}(x)$ is a stationary solution of (5).

Suppose that $\bar{u}^T$ is the unique solution of (3) and converges to $\bar{u}$ as $T \to \infty$. Then by the stationarity of (X^ℓ, dB),

$$\bar{u}^T(s-t, x) = \bar{u}^{T+t}(s, x)$$

holds. Hence tending T to ∞, we have

$$\bar{u}(s-t, x) = \bar{u}(s, x) \quad \text{for } t \leq s.$$

This completes the proof of Theorem 3.

4. Proof of Theorem 5. Let $\mathcal{B}$ be the space of all bounded R^m-valued functions defined on $(-\infty, \infty) \times R^n$, endowed with the usual norm, $|||\cdot|||$, of supremum. We define $F\bar{u}$ by the right side of (5)

for $\bar{u} \in \mathcal{B}$ and $s \in R^1$. Then using the similar evaluations as in [1] under conditions (C 2) and (C 4), we can show that

(8) $\|F\bar{u}_1 - F\bar{u}_2\| \le \lambda\|\bar{u}_1 - \bar{u}_2\|$

with a constant $\lambda < 1$. Hence we have a unique solution $\bar{u}$ of (5).

Putting $\bar{u}_\theta(t, x) = \bar{u}(t+\theta, x)$, we see that

$$E_{(s+\theta,\ x)}\int_{s+\theta}^{\infty} f^{\ell}(X^{\ell}(t),\ \bar{u}_{-\theta}(t,\ X^{\ell}(t))J(\ell,\ t,\ s+\theta,\ \bar{u}_{-\theta})dt$$

is independent of θ, by the stationarity of (X^{ℓ}, dB).

Therefore

$$u_{-\theta}(s+\theta,\ x) = u^{\ell}(s,\ x)$$

$$= E_{(s+\theta,\ x)}\int_{s+\theta}^{\infty} f^{\ell}(X^{\ell}(t),\ \bar{u}_{-\theta}(t,\ X^{\ell}(t))J(\ell,\ t,\ s+\theta,\ \bar{u}_{-\theta})dt$$

This means that $\bar{u}_{-\theta}$ is a solution of (5). So, $\bar{u}_{-\theta} = \bar{u}$, namely $\bar{u}$ is stationary.

We shall now evaluate $u^{T\ell}(s, x) - u^{\ell}(s, x)$. For simplicity we drop the suffix ℓ.

$$u^T(s,\ x)-u(s,\ x) = E_{(sx)}\int_T^{\infty} f(X,\ \bar{u})J(t,\ s,\ \bar{u})dt$$

$$+ E_{(sx)}h(X(T))J(T,\ s,\ \bar{u}^T) + E_{(sx)}\int_s^T (f(X,\ \bar{u}) - f(X,\ \bar{u}^T))J(t,\ s,\ \bar{u})dt$$

$$+ E_{(sx)}\int_s^T f(X,\ \bar{u}^T)(J(t,\ s,\ \bar{u}) - J(t,\ s,\ \bar{u}^T))dt.$$

We may assume $A > 1$ and in the following K_i stands for a constant which is independent of s, x, T and A. The absolute values of the first and second terms are less than $K_1 e^{-A(T-s)}$. Putting $\rho_T(s) = \sup_x \|\bar{u}(s, x) - \bar{u}^T(s, x)\|$, we see the following estimates,

$$|\text{the third term}| \le K_2\int_s^T \rho_T(t)e^{-A(t-s)}ds$$

$$|\text{the fourth term}| \le K_3\int_s^T \rho_T^2(t)e^{-\frac{A}{2}(t-s)}dt.$$

Hence we have

$$\|\bar{u}(s, x) - \bar{u}^T(s, x)\|^2 \leq K_4(e^{-A(T-s)} + \int_s^T \rho_T^2(t)e^{-\frac{A}{2}(t-s)} dt).$$

namely,

$$\rho_T^2(s) \leq K_4(e^{-A(T-s)} + \int_s^T \rho_T^2(t)e^{-\frac{A}{2}(t-s)} dt).$$

Since $\rho_T(s)$ is bounded in (T, s), we get

$$\rho_T^2(s) \leq e^{(K_4-\frac{A}{2})(T-s)} \times \text{bounded function, if } K_4 < \frac{A}{2}.$$

Therefore $\bar{u}^T$ converges to $\bar{u}$ uniformly on $[0\ s_0] \times R^n$.

The last part of Theorem 5 is proved as follows; Put $\bar{u}_0(s,x) \equiv 0$ and $\bar{u}_{n+1} = F\bar{u}_n$. Then $\bar{u}_n$ is independent of s and $\frac{\partial \bar{u}_n}{\partial x_i}$ is bounded and continuous by the assumption of smoothness of coefficients. Moreover setting $\rho_n = \sup_{i,\ell,x} |\frac{\partial u_n^\ell}{\partial x_i}(x)|$, we have $\rho_{n+1} \leq \frac{K_5}{A}(\rho_n + 1)$.

Hence, if $K_5 < A$, then $\rho_{n+1} \leq \sum_{j=i}^{n+1} (\frac{K_5}{A})^j \leq \frac{1}{1 - \frac{K_5}{A}}$, $n = 0, 1, \cdots,$ hold. On the other hand, $\bar{u}_n$ tends to $\bar{u}$ uniformly by virtue of (8). Therefore $\bar{u}$ is Lipschitz continuous. This completes the proof of Theorem 5.

The following simple example shows that (C 4) is necessary in some sense, for the uniqueness of solution of (5).

Suppose that $m = 1$, $\alpha_{ii}(x) = 1$, $\alpha_{ij}(x) = 0$, $(i \neq j)$, $b_i = \gamma_i = 0$ and $c(x, v) = A$. Let $f(x, v)$ be independent of x and satisfy (C 2). If $f(0) = 0$ and $f(1) = -A$, then two constant functions 1 and 0 satisfy (5). Moreover for $h(x) = 1$, 1 is the unique solution of (3) and for $h(x) = 0$, 0 is the unique solution. This function $f(v)$ can not have a Lipschitz constant K which is less than A. On the other hand, if $K < A$, then (5) has a unique solution.

References

[1] M. I. Freidlin: Quasilinear parabolic equations and measures in function space, Functional Anal. Appl. 1 (1967), 234-240.

[2] K. Ito: On a formula concerning stochastic differentials, Nagoya Math. Jour. 3 (1951), 55-65.

[3] H. P. McKean: Stochastic integrals, Academic Press, 1969.

[4] I. M. Sonin: On a class of degenerate diffusion processes, Th. Prob. Appl. 12 (1967) 490-496.

Department of Mathematics
Kobe University
Kobe, Japan.

ON WHITE NOISE AND INFINITE DIMENSIONAL ORTHOGONAL GROUP

Hisao Nomoto

Introduction

The purpose of this report is to consider some properties of white noise related to the infinite dimensional orthogonal group $O(H)$ on a real Hilbert space. Since the group $O(H)$ is an analogue of the finite dimensional orthogonal group $O(n)$ of R^n, it seems to be reasonable to inquire whether it possess similar properties to the group $O(n)$. In particular, we shall consider about the invariant measure of the group $O(H)$. In [3], D. Shale studied invariant integrations over infinite dimensional manifolds, and in [4] H. Shimomura considered a construction of the invariant measure of $O(H)$. On the other hand, Y. Yamasaki [5] pointed out that the group $O(H)$ fails to possess invariant finite measure over the σ-algebra for which every function $g \to (gx, y)$ $(x, y \in H)$ is measurable, and constructed an $O(H)$ - invariant measure over a linear space which includes all bounded linear operators of H. Our approach for this problem is to consider the projective sequence of Haar measures of finite dimensional orthogonal groups associated with the projective sequence of finite dimensional Gaussian measure over the general linear groups, and we shall obtain results essentially the same to [5] and [6].

1. White noise. Let E be an infinite dimensional real nuclear space and H be its completion by a continuous Hilbertian norm $\| \|$ of E. Then we have the relation

$$E \subset H \subset E^*$$

where E^* is the conjugate space of E. Consider a function $C(\xi)$ on E defined by

$$C(\xi) = e^{-\|\xi\|^2/2}, \quad \xi \in E. \tag{1}$$

Then, by Minlos' theorem, there exists a unique probability measure μ on E^* such that

(2) $$\int_{E^*} e^{i\langle x,\xi\rangle} d\mu(x) = e^{-\|\xi\|^2/2},$$

where $\langle x, \xi\rangle$ denotes the canonical bilinear form. The measure μ is defined on the σ-algebra generated by all cylinder sets of E^*. We call μ a Gaussian measure or white noise.

Let $O(H)$ be the group formed by all linear and orthogonal operators acting on H. Since, as was explained in [5], the group $O(H)$ has no finite invariant measure, we shall consider the projective limit measure of Haar measures of finite dimensional orthogonal groups $O(K_n)$ $(n = 1, 2, \dots)$, where K_n is a finite dimensional subspace of H such that $K_n \subset K_{n+1}$ and $\bigcup_{n=1}^{\infty} K_n$ is dense in H. For this purpose, we will avail of white noise which is realized on the projective limit space of a sequence of general linear groups.

2. Invariant measure of the orthogonal groups

Let μ_n be the standard Gaussian measure on R^{n^2}:

(3) $$\mu_n(dx) = \left(\frac{1}{2\pi}\right)^{\frac{n^2}{2}} \exp\{-\tfrac{1}{2}\|x\|^2\}dx.$$

Let $GL(n)$ be the n-dimensional general linear group over R. Then, considering $GL(n) \subset R^{n^2}$, we have $\mu_n(GL(n)) = 1$.

Let

(4) $$GL(n) = O(n) \times T(n)$$

be an Iwasawa decomposition of the group $GL(n)$, where $O(n)$ being the group formed by all $n \times n$-orthogonal matrices and $T(n)$ the group formed by all matrices $t = [t_{ij}]_{1\le i,\, j\le n}$ with $t_{ii} > 0$, $t_{ij} = 0$ $(i > j)$.

For any matrix $a^{(n)} \in GL(n)$ with

(5) $$a^{(n)} = g^{(n)} t^{(n)}, \quad g^{(n)} \in O(n), \quad t^{(n)} \in T(n)$$

we define the mapping $\phi_n : GL(n) \to O(n)$ by

$$\phi_n(a^{(n)}) = g^{(n)}.$$

Then ϕ_n induces the normalized Haar measure $\nu_n = \phi_n \cdot \mu_n$ on $O(n)$.

Let $p_{n,m}$ $(n < m)$ be the projection defined by

$$p_{n,m}(a^{(m)}) = a^{(n)}$$

where $a^{(m)} = [a_{ij}]_{1 \leq i,\, j \leq m}$ and $a^{(n)} = [a_{ij}]_{1 \leq i,\, j \leq n}$.

Then, in symbol, we obtain two projective sequences of probability spaces such that

$$(6) \qquad \begin{array}{ccccccc} \longleftarrow & [GL(n), \mu_n] & \xleftarrow{p_{n,m}} & [GL(m), \mu_m] & \xleftarrow{p_m} & [GL(\infty), \mu_\infty] \\ & \phi_n \big\downarrow & & \big\downarrow \phi_m & & \\ \longleftarrow & [O(n), \nu_n] & \xleftarrow{\pi_{n,m}} & [O(m), \nu_m] & \xleftarrow{\pi_m} & [O(\infty), \nu_\infty] \end{array}$$

The precise meanings of (6) are as follows.

We set $G_n = \{a^{(n)} \in GL(n)$: all principal minor parts of $a^{(n)}$ are not zero$\}$ $(\mu_n(G_n) = 1)$ and

$$O_n = \phi_n(G_n).$$

Let $g^{(n)}$ be in O_n and assume $g^{(n)} = \phi_n(a^{(n)})$, then $a^{(n)} = g^{(n)} t^{(n)}$ for some $t^{(n)} \in T(n)$ and it is easy to obtain

$$\begin{aligned} a^{(n-1)} &= p_{n-1,n}(a^{(n)}) \\ &= p_{n-1,n}(g^{(n)}) p_{n-1,n}(t^{(n)}). \end{aligned}$$

Since $a^{(n-1)} \in G_{n-1}$ and $t^{(n-1)} = p_{n-1,n}(t^{(n)}) \in T(n-1)$, $g^{(n-1)} = p_{n-1,n}(g^{(n)}) \in G_{n-1}$ but not necessary $g^{(n-1)} \in O_{n-1}$. So, we considere the decomposition (5) of $g^{(n-1)}$:

$$g^{(n-1)} = h^{(n-1)} s^{(n-1)}.$$

Then $h^{(n-1)}$ is uniquely determined by $g^{(n)}$. Thus we can define the mapping $\pi_{n-1,n} : O_n \to O_{n-1}$ by

$$\pi_{n-1,n}(g^{(n)}) = h^{(n-1)}$$

and we set

$$\pi_{n,m} = \pi_{n,n+1} \cdots \pi_{m-1,m} \quad (n < m).$$

Then we have the following theorem.

Theorem 1. $[(G_n, \mu_n) : p_{n,m}]$ and $[(O_n, \nu_n) : \pi_{n,m}]$ are projective sequences of probability spaces which are related so as

$$\phi_n p_{n,m} = \pi_{n,m} \phi_m$$

$$\nu_n = \phi_n \cdot \mu_n .$$

Therefore, there exist projective limit probability spaces in the sense of Bochner [1],

$$(G_\infty, \mu_\infty) = \varprojlim (G_n, \mu_n)$$

and

$$(O_\infty, \nu_\infty) = \varprojlim (O_n, \nu_n)$$

By definition of projective limit, G_∞ is a sequential space such that

$$G_\infty : a^{(\infty)} = (a^{(n)} : n \geqq 1), \; a^{(n)} \in G_n, \; a^{(n)} = p_{n,m}(a^{(m)}) \quad (n < m).$$

So, $a^{(\infty)}$ can be identified with an infinite matrix as follows.

(7) $$a^{(\infty)} = [a_{ij}]_{1 \leqq i, j < \infty}, \quad a^{(n)} = [a_{ij}]_{1 \leqq i, j \leqq n}.$$

Now, we define random variables $G_{ij}^{(n)}$ on the space (O_∞, ν_∞) by

$$G_{ij}^{(n)}(g^{(\infty)}) = \sqrt{n}\, g_{ij}^{(n)}, \quad 1 \leq i, j \leq n$$

where $g^{(\infty)} = (g^{(n)} : n \geqq 1) \in O_\infty$ and $g^{(n)} = [g_{ij}^{(n)}]_{1 \leqq i, j \leqq n}$.

Then we have the following theorem.

Theorem 2. The limit

$$\lim_{n \to \infty} G_{ij}^{(n)}(g^{(\infty)}) = G_{ij}(g^{(\infty)}) \quad \text{(a.e.)}$$

exists and $\{G_{ij}\}$ $(1 \leqq i, j < \infty)$ is an independent Gaussian system.

Proof. 1). If $a^{(\infty)} \in G_\infty$ has the expression (7), then defining

(8) $$A_{ij}(a^{(\infty)}) = a_{ij}$$

we have independent Gaussian random variables $\{A_{ij}\}$ on G_∞. Furthermore, let us introduce another random variables on G_∞ by

(9) $$B_{ij}^{(n)}(a^{(\infty)}) = \sqrt{n}\, g_{ij}^{(n)}, \quad 1 \leqq i, j \leqq n,$$

where

(10) $$a^{(n)} = g^{(n)}t^{(n)}, \quad g^{(n)} = [g_{ij}^{(n)}]_{1\leqq i, j\leqq n}, \quad t^{(n)} = [t_{ij}^{(n)}]_{1\leqq i, j\leqq n}.$$

Then, by virtue of theorem 1, to prove our theorem, it is enough to show that

(11) $$\lim_{n\to\infty} B_{ij}^{(n)}(a^{(\infty)}) = A_{ij}(a^{(\infty)}) \quad \text{(a.e.)}.$$

2). In (10), since $g^{(n)} \in O(n)$ and $t^{(n)} \in T(n)$, we have following relations:

(12) $$a_{ij} = \sum_{p=1}^{j} g_{ip}^{(n)} t_{pj}^{(n)}$$

(13) $$\sum_{i=1}^{n} a_{ip}a_{iq} = \sum_{i=1}^{p} t_{ip}^{(n)} t_{iq}^{(n)}, \quad 1 \leqq p \leqq q \leqq n$$

In particular,

(14) $$a_{i1} = g_{i1}^{(n)} t_{11}^{(n)}$$

and

(15) $$t_{11}^{(n)2} = \sum_{i=1}^{n} a_{i1}^2 = \sum_{i=1}^{n} A_{i1}^2(a^{(\infty)}).$$

Therefore, using the strong law of large numbers, we obtain

$$\lim_{n\to\infty} \frac{t_{11}^{(n)2}}{n} = \lim_{n\to\infty} \frac{1}{n} \sum_{i=1}^{n} A_{i1}^2(a^{(\infty)}) = 1 \quad \text{(a.e.)},$$

so that

(16) $$\lim_{n\to\infty} \frac{t_{11}^{(n)}}{\sqrt{n}} = 1 \quad \text{(a.e.)}$$

since $t_{11}^{(n)} > 0$. Thus we get from (14) and (16)

(17) $$\lim_{n\to\infty} \sqrt{n}\, g_{i1}^{(n)} = a_{i1} \quad \text{(a.e.)}.$$

Now, assume that already we have shown that

(18) $$\lim_{n\to\infty} \frac{t_{ij}^{(n)}}{\sqrt{n}} = 0, \quad 1 \leq i < j \leq k \quad \text{(a.e.)},$$

(19) $$\lim_{n\to\infty} \frac{t_{jj}^{(n)}}{\sqrt{n}} = 1, \quad 1 \leq j \leq k \quad \text{(a.e.)}$$

and

(20) $$\lim_{n\to\infty} \sqrt{n}\, g_{ij}^{(n)} = a_{ij}, \quad 1 \leq j \leq k \quad \text{(a.e.)}$$

hold for k. Under these assumptions we will show that above relations hold for $k+1$.

Applying the strong law of large numbers to relations (13) together with (18) and (19), we can easily obtain the following formulas successively.

To begin with

(21) $$\lim_{n\to\infty} \frac{t_{i,k+1}^{(n)}}{\sqrt{n}} = 0, \quad 1 \leq i \leq k \quad \text{(a.e.)}$$

and in the next place

(22) $$\lim_{n\to\infty} \frac{t_{k+1,k+1}^{(n)}}{\sqrt{n}} = 1 \quad \text{(a.e.)}.$$

Therefore, from (12), (20), (21) and (22) we get

$$\begin{aligned} a_{i,k+1} &= \lim_{n\to\infty} \sum_{p=1}^{k+1} g_{ip}^{(n)} t_{p,k+1}^{(n)} \\ &= \lim_{n\to\infty} \sqrt{n}\, g_{i,k+1}^{(n)} \frac{t_{k+1,k+1}^{(n)}}{\sqrt{n}} + \lim_{n\to\infty} \sum_{p=1}^{k} \sqrt{n}\, g_{i,p}^{(n)} \frac{t_{p,k+1}^{(n)}}{\sqrt{n}} \\ &= \lim_{n\to\infty} \sqrt{n}\, g_{i,k+1}^{(n)} . \end{aligned}$$

This proves the theorem.

References

[1] S. Bochner, Harmonic Analysis and the Theory of Probability. Univ. of Calif. Press (1955).

[2] R.A. Minlos, Generalized random processes and their extension to measures, Trudy Moskow. Math. Obsc. 8(1959), 497-518.

[3] D. Shale, Invariant integration over the infinite dimensional orthogonal group and related spaces, Trans. Amer. Math. Soc., 124(1966), 148-157.

[4] H. Shimomura, Invariant measure on the ∞-dimensional orthogonal group. A report submitted to Kyoto Univ. (1970). (in Japanese)

[5] Y. Yamasaki, Invariant measure of the infinite dimensional rotation group, Publ. RIMS, Kyoto Univ., 8(1972/73), 131-140.

[6] " , Projective limit of Haar measures on O(n), Publ. RIMS, Kyoto Univ., 8(1972/73), 141-149.

[7] H. Yoshizawa, Rotation group of Hilbert space and its application to Brownian motion, Proceedings of the International Conference on Functional Analysis and Related Topics, Tokyo, (1969), 414-423.

Nagoya University
Nagoya, JAPAN

ON MOMENT INEQUALITIES AND IDENTITIES FOR STOCHASTIC INTEGRALS

A.A.Novikov

In this paper, we give new conditions, which provide the validity of moment inequalities and identities for stochastic integrals with respect to a Wiener process. In a certain sense these conditions are non-improvable.

1. Let $(\Omega,\mathcal{F},P)$ be a probability space, $\{\mathcal{F}(t)\}$, $t\geq 0$, a non-decreasing family of sub- σ -algebras $\mathcal{F}$ and $W(t)=(w_i(t))$, $1\leq i\leq m$, a m -vector standard Wiener process with respect to $\{\mathcal{F}(t)\}$. We shall denote by $H_2[0,T]$ the class of random functions $f(t,\omega)=(f_{ij}(t,\omega))$, $1\leq j\leq n$, which are jointly measurable, adapted to $\{\mathcal{F}(t)\}$ and for which $\int_0^T|f(t,\omega)|^2dt<\infty$ a.s. (here $|f(t,\omega)|^2\equiv\sum_{i,j}f^2_{ij}(t,\omega)$). For random functions from $H_2[0,T]$, an n-vector stochastic integral $\int_0^T f(t,\omega)\,dW(t)$ is defined ([1],[2]).

Theorem 1. Let $f(t,\omega)\in H_2[0,T]$ and

$$E\Big(\int_0^T|f(t,\omega)|^2dt\Big)^{P/2}<\infty.$$

Then there exist positive constants A_p and B_p depending only on p, such that

$$A_pE\Big(\int_0^T|f(t,\omega)|^2dt\Big)^{P/2}\leq E\Big|\int_0^T f(t,\omega)dW(t)\Big|^P\leq B_pE\Big(|f(t,\omega)|^2dt\Big)^{P/2},$$

where the left inequality holds for $p>1$ and the right inequality holds for $p>0$.

The method of proving this theorem (see [3]) consists in

applying Ito's formula to a specially chosen function

$$\left(\delta + c\int_0^T |f(t,\omega)|^2 dt + \left|\int_0^T f(t,\omega)\,dw(t)\right|^2\right)^{p/2},$$

where δ and c are some auxiliary positive constants, and then some simple transformations. Earlier Zakai [4] used a similar method for deducing moment inequalities of another kind. Theorem 1 for $p>1$ can be obtained also from results of Millar [5], where an analogous assertion was proved by using martingale transformations of Burkholder [6].

Insignificantly modifyng the proof of Theorem 1 in [3] we obtained the following values for the constants:

$$A_p = \begin{cases} p/2\,(p-1)(2/p-1)^{1-p/2}, & 1<p\le 2; \\ (2/p)^{p/2}, & p\ge 2; \end{cases} \qquad B_p = \begin{cases} (2/p)^{p/2}, & 0<p\le 2, \\ (p)^{p/2}-(3-p)^{p/2}, & 2\le p<3, \\ [p/2\,(p-1)]^{p/2}, & p\ge 3. \end{cases}$$

Theorem 1 gives the left inequality only for $p>1$. The next example shows that this circumstance is not caused by the method used, but is an effect of specific properties of stochastic integrals. Let $f(t,\omega)=\chi\{\tau_b \ge t\}$, where $\chi\{\ \}$ is the indicator function, and $\tau_b \equiv \inf\{t\ge 0: w(t)\le b\sqrt{t}-1\}$, $b>0$, be a stopping time for a Wiener process. For the chosen function $\int_0^\infty f^2(t,\omega)\,dt=\tau_b$ and $\int_0^\infty f(t,\omega)\,dw(t)=w(\tau_b)=b\sqrt{\tau_b}-1$ a.s.. It is easy to show that $E\tau_b^{1/2}=1/b$ ([7]). So, in case $p=1$ and $T=\infty$, the left inequality of Theorem 1 would be equivalent to the next: $A_1 E\tau_b^{1/2}=A_1/b\le E|w(\tau_b)|$. Since b is an arbitary positive number and $E|w(\tau_b)|\le 2$, the constant A_1 cannot be positive.

2. The moment inequalities considered above allow to prove the next assertion about moment identities (see [7]), which is formulated below for scalar stochastic integrals.

Theorem 2. Let $f(t,\omega) \in H_2[0,T]$ and

$$E\Big(\int_0^T |f(t,\omega)|^2 dt\Big)^{n/2} < \infty .$$

Then

$$(1)\quad E\left\{\Big(\int_0^T |f(t,\omega)|^2 dt + \alpha\Big)^{n/2} H_{e_n}\left(\frac{\int_0^T f(t,\omega)dw(t)+\beta}{\big(\int_0^T |f(t,\omega)|^2 dt+\alpha\big)^{1/2}}\right)\right\} = \alpha^{n/2} H_{e_n}\big(\beta/\sqrt{\alpha}\big).$$

Here $H_{e_n}(z)$ is the Hermite polinomial $\big(He_n(z) = (-1)^n \exp(z^2/2) \cdot \frac{d^n}{dz^n}\exp(-z^2/2),\ n=0,1,\ldots\big)$ and α, β are any constants.

This theorem in case $n=1$ asserts that the expectation of the stochastic integral equals zero under the condition

$$(2)\qquad E\Big(\int_0^T |f(t,\omega)|^2 dt\Big)^{1/2} < \infty$$

which is not so strong as the well-known condition $E\int_0^T |f(t,\omega)|^2 dt < < \infty$.

In particular, if τ is any stopping time for a Wiener process, then $Ew(\tau)=0$ provided $E\tau^{1/2} < \infty$. This result was obtained also by Burkholder and Gundy [8], who used a different method.

In general condition (2) cannot be dispensed with. Indeed, if in the example considered above we put $b=0$, then $Mw(\tau_0)=1$ and $M\tau_0^{1/2-\delta}$ for any $\delta>0$, but, of course, $M\tau_0^{1/2}=\infty$.

The formula (1) can be useful when studying properties of some stopping times. For example, define $\tau(a,b,c) \equiv \inf\{t \geq 0:$

$w(t)+a \leq b\sqrt{t+c}\ \}$, where $c \geq 0$ and $b\sqrt{c} < a$, and let z_n be the largest root of the equation $He_n(z)=0$. Then choosing suitable constants α , β and function $f(t,\omega)$, we find out that, if $b > z_n$, then

$$E(\tau(a,b,c)+c)^{n/2} = \frac{c^{n/2} He_n(a/\sqrt{c})}{He_n(b)}$$

and, if $b \leq z_n$, then $E\tau^{n/2}(a,b,c) = \infty$.

3. The moment identities (1) are closely connected with the considered below exponential identity (4), playing a fundamental role in some aspects of the theory of stochastic differential equations. In the following theorem we give one new sufficient condition, which provides the validity of (4) for scalar stochastic integrals.

Theorem 3. Let $f(t,\omega) \in H_2[0,T]$ and

(3) $$E \exp\left(1/2 \int_0^T |f(t,\omega)|^2 dt\right) < \infty .$$

Then

(4) $$E \exp\left\{\int_0^T f(t,\omega)\, dw(t) - 1/2 \int_0^T |f(t,\omega)|^2 dt\right\} = 1 .$$

The problem of finding sufficient conditions for the validity of (4) was raised first by Girsanov [9].

Theorem 3 with $(1+\delta)$, $\delta > 0$, replacing $1/2$ in the condition (3) can be found in [1]. McKean [2] considered the case, when $f(t,\omega)$ is pure imaginary.

Proof. We shall give a proof of Theorem 3 for one special case. The general case is similar in principle, although the details are somewhat more complicated.

Let $f(t,\omega)=\chi\{\tau \geq t\}$, where τ is some stopping time for a Wiener process. In this case we must show that, if

$$E\exp(1/2\,\tau)<\infty, \tag{5}$$

then

$$E\exp\{1/2\,w(\tau)-1/2\,\tau\}=1. \tag{6}$$

Put $\tau_a \equiv \inf\{t\geq 0: w(t)\leq t-a\}$, $0<a<\infty$. The distribution of this random variable is well-known and it is easy to calculate that $E\exp(1/2\,\tau_a)=\exp(a)<\infty$. Since $w(\tau_a)-1/2\,\tau_a = 1/2\,\tau_a - a$ a.s., it follows that $E\exp\{w(\tau_a)-1/2\,\tau_a\}=1$. Since the process $\{\exp(w(t)-1/2\,t), \mathcal{F}(t)\}$ is a martingale, we have by Doob's theorem

$$E\exp\{w(\tau\wedge\tau_a)-1/2\,\tau\wedge\tau_a\}=1,$$

where $\tau\wedge\tau_a$ is the minimum of τ and τ_a . This relation can be rewritten in the next equivalent form

$$E\chi\{\tau_a\leq\tau\}\exp\{1/2\,\tau_a-a\}+E\chi\{\tau<\tau_a\}\exp\{w(\tau)-1/2\,\tau\}=1. \tag{7}$$

By the condition (5)

$$E\chi\{\tau_a\leq\tau\}\exp\{1/2\,\tau_a-a\}\leq E\exp\{1/2\,\tau-a\}\to 0$$

as $a\to\infty$. Since $P\{\tau_a\to\infty\}=1$ as $a\to\infty$, in the second term of (7) we can monotonic convergence theorem. The proof is completed.

Now we show that, generally speaking, condition (3) cannot be dispensed with. Indeed, let $\tau_c \equiv \inf\{t \geq 0 : w(t) \leq ct - 1\}$, $0 < c < 1$. It is easy to find that $E\exp(1/2\, c^2 \tau_c) = \exp(c) < \infty$, but $E\exp(1/2\, \tau_c) = \infty$. On the other hand, direct calculations give

$$E\exp\{w(\tau_c) - 1/2\,\tau_c\} = \exp\{2(c-1)\} < 1.$$

Therefore, choosing a suitable constant c, for any $\delta > 0$, we can construct such a stopping time τ, for which $E\exp\{(1/2 - \delta)\tau\} < \infty$, but the identity (6) fails to hold and, of course, $E\exp\{1/2\,\tau\} = \infty$.

REFERENCES

[1] И.И.Гихман, А.В.Скороход, Стохастические дифференциальные уравнения, Киев, "Наукова думка", 1968 г.

[2] H.P.McKean, Stochastic integrals, AP, New-York, 1969.

[3] А.А.Новиков, О моментных неравенствах для стохастических интегралов, Теория вероят. и ее примен., XVI, 3(1971),548-550.

[4] M.Zakai, Some moment inequalities for stochastic integrals and for solutions of stochastic differential equations, Israel J. Math., 5, 3(1967), 170-176.

[5] P.W.Millar, Martingale integrals, Trans. Amer. Math. Soc., 135, 1(1968), 145-166.

[6] D.L.Burkholder, Martingale transforms, Ann. Math. Statist., 37, 6(1966), 1494-1505.

[7] А.А.Новиков, О моментах остановки винеровского процесса, Теория вероят. и ее примен., XVI, 3(1971), 458-465.

[8] D.L.Burkholder, R.F.Gundy, Extrapolation and interpolation of quasi-linear operators on martingales, Acta Math., 124, 3-4(1970), 249-304.

[9] И.В.Гирсанов, О преобразовании одного класса случайных процессов с помощью абсолютно-непрерывной замены меры, Теория вероят. и ее примен., У, 3(I960), 3I4-330.

**Steklov Mathematical Institute
of the Academy of Sciences of the USSR
Moscow**

ON A MARKOVIAN PROPERTY OF GAUSSIAN PROCESSES

Yasunori Okabe

§1. Introduction

In this talk(*) we shall consider real-valued L^2-continuous Gaussian processes $\mathbb{X} = (X(x) \; ; \; x \in \mathbb{R}^d)$ on a probability space $(\Omega, \mathbb{F}, P)$ with mean zero and correlation functions $R = R(x, y)$ whose time spaces are the d-dimensional Euclidean space $\mathbb{R}^d$ and discuss a new Markovian property more general than the usual one in Markov processes. The former concept was given by K. Urbanik [13] and H. P. McKean, Jr. [7]. Roughly speaking, $\mathbb{X}$ has the Markovian property in a domain D in $\mathbb{R}^d$ if and only if, conditioned by the knowledge in an arbitrary small neighbourhood of the boundary ∂D, the future (the information in the exterior of D) is independent of the past (the one in the interior of D).

We shall give the precise definition of Markovian property. For any open set D in $\mathbb{R}^d$ we define a sub-σ-field $\mathbb{F}(D)$ of $\mathbb{F}$ by

(1.1) $\mathbb{F}(D)$ = the smallest σ-field for which all $X(x)$ $(x \in D)$ are measurable.

Then, we define the following sub-σ-fields of $\mathbb{F}(\mathbb{R}^d)$ for each open set D in $\mathbb{R}^d$.

Definition 1.1. For each open set D in $\mathbb{R}^d$ we put

$$\mathbb{F}^+(D) = \bigcap_{n=1}^{\infty} \mathbb{F}((D^c)_n) \qquad \text{(the future)},$$

(*) Adding to the content of this proceeding, I talked that for any stationary Gaussian process with Markovian property there exists an infinite-dimensional simple Markov process whose infinitesimal generator characterizes the original stationary Gaussian process completely. The detailed content will be published elsewhere.

$$\mathbb{F}^-(D) = \bigcap_{n=1}^{\infty} \mathbb{F}(D_n) \qquad \text{(the past)},$$

$$\partial\mathbb{F}(D) = \bigcap_{n=1}^{\infty} \mathbb{F}((\partial D)_n) \qquad \text{(the germ)},$$

where for any set B in $\mathbb{R}^d$ and any $n \in \mathbb{N}$ B_n denotes the set $\{x \in \mathbb{R}^d ;$ distance $(x, B) < \frac{1}{n}\}$.

Now, following H. P. McKean, Jr. [7], we can state the definition of the Markovian property.

Definition 1.2. We say that the process $\mathbf{X}$ has the Markovian property in an open set D in $\mathbb{R}^d$ if

$$(1.2) \qquad \mathbb{F}^-(D) \underset{\partial\mathbb{F}(D)}{\perp\!\!\!\perp} \mathbb{F}^+(D),$$

which means that the future field $\mathbb{F}^+(D)$ is independent of the past field $\mathbb{F}^-(D)$, conditioned by the knowledge of the germ field $\partial\mathbb{F}(D)$.

It was P. Lévy's Brownian motion that gave rise to H. P. McKean's study [7]. It is a Gaussian process with continuous paths whose correlation function R is given by

$$(1.3) \qquad R(x, y) = \frac{1}{2}(|x| + |y| - |x - y|) \quad (x, y \in \mathbb{R}^d).$$

In the sequel, we shall assume that d is odd :

$$(1.4) \qquad d = 2p - 1.$$

Then, it is easy to see that for any $y \in \mathbb{R}^d$

$$(1.5) \qquad \Delta_x^p R(x, y) = c_1 \cdot (\delta(x - y) - \delta(x)),$$

where c_1 is a positive constant and Δ denotes the Laplacian.

At this point, for any Gaussian process $\mathbf{X}$, we shall rephrase the Markovian property (1.2) by means of the linear manifolds. For any open set

D in $\mathbb{R}^d$ we define a closed subspace $\mathbb{M}(D)$ of $L^2(\Omega, \mathcal{F}, P)$ by

(1.6) $\quad \mathbb{M}(D)$ = the closed linear hull of $\{X(x) ; x \in D\}$.

Similarly as Definition 1.1, we give

<u>Definition 1.3.</u> For each open set D in $\mathbb{R}^d$ we set

$$\mathbb{M}^+(D) = \bigcap_{n=1}^{\infty} \mathbb{M}((D^c)_n) \qquad \text{(the future)},$$

$$\mathbb{M}^-(D) = \bigcap_{n=1}^{\infty} \mathbb{M}(D_n) \qquad \text{(the past)},$$

$$\partial\mathbb{M}(D) = \bigcap_{n=1}^{\infty} \mathbb{M}((\partial D)_n) \qquad \text{(the germ)}.$$

Let $\mathcal{P}_D$ be the projection of $\mathbb{M}(\mathbb{R}^d)$ onto $\mathbb{M}^-(D)$. Then, it is clear that for each open set D in $\mathbb{R}^d$

(1.7) $$\partial\mathbb{M}(D) \subset \mathbb{M}^+(D) \cap \mathbb{M}^-(D) \subset \mathcal{P}_D\mathbb{M}^+(D).$$

Since $\mathbb{X}$ is a Gaussian process, it can be proved ([7]) that (1.2) is equivalent to

(1.8) $$\mathcal{P}_D\mathbb{M}^+(D) = \partial\mathbb{M}(D).$$

Now, we shall consider Lévy's Brownian motion again. It follows from (1.5) that for any open set D and any $y \in D$

(1.9) $$\Delta_x^p R(x, y) = 0 \quad \text{in} \quad (\overline{D} \cup 0)^c.$$

This implies that $R(\cdot, y)$ $(y \in D)$ is a solution of the exterior Dirichlet problem for the differential operator Δ^p. Representing the projection $\mathcal{P}_D X(x)$ $(x \notin \overline{D})$ by means of the Green function for the exterior problem, H. P. McKean, Jr. [7] showed

Theorem 1.1. ([7]) Lévy's Brownian motion with an odd-dimensional time parameter satisfies (1.8) for any bounded domain D containing the origin with smooth boundary.

G. M. Molchan [8] gave an alternative proof of Theorem 1.1 with the aid of Hilbert spaces with reproducing kernels. Before explaining the idea in [8], we shall give a condition equivalent to (1.8) by means of a functional space. Let X be any Gaussian process with $R(x, y)$ correlation function. We denote by $\mathcal{H}$ the reproducing kernel Hilbert space whose reproducing kernel is R. It is known ([1]) that the space $\mathcal{H}$ is a subspace of the space of continuous functions on $\mathbb{R}^d$ containing $R(\cdot, y)$ $(y \in \mathbb{R}^d)$ as its element and it has the next properties:

(1.10) $\quad (u, R(\cdot, y))_{\mathcal{H}} = u(y)$ for any $u \in \mathcal{H}$ and any $y \in \mathbb{R}^d$;

(1.11) $\quad$ {finite linear combinations $\sum_{j=1}^{n} c_j R(\ , y_j)$; $c_j \in \mathbb{R}$, $y_j \in \mathbb{R}^d$, $n \in \mathbb{N}$} is dense in $\mathcal{H}$.

Next, we define for any open set D in $\mathbb{R}^d$ the closed subspace $\mathcal{H}(D)$ of $\mathcal{H}$ by

(1.12) $\quad \mathcal{H}(D)$ = the closed linear hull of $\{R(\cdot,y) \ ; \ y \in D\}$.

Then, similarly as Difinition 1.3, we give

Definition 1.4. For each open set D in $\mathbb{R}^d$ we denote

$$\mathcal{H}^+(D) = \bigcap_{n=1}^{\infty} \mathcal{H}((D^c)_n) \qquad \text{(the future)},$$

$$\mathcal{H}^-(D) = \bigcap_{n=1}^{\infty} \mathcal{H}(D_n) \qquad \text{(the past)},$$

$$\partial\mathcal{H}(D) = \bigcap_{n=1}^{\infty} \mathcal{H}((\partial D)_n) \qquad \text{(the germ)}.$$

By (1.10) and (1.12), it is clear that

(1.13) $\quad$ the orthogonal complement of $\mathcal{H}(D) = \{u \in \mathcal{H}\ ;\ u = 0 \text{ in } D\}$.

Since there exists a unitary operator from $\mathbf{M}(\mathbb{R}^d)$ onto $\mathcal{H}$ under which $X(y)$ $(y \in \mathbb{R}^d)$ corresponds to $R(\cdot, y)$, it follows from (1.8) that (1.2) is equivalent to

(1.14) $$\mathcal{P}_D \mathcal{H}^+(D) = \partial\mathcal{H}(D),$$

where $\mathcal{P}_D$ denotes the projection of $\mathcal{H}$ onto $\mathcal{H}^-(D)$.

Again, we shall return to Lévy's Brownian motion. By using the uniqueness for the interior and exterior Dirichlet problem, G. M. Molchan [8] characterized the spaces $\mathcal{H}(D)$ as follows.

Theorem 1.2. ([8]) *For any bounded domain* D *with smooth boundary,* $\mathcal{H}(D)$ *is equal to the subspace of* $u \in W^p_{L^2}(\mathbb{R}^d)$ *such that*

(i) $u(0) = 0$,

(ii) $\Delta^p u = 0$ *in* $(\overline{D} \cup 0)^c$,

(iii) *if* p *is even* $2k$, *then* $\Delta^k u \in L^2(\mathbb{R}^d)$,
or if p *is odd* $2k+1$, *then* $\Delta^k u \in L^2_{loc}(\mathbb{R}^d) \cap W^1_{L^2}(\mathbb{R}^d)$.

As an application of Theorem 1.2, G. M. Molchan [8] proved Theorem 1.1.

§2. The purpose of this talk

As we have seen in section one, it is the interior and exterior Dirichlet problem in the theory of differential equations upon which the studies in H. P. McKean, Jr. [7] and G. M. Molchan [8] depend. After characterizing the spaces $\mathcal{H}(D)$ by means of Sobolev spaces, G. M. Molchan [8] showed the Markovian property of Lévy's Brownian motion with an odd-dimensional time parameter space R^d $(d = 2p - 1)$. In this case, it is further proved in [7] and [8] that Lévy's Brownian motion has a p-ple Markovian property in some sense.

Recently, G. M. Molchan [9] and L. D. Pitt [11] considered (stationary) Gaussian processes X more general than Lévy's Brownian motion whose correlation functions are fundamental solutions of uniformly elliptic and self-adjoint differential operators with smooth variable coefficients, and showed, using the same idea as in G. M. Molchan [8], that X has the Markovian property in any bounded domain with smooth boundary and moreover that X has a finite multiple Markovian property in some sense. In general, it is not easy to characterize the reproducing kernel Hilbert spaces by means of Sobolov spaces.

However, this procedure is unnecessary as far as the Markovian property (1.2) is concerned. We shall mention this point in §3. In fact, in [4] and [10], with the aid of the theory of M. Sato's hyperfunctions [12], we have treated the case of what is called an infinitely multiple Markovian property and showed that a stationary Gaussian process whose spectral density is the inverse of an entire function of infra-exponential type has the Markovian property in any bounded open convex set under the additional assumption, where the subspaces $\mathcal{H}(D)$ are characterized in the total space $\mathcal{H}(R^d)$ from the point of view of Fourier hyperfunctions ([2], [12]). For the purpose of explaining the fundamental feature of [4] and [10], we shall give in §4 an another proof the result of H. P. McKean, Jr. [7] and G. M. Molchan [8] about Lévy's Brownian motion, following the idea in [4] and [10].

We give one more remark here. It is announced in [11] that if a stationary Gaussian process has the Markovian property, then its spectral density is the inverse of an entire function of infra-exponential type. By using the theory of ultra-distributions [3], S. Kotani [5] showed this fact under the same assumptions as in [4].

§3. An infinitely multiple Markovian property

At first, we shall consider any purely non-deterministic stationary Gaussian process $\mathbb{X} = (X(t) ; t \in \mathbb{R}^1)$ whose time parameter space is $\mathbb{R}^1$. Then, it is well known ([10]) that $\mathbb{X}$ has the following canonical representation : for any $t \in \mathbb{R}$

(3.1) $$X(t) = \int_{-\infty}^{t} E(t - s)dB(s),$$

(3.2) $$\mathbf{F}((-\infty, t)) = \sigma(dB(s) ; s < t),$$

where E is a real L^2-function vanishing in the negative axis and $(B(t) ; t \in \mathbb{R})$ is a Brownian motion.

We denote by h the inverse Fourier transformation :

(3.3) $$h(\lambda) = \frac{1}{2\pi}\int_0^{\infty} e^{i\lambda t}E(t)dt.$$

Then, N. Levinson - H. P. McKean, Jr. [6] showed

<u>Theorem 3.1.</u> ([6]) <u>The process</u> $\mathbb{X}$ <u>has the Markovian property in</u> $(-\infty, 0)$ <u>if and only if</u>

(3.4) $h(-\cdot)$ <u>is an inverse of an entire function</u> P <u>of infra-exponential type.</u>

We recall that an entire function f on $\mathbb{C}^d$ is said to be of infra-exponential type if for any $\varepsilon > 0$ there exists some constant $c_\varepsilon > 0$ such that for any $z \in \mathbb{C}^d$

(3.5) the estimate $|f(z)| \leq c_\varepsilon e^{\varepsilon|z|}$ holds.

Let P be any entire function of such a type expanded in the form

(3.6) $$P(\cdot) = \sum_{n=0}^{\infty} c_n(-i\cdot)^n \text{ in } \mathbb{C}.$$

Then, formally, we can associate a differential operator $P(\frac{1}{i}\frac{d}{dt})$ of infinite order :

(3.7) $$P(\frac{1}{i}\frac{d}{dt}) = \sum_{n=0}^{\infty} c_n(-\frac{d}{dt})^n.$$

Generally, this operator does not make sense in the space of L. Schwartz's distributions. However, in the space of M. Sato's hyperfunctions more general than the former, we can regard this operator as a local operator ([12]). Based upon this point, I gave an alternative proof of Theorem 3.1 from the point of view of operator-theory ([10]). In doing so, it is not necessary to characterize the reproducing kernel Hilbert space by means of Sobolev spaces. It is indeed impossible because we treat the case of what is called an infinitely multiple Markovian property.

Next, we shall consider a stationary Gaussian process $\mathbf{X} = (X(x)\ ;\ x \in \mathbb{R}^d)$ with the spectral density Δ whose parameter space is $\mathbb{R}^d$. Then, following the idea in [10], S. Kotani - Y. Okabe [4] proved

Theorem 3.2. ([4]) Let's assume that the spectral density Δ satisfies the following conditions :

(i) Δ is an inverse of an entire function of infra-exponential type ;

(ii) There exists a positive number t_0 and a non-negative increasing continuous function $T(t)$ $(t \in [t_0, \infty))$ such that

(a) $$\Delta^{-1}(x) \leq e^{T(|x|)} \quad \text{for any } x \in \mathbb{R}^d,\ |x| \geq t_0,$$

(b) $$\int_{t_0}^{\infty} \frac{T(t)}{1+t^2}\, dt < \infty.$$

Then, $\mathbf{X}$ has the Markovian property in any bounded open convex set in $\mathbb{R}^d$.

Remark 3.1. It follows from (3.1) and (3.3) that (3.4) in Theorem 3.1

implies that the spectral density of X is an inverse of an entire function of infra-exponential type.

Remark 3.2. The condition (ii) in Theorem 3.2 is stronger than the purely non-deterministicness in the one dimensional case. In fact, it implies that

$$(3.8) \qquad \bigcap_{t>0} \mathcal{H}(x \in \mathbb{R}^d \ ; \ |x| > t) = \{0\}.$$

§4. Lévy's Brownian motion with multi-dimensional time

For the purpose of explaining the idea in [4] and [10], we shall consider Lévy's Brownian motion whose time parameter space is an odd-dimensional Euclidean space $\mathbb{R}^d$ and prove the next theorem, which is stronger than Theorem 1.1 in the sense that now the Markovian property holds in any open set in $\mathbb{R}^d$.

Theorem 4.1. Let's assume that $P = \frac{d+1}{2}$ is even. Then, Lévy's Brownian motion whose time space is $\mathbb{R}^d$ has the Markovian property in any open set in $\mathbb{R}^d$.

It follows from (1.3) that

$$(4.1) \qquad R(x, y) = c_2{}^2 \int_{\mathbb{R}^d} \frac{e^{ix\cdot\lambda} - 1}{|\lambda|^p} \, \frac{e^{-iy\cdot\lambda} - 1}{|\lambda|^p} \, d\lambda,$$

where c_2 is a positive constant. For any $x \in R^d$ we define an L^2-function $h(x, \lambda)$ by

$$(4.2) \qquad h(x, \lambda) = c_2 \, \frac{e^{ix\cdot\lambda} - 1}{|\lambda|^p} .$$

Further, we denote by $E(x, y)$ the Fourier transformation of $h(x, \lambda)$:

$$(4.3)\qquad E(x, y) = \int_{\mathbb{R}^d} e^{-iy\cdot\lambda} h(x, \lambda)d\lambda.$$

Then, it follows from (4.1), (4.2) and (4.3) that

$$(4.4)\qquad R(x, y) = (h(x, \cdot), h(y, \cdot))_{L^2} = (2\pi)^{-2d}(E(x, \cdot), E(y, \cdot))_{L^2}.$$

By using the uniqueness of Fourier transformations, it is easy to see that

(4.5) if an L^2-function f satisfies

$$(h(x, \cdot), f)_{L^2} = 0 \quad \text{for any} \quad x \in \mathbb{R}^d,$$

then $f = 0$.

Therefore, by (1.10), (1.11), (4.4) and (4.5), we have

<u>Lemma 4.1.</u> <u>There exists uniquely a unitary operator</u> K <u>from</u> $\mathcal{H}$ <u>onto a real</u> L^2<u>-space</u> $L^2(\mathbb{R}^d)$ <u>such that</u>

$$(4.6)\qquad K(R(\cdot, y)) = (2\pi)^{-d}E(y, \cdot) \quad \text{for any} \quad y \in \mathbb{R}^d.$$

<u>Moreover, the next relation holds</u> :

$$(4.7)\qquad u(x) = (2\pi)^{-d}(E(x, \cdot), Ku)_{L^2} \quad \underline{\text{for any}} \quad u \in \mathcal{H} \ \underline{\text{and any}} \quad x \in \mathbb{R}^d.$$

Noting (4.2) and (4.3), we see that

$$(4.8)\qquad \begin{aligned} \Delta_x^{\frac{p}{2}}E(x, y) &= c_2(2\pi)^d\delta(y - x), \\ \Delta_x^{\frac{p}{2}}E(x, y) &= c_2(2\pi)^d(\delta(x - y) - \delta(y)). \end{aligned}$$

Combining (4.7) and (4.8), we find that

$$(4.9)\qquad \Delta^{\frac{p}{2}}u = c_2 Ku \quad \text{for any} \quad u \in \mathcal{H}.$$

In particular, it follows from (4.6), (4.7), (4.8) and (4.9) that

$$\Delta_x^p R(x, Y) = c_2^{\,2}(\delta(x - y) - \delta(x)). \tag{4.10}$$

Now, we shall prove the fundamental Lemma 4.2 in our talk.

Lemma 4.2. Let u be any element of $\mathcal{H}$ and D be any open set in $\mathbb{R}^d$.

(i) If u belongs to $\mathcal{H}(D)$, then there exist a distribution $T \in \mathcal{D}(\mathbb{R}^d)'$ and some constant c such that

$$\Delta^p u = T + c\cdot\delta, \quad \text{supp } T \subset \overline{D}. \tag{4.11}$$

Conversely,

(ii) if u satisfies (4.11), then u belongs to $\mathcal{H}^-(D)$.

Proof. (i) Assume that u belongs to $\mathcal{H}(D)$. Then, it follows from (1.12) that there exists a sequence $(u_n)_{n=1}^{\infty}$ in $\mathcal{H}$ convergent to u in $\mathcal{H}$ such that each u_n has the form

$$u_n = \sum_{j=1}^{\ell_n} c_j^n R(\cdot, y_j^n), \quad y_j^n \in D.$$

Since the convergence in $\mathcal{H}$ implies the uniform convergence on any compact set, it can be seen by (4.10) that

$$\lim_{n\to\infty} c_2^{\,2} \sum_{j=1}^{\ell_n} c_j^n(\delta(\cdot - y_j^n) - \delta(\cdot)) = \Delta^p u \quad \text{in } \mathcal{D}(\mathbb{R}^d)'. \tag{4.12}$$

If the origin $0 \in \overline{D}$, then (4.12) implies that the support of $\Delta^p u$ is contained in $\overline{D}$. Let's consider the case where $0 \notin \overline{D}$. Then, taking a function in $\mathcal{D}(\mathbb{R}^d)$ which is zero in D and is one at the origin, we can see by (4.12) that $-c_2^{\,2} \sum_{j=1}^{\ell_n} c_j^n$ is convergent to some $c \in \mathbb{R}$. Therefore, in this case, it follows from (4.12) that u satisfies (4.11). Next, we shall show (ii).

Let's consider any $n \in N$ and any $v \in \mathcal{H}(D_n)^{\perp}$ (the orthogonal complement of $\mathcal{H}(D_n)$). At first, we note that there exist an open set U and some $\delta > 0$ such that

(4.13) $$\overline{D} \subset U \subset D_n,$$

(4.14) $$\text{If } x \in U, \ |y| < \delta, \text{ then } x - y \in D_n.$$

Take any $\varphi \in C_0^\infty(|x| < \delta)$ and any $\psi \in C_0^\infty(\mathbb{R}^d)$, $\psi(0) = 1$. Then, we define $\psi_m \in C_0^\infty(\mathbb{R}^d)$ by $\psi_m(\cdot) = \psi(\frac{\cdot}{m})$ $(m \in \mathbb{N})$.

It follows from (1.13) and (4.14) that

$$\psi_m(\varphi * v) = 0 \text{ in } U.$$

Hence, by (4.11) and (4.13), we get

$$<\Delta^p u, \ \psi_m(\varphi * v)> = c \ \varphi * v(0).$$

On the other hand, noting (4.9), we find that

$$\lim_{m\to\infty} <\Delta^p u, \ \psi_m(\varphi * v)> = c_2{}^2 \int_{R^d} Ku\varphi * Kv \, dx.$$

So that,

(4.15) $$c_2{}^2 \int_{\mathbb{R}^d} Ku\varphi * Kv \, dx = c \ \varphi * v(0).$$

Letting φ tend to a δ-function in (4.15), we see that

$$c_2{}^2 (Ku, Kv)_{L^2} = cv(0).$$

Since $R(\ , 0) = 0$ ((1.3)), it follows from (1.10) that $v(0) = 0$.

Thus,

$$(Ku, Kv)_{L^2} = 0.$$

By Lemma 4.1, this implies that $(u, v) = 0$. Since n is arbitrary and

v is any element of $\mathcal{H}(D_n)^{\perp}$, this yields that u belongs to $\mathcal{H}^-(D)$.

(Q. E. D.)

After these preparations, we can prove Theorem 4.1. Let D be any open set in $\mathbb{R}^d$. By (1.14), it suffices to show that for any $u \in \mathcal{H}^+(D)$

$$v = \mathcal{P}_D u \in \partial\mathcal{H}(D). \tag{4.16}$$

We set $w = u - v$. Since w is orthogonal to $\mathcal{H}(D)$, it follows from (1.13) that $w = 0$ in D and so

$$\Delta^p w = 0 \quad \text{in} \quad D. \tag{4.17}$$

Take any $n \in N$ and fix it. By Definition 1.4,

$$u \in \mathcal{H}((D^c)_m) \quad \text{and} \quad v \in \mathcal{H}(D_m) \quad \text{for any} \quad m \geq n.$$

By using Lemma 4.2, we see that there exist distributions $T_m^{(j)}$ and constants $c_m^{(j)}$ $(j = 1,\ 2,\ m \geq n)$ such that

$$\Delta^p u = T_m^{(1)} + c_m^{(1)}\delta, \quad \operatorname{supp} T_m^{(1)} \subset \overline{(D^c)_m}, \tag{4.18}$$

$$\Delta^p v = T_m^{(2)} + c_m^{(2)}\delta, \quad \operatorname{supp} T_m^{(2)} \subset \overline{D_m}. \tag{4.19}$$

Therefore, considering whether the origin belongs to ∂D or not and then noting (4.17), we can take m larger than n such that

$$\operatorname{supp} T_m^{(2)} \subset (\partial D)_m. \tag{4.20}$$

Hence, it follows from Lemma 4.2 (ii), (4.19) and (4.20) that

$$v \in \mathcal{H}^-((\partial D)_m) \subset \mathcal{H}((\partial D)_n).$$

Since n is arbitrary, this implies (4.16). Thus, we have proved Theorem 4.

References

[1] N. Aronszajn : Theory of reproducing kernels, Trans. Amer. Math. Soc. 68 (1950), 337-404.

[2] T. Kawai : On the theory of Fourier hyperfunctions and its applications to partial differential equations with constant coefficients, J. Fac. Sci. Univ. Tokyo Sect. IA 17 (1971), 467-517.

[3] H. Komatsu : On ultra-distributions, Katada Symposium, 1971, to appear in Lecture notes in Math., Springer-Verlag.

[4] S. Kotani and Y. Okabe : On a Markovian property of stationary Gaussian process with a multi-dimensional parameter, Katada Symposium, 1971, to appear in Lecture notes in Math., Springer-Verlag.

[5] S. Kotani : On a Markovian property of stationary Gaussian processes with a multi-dimensional parameter, II, Proceedings of the Second Japan-USSR Symposium on Probability Theory.

[6] N. Levinson and H. P. McKean, Jr. : Weighted trigonometrical approximation on $\mathbb{R}^1$ with application to the germ field of a stationary Gaussian noise, Acta Math. 112 (1964), 99-143.

[7] H. P. McKean, Jr. : Brownian motion with a several dimensional time, Theor. Probability Appl. 8 (1963), 357-378.

[8] G. M. Molchan : On some problems concerning Brownian motion in Lévy's sense, Theor. Probability Appl. 12 (1967), 682-690.

[9] G. M. Molchan : Characterization of Gaussian fields with Markov property, Dokl. Akad. Nauk USSR 197 (1971), 784-787. (in Russian).

[10] Y. Okabe : Stationary Gaussian processes with Markovian property and M. Sato's hyperfunctions, to appear in J. Math. Soc. Japan.

[11] L. D. Pitt : A Markov property for Gaussian processes with a multidimensional parameter, Arch. Rational Mech. Anal. 43 (1971), 367-395.

[12] M. Sato : Theory of hyperfunctions I, II, J. Fac. Sci. Univ. Tokyo Sect. I. 8 (1959), 139-193, 387-437.

[13] K. Urbanik : Generalized stationary processes of Markovian character, Studia Math. 21 (1962), 261-282.

Department of Mathematics
Faculty of Science
Osaka University

Current Address
Department of Mathematics
Faculty of Science
Nagoya University

THE LOG LOG LAW FOR CERTAIN DEPENDENT RANDOM SEQUENCES

Hiroshi Oodaira

1. Introduction

The purpose of this paper is to extend Strassen's law of the iterated logarithm for independent identically distributed (i.i.d.) random variables [7] to certain classes of dependent random sequences.

We first consider the following stationary random sequence:

$$(1) \qquad X_j = \sum_{k=-\infty}^{\infty} c_{k-j}\xi_k \ , \ j = 1, 2, \ldots , \text{ with } \sum_{k=-\infty}^{\infty} c_k^2 < \infty ,$$

where ξ_k , $k = 0, \pm 1, \ldots$, are i.i.d. random variables with $E\xi_k = 0$ and $E\xi_k^2 = 1$. In a recent paper [2] Yu. A. Davydov proved the following result. Set $S_0 = 0$ and $S_n = \sum_{j=1}^{n} X_j$. Define a random sequence $\{g_n(t), 0 \le t \le 1, n = 1, 2, \ldots\}$ in the space $C[0,1]$ of all continuous functions on $[0,1]$ vanishing at the origin with the sup norm $\|\cdot\|$, by

$$g_n(t) = s_n^{-1/2}\{S_{[nt]} + (nt-[nt])(S_{[nt]+1}-S_{[nt]})\} ,$$

where $s_n^2 = ES_n^2$. If $E|\xi_k|^{2i} < \infty$, $i \ge 2$, and $s_n^2 \sim n^{\gamma}$, $2/(i+2) < \gamma \le 2$, then the sequence of probability measures corresponding to $\{g_n(t)\}$ converges weakly to the probability measure corresponding to the Gaussian process with mean zero and covariance kernel

$$(2) \qquad \Gamma_{\gamma}(s,t) = (1/2)\{s^{\gamma}+t^{\gamma}-|s-t|^{\gamma}\} , \ 0 < \gamma \le 2 , \ 0 \le s,t \le 1.$$

A question naturally arises as to whether Strassen's log log law can be extended to the stationary sequence (1) considered by Davydov. A partial answer is given by the following theorem 1. Define a random sequence $\{f_n(t), 0 \le t \le 1, n = 3, 4, \ldots\}$ in $C[0,1]$ by

$$(3) \qquad f_n(t) = (2s_n^2 \log_2 n)^{-1/2}\{S_{[nt]} + (nt-[nt])(S_{[nt]+1} - S_{[nt]})\} ,$$

where $\log_2 n = \log\log n$.

Theorem 1. Suppose that

(a) $E|\xi_k|^{2i} < \infty$, $i \ge 2$,

(b) $s_n^2 = ES_n^2 \sim n^{\gamma}$, $2/i < \gamma \le 2$,

(c) $\sum_{k=0}^{n} |c_k| = O(n^{\beta})$, $\sum_{k=-n}^{0} |c_k| = O(n^{\beta})$, with $\beta < (\gamma/2)-(1/4)$,

(d) $\sum_{k=n^{\delta}+1}^{\infty} (\sum_{j=1}^{n} c_{k-j})^2 = O(n^{\lambda(\delta)})$, $\sum_{k=-\infty}^{-n^{\delta}-1} (\sum_{j=1}^{n} c_{k-j})^2 = O(n^{\lambda(\delta)})$,

with $\lambda(\delta) < \gamma - (2/i)$ for some $\delta < (\gamma/2)/(\beta+(1/4))$.
Then, with probability one, the sequence of functions $\{f_n(t)\}$ is relatively compact and the set of its limit points is contained in the unit ball of the reproducing kernel Hilbert space (RKHS) with kernel (2).

Example. Let $c_k = 0$ for $k \ge 0$, $= (-k)^{-\alpha}$ for $k \le -1$, with $1/2 < \alpha < 1$. If $i > 2/(2\alpha-1)$, then conditions (a)-(d) are satisfied and the conclusion of Theorem 1 holds with $\gamma = 3-2\alpha$.

Next we consider the following random sequence:

(4) $$S_0 = 0,\ S_n = \sum_{k=0}^{n-1} (n-k)^{\alpha}\xi_k\ ,\quad \alpha \geq 0\ ,$$

where ξ_k , $k = 0, 1, \ldots$, are i.i.d. with $E\xi_k = 0$, $E\xi_k^2 = 1$ and $E|\xi_k|^{2+\delta} < \infty$ for some $\delta > 0$. We define $f_n(t)$ for (4) as in (3). Then we have

<u>Theorem 2</u>. <u>With probability one the sequence</u> $\{f_n(t)\}$ <u>is relatively compact and the set of its limit points coincides with the unit ball of the RKHS with kernel</u>

(5) $$\Gamma(s,t) = (2\alpha+1) \int_0^{s\wedge t} (s-\lambda)^{\alpha}(t-\lambda)^{\alpha}d\lambda\ ,\quad 0 \leq s,t \leq 1.$$

Similarly, if

(6) $$S_0 = 0,\ S_n = \sum_{k=0}^{n-1} \left(1-\frac{k}{n}\right)^{\alpha}\xi_k\ ,\quad \alpha \geq 0,$$

then

<u>Theorem 3</u>. <u>With probability one the sequence</u> $\{f_n(t)\}$ <u>for</u> (6) <u>is relatively compact and the set of its limit points is the unit ball of the RKHS with kernel</u>

$$\Gamma(s,t) = (2\alpha+1) \int_0^{s\wedge t} \left(1-\frac{\lambda}{s}\right)^{\alpha}\left(1-\frac{\lambda}{t}\right)^{\alpha}d\lambda\ .$$

As an immediate corollary to Theorem 3 we have

$$\lim_n \sup S_n/([2/(2\alpha+1)]n \log_2 n)^{1/2} = 1 \text{ a.s. } .$$

This is a version of the log log law, obtained by V. F. Gaposhkin [3], for uniformly bounded ξ_k with variable distributions.

The method of proof is essentially that of V. Strassen [7]. We imbed ξ_k in Brownian motion by Skorohod's representation theorem, and, using Strassen's and Breiman's results [8], [1], we reduce the problem to that of Gaussian processes. The proof will be completed by applying the following theorems which are slight modifications of the ones proved in our earlier paper [5] (see also [4]).

Theorem 4. *Let* $X(t)$, $t \geq 0$, *be a real separable Gaussian process with* $X(0) = 0$, *mean zero and continuous covariance kernel* $R(s,t)$. *Define a sequence of functions* $\{f_n^*(t), n = 3, 4, \ldots\}$ *in* $C[0,1]$ *by*

$$f_n^*(t) = (2R(n,n)\log_2 n)^{-1/2} X(nt) , \quad 0 \leq t \leq 1.$$

Suppose that, for any $T > 0$, *there exist a positive function* $v(r) \uparrow \infty$, $r \geq 0$, *a positive monotone nondecreasing function* $g(x,T)$, $x \geq 0$, *and a covariance kernel* $\Gamma(s,t)$, $0 \leq s,t \leq 1$, *such that*

(a) $$\sup_{0\leq s,t\leq 1} |v^{-1}(r)R(rs,rt) - \Gamma(s,t)| \to 0, \quad r \to \infty ,$$

(b) $$|R(rs,rs)-2R(rs,rt)+R(rt,rt)| < v(r)g(|s-t|,T)$$

for all $r \geq 0$ *and* $0 \leq s,t \leq T$,

(c) $$\{g(1,T)\}^{-1/2} \int_1^\infty g^{1/2}(e^{-u^2},T)du \leq C < \infty ,$$

(d) $v(T)/g(1,T) \uparrow \infty$ *as* $T \to \infty$,

(e) $\Gamma(t,t)$ *is monotone increasing and* $\Gamma(s,t)$ *is strictly positive definite, and*

(f) $\Gamma(1,1) = 1$.

<u>Then, with probability one, the sequence of functions</u> $\{f_n^*(t)\}$ <u>is relatively compact and the set of its limit points is contained in the unit ball of the RKHS with kernel</u> $\Gamma(s,t)$.

Let $L(X,t)$ denote the closed linear manifold spanned by $\{X(s), 0 \le s \le t\}$ and let $L^*(X,r\delta) = \bigcap_{h>0} L(X,r\delta+h)$ for $0 \le \delta < 1$ and let $L'(X,r\delta)$ be its orthogonal complement in $L(X,r)$. Let $X^*_{r\delta}(t)$ and $X'_{r\delta}(t)$ denote the projections of $X(t)$, $0 \le t \le r$, on $L^*(X,r\delta)$ and $L'(X,r\delta)$ respectively. Set

$$R^*_{r\delta}(s,t) = EX^*_{r\delta}(s)X^*_{r\delta}(t) \ , \quad 0 \le s,t \le r \ ,$$

$$R'_{r\delta}(s,t) = EX'_{r\delta}(s)X'_{r\delta}(t) \ , \quad 0 \le s,t \le r \ .$$

<u>Theorem 5.</u> <u>Suppose that, in addition to conditions</u> (b), (c), (d) <u>and</u> (f) <u>of Theorem</u> 4, <u>there exist, for each</u> $0 \le \delta < 1$, <u>kernels</u> $\Gamma^*_\delta(s,t)$ <u>and</u> $\Gamma'_\delta(s,t)$, $0 \le s,t \le 1$, <u>such that</u>

(a') $$\sup_{0\le s,t\le 1} |v^{-1}(r)R^*_{r\delta}(rs,rt) - \Gamma^*_\delta(s,t)| = o((\log r)^{-1}), \text{ and}$$

$$\sup_{0\le s,t\le 1} |v^{-1}(r)R'_{r\delta}(rs,rt) - \Gamma'_\delta(s,t)| = o((\log r)^{-1}),$$

(b') $H(\Gamma) = H(\Gamma^*_\delta) \oplus H(\Gamma'_\delta)$, <u>where</u> $\Gamma(s,t) = \Gamma^*_\delta(s,t) + \Gamma'_\delta(s,t)$,

(c') $\Gamma^*_\delta(t,t) \to 0$ <u>as</u> $\delta \to 0$, <u>uniformaly in</u> $0 \le t \le 1$, <u>and</u>

(d') $\Gamma'_\delta(t,t)$ <u>is monotone increasing and</u> $\Gamma'_\delta(s,t)$, $\delta \le s,t \le 1$, <u>is strictly positive definite</u>.

<u>Then, with probability one, the set of limit points of</u> $\{f_n^*(t)\}$ <u>contains the unit ball of the RKHS with kernel</u> $\Gamma(s,t)$.

2. Proof of Theorem 1.

By Skorohod's representation theorem, $\{\xi_k, k = 1, 2, \ldots\}$ and $\{\xi_k, k = 0, -1, \ldots\}$ can be imbedded respectively in independent Brownian motions $B(t)$, $B'(t)$, $t \geq 0$, with $B(0) = B'(0) = 0$, in such a way that the sequences $\{B(T_k)-B(T_{k-1}), k = 1, 2, \ldots\}$ and $\{B'(T'_{k+1})-B'(T'_k), k = 0, -1, \ldots\}$ have the same distributions as $\{\xi_k, k = 1, 2, \ldots\}$ and $\{\xi_k, k = 0, -1, \ldots\}$ respectively, where T_k and T'_k are sums of nonnegative i.i.d. random variables with $T_0 = T'_0 = 0$. We shall write $B(t) = -B'(-t)$ for $t \leq 0$ and $T_k = -T'_{-k}$, so that $B(T_k)-B(T_{k-1}) = B'(T'_{1-k})-B'(T'_{-k})$ for $k = 0, -1, \ldots$. Since

$$S_n = \sum_{j=1}^{n} X_j = \sum_{k=-\infty}^{\infty} \left(\sum_{j=1}^{n} c_{k-j}\right)\xi_k ,$$

the random variables

$$\sum_{k=-\infty}^{\infty} \left(\sum_{j=1}^{n} c_{k-j}\right)[B(T_k)-B(T_{k-1})] , \quad n = 1, 2, \ldots , \tag{7}$$

are distributed in the same way as S_n. It is hence sufficient to prove the theorem for (7). We shall denote (7) by S_n as well.

Set

$$S_{n,N} = \sum_{k=-N}^{N} \left(\sum_{j=1}^{n} c_{k-j}\right)[B(T_k)-B(T_{k-1})] \quad \text{and} \quad S'_{n,N} = S_n - S_{n,N} .$$

Then, by assumption (d),

$$ES'^{\,2}_{n,n^\delta} = \sum_{k=-\infty}^{-n^\delta-1} \left(\sum_{j=1}^{n} c_{k-j}\right)^2 + \sum_{k=n^\delta+1}^{\infty} \left(\sum_{j=1}^{n} c_{k-j}\right)^2 \leq Cn^{\lambda(\delta)} ,$$

and it is easy to see that $E(S'_{n,n^\delta})^{2i} \leq C'(E(S'_{n,n^\delta})^2)^i$. Note that

$ES^2_{n,n^\delta} \sim s^2_n$. Let

$$f_{n,n^\delta}(t) = (2s_n^2 \log_2 n)^{-1/2}\{S_{[nt],n^\delta}+(nt-[nt])(S_{[nt]+1,n^\delta} - S_{[nt],n^\delta})\} .$$

Lemma 1. $P(\ \lim_n \sup\ ||f_n - f_{n,n^\delta}|| \geq \varepsilon\ ) = 0$ for any $\varepsilon > 0$.

Proof. Since f_n and f_{n,n^δ} are piecewise linear, it suffices to show that

$$P(\ \lim_n \sup\ \{\ \max_{1\leq k\leq n} |S'_{k,k^\delta}| \geq \varepsilon(2s_n^2\log_2 n)^{1/2}\}\ ) = 0.$$

We have

$$\begin{aligned} P(\ \max_{1\leq k\leq n} |S'_{k,k^\delta}| \geq \varepsilon(2s_n^2\log_2 n)^{1/2}) &\leq \sum_{k=1}^{n} P(|S'_{k,k^\delta}| \geq \varepsilon(2s_n^2\log_2 n)^{1/2}) \\ &\leq \sum_{k=1}^{n} \{E(S'_{k,k^\delta})^{2i}/\varepsilon^{2i}(2s_n^2\log_2 n)^i\} \\ &\leq (C_1/s_n^{2i}) \sum_{k=1}^{n} (E(S'_{k,k^\delta})^2)^i \\ &\leq (C_2/s_n^{2i}) \sum_{k=1}^{n} k^{i\lambda(\delta)} \\ &\leq C_3 n^{i(\lambda(\delta)-\gamma)+1} \quad \text{for sufficiently large } n. \end{aligned}$$

Since $i(\lambda(\delta)-\gamma)+1 < -1$ by assumption (d), the Borel-Cantelli lemma completes the proof.

Set now

$$S^*_{n,n^\delta} = \sum_{k=-n^\delta}^{n^\delta} \left(\sum_{j=1}^{n} c_{k-j} \right) [B(k)-B(k-1)] \ , \quad \text{and}$$

$$f^*_{n,n^\delta}(t) = (2s_n^2 \log_2 n)^{-1/2} \{ S^*_{[nt],n^\delta} + (nt-[nt])(S^*_{[nt]+1,n^\delta} - S^*_{[nt],n^\delta}) \} \ .$$

Then we have

<u>Lemma 2</u>. $P(\lim_n \|f_{n,n^\delta} - f^*_{n,n^\delta}\| = 0) = 1$.

<u>Proof</u>. $|S_{n,n^\delta} - S^*_{n,n^\delta}|$

$$= \left| \sum_{k=-n^\delta}^{n^\delta} \left(\sum_{j=1}^{n} c_{k-j} \right) \{ [B(T_k)-B(k)] - [B(T_{k-1})-B(k-1)] \} \right|$$

$$= \left| \left(\sum_{j=1}^{n} c_{n^\delta - j} \right) [B(T_{n^\delta}) - B(n^\delta)] + \sum_{k=-n^\delta}^{n^\delta} (c_{k-n} - c_k) [B(T_k) - B(k)] - \left(\sum_{j=1}^{n} c_{-n^\delta - j} \right) [B(T_{-n^\delta - 1}) - B(-n^\delta - 1)] \right|$$

$$\le \left\{ \left| \sum_{j=1}^{n} c_{n^\delta - j} \right| + \sum_{k=1}^{n^\delta} |c_{k-n} - c_k| \right\} \max_{1 \le k \le n^\delta} |B(T_k) - B(k)| + \left\{ \sum_{k=-n^\delta}^{-1} |c_{k-n} - c_k| + \left| \sum_{j=1}^{n} c_{-n^\delta - j} \right| \right\} \max_{-n^\delta - 1 \le k \le -1} |B(T_k) - B(k)| \ .$$

Now, by assumption (c) and a theorem of Strassen ([8], p.320, Theorem 1.5),

$$\le C_1 (n^\beta + n^{\delta\beta}) n^{\delta/4} (\log n^\delta)^{1/2} (\log_2 n^\delta)^{1/4} + C_1 (n^\delta + n)^\beta n^{\delta/4} (\log n^\delta)^{1/2} (\log_2 n^\delta)^{1/4}$$

$$\leq \begin{cases} C_2 n^{\delta(\beta+(1/4))}(\log n^{\delta})^{1/2}(\log_2 n^{\delta})^{1/4} & \text{for } \delta \geq 1, \\ C_2 n^{\beta+(1/4)}(\log n)^{1/2}(\log_2 n)^{1/4} & \text{for } \delta \leq 1. \end{cases}$$

We thus have

$$\max_{1\leq k\leq n} |S_{k,k^{\delta}} - S^*_{k,k^{\delta}}|/(2s_n^2 \log_2 n)^{1/2}$$

$$\leq \begin{cases} C_3 n^{\delta(\beta+(1/4))-(\gamma/2)}(\log n^{\delta})^{1/2}(\log_2 n^{\delta})^{1/4} & \text{for } \delta \geq 1, \\ C_3 n^{\beta+(1/4)-(\gamma/2)}(\log n)^{1/2}(\log_2 n)^{1/4} & \text{for } \delta \leq 1. \end{cases}$$

By assumption (d), $\delta(\beta+(1/4))-(\gamma/2) < 0$ for $\delta \geq 1$, and by assumption (c), $\beta+(1/4)-(\gamma/2) < 0$ for $\delta \leq 1$. Hence

$$\max_{1\leq k\leq n} |S_{k,k^{\delta}} - S^*_{k,k^{\delta}}|/(2s_n^2 \log_2 n)^{1/2} \to 0 \quad \text{a.s.} \quad \text{as} \quad n \to \infty.$$

Since $f_{n,n^{\delta}}$ and $f^*_{n,n^{\delta}}$ are piecewise linear, this proves the lemma.

Let

$$S_n^* = \sum_{k=-\infty}^{\infty} \left(\sum_{j=1}^{n} c_{k-j}\right)[B(k)-B(k-1)] ,$$

$$Y(t) = S^*_{[t]} + (t-[t])(S^*_{[t]+1} - S^*_{[t]}) \quad \text{for} \quad t \geq 0, \text{ and}$$

$$f_n^*(t) = (2s_n^2 \log_2 n)^{-1/2} Y(nt) \quad \text{for} \quad 0 \leq t \leq 1, \; n = 3, 4, \ldots .$$

Then, just as Lemma 1, we have

Lemma 3. $P(\limsup_n ||f_n^* - f_{n,n^\delta}^*|| \geq \varepsilon) = 0$ for any $\varepsilon > 0$.

From Lemmas 1-3 it follows that

$$P(\lim_n ||f_n - f_n^*|| = 0) = 1,$$

and so it suffices to prove that the conclusion of the theorem holds for $\{f_n^*(t)\}$. Since $Y(t)$ is Gaussian, we need only to verify that conditions (a)-(f) of Theorem 4 are satisfied for $Y(t)$.

Put $R(s,t) = EY(s)Y(t)$. Then

$$R(s,t) = (1/2)\{EY^2(s)+EY^2(t)-E|Y(s)-Y(t)|^2\}.$$

An elementary calculation shows that

$$|EY^2(rt)/r^\gamma - t^\gamma| \leq C_1 r^{-\gamma/2}, \text{ and}$$

$$|E|Y(rs)-Y(rt)|^2/r^\gamma - |s-t|^\gamma| \leq C_2 r^{-\gamma/2},$$

and so

$$\sup_{0\leq s,t\leq 1} |r^{-\gamma}R(rs,rt)-(1/2)\{s^\gamma+t^\gamma-|s-t|^\gamma\}| \to 0 \quad \text{as} \quad r \to \infty.$$

We have also

$$E|Y(rs)-Y(rt)|^2 \leq r^\gamma C_3 |s-t|^\gamma \quad \text{for all} \quad r, s, t \geq 0.$$

Thus conditions (a)-(f) of Theorem 4 are satisfied with $v(r) = r^\gamma$, $g(x,T) = C_3|x|^\gamma$ and $\Gamma(s,t) = (1/2)\{s^\gamma+t^\gamma-|s-t|^\gamma\}$. This completes the proof.

Remark. If $\gamma = 1$, it may be shown that conditions of Theorem 5 are also fulfilled, and the set of limit points of $\{f_n(t)\}$ is the unit ball of the RKHS associated with Brownian motion. In this connection, see also [6].

3. Proof of Theorem 2.

Again, by Skorohod's representation theorem, ξ_k, $k = 0, 1, \ldots$, can be imbedded in a Brownian motion $B(t)$, $t \geq 0$, so that $B(T_{k+1}) - B(T_k)$, $k = 0, 1, \ldots$, are distributed in the same way as ξ_k, $k = 0, 1, \ldots$, where $T_0 = 0, T_1, T_2, \ldots$, are sums of nonnegative i.i.d. random variables. It is hence sufficient to prove the theorem for $\{f_n(t)\}$ with

$$S_0 = 0, \quad S_n = \sum_{k=0}^{n-1} (n-k)^{\alpha} [B(T_{k+1}) - B(T_k)], \quad n = 1, 2, \ldots .$$

Let

$$S_0^* = 0, \quad S_n^* = \sum_{k=0}^{n-1} (n-k)^{\alpha} [B(k+1) - B(k)], \quad n = 1, 2, \ldots, \text{ and}$$

$$f_n^*(t) = (2s_n^2 \log_2 n)^{-1/2} \{S_{[nt]}^* + (nt - [nt])(S_{[nt]+1}^* - S_{[nt]}^*)\}$$
$$\text{for } 0 \leq t \leq 1.$$

Note that $s_n^2 = ES_n^{*2} = \sum_{k=0}^{n-1} (n-k)^{2\alpha} \sim (2\alpha+1)^{-1} n^{2\alpha+1}$. In Lemma 4 the following result of L. Breiman [1] will be used:

$$|B(T_k) - B(k)| = o(k^{1/(2+\delta)} (\log k)^{1/2}) \quad \text{a.s.}, \; 0 < \delta < 2.$$

Lemma 4. $P(\lim_n ||f_n - f_n^*|| = 0) = 1.$

Proof. Since f_n and f_n^* are piecewise linear, it suffices to show that

$$(2s_n^2 \log_2 n)^{-1/2} \max_{1\le j\le n} |S_j - S_j^*| \to 0 \quad \text{a.s.} \quad \text{as} \quad n \to \infty.$$

If j ($\le n$) is sufficiently large,

$$|S_j - S_j^*| = |\sum_{k=0}^{j-1} (j-k)^{\alpha}\{[B(T_{k+1})-B(k+1)]-[B(T_k)-B(k)]\}|$$

$$= |\sum_{k=1}^{j} \{(j-k+1)^{\alpha}-(j-k)^{\alpha}\}[B(T_k)-B(k)]|$$

$$\le C_1 j^{\alpha} j^{1/(2+\delta)} (\log j)^{1/2}$$

$$\le C_1 n^{\alpha} n^{1/(2+\delta)} (\log n)^{1/2}$$

and hence

$$(2s_n^2 \log_2 n)^{-1/2} \max_{1\le j\le n} |S_j - S_j^*| \le C_2 n^{-1/2} n^{1/(2+\delta)} (\log n)^{1/2} \to 0 \quad \text{a.s.}$$

The lemma is proved.

By Lemma 4 it is enough to prove the theorem for $\{f_n^*(t)\}$. Let

$$Y(t) = S^*_{[t]} + (t-[t])(S^*_{[t]+1} - S^*_{[t]}) \quad \text{for} \quad t \ge 0.$$

Then $Y(t)$ is a Gaussian process with $EY(t) = 0$, $EY^2(n) = s_n^2$, and we have

$$f_n^*(t) = (2s_n^2 \log_2 n)^{-1/2} Y(nt) \ , \quad 0 \le t \le 1.$$

By an elementary calculation we see that the covariance kernel $R(s,t) = EY(s)Y(t)$ satisfies the following relations:

$$\sup_{0\le s,t\le 1} \left| (2\alpha+1) r^{-(2\alpha+1)} R(rs,rt) - (2\alpha+1) \int_0^{s\wedge t} (s-\lambda)^{\alpha}(t-\lambda)^{\alpha} d\lambda \right| = O(r^{-1})$$

and

$$E|Y(rs)-Y(rt)|^2 = \begin{cases} C_1 r^{2\alpha+1} |s-t|^{2\alpha} T & \text{if } 0 \le \alpha \le 1, \\ C_2 r^2 |s-t|^2 T^{2\alpha-1} & \text{if } \alpha \ge 1, \end{cases}$$

$$\text{for all } r \ge 0, \ 0 \le s,t \le T.$$

Thus conditions (a)-(f) of Theorem 4 are satisfied with $v(r) = (2\alpha+1)^{-1} r^{2\alpha+1}$, $g(x,T) = C_3|x|^{2\alpha} T$ for $0 \le \alpha \le 1$, $= C_3|x|^2 T^{2\alpha-1}$ for $\alpha \ge 1$ and covariance kernel (5).

Now, for $0 \le \delta < 1$ and $0 \le t \le 1$,

$$Y^*_{r\delta}(rt) = \begin{cases} Y(rt) & \text{for } t \le \delta, \\ \sum_{k=0}^{r\delta-1} (r\delta-k)^{\alpha}[B(k+1)-B(k)] & \text{for } t > \delta \text{ and } r\delta = [r\delta], \\ \sum_{k=0}^{[r\delta]} [([rt]-k)^{\alpha}+(rt-[rt])\{([rt]+1-k)^{\alpha}-([rt]-k)^{\alpha}\}] \cdot [B(k+1)-B(k)] & \text{for } t > \delta \text{ and } r\delta > [r\delta]. \end{cases}$$

Let

$$R_{r\delta}(rs,rt) = EY^*_{r\delta}(rs)Y^*_{r\delta}(rt) \ ,$$

$$\Gamma^*_{\delta}(s,t) = (2\alpha+1) \int_0^{\delta\wedge s\wedge t} (s-\lambda)^{\alpha}(t-\lambda)^{\alpha} d\lambda \ , \quad \text{and}$$

$$\Gamma'_\delta(s,t) = \Gamma(s,t) - \Gamma^*_\delta(s,t) .$$

Then we have

$$\sup_{0\le s,t\le 1} |(2\alpha+1)r^{-(2\alpha+1)}R_{r\delta}(rs,rt) - \Gamma^*_\delta(s,t)| = O(r^{-1}) ,$$

$$\sup_{0\le t\le 1} \Gamma^*_\delta(t,t) \to 0 \quad \text{as} \quad \delta \to 0 ,$$

and we see that conditions (a')-(d') of Theorem 5 are satisfied. This completes the proof.

The proof of Theorem 3 is quite similar to that of Theorem 2 and is, hence, omitted.

References

[1] L. Breiman: On the tail behavior of sums of independent random variables, Z. Wahrscheinlichkeitstheorie verw. Geb. 9 (1967), 20-25.

[2] Yu. A. Davydov: The invariance principle for stationary processes, Theory of Prob. and its Appl. 15 (1970), 487-498.

[3] V. F. Gaposhkin: The law of the iterated logarithm for Cesaro's and Abel's methods of summation, Theory of Prob. and its Appl. 10 (1965), 411-420.

[4] H. Oodaira: On Strassen's version of the law of the iterated logarithm for Gaussian processes, Z. Wahrscheinlichkeitstheorie verw. Geb. 21 (1972), 289-299.

[5] H. Oodaira: The law of the iterated logarithm for Gaussian processes, (to appear).

[6] H. Oodaira and K. Yoshihara: Note on the law of the iterated logarithm for stationary processes satisfying mixing conditions,

Kodai Math. Sem. Rep. 23 (1971), 335-342.

[7] V. Strassen: An invariance principle for the law of the iterated logarithm, Z. Wahrscheinlichkeitstheorie verw. Geb. 3 (1964), 211-226.

[8] V. Strassen: Almost sure behavior of sums of independent random variables and martingales, Proc. Fifth Berkeley Symp. Math. Stat. Prob. Vol. 2, Univ. of California Press, (1967), 315-343.

Department of Applied Mathematics
Faculty of Engineering
Yokohama National University
Minami-Ku, Yokohama
Japan

THE CONCENTRATION FUNCTIONS OF SUMS OF INDEPENDENT RANDOM VARIABLES

B. A. Rogozin

1. One-dimensional case. Let $\xi_1, \dots, \xi_n, \dots$ be a sequence of independent random variables and $S_n = \sum_{k=1}^{n} \xi_k$. The concentration function of a random variable ξ is a function of the positive variable l defined by

$$Q(\xi, l) = \sup_{-\infty < x < \infty} P\{x \le \xi \le x + l\}.$$

We distinguish three types of results concerned with estimation of $Q(S_n, L)$:

i) Estimation of $Q(S_n, L)$ in the case when $\max_{1 \le k \le n} Q(\xi_k, L)$ is not small and $\min_{1 \le k \le n} Q(\xi_k, l_k)$ is not near 1.

ii) Estimation of $Q(S_n, L)$ in the case when $\max_{1 \le k \le n} Q(\xi_k, L)$ is small.

iii) Estimation of $Q(S_n, L)$ in the case when $\min_{1 \le k \le n} Q(\xi_k, l_k)$ is near 1.

In the paper, for the first two types of estimation, we only indicate the references.

Estimations of the first type can be found in the fundamental papers of Doeblin and Levy [2] , Levy [13] , Doeblin [3]and Kolmogorov [9] , [10] , and also in papers of Rogozin [20] , [21] . Some more precise estimations of $Q(S_n, L)$ of the first type were obtained by Kolmogorov [11] , Le Cam [12] , Esseen [6] , [7]. In [6] , [7] the characteristic function is used; in particular, the estimations of $Q(\xi, l)$ depend on the characteristic function of ξ (see also Rosén [22] , Petrov [18]). We note that, for some important special classes of distributions of ξ_K, $K=1,\dots,n$, the estimations of the first type are obtained in papers of Littlewood and Offord [14] , Offord [16], Erdös [5], Chung and Erdös [1], and also in the recent papers of Petrov[18] and Dunnage [4].

Estimations of the second type can be found under the hypothesis stated by Prokhorov [19] concerning $Q(S_n,\ell)$ provided that random variables $\xi_1,\xi_2,\dots,\xi_n$ are equally distributed and $Q(\xi_1,\ell)\le d$. The paper of Kesten [8] contains very interesting and general estimations of $Q(S_n,L)$ of the second type.

An estimation of the third type was proposed in the footnote to [21]. The proof of the estimation is not published yet. Here we would like to indicate the changes which are necessary to obtain the estimation:

$$Q(S_n,L)\le 2\Phi\left(\frac{C_1 L}{\sqrt{s}}\right)-1+\frac{C_2\max\limits_{1\le K\le n}\ell_K}{\sqrt{s}}, \tag{1}$$

where $\Phi(x)$ is the distribution function of the standard normal law, C_1 and C_2 are positive absolute constants and $s=\sum_{K=1}^{n}\ell_K^2(1-Q(\xi_K,\ell_K))$.

To prove (1) we use, instead of Lemma of [21] , the following one.

Lemma 1. Let $P\{\xi_K=\ell'_K\}=P\{\xi_K=-\ell'_K\}=\frac{1}{2}$ and $\ell_K\le\ell'_K$, $K=1,2,\dots,n$. There exists an absolute constant C_3 such that

$$Q(S_n,L)\le 2\Phi\left(\frac{\sqrt{2}L}{\sqrt{s}}\right)-1+\frac{C_3\max\limits_{1\le K\le n}\ell_K}{\sqrt{s}},$$

where $s=\sum_{K=1}^{n}4\ell_K^2$.

The proof of the lemma is similar to that of Lemma in [21]: we use, instead of Lemma 1 of [20] , the following result.

Lemma 2. Let $P\{\xi_K=\ell'_K\}=P\{\xi_K=-\ell'_K\}=\frac{1}{2}$ and $0<\ell\le\ell'_K$, $K=1,2,\dots n$. There exists an absolute constant C_4 such that

$$Q(S_n,L)\le 2\Phi\left(\frac{L}{2\ell\sqrt{n}}\right)-1+\frac{C_4}{\sqrt{n}}.$$

This inequality is a consequence of Theorem 3 of [5].

The proof of (1) is now similar to that of Theorem of [21].

2. Multidimensional case. A multidimensional generalization of the concentration function was given by Sevastyanov [24]. A vector $(x_1,\dots,x_N)$ in R^N will be denoted by x. The inner product of two vectors x and y in R^N is defined as $(x,y)=\sum_{k=1}^{N} x_k y_k$ and the norm of x as $|x|=\sqrt{(x,x)}$. By $\mathcal{M}_m$, $m=1,2,\dots,N$, we denote the class of all m-dimensional subspaces of R^N, and by λ_M - the Lebesgue measure on $M\in\mathcal{M}_m$.

The concentration function $Q_m(\xi,v)$, $m=1,2,\dots,N$, of a random vector ξ is a function of a positive variable v defined by

$$Q_m(\xi,v)=\sup_{A\in C_{m,v}} P\{\xi\in A\},\quad m=1,\dots,N,$$

where $C_{m,v}$ is the class of all convex closed sets $A\subset R^N$ with $d_m(A)\leq v$ where the m-dimensional diameter $d_m(A)$ of $A\subset R^N$ is defined by

$$d_m(A)=\sup_{x\in R^N}\ \sup_{M\in\mathcal{M}_m} \lambda_M(M\Pi(A+x)).$$

Let $\xi_1,\dots,\xi_n$ be a sequence of the independent random vectors and $S_n=\sum_{k=1}^{n}\xi_k$.

In [24], a hypothesis concerning the estimation of concentration functions of sums of independent identically distributed random vectors was formulated. This hypothesis has been proved by Sazonov [23] under certain weak additional conditions on the common probability distributions of $\xi_1,\dots,\xi_n$. In [23], generalizations of a combinatorial result of Sperner [25] obtained in [15] were used.

Estimations of the concentration function of S_n defined relative to N-dimensional intervals are given in [6] and [4], and general estimations of the concentration functions of S_n defined relative to spheres are obtained in [7]. Recently estimations of $Q_1(S_n,v)$ were obtained by Paulauskas [17].

In this paper, using the results of [7] and [23], we obtain some estimations of $Q_m(S_n, v)$, $m=1,2,\dots N$, from which the estimations of the hypothesis in [24] follow immediately.

Let ξ'_K be a random vector independent of ξ_K with the same distribution. We denote by $\overline{P}_K$ the probability distribution of $\xi_K - \xi'_K$ and put

$$\mathcal{X}_K(t) = \inf_{|x|=1} \int_{\{|y|\le t\}} (x,y)^2\, \overline{P}_K(dy), \quad K=1,2,\dots,n.$$

Define

$$\mathcal{Y}_n(u) = \sup_{t\ge u} \frac{u^2}{t^2} \sum_{K=1}^{n} \mathcal{X}_K(t).$$

Theorem.

$$(2) \qquad Q_K(S_n, v) \le C(N) \left(\rho_K(v) \vee u\right)^{N-K+1} \mathcal{Y}_n(u)^{-\frac{N-K+1}{2}},$$

where $\rho_K(v) = k\,(v \vee 1)$ and $u \vee t = \max\{u,t\}$.

Proof. Let $e_1, e_2, \dots, e_m$ be an orthogonal basis of $M \in \mathcal{M}_m$ and

$$\Sigma_t(M) = \{x:\ x \in R^N,\ \sum_{K=1}^{m} (x, e_K)^2 \le t^2\}.$$

Applying Theorem 6.2 from [7] to $P\{S_n \in \Sigma_t(M) + z\}$ we obtain:

$$(3) \qquad \sup_{z\in M} P\{S_n \in \Sigma_t(M) + z\} \le C(m) \cdot \frac{(t \vee v)^m}{\left(\sup_{u\ge v} \frac{v^2}{u^2} \sum_{K=1}^{n} \mathcal{X}_K(u,M)\right)^{\frac{m}{2}}},$$

where

$$\mathcal{X}_K(t,M) = \inf_{\substack{|z|=1\\ z\in M}} \int_{\{y\in \Sigma_t(M)\}} (z,y)^2\, \overline{P}_K(dy), \quad \left(\mathcal{X}_K(t) = \mathcal{X}_K(t, R^N)\right).$$

Since $\mathcal{X}_K(u,M) \ge \mathcal{X}_K(u)$, it follows from (3) that

$$(4) \qquad \sup_{z\in M} P\{S_n \in \Sigma_t(M) + z\} \le C(N)(t \vee v)^m\, \mathcal{Y}_n(v)^{-\frac{m}{2}}.$$

Denote by $C'_{m,v}$ the class of all convex closed sets $A \subset R^N$ for which $d'_m(A) = \min_{M\in\mathcal{M}_m} d_1(P_M A) \le v$ and by $P_M A$ the orthogonal projection of A on M. Using (4), we have

$$\sup_{A\in C'_{m,v}} P\{S_n\in A\}\le C(N)(v\vee u)^m \varphi_n(u)^{-\frac{m}{2}}.$$

Inequality (2) can be proved by induction. Applying inequality (5) with $m=N$, we obtain

$$Q_1(S_n,v)=\sup_{A\in C'_{N,v}} P\{S_n\in A\}\le C(N)(v\vee u)^N \varphi_n(u)^{-\frac{N}{2}}.$$

Now we can show that inequality (2) with $k=i+1$ follows from (2) with $k=i$. For every set $A\in C_{i+1,\,v}$, we have either $d_i(A)\le 1$ or $d_i(A)>1$. If $d_i(A)\le 1$, then, from (2) with $k=i$, it follows that

(6) $$P\{S_n\in A\}\le C(N)(\rho_i(1)\vee u)^{N-i+1}\varphi_n(u)^{-\frac{N-i+1}{2}}.$$

And from (6) it follows immediately that

(7) $$P\{S_n\in A\}\le C(N)(\rho_i(1)\vee u)^{N-i}\varphi_n(u)^{-\frac{N-i}{2}}.$$

Since $C(N)>1$, inequality (7) is evident provided

$$(\rho_i(1)\vee v)/\sqrt{\varphi_n(u)}\ge 1.$$

If $d_i(A)>1$, then there exists a subspace $M'\in\mathcal{M}_i$; and a vector $x\in R^N$ such that $d_i(A\cap(M'+x))>1$. And since $d_{i+1}(A)\le v$ it follows that $\sup_{z\in A}\inf_{y\in M'+x}|z-y|\le\rho_{i+1}(v)$ and hence $d_1(P_{M''}A)\le\rho_{i+1}(v)$, where M'' is the orthogonal complement to M'. Applying the inequality (5) with $m=N-i$, we have

(8) $$P\{S_n\in A\}\le C(N)(\rho_{i+1}(v)\vee u)^{N-i}\varphi_n(u)^{-\frac{N-i}{2}}.$$

From (7) and (8) we immediately obtain (2) with $k=i+1$.

References

[1] K.L. Chung and P. Erdös: Probability limit theorems assuming only the first moment, Mem.Amer.Math.Soc., (1951).

[2] W. Doeblin et P. Lévy: Sur les sommes de variables aléatoires à dispersions bornés inferieurement, C.R.Acad.Sci. Paris, 902 (1936), 2027-2029.

[3] W. Doeblin: Sur les sommes d'un grand nombre des variables aléatoires independantes, Bull.Sci.Math., 63(1939), 23-64.

[4] J.E.A. Dunnage: Inequalities for concentration of sums of independent random variables, Proc.London Math.Soc., (3) 23(1971), 489-514.

[5] P.Erdös: On a lemma of Littlewood and Offord, Bull.Amer.Math. Soc., 51(1945), 898-902,

[6] C.-G. Esseen: On the Kolmogorov-Rogozin inequality for the concentration function, Z. Wahrscheinlichkeitstheorie verw.Geb., 5(1966), 210-216.

[7] C.-G.Esseen: On the concentration function of a sum of independent random variables, Z. Wahrscheinlichkeitstheorie verw. Geb., 9(1968), 290-308.

[8] H. Kesten: A sharper form of the Doeblin-Levy-Kolmogorov-Rogozin inequality for concentration functions, Math.Scand.,25(1969), 133-144.

[9] A. Kolmogorov: Two uniform limit theorems for sums of independent random variables, Theor.Prob.Appl.,1(1956), 384-394.

[10] A. Kolmogorov: Sur les proprietes des fonctions de concentrations de M.P.Lévy, Ann.Inst.Henri Poincaré, 16(1958), 27-34.

[11] A. Kolmogorov: On the approximation of distributions of sums of independent summands by infinitely divisible distributions, Sankhya, Ser A 25 (1963), 159-174.

[12] L.LeCam: On the distribution of sums of independent random variables, Bernoulli, Bayes, Laplace, ed. by J.Neyman and L. LeCam, Berlin-Heidelberg-New York: Springer, 1965, 179-220.

[13] P. Lévy: Theorie de l'addition des variables aleatoires, 2 ed., Gauthier-Villars, Paris 1954.

[14] J.E. Littlewood and A.C. Offord: On the number of real roots of a random algebraic equation, III, Recueil Math., 5, 12(1943) 277-286.

[15] Л.Д.Мешалкин, Обобщение теоремы Шпернера о числе подмножеств конечного множества, Теория вероят. и ее примен., 8(1963), 219-220.

[16] A.C. Offord: An inequality for sums of independent random variables, Proc.London Math.Soc.,(2) 48 (1945), 467-477.

[17] В.И.Паулаускас, О функциях концентрации случайных векторов, ДАН СССР, 204(1972), 791-794.

[18] В.В.Петров, Об оценке функций концентраций суммы независимых случайных величин, Теория вероят. и ее примен., 15(1970), 718-721.

[19] Ю.В.Прохоров, Экстремальные задачи в предельных теоремах, Труды 6 Всесоюзного совещания по теории вероятностей и математической статистике, Вильнюс, 1962, 77-84.

[20] B.A. Rogozin: An estimate for concentration functions, Theor. Probab.Appl., 6(1961), 94-97.

[21] B.A. Rogozin: On the increase of dispersion of sums of independent random variables, Theor.Probab.Appl.,6(1961),97-99.

[22] B.Rosén: On the asymptotic distribution of sums of independent random variables, Ark.Mat., 4(1961), 323-332.

[23] В.В.Сазонов, О многомерных функциях концентрации, Теория вероят. и ее примен., 11 (1966), 683-690.

[24] Б.А.Севастьянов, О многомерных функциях концентрации, Теория вероят. и ее примен., 8 (1963), 124-125.

[25] E.Sperner: Ein Satz über Untermengen einer endlichen Menge, Math.Z., 27(1928), 544-548.

Institute of Mathematics of the Academy of Sciences of the USSR, Novosibirsk

ON LIE GROUP STRUCTURE OF SUBGROUPS OF O(S)

Hiroshi Sato

Let S be a real nuclear space of all rapidly decreasing functions on the real line. We say a linear homeomorphism g of S onto itself is a rotation if it satisfies

$$\int_{-\infty}^{+\infty} |\xi(x)|^2 dx = \int_{-\infty}^{+\infty} |g\xi(x)|^2 dx$$

for every ξ in S and denote by $O(S)$ the group of all rotations. Evidently $O(S)$ is a subgroup of the orthogonal group O_∞ of the Hilbert space $L^2(R^1)$.

Every g in $O(S)$ induces an automorphism on the probability space of the Gaussian White Noise and consequently every one-parameter subgroup induces a flow. So we are interested in the investigation of $O(S)$. Especially we are interested in constructing one-parameter subgroups and clarifying their probabilistic meanings. For these purposes the theory of Lie group may be useful, that is, if we can find an infinite dimensional Lie group structure in $O(S)$ or its subgroup, then every element in the associated Lie algebra induces a one-parameter subgroup through the exponential map and consequently, a flow. It is very nice if we can find probabilistic meanings of elements in the associated Lie algebra.

T. Hida has investigated Lie subgroups of the unitary group consisting of several one-parameter subgroups whose probabilistic meanings are recognized, while H. Nomoto has investigated the Haar measure on O_∞.

In this paper, we will investigate the Lie group structure of subgroups of $O(S)$. Of course $O(S)$ is a subgroup of O_∞ and O_∞ is a Banach Lie group modeled on the Banach space of all continuous skew-symmetric linear operators on $L^2(R^1)$. But in concordance with this structure, O_∞ does not include important one-parameter subgroups. For example, O_∞ does not include the shift

$$S_t : \xi(x) \rightarrow \xi(x-t), \quad \xi \in S, \qquad t \in R^1.$$

We therefore try to find another structure which is more suitable for our purpose.

(I). It is well-known that S is isomorphic to $\mathscr{s}$, the space of all rapidly decreasing sequences with a countable system of norms defined by

$$\| x \|_p = [\sum_{n=1}^{+\infty} n^p x_n^2]^{\frac{1}{2}}, \qquad p = 1,2,3,\ldots,$$

for $x = \{x_n\}_{n=1}^{+\infty}$ in $\mathscr{s}$.

Let $o(\mathscr{s})$ be the group of all orthogonal matrices that induce homeomorphisms of $\mathscr{s}$. Then we can explicitly determine $o(\mathscr{s})$ as follows :

<u>Theorem 1.</u> <u>An orthogonal matrix</u> $\underline{a} = (a_{nm})_{n,m=1}^{+\infty}$ <u>belongs to</u> $o(\mathscr{s})$ <u>if and only if for every non-negative integer</u> p, <u>there exists a non-negative integer</u> $q = q(p)$ <u>such that</u>

$$(\ast) \qquad \sum_{m,n=1}^{+\infty} \frac{n^p}{m^q}(a_{mn}^2 + a_{nm}^2) < +\infty.$$

<u>Proof</u>. Let $\underline{a} = (a_{mn})$ be an orthogonal matrix, i.e.,

$$\sum_{k=1}^{\infty} a_{mk} a_{nk} = \sum_{k=1}^{\infty} a_{km} a_{kn} = \delta_{mn}, \qquad m,n = 1,2,3,\ldots .$$

Suppose $\underline{a}$ is in $o(\mathscr{s})$ and put $e^m = \{\delta_{mj}\}_{j=1}^{\infty}$, $m = 1,2,3,\ldots$. Since $\underline{a}$ is a homeomorphism of $\mathscr{s}$, for every positive integer p there exist a positive integer $q = q(p)$ and a positive constant c such that

$$\| \underline{a} x \|_p \leq c \| x \|_q$$

for every x in $\mathscr{s}$. Substituting x by e^m, we have

$$\sum_{n=1} n^p a_{nm}^2 = \| ae^m \|_p^2$$

$$\leq c^2 \| e^m \|_q^2 = c^2 m^q, \qquad m = 1,2,3,\ldots \ .$$

On the other hand, since $\underline{a}^{-1}$ is also a homeomorphism of $\mathcal{S}$ and given by the transposed matrix of $\underline{a} = (a_{nm})$, we can prove (★) immediately.

Conversely, assume that (★) is true for an orthogonal matrix $a = (a_{nm})$. Then there is no difficulty in proving that $\underline{a}$ is a homeomorphism of $\mathcal{S}$.

Corollary 1. An infinite permutation π belongs to $o(\mathcal{S})$ if and only if for every non-negative integer p, there exists a non-negative integer $q = q(p)$ such that

$$\sum_{n=1}^{\infty} \frac{\pi(n)^p + \pi^{-1}(n)^p}{n^q} < + \infty .$$

Proof. Since a permutation π is given by an orthogonal matrix $(\delta_{m\pi(n)})_{n,m=1}$ and (★) is equal to

$$\sum_{n,m=1}^{\infty} \frac{n^p}{m^q}(\delta^2_{m\pi(n)} + \delta^2_{n\pi(m)})$$

$$= \sum_{m=1}^{\infty} \frac{\pi^{-1}(m)^p + \pi(m)^p}{m^q} ,$$

Corollary 1 is derived from Theorem 1 immediately.

Corollary 2. Let $\underline{a} = (a_{mn})$ be an orthogonal matrix and $g(n)$ be a positive integer-valued function. Assume that for every positive integer p, there exists a positive integer $r = r(p)$ such that

$$\sum_{n=1}^{\infty} n^p g(n)^{-r} < + \infty ,$$

and assume for every n

$$a_{mn} = a_{nm} = 0, \qquad \text{if } m \leq g(n).$$

Then $\underline{a}$ is in $o(\mathcal{S})$.

Proof. For every positive integer p, put

$$q = \max(p+3,\ r(p+1)).$$

Then we have

$$\sum_{m,n=1}^{\infty} \frac{n^p}{m^q}(a_{mn}^2 + a_{nm}^2)$$

$$= \sum_{n=1}^{\infty} n^p \sum_{m=g(n)}^{\infty} \frac{1}{m^q}(a_{mn}^2 + a_{nm}^2)$$

$$\leq 2\sum_{n=1}^{\infty} n^p \Big(\sum_{m=g(n)}^{n} \frac{1}{m^q} + \sum_{m=n+1}^{\infty} \frac{1}{m^q} \Big)$$

$$\leq 2\sum_{n=1}^{\infty} n^p \Big(\frac{n}{g(n)^q} + \frac{1}{q-1}\frac{1}{n^{q-1}} \Big)$$

$$\leq 2\sum_{n=1}^{\infty} \frac{n^{p+1}}{g(n)^q} + 2\sum_{n=1}^{\infty} \frac{n^{p+1}}{n^q} < +\infty .$$

However $o(\mathcal{S})$ is still not sufficient for our purpose, since the shift and other important one-parameter subgroups do not have simple forms in $o(\mathcal{S})$. So we must investigate a class of specified subgroups of $O(S)$, which is derived from the change of variables.

(II). Let $\mathfrak{A}$ be the collection of all real odd functions which have slowly increasing derivatives of all orders and for every h in $\mathfrak{A}$ define a transformation $\tilde{g}(h)$ by

$$\tilde{g}(h)\xi = T_h * \xi, \qquad \xi \in S,$$

where T_h is the inverse Fourier transform of $e^{ih(x)}$ and $*$ is the convolution in the distribution sense.

Theorem 2. $G_0 = \{\tilde{g}(h) ; h \in \mathfrak{a}\}$ is the maximal abelian subgroup of $O(S)$ that contains the shift. (H. Sato [1]).

In fact the shift S_t is given by $\tilde{g}(tx)$.

Every one-parameter subgroup of G_0 is derived from $\mathfrak{a}$ and conversely, every h in $\mathfrak{a}$ induces a one-parameter subgroup $\{\tilde{g}(th) ; t \in R^1\}$ of G_0 and we can calculate the spectral type of $\tilde{g}(th)$ as a one-parameter orthogonal group .

Proposition 1. Let $E(\Delta)$ be the spectral measure of the shift as a one-parameter orthogonal group. Then the spectral measure $F(\Delta)$ of $\tilde{g}(th)$ is given by

$$F(\Delta) = E(h^{-1}(\Delta)).$$

(III). We can find another Lie group structure in a subgroup of $O(S)$. For every u in S, define a transformation $\hat{g}(u)$ of S by

$$\hat{g}(u) : \xi(x) \to \exp\tfrac{1}{2}u(x)\xi(f_u(x)), \quad \xi \in S,$$

where

$$f_u(x) = \int_0^x e^{u(y)}dy, \qquad x \in R^1.$$

Then the collection

$$G = \{ \hat{g}(u) ; u \in S \}$$

is a subgroup of O(S). Since the correspondence between u and $\hat{g}(u)$ is bijective, we introduce the topology of S into G through the map $\hat{g}$. Then G is a topological group in this topology. Considering S is a countably normed space with an increasing sequence of norms

$$\|\xi\|_k = \sup_{0 \leq q \leq k} \sup_{-\infty < x < +\infty} (1+x^{2k}) |\xi^{(q)}(x)|$$

$k = 0,1,2,\ldots$, we define an I.L.B.-system $\{S^k ; k = 0,1,2,\ldots\}$ where S^k is the $\| \|_k$-completion of S.

Theorem 3. *The topological group* G *is an I.L.B. (inverse limit Banach). Lie group modeled on the I.L.B.-system* $\{S^k\}$. (H. Sato [3]).

Proposition 2. The corresponding bracket product on S, which is the tangent space of the identity of G, is given by

$$[u,v] = F_u v' - u' F_v,$$

for every u and v in S, where

$$F_u(x) = \int_0^x u(y)dy, \qquad x \in R^1.$$

The explicit formula of the exponential map is also obtained. To begin with, we define a function $h(t, x ; u)$ as follows. For every u in S, define

$$\Lambda = \{ x \in R^1 ; F_u(x) \neq 0 \}.$$

Then Λ is an at most countable union of open intervals

$$\Lambda = \bigcup_n (a_n, b_n).$$

Define a function φ_n on each interval (a_n, b_n) by

$$\varphi_n(x) = \int_{c_n}^{x} \frac{dy}{F_u(y)}, \qquad a_n < x < b_n,$$

where c_n is an arbitrary fixed point in (a_n, b_n) and define

$$h(t, x ; u) = \begin{cases} x, & x \in \Lambda, \quad t \in R^1, \\ \varphi_n^{-1}(\varphi_n(x)+t), & a_n < x < b_n, \quad t \in R^1. \end{cases}$$

Proposition 3. The exponential map of the tangent space S into G is given by

$$\exp tu = \hat{g}\left(\int_0^t u(h(r, x ; u))dr\right)$$

for every u in S.

References

[1] H. Sato : A family of one-parameter subgroups of $O(S_r)$ arising from the variable change of the White Noise. Publ. R.I.M.S., Kyoto Univ., 5(1969), 165-191.

[2] H. Sato : One-parameter subgroups and a Lie subgroup of an infinite dimensional rotation group. J. Math. Kyoto Univ., 11(1971), 253-300.

[3] H. Sato : An I.L.B. Lie subgroup of O(S). (to appear).

[4] H. Omori : Theory of infinite dimensional Lie group. Lecture note at Tokyo Metropolitan Univ., (1972). (in Japanese).

Department of Mathematics
Kyushu University
Fukuoka, Japan

ON THE MULTIDIMENSIONAL CENTRAL LIMIT THEOREM WITH A WEAKENED CONDITION ON MOMENTS

V.V.Sazonov

1. Introduction and notation. In [1] , M.Katz constructed the following estimate of the rate of convergence in the central limit theorem. Let g be a function belonging to the class $\mathcal{G}$ (the definition of $\mathcal{G}$ see below in the list of notation), e.g. $g(a)=|a|^{\delta}$, $0<\delta\leq 1$, or $g(a)=(\log(1+|a|))/\log 2$. Let $\xi_1, \xi_2, \dots$ be a sequence of real independent identically distributed random variables such that $E\xi_1=0$, $E\xi_1^2=1$. Then there exists an absolute constant c such that

$$\sup_x \left| P\left(\sum_1^n \xi_i/\sqrt{n} \leq x\right) - \frac{1}{\sqrt{2\pi}} \int_{-\infty}^{x} e^{-t^2/2} dt \right| \leq c \, \frac{E\xi_1^2 g(\xi_n)}{g(\sqrt{n})}$$

The classical Berry-Esseen theorem is a special case of this estimate and is obtained by putting $g(a)=|a|$. A Bikelis in [2] extended this estimate to differently distributed random variables with values in R^k.

The aim of the present paper is to improve the Bikelis estimate in the case of identically distributed summands by expressing its dependence on the distribution P of ξ_1 in terms of the variation of the signed measure $P-Q$ where Q is the limit normal distribution.

Both Katz and Bikelis used in their papers the method of characteristic functions. We use a different method, the so-called method of convolutions (see [3] - [7]).

An attempt to prove an estimate of the type considered in the present paper was also made by V.Paulauskas in [8] . However he considered only the special case $g(a)=|a|^{\delta}$, $0<\delta<1$, and even in this case his estimate is much weaker than that of the present paper (see Remark 1 at the end of the paper).

In the sequel, the following notation will be used. For $x=(x_1,\dots,x_k)$, $y=(y_1,\dots,y_k)\in R^k$ we put as usual $(x,y)=\sum_1^k x_i y_i$, $|x|=(x,x)^{1/2}$. By $\mathcal{G}$ we denote the class of all non-negative even functions $g(x)$ on R such that $g(x)\neq 0$ when $x\neq 0$, $g(1)=1$ and

$$g(x_1)\leq g(x_2),\quad x_1/g(x_1)\leq x_2/g(x_2)\quad,\ \text{when}\quad 0\leq x_1\leq x_2.$$

For any function $g\in\mathcal{G}$, denote by $\mathcal{P}_g$ the class of all non-degenerate Borel probability distributions P on R^k such that

$$\int_{R^k} x\,dP=0\,,\quad \int_{R^k}|x|^2 g(|x|)\,dP<\infty$$

If μ is a signed measure, then $|\mu|$ stands for its variation. For any non-degenerate probability distribution P on R^k such that $\int_{R^k} x\,dP=0$, $\int_{R^k}|x|^2 dP<\infty$, and for any $z>0$, $g\in\mathcal{G}$, we put

$$\nu_{z,g}(P)=\nu_{z,g}=\int_{R^k}(\Delta^{-1}x,x)^{z/2}\,g((\Delta^{-1}x,x)^{1/2})\,|P-Q|(dx),$$

$$\beta_{z,g}(P)=\beta_{z,g}=\int_{R^k}(\Delta^{-1}x,x)^{z/2}\,g((\Delta^{-1}x,x)^{1/2})\,P(dx),$$

where Q is the normal distribution with the same first and second moments as those of P.

The convolution of measures μ_1,μ_2 is denoted by $\mu_1*\mu_2$; if μ is a measure, then μ^n will denote the n-time convolution of μ with itself; δ_x denotes the probability distribution concentrated at a point $x \in R^k$. By N_T, $T>0$ (resp. φ_T, $T>0$), we denote the normal distribution (resp. the density function of the normal distribution) with zero mean and variance-covariance matrix $T^{-2}I$, where I is the $(k\times k)$-identity matrix.

$\mathcal{C}$ stands for the class of all universally measurable convex subsets of R^k. For any $\ell>0$, we put $S_\ell = \{x : x \in R^k, |x| \le \ell\}$.

Finally, by $C, C(k)$ we denote absolute constants and constants depending only on the dimension k, respectively. The same symbol C (resp. $C(k)$) may stand for different constants (resp. different constants depending only on the dimension k).

2. The main result

Theorem. Let $\xi_1, \xi_2, \dots$ be a sequence of independent random variables with values in R^k with the same non-degenerate distribution P having zero mean and finite variance-covariance matrix Δ. Denote by P_n the distribution of the normalized sum $n^{-1/2}\sum_1^n \xi_i$, and let Q be the normal distribution with the same first and second moments as those of P. Then, for any $g \in \mathcal{G}$

$$(1)\qquad \sup_{E\in\mathcal{C}} |P_n(E)-Q(E)| \le c(k)\frac{\bar{\nu}_{2,g}}{g(\sqrt{n})}, \quad n=1,2,\dots,$$

where

$$\bar{\nu}_{2,g} = |P-Q|\,(x:(\Delta^{-1}x,x)\le 1) + \int_{(\Delta^{-1}x,x)\ge 1} (\Delta^{-1}x,x)\, g((\Delta^{-1}x,x)^{1/2})\,|P-Q|\,(dx).$$

The constant $c(k) \le Ck^{5/2}$

Proof. Without loss of generality we may suppose that $P \in \mathcal{P}_g$. Just as in the proof of Lemma 2 in [6], one may show that it suffices to prove the theorem for $\Delta = I$. It is this case we consider in the sequel.

We have obviously

$$| P(E)-N_1(E)| \leq \frac{1}{2} | P-N_1| (k) \leq$$

$$\leq \frac{1}{2} \Big(|P-N_1| (S_1) + \int_{S_i^c} |x|^2 g(|x|) \, |P-N_1| (dx) = \frac{1}{2} \bar{\nu}_{2,g}$$

and so (1) holds true for $n = 1$ with $c(k) = 1/2$. We shall show that, if (1) holds for all values of n less than some fixed value with the constant $c(k)$ which will be indicated later, then it also holds for this fixed value of n with the same constant $c(k)$ In the sequel, n will be a fixed integer $\geqslant 2$.

Denote by $P_{(n)}$ the distribution of $n^{-1/2}\xi_1$, and put

$$H_i = \begin{cases} \delta_0 , & i=0 \\ P_{(n)}^i - N_{n^{1/2}}^i , & i=1,\dots,n \end{cases}$$

Noting that $P_n = P_{(n)}^n$, $N_1 = N_{n^{1/2}}^n$, we have for all $T > 0$ (cf. [5] , p. 186)

(2)
$$(P_n - N_1) * N_T = H_n * N_T =$$

$$= H_1 * \sum_{i=0}^{n-1} (P_{(n)}^i * N_{n^{1/2}}^{n-i-1}) * N_T = \Big(\sum_{i=1}^{n-1} U_i + n U_0 \Big) * H_1$$

where

$$U_i = H_i * N_{\tau_i} , \quad \tau_i = \left(\frac{n-i-1}{n} + \frac{1}{T^2} \right)^{-1/2}, \quad i=0,1,\dots,n-1.$$

Let E be an arbitrary fixed set belonging to $\mathcal{C}$. Put

$$f_i(x) = H_i(E+x), \quad g_i(x) = U_i(E+x), \quad i = 0, 1, \ldots, n-1,$$

and note that (cf. [5], p. 186)

$$g_i(x) = \int_{R^k} f_i(z)\, \varphi_{\tau_i}(x-z)\, dz \tag{3}$$

Denoting

$$h_i(x) = g_i(-x) - g_i(0) + \sum_{j=1}^{k} \frac{\partial g_i}{\partial x_j}(0)\, x_j - \frac{1}{2} \sum_{j,z=1}^{k} \frac{\partial^2 g_i}{\partial x_j \partial x_n}(0)\, x_j x_z \tag{4}$$

and using the fact that the first and second corresponding moments of the distributions $P_{(n)}$ and $N_{n^{1/2}}$ are the same, we obtain

$$|(U_i * H_1)(E)| = \left| \int_{R^k} g_i(-x)\, H_1(dx) \right| = \tag{5}$$

$$= \left| \int_{R^k} h_i(x)\, H_1(dx) \right| \le \sup_{\substack{x \\ i = 0,1,\ldots,n-2.}} \frac{g(\sqrt{n})\, |h_i(x)|}{|x|^2 g(|x|\sqrt{n})} \int_{R^k} \frac{|x|^2 g(|x|\sqrt{n})}{g(\sqrt{n})} |H_1|(dx).$$

Further we have obviously

$$\int_{R^k} \frac{|x|^2 g(|x|\sqrt{n})}{g(\sqrt{n})} |H_1|(dx) = \frac{\nu_{2,g}}{n g(\sqrt{n})} \le \frac{\bar{\nu}_{2,g}}{n g(\sqrt{n})} \tag{6}$$

and by the properties of g

$$\frac{g(\sqrt{n})}{|x|^2 g(|x|\sqrt{n})} \le \begin{cases} \dfrac{1}{|x|^3}, & |x| \le 1 \\ \dfrac{1}{|x|^2}, & |x| \ge 1. \end{cases} \tag{7}$$

Moreover, by (3) and (4) we have

$$|h_i(x)| =$$

$$(8) \qquad = \Big| \int_{R^k} f_i(z)\Big[\varphi_{\tau_i}(-x+z)-\varphi_{\tau_i}(z)-\sum_{j=1}^{k}\frac{\partial\varphi_{\tau_i}}{\partial x_j}(z)x_j-\frac{1}{2}\sum_{j,z=1}^{k}\frac{\partial^2\varphi_{\tau_i}}{\partial x_j\,\partial x_z}(z)x_j x_z\Big]dz\Big| \le$$

$$\le \sup_z |f_i(z)|\Big(2+k^{1/2}|z|\max_j\int_{R^k}\Big|\frac{\partial\varphi_{\tau_i}}{\partial x_j}(x)\,dz+\frac{1}{2}k|x|^2\max\int\frac{\partial^2\varphi_{\tau_i}}{\partial x_j\,\partial x_z}(z)\Big|dz\Big)$$

and, on the other hand, using Taylor's formula, by (3) and (4), we obtain

$$(9) \qquad |h_i(x)| \le \frac{1}{6}\sum_{i,z,j=1}^{k}\Big|\frac{\partial^3 g_i(-\vartheta x)}{\partial x_j\,\partial x_z\,\partial x_j}\Big|\,|x_j x_z x_s| \le$$

$$\le \frac{1}{6}k^{3/2}|x|^3 \sup_z |f_i(z)| \max_{j,z,s}\int_{R^k}\Big|\frac{\partial^3\varphi_{\tau_i}}{\partial x_j\,\partial x_z\,\partial x_i}(z)\Big|\,dz$$

(here $|\vartheta|\le 1$). Now note that by the inductive hyposisis

$$(10) \qquad \sup_z |f_i(z)| \le M_i = \begin{cases} c(k)\,\bar{\nu}_{2,g}/g(\sqrt{i}), & i=1,\dots,n-1 \\ 1, & i=0 \end{cases}$$

and that, as a simple calculation shows, for any $\tau>0$,

$$(11) \qquad \int_{R^k}\Big|\frac{\partial\varphi_\tau}{\partial x_j}\Big|dx \le c\tau, \quad \int_{R^k}\Big|\frac{\partial^2\varphi_\tau}{\partial x_j\,\partial x_z}\Big|dx \le c\tau^2, \quad \int_{R^k}\Big|\frac{\partial^3\varphi_\tau}{\partial x_j\,\partial x_z\,\partial x_i}\Big|dx \le c\tau^3$$

Combining (7), (9), (10), (11), we get

$$(12)\qquad \sup_{|x|\le 1} \frac{g(\sqrt{n})\,|h_i(x)|}{|x|^2 g(|x|\sqrt{n})} \le c k^{3/2} \tau_i^3 M_i ,$$

and from (7), (8), (10), (11), we obtain

$$(13)\qquad \sup_{|x|\geq 1} \frac{g(\sqrt{n})\,|h_i(x)|}{|x|^2 g(|x|\sqrt{n})} \le c\left(1+k^{1/2}\tau_i + k\tau_i^2\right) M_i$$

From now on we shall suppose (unless otherwise specified) that $T \geq 1$. Under this condition, we have $\tau_i \geq 2^{-1/2}$, $i=0,1,\dots,n-2$, and so from (12) and (13), it follows that

$$(14)\qquad \sup_{x} \frac{g(\sqrt{n})\,|h_i(x)|}{|x|^2 g(|x|\sqrt{n})} \le c k^{3/2} \tau_i^3 M_i .$$

From (5), (6), (10) and (14), we now deduce

$$(15)\qquad |U_i * H_1(E)| \le \begin{cases} \dfrac{c k^{3/2}\, c(k)\, \bar{\nu}_{2,g}^{-2}}{n g(\sqrt{n})} \dfrac{\tau_i^3}{g(\sqrt{i})}, & 0 < i \le n-2 \\[2ex] c k^{3/2} \dfrac{\bar{\nu}_{2,g}}{n g(\sqrt{n})}, & i = 0 \end{cases}$$

Further by the properties of g , we have

$$(16)\qquad \sum_{i=1}^{n-2} \frac{\tau_i^3}{g(\sqrt{i})} = \sum_{i=1}^{n-2} \frac{1}{g(\sqrt{i})\left(\frac{n-i-1}{n} + \frac{1}{T^2}\right)^{3/2}} \le$$

$$\le \int_1^{n-1} \frac{dx}{g(\sqrt{x-1})\left(\frac{n-x-1}{n} + \frac{1}{T^2}\right)^{3/2}} \le \frac{\sqrt{n-2}}{g(\sqrt{n-2})} \int_1^{n-1} \frac{dx}{\sqrt{x-1}\left(\frac{n-x-1}{n} + \frac{1}{T^2}\right)^{3/2}}$$

(cf. (23) in [5]).

Let us estimate now $|(U_{n-1} * H_1)(E)|$. Denoting

$$h(x) = g_{n-1}(-x) - g_{n-1}(0),$$

we obtain similarly to (5)

$$(17) \qquad |(U_{n-1} * H_1)(E)| \le \sup_x \frac{g(\sqrt{n})|h(x)|}{g(\sqrt{n}|x|)} \int_{R^k} \frac{g(|x|\sqrt{n})}{g(\sqrt{n})} |H_1|(dx)$$

Now we have obviously

$$(18) \qquad \int_{R^k} \frac{g(|x|\sqrt{n})}{g(\sqrt{n})} |H_1|(dx) = \frac{\nu_{0,g}}{g(\sqrt{n})} \le \frac{\bar{\nu}_{2,g}}{g(\sqrt{n})},$$

and by the properties of g,

$$(19) \qquad \frac{g(\sqrt{n})}{g(|x|\sqrt{n})} \le \begin{cases} 1/|x|, & |x| \le 1 \\ 1, & |x| \ge 1 \end{cases}$$

Similarly to (8) and (9), we have

$$(20) \qquad |h(x)| = \Big| \int_{R^k} f_{n-1}(z) \big(\varphi_{\tau_{n-1}}(-x-z) - \varphi_{\tau_{n-1}}(-z) \big) dz \Big| \le$$

$$\le 2 \sup_z |f_{n-1}(z)|,$$

$$(21) \qquad |h(x)| \le \sum_{j=1}^{n} \Big| \frac{\partial g_{n-1}(-\vartheta x)}{\partial x_j} \Big| |x_j| \le$$

$$\le k^{1/2} |x| \sup_z |f_{n-1}(z)| \int_{R^k} \Big| \frac{\partial \varphi_{\tau_{n-1}}}{\partial z} \Big| dz$$

(here $|\vartheta| \le 1$). From (19)-(21) and (10), we derive

$$(22) \qquad \sup_{|x|\ge 1} \frac{g(\sqrt{n})\,|h(x)|}{g(|x|\sqrt{n})} \le C(k)\,\frac{\bar{\nu}_{2,g}}{g(\sqrt{n})},$$

$$(23) \qquad \sup_{|x|\le 1} \frac{g(\sqrt{n})\,|h(x)|}{g(|x|\sqrt{n})} \le ck^{1/2} c(k)\,\frac{\bar{\nu}_{2,g}}{g(\sqrt{n})}\,T.$$

By (17), (18), (22) and (23), we have

$$(24) \qquad |(U_{n-1} * H_1)(E)| \le ck^{1/2} C(k)\,\frac{\bar{\nu}_{2,g}^2}{g^2(\sqrt{n})}\,T.$$

Combining (2), (15), (16) and (24), we obtain for all $T \ge 1$

$$(25) \qquad \sup_{E\in\mathcal{C}} |(P_n - N_1) * N_T(E)| \le ck^{3/2}\,\frac{\bar{\nu}_{2,g}}{g(\sqrt{n})}\left(1 + c(k)\,\frac{\bar{\nu}_{2,g}}{g(\sqrt{n})}\,T\right)$$

We shall use now Lemma 1 from [6] . By this Lemma, from (25) it follows that for all $T \ge 1$

$$(26) \qquad \sup_{E\in\mathcal{C}} |P_n(E) - N_1(E)| \le ck^{3/2}\,\frac{\bar{\nu}_{2,g}}{g(\sqrt{n})}\left(1 + c(k)\,\frac{\bar{\nu}_{2,g}}{g(\sqrt{n})}\,T\right) + ckT^{-1}$$

If

$$(27) \qquad \frac{g(\sqrt{n})}{k^{1/4}\,c^{1/2}(k)\,\bar{\nu}_{2,g}} \ge 1,$$

take in (26) to be equal to the left hand side of (27). We then get

$$(28) \qquad \sup_{E\in\mathcal{C}} |P_n(E) - N_1(E)| \le c(k)\left\{c\left(\frac{k^{3/2}}{c(k)} + \frac{k^{5/4}}{c^{1/2}(k)}\right)\right\}\frac{\bar{\nu}_{2,g}}{g(\sqrt{n})}$$

Supposing that

(29) $$c(k) \geq c k^{5/2}$$

with an appropriate C, we see that the left hand side in (28) is less or equal to $c(k)\,\bar{\nu}_{2,g}\, g^{-1}(\sqrt{n})$.

If (27) is not satisfied, using (29) we have

$$c(k)\,\bar{\nu}_{2,g}\, g^{-1}(\sqrt{n}) \geq c^{1/2}(k)\, k^{-1/4} \geq 1$$

and (1) follows.

The theorem is proved.

3. Remarks. 1. The quantity $\bar{\nu}_{2,g}$ may be majorized by a function of the "pseudomoment" $\nu_{2,g}$. Namely we shall show that

(30) $$\bar{\nu}_{2,g} \leq (1 + Ck^{-1})\, \tilde{\nu}_{2,g}$$

where

$$\tilde{\nu}_{2,g} = \max\left\{\nu_{2,g},\, (h^{-1}(\nu_{2,g}))^{k}\right\}$$

and

$$h(a) = a^{k+2} g(a), \quad a \geq 0.$$

It suffices to prove (30) when $P \in \mathcal{P}_g$ and when the variance-covariance matrix of P is the identity matrix (cf. the beginning of the proof of Lemma 2 in [6]).

Note that (30) is a consequence of the following inequality: for any probability distribution P' on R^k

(31) $$|P' - N_1|(R^k) \leq ck^{-1}\left\{h^{-1}\left(\int_{R^k} |x|^2 g(|x|)\, |P' - N_1|(dx)\right)\right\}^{k}$$

Indeed, according to (31)

$$\bar{\nu}_{2,g} = |P-N_1|(R^k) + \nu_{2,g} \le ck^{-1}\{h^{-1}(\nu_{2,g})\}^k + \nu_{2,g} \le (1+ck^{-1})\tilde{\nu}_{2,g}$$

It remains to prove (31). Just as in [6] for $g(a)=|a|$ (see Remark 1 in [6]), one may show that, if $P \ne N_1$,

$$(32)\qquad \frac{|P-N_1|(R^k)}{\left(h^{-1}\left(\int_{R^k} |x|^2 g(|x|)\,|P-N_1|(dx)\right)\right)^k} \le \sup_a \frac{2N_1(S_a)}{\left(h^{-1}\left(\int_{S_a} |x|^2 g(|x|) N_1(dx)\right)\right)^k} .$$

Denote by J_a the quantity which supremum is taken of in the right hand side of (32). To estimate (32) we consider separately two cases $a \le 1$ and $a > 1$. When $a \le 1$, since the function $g(z)/z$ is non-increasing, we have

$$(33)\qquad \int_{S_a} |x|^2 g(|x|) N_1(dx) = \frac{2}{2^{k/2}\Gamma(k/2)} \int_0^a g(z) z^{k+1} e^{-z^2/2} dz \ge K g(a) a^{k+2}$$

where

$$(34)\qquad K = K(a,k) = \left(2^{k/2-1} e^{1/2}\Gamma(k/2)(k+3)\right)^{-1} < 1 .$$

Since g is non-decreasing, using (34) we can continue inequality (33) as follows

$$(35)\qquad \ge g\left(K^{1/(k+2)}a\right)\left(K^{1/(k+2)}a\right)^{k+2} = h\left(K^{1/(k+2)}a\right)$$

From (35) and the obvious inequality

$$N_1(S_a) \leq 2^{1-k/2}\Gamma^{-1}(k/2)k^{-1}a^k,$$

we get

(36) $$J_a \leq Ck^{-1}, \quad a \leq 1$$

Let now $a > 1$. It is easily seen that

(37) $$I_a = \frac{2}{2^{\frac{k}{2}}\Gamma(\frac{k}{2})}\int_0^a r^{k+2}g(r)e^{-r^2/2}dr \geq c\frac{2}{2^{\frac{k}{2}}\Gamma(\frac{k}{2})}\int_0^a r^{k+1}e^{-\frac{r^2}{2}}dr = cI'_a$$

Moreover, the properties of g and the definition of h imply

(38) $$h^{-1}(a) \geq \min(a^{1/(k+2)}, a^{1/(k+3)})$$

From (37) and (38) we have

$$(h^{-1}(I_a))^k \geq c(I'_a)^{k/(k+2)}\min(1, (I'_a)^{-k/(k+2)(k+3)})$$

But $I'_a \leq k$ and so

(39) $$(h^{-1}(I_a))^k \geq c(I'_a)^{k/(k+2)}$$

Since

$$N_1(S_a)(I'_a)^{k/(k+2)}$$

as a function of a is non-increasing, we obtain from (39)

(40) $$J_a \leq cN_1(S_1)(I')^{-k/(k+2)} \leq ck^{-1}, \quad a \geq 1$$

Combining (32), (36) and (40), we get (31).

2. It is not difficult to show that

$$\nu_{2,g} \leq c\beta_{2,g}$$

This inequality together with (30) implies

$$\bar{\nu}_{2,g} \leq c\beta_{2,g}$$

REFERENCES

[1] M.L.Katz, Note on the Berry-Esseen theorem. Ann.Math.Stat., 34, 3 (1963), 1107-1108

[2] А.Бикялис, О центральной предельной теореме в R^k, I, Литовский математический сборник, 9, I (I97I), 27-58

[3] H.Bergström, On the central limit theorem in the space R^k, $k>1$, Skand. Aktuarietidskrift, 1-2 (1945), 106-127

[4] H.Bergström, On the central limit theorem in , Z. Wahrscheinlichkeitstheorie verw. Geb., 14 (1969), 113-126

[5] V.Sazonov, On the multi-dimensional central limit theorem, Sankhya, Ser. A, 30, 2 (1968), 181-204

[6] V.Sazonov , On the estimation of the rate of convergence in the multi-dimensional central limit theorem, to be published in Proc. of the VI Berkeley Symposium on Math. Statist. and Probab.

[7] В.Паулаускас, Об одном усилении теоремы Ляпунова, Литовский математический сборник, 9, 2 (I969), 323-328

[8] В.Паулаускас, Одна оценка скорости сходимости с использованием псевдомоментов, Литовский математический сборник, 9, 2 (I97I), 3I7-327

Steklov Mathematical Institute
of the Academy of Sciences of the USSR
Moscow

STATISTICS OF DIFFUSION TYPE PROCESSES

A.N.Shiryayev

I. Effective filtering examples

1. Let $(\Omega, \mathcal{F}, P)$ be a complete probability space, $(\mathcal{F}_t)$, $t \geq 0$, a non-decreasing right continuous family of sub-σ - algebras of $\mathcal{F}$ and $W = (W_t, \mathcal{F}_t)$ a standard Wiener process. An observation process, $\xi = (\xi_t, \mathcal{F}_t)$ is supposed to be of diffusion type with the differential

(1) $$d\xi_t = A(t,\omega)dt + B(t,\xi)dW_t, \quad \xi_0 = 0,$$

where the stochastic process $A = (A(t,\omega), \mathcal{F}_t)$ "contains" the unobservable part of the signal (e.g., $A(t,\omega)$ is an unknown signal) which is to be estimated from observations ξ, and $B(t,\xi)$ is $\mathcal{F}_t^{\xi} = \sigma(\omega : \xi s, s \leq t)$ measurable for each $t \geq 0$.

In the last decade, for processes of type (1), considerable advances have been made in optimal (least squares) non-linear filtering (cf. [1]-[8]). Here the main result is formulated as follows.

Let $\pi_t(g) = M[g(t,\omega)|\mathcal{F}_t^{\xi}]$ be an optimal least squares estimate of the fuction $g = g(t,\omega)$, $Mg^2(t,\omega) < \infty$, $t \geq 0$, by observations ξ_s, $s \leq t$. If the function $f = f(t,\omega)$ to be estimated is such that there exists a function $F(t,\omega)$ with the property that the process $x_t = f(t,\omega) - f(0,\omega) - \int_0^t F(s,\omega)ds$ is a square integrable martingale (under some natural assumptions; for more detail see [7], [8]), then

(2) $$\pi_t(f) = \pi_0(f) + \int_0^t \pi_s(F)ds + \int_0^t \left\{\pi_s(D) + [\pi_s(fA) - \pi_s(f)\pi_s(A)]B^{-1}(s,\xi)\right\} d\overline{W}_s,$$

where $\overline{W} = (\overline{W}_t, \mathcal{F}_t^{\xi})$, $t \geq 0$, is a Wiener process,

$$\overline{W}(t) = \int_0^t \frac{d\xi_s - \pi_s(A)ds}{B(s,\xi)}$$

with $\langle x, W\rangle_t = \int_0^t D_s\, ds$. (Equation (2) is called the optimal non-linear filtering equation.)

To illustrate possibilities of application of equation (2) let us consider a particular case of process (1):

$$(3) \qquad d\xi_t = [A_0(t,\xi) + \theta A_1(t,\xi)]dt + B(t,\xi)dw_t ,$$

where $\theta = \theta(\omega)$ is a random variable with finite moments independent of the Wiener process W. Putting $f(t,\omega) \equiv 0$ in (2) we find the stochastic differential of $\pi_t(\theta) = M(\theta | \mathcal{F}_t^\xi)$:

$$d\pi_t(\theta) = [A_0(t,\xi)\pi_t(\theta) + A_1(t,\xi)\pi_t(\theta^2)]B^{-1}(t,\xi)dt -$$

$$(4) \qquad - [A_0(t,\xi)\pi_t(\theta) + A_1(t,\xi)\pi_t^2(\theta)]B^{-1}(t,\xi)[d\xi_t - \pi_t(\theta)dt]$$

$$(F(t,\omega) \equiv 0, \; D_t \equiv 0).$$

From (4), it is seen that the conditional moment $\pi_t(\theta) = M(\theta | \mathcal{F}_t^\xi)$ can not be found unless the second conditional moment $\pi_t(\theta^2) = M(\theta^2 | \mathcal{F}_t^\xi)$ is known etc. In view of this "non-closeness" (lower moments are determined by higher ones) the cases when filtering equations can be effectively solved are very important.

Before stating our main result (Theorem 2) let us consider the process with differential (3). This example will illustrate the idea used in the general situation.

Theorem 1. *Let the distribution* $P(\theta \leq x | \xi_0)$ *is normal*, $N(m, \gamma)$, *and*

$$P\left(\int_0^t [A_0^2(s,\xi) + B^2(s,\xi)]ds < \infty \right) = 1, \quad t \geq 0,$$

$$|A_1(t,\xi)| \le C_1 < \infty \quad , \qquad P\text{ - a.s.,}$$

$$B^2(t,\xi) \geq C_2 > 0 \quad , \qquad P\text{ - a.s.}$$

Then, for any $t > 0$, a posteriori distribution $P(\theta \le x | \mathcal{F}_t^{\xi})$ is also normal, $\mathcal{N}(m_t, \gamma_t)$, with $m_t = M(\theta | \mathcal{F}_t^{\xi}), \gamma_t = M[(\theta - m_t)^2 | \mathcal{F}_t^{\xi}]$ satisfying the equations

(5) $$dm_t = \gamma_t \frac{A_1(t,\xi)}{B^2(t,\xi)} [d\xi_t - (A_0(t,\xi) + A_1(t,\xi) m_t) dt], \quad m_0 = m,$$

(6) $$\dot{\gamma}_t = -\gamma_t^2 \frac{A_1(t,\xi)}{B^2(t,\xi)} , \quad \gamma_0 = \gamma,$$

the solutions of which are given by

(7) $$m_t = m \frac{\gamma_t}{\gamma} + \gamma_t \int_0^t \frac{A_1(s,\xi)}{B^2(s,\xi)} (d\xi_s - A_0(s,\xi) ds),$$

(8) $$\gamma_t = \frac{\gamma}{1 + \gamma \int_0^t \frac{A_1^2(s,\xi)}{B^2(s,\xi)} ds}$$

Remark. It is important to emphasize that, although the pair (θ,ξ) is not Gaussian, the conditional distribution $P(\theta \le x | \mathcal{F}_t^{\xi})$, $t \geq 0$, is. It follows from the fact that the initial distribution $P(\theta \le x | \xi_0)$ is normal and the drift co-efficient in (3) is linear in θ .

The proof of the theorem is simple. (In case of a Markov (θ,ξ)-process see also [2].) Let μ_W and μ_ξ be the Wiener measure and measure corresponding to the process ξ . Let μ_{ξ^α} be the measure of process ξ^α with the differential

$$d\xi_t^\alpha = [A_0(t,\xi^\alpha) + \alpha A_1(t,\xi^\alpha)]dt + B(t,\xi^\alpha)dW_t ,$$

where $-\infty < \alpha < \infty$.

Then (cf. [2], [8], [9]) the Radon-Nikodym derivative $\frac{d\mu_{\xi^\alpha}}{d\mu_\xi}(t,\xi)$ of μ_{ξ^α} relative to μ_ξ on the interval $[0,t]$ is

$$\frac{d\mu_{\xi^\alpha}}{d\mu_\xi}(t,\xi) = exp\Big[\int_0^t (\alpha - m_\alpha)\frac{A_1(s,\xi)}{B^2(s,\xi)}[d\xi_t - (A_0(s,\xi) + m_s A_1(s,\xi))ds] -$$

$$- \frac{1}{2}\int_0^t (\alpha - m_s)^2 \frac{A_1^2(s,\xi)}{B^2(s,\xi)} ds\Big]$$

and hence the density

$$\varrho_\alpha(t) = \frac{\partial P(\theta \le \alpha | \mathcal{F}_t^\xi)}{\partial \alpha} = \varrho_\alpha(0)\frac{d\mu_{\xi^\alpha}}{d\mu_\xi}(t,\xi) = \frac{1}{\sqrt{2\pi\gamma}} exp\Big(-\frac{(\alpha - m)^2}{2\gamma}\Big) \times$$

$$\times exp\Big[\int_0^t (\alpha - m_s)\frac{A_1(s,\xi)}{B^2(s,\xi)}\big[d\xi_s - \big(A_0(s,\xi) + m_s A_1(s,\xi)\big)ds\big] - \frac{1}{2}\int_0^t (\alpha - m_s)\frac{2A_1^2(s,\xi)}{B^2(s,\xi)} ds.$$

It is seen that $\varrho_\alpha(t)$ is the density of a Gaussian distribution with parameteres depending on ξs, $s \le t$.

Formulas (5) and (6) follow immediately from (2) where one should put $f(t,\omega) \equiv 0$, $f(t,\omega) \equiv \theta^2$ and take into account that, since $P(\theta \le x | \mathcal{F}_t^\xi)$ is normal,

$$M[\theta^2(\theta - m_t) | \mathcal{F}_t^\xi] = 2m_t\gamma_t \qquad \text{(P- a.s.)}$$

2. The example considered is a particular case of the following general one.

Suppose that the "unobservable" process $\theta=(\theta_t)$, $t\geq 0$, and "observation" process $\xi=(\xi_t)$, $t\geq 0$, have the stochastic differentials

$$\begin{aligned} d\theta_t &= [a_0(t,\xi)+a_1(t,\xi)\theta_t]dt+b_1(t,\xi)dW_1(t)+b_2(t,\xi)dW_2(t) \\ d\xi_t &= [A_0(t,\xi)+A_1(t,\xi)\theta_t]dt+B(t,\xi)dW_2(t), \end{aligned} \tag{9}$$

where $W_1=(W_1(t))$, $W_2=(W_2(t))$, $t\geq 0$, are independent Wiener processes.

<u>Theorem 2. Let the conditional distribution</u> $P(\theta_0\leq x|\xi_0)$ <u>be normal</u> $N(m,\gamma_0)$, $|m_0|<\infty$, $0\leq\gamma_0<\infty$. <u>Let also</u>

(i) $P\left(\int_0^t[a_0^2(s,\xi)+A_0^2(s,\xi)+b_1^2(s,\xi)+b_2^2(s,\xi)]ds<\infty\right)=1,\ t\geq 0,$

(ii) $|A_1(t,\xi)|\leq C_1<\infty,\quad \left|a_0(t,\xi)-\frac{b_2(t,\xi)}{B(t,\xi)}A_0(t,\xi)\right|\leq C_1<\infty,$

$$\left|a_1(t,\xi)-\frac{b_2(t,\xi)}{B(t,\xi)}A_1(t,\xi)\right|\leq C_1<\infty,\quad P\text{-a.s.},$$

(iii) $b^2(t,\xi)\geq C_2>0$, P - a.s.,

(iv) <u>any two continuous solutions of the equation</u> $\eta_t=\xi_0+\int_0^t B(s,\eta)dW_2(s)$ <u>have the same finite-dimensional distributions.</u>

<u>Then the stochastic process</u> (θ_t,ξ_t), $t\geq 0$ <u>is conditionally Gaussian, i.e., for any</u> $0\leq t_1\leq\cdots\leq t_n\leq t$ <u>the distri-</u>

bution $P(\theta_{t_1} \le x_1, \dots, \theta_{t_n} \le x_n \mid \mathcal{F}_t^{\xi})$ is (P- a.s.) normal.

Moreover, $m_t = M(\theta_t \mid \mathcal{F}_t^{\xi})$ and $\gamma_t = M[(\theta_t - m_t)^2 \mid \mathcal{F}_t^{\xi}]$ satisfy the closed system of equations

$$(10) \quad dm_t = [a_0 + a_1 m_t] + \frac{b_2 B + \gamma_t A_1}{B^2}[d\xi_t - (A_0 + A_1 m_t)dt],$$

$$(11) \quad \dot{\gamma}_t = 2a_1\gamma_t + (b_1^2 + b_2^2) - \frac{(b_2 B + \gamma_t A_1)^2}{B^2}$$

$$(m_0 = m, \quad \gamma_0 = \gamma; \quad a_0 = a_0(t,\xi),\ a_1 = a_1(t,\xi), \dots).$$

This theorem contains, as particular cases, many of the results obtained in the filtering theory. The most known is the Kalman-Bucy case, [10] , where the co-efficients in (9) depend only on t. The method of [10] in obtaining the equations for m_t, γ_t was based on the Wiener-Hopf method and enabled to consider non-stationary problems but did not work in case of the co-efficients in (9) depending on the whole past of the observation process that is most important in problems of control by incomplete data.

According to Theorem 2 the a posteriori mean m_t and variance γ_t satisfy the system of equations (10), (11). The first one is a stochastic differential equation, and the second is a Riccati equation with random co-efficients. An important question arises whether a solution of these equations is unique (for equation (11) - in the class of non-negative functions).

Theorem 3. Under the assumptions of Theorem 2 the system of equations (10), (11) has a unique solution.

A full proof of Theorems 2 and 3 will be published soon, [8]. Here we merely note that the proof of Theorem 2 employs a similar idea as the one of Theorem 1, i.e. in case $\theta_t \equiv \theta$.

3. Let us make use of Theorem 1 to simply prove a result of information theory [11] , [12] , [13] on the structure of optimal coding and encoding in transmission of a Gaussian random variable θ through an information channel with white noise and noise-free feedback.

Let $\theta=\theta(\omega)$ be a Gaussian random variable, $N(m,\gamma)$, to be transmitted through an information channel with white noise. Assume that noise-free feeback is possible. The corresponding model in case of continuous time is constructed as follows. Let $\xi=(\xi t)$, $t\geq 0$, be an unput message,

(12) $$d\xi_t=[A_0(t,\xi)+\theta A_1(t,\xi)]dt+B\,dW_t\,,\quad \xi_0=0.$$

A choice of functions (A_0,A_1) gives coding and their dependence on ξ means that noise-free feedback is possible. Coding is assumed to be such that, at each moment $t\geq 0$,

(13) $$M[A_0(t,\xi)+\theta A_1(t,\xi)]^2\leq P,$$

where P is a given constant - an "energy characteristic" of the transmitter.

At each moment $t\geq 0$, an estimate $\hat{\theta}_t=\hat{\theta}(t,\xi)$ by the data ξs, $s\leq t$, is to be chosen to minimize the encoding error $M(\theta-\hat{\theta}_t)^2$. Hence the problem is to find a coding (A_0^*,A_1^*) satisfying condition (13) and encoding $\theta_t^*=\theta^*(t,\xi)$, $t\geq 0$, such that, for any $t\geq 0$,

$$\Delta(t)=\inf_{\hat{\theta},\,(A_0,A_1)} M(\theta-\hat{\theta}_t)^2=M(\theta-\theta_t^*)^2$$

If a coding (A_0,A_1) has been chosen, the best encoding estimate is clearly $\theta_t^*=m_t=M(\theta/\mathcal{F}_t^{\xi})$. Thus the problem of optimal encoding is simple.

For any pair (A_0, A_1) in view of (8) $\gamma_t = M[(\theta - m_t)^2 | \mathcal{F}_t^\xi]$ is given by the formulae

$$(14) \qquad \gamma_t = \frac{\gamma}{1 + \gamma \int_0^t \frac{A_1^2(s,\xi)}{B^2} ds}$$

and

$$(15) \qquad m_t = \gamma_t \left[\frac{m}{\gamma} + \int_0^t \frac{A_1(s,\xi)}{B} (d\xi_s - A_0(s,\xi) ds) \right].$$

Let us rewrite the restriction (13) in another form:

$$(16) \qquad P \geq M[A_0(t,\xi) + \theta A_1(t,\xi)]^2 =$$

$$= M[A_0(t,\xi) + m_t A_1(t,\xi)]^2 + M[\gamma_t \cdot A_1^2(t,\xi)].$$

It follows from (16) in particular that

$$P \geq M[\gamma_t A_1^2(t,\xi)],$$

what, together with (14), gives the inequality

$$(17) \qquad P \geq M \frac{\gamma A_1^2(t,\xi)}{1 + \int_0^t \frac{\gamma A_1^2(s,\xi)}{B^2} ds}.$$

Suppose now that the coding under consideration is such that $A_1(t,\xi)$ does not depend on ξ. Then

$$(18) \qquad u(t) \leq P + \frac{P}{B^2} \int_0^t u(s) ds$$

where $u(t) = \gamma A_1^2(t)$. By the Granuol-Bellman inequality,

$$u(t) \leq P \exp\left(\frac{P}{B^2} t\right),$$

i.e.

(19) $$A_1^2(t) \leq \frac{P}{\gamma} \exp\left(\frac{P}{B^2} t\right).$$

Therefore

(20) $$M\gamma_t = \gamma_t = \frac{\gamma}{1+\gamma \int_0^t \frac{A_1^2(s)}{B^2} ds} \geq \gamma e^{-\frac{P}{B^2}t},$$

and hence

(21) $$\Delta(t) = \inf_{\hat{\theta}, (A_0, A_1)} M(\theta - \hat{\theta}_t)^2 = \inf_{(A_0, A_1)} M(\theta - m_t)^2 = \inf_{(A_0, A_1)} \gamma_t \geq \gamma e^{-\frac{P}{B^2}t}$$

Now note that (19)-(21) turn into equalities if

$$A_1^2(t) = \frac{P}{\gamma} \exp\left(\frac{P}{B^2} t\right),$$

$$A_0(t,\xi) = -m_t A_1(t).$$

Thus the following result has been proved (compare with [11], [12], [13]).

Theorem 4. If $A_1(t,\xi) = A_1(t)$ (does not depend on ξ), then an optimal coding is given by

$$A_1^*(t) = \sqrt{\frac{P}{\gamma}} \exp\left(\frac{P}{2B^2} t\right)$$

$$A_0^*(t,\xi) = -m_t \cdot A_1^*(t).$$

The signal transmitted, $\xi = (\xi_t),\ t \geq 0$, and an optimal encoding $\theta^* \equiv m = (m_t),\ t \geq 0$, are defined by the equations

$$d\xi_t = \sqrt{\frac{P}{\gamma}}\, exp\left(\frac{P}{2B^2}t\right)(\theta - m_t)dt + B\,dW_t$$

$$dm_t = \frac{\sqrt{P\gamma}}{B^2}\, exp\left(-\frac{P}{2B^2}t\right)d\xi_t$$

The encoding error

$$\Delta(t) = \gamma e^{-\frac{P}{B^2}t}$$

Remark. By the same method one can show that, if no feedback is used in coding, i.e. $A_0 = A_0(t)$, $A_1 = A_1(t)$, then optimal $A_0^* = -mA_1^*$, $A_1^* = \sqrt{\frac{P}{\gamma}}$, and the error is $\gamma(1+\frac{P}{B^2}t)^{-1}$.

2. Sequential estimation of parameters in diffusion type processes

1. Let us consider the process $\xi = (\xi_t)$, $t \geq 0$ of diffusion type with the differential

(22) $$d\xi_t = \lambda A(t,\xi)dt + dW_t\ , \quad \xi_0 = 0,$$

where is an unknown parameter, $-\infty < \lambda < \infty$, and $A(t,\xi)$ is $\mathcal{F}_t^\xi$ -measurable for each $t \geq 0$. We shall suppose that for any continuous function $x = (x_t)$, $t \geq 0$, $x_0 = 0$, there exists $\int_0^t |A(s,x)|ds < \infty$ for $t \geq 0$

$$P_\lambda\left(\int_0^t A^2(s,\xi)ds < \infty\right) = 1\ , \quad P_0\left(\int_0^t A^2(s,\xi)ds < \infty\right) = 1$$

where the index λ shows that the ξ -process is considered with this fixed λ .

According to [9] , it follows from these conditions that the Wiener measure μ_W is equivalent, for each λ, $-\infty < \lambda < \infty$, to the measure μ_ξ^λ of the process ξ and

(23) $$\frac{d\mu_{\xi^\lambda}}{d\mu_\xi}(t,\xi) = \frac{d\mu_{\xi^\lambda}}{d\mu_{\xi^0}}(t,\xi) = exp\left[\lambda\int_0^t A(s,\xi)d\xi_s - \frac{\lambda^2}{2}\int_0^t A^2(s,\xi)ds\right].$$

The most simple and usual estimator of the parameter λ by observations ξ_s, $s \le t$, is the maximum likelihood estimator $\lambda_t(\xi)$ which, in view of (23), is

(24) $$\lambda_t(\xi) = \frac{\int_0^t A(s,\xi)d\xi_s}{\int_0^t A^2(s,\xi)ds}$$

Denote by M_λ the mathematical expectation relative to the measure P_λ. Then, under some natural assumptions about $A(t,\xi)$ (say, if $0 < C \le |A(t,\xi)| \le C < \infty$, one can easily deduce

(25) $$M_\lambda \lambda_t(\xi) = \lambda + \frac{\partial}{\partial\lambda} M_\lambda \left[\int_0^t A^2(s,\xi)ds\right]^{-1}$$

This formulae shows that, in a general case, the maximum likelihood estimator is biased, and the bias is $\frac{\partial}{\partial\lambda} M_\lambda \left[\int_0^t A^2(s,\xi)ds\right]^{-1}$

R.Sh.Lipcer and the author proposed to consider sequential maximum likelihood estimator which turned out to possess many nice properties. These estimators are constructed as follows.

Let H be a non-negative number. Define a Markov time

(26) $$\tau(H) = inf\left\{t : \int_0^t A^2(s,\xi)ds \geqslant H\right\}$$

If, for any λ, $P_\lambda\left(\int_0^\infty A^2(s,\xi)ds = \infty\right) = 1$, then obviously $P_\lambda(\tau(H) < \infty) = 1.$

We call

$$\tilde{\lambda}(H) = \lambda_{\tau(H)}(\xi),$$

a sequential maximum likelihood estimator of λ . Then from (24) and (26) we get

(27) $$\tilde{\lambda}(H)=\frac{1}{H}\int_0^{\tau(H)} A(s,\xi)\,d\xi_s$$

From (27) and also from (25) with $\tau(H)$ instead of t , one can see that, for any $H>0$, the estimator $\tilde{\lambda}(H)$ is unbiased: $M_\lambda\tilde{\lambda}(H)\equiv\lambda$. It is also clear that $D_\lambda\tilde{\lambda}(H)\equiv\frac{1}{H}$ and

$$\sqrt{H}\,(\tilde{\lambda}(H)-\lambda)=\frac{1}{\sqrt{H}}\int_0^{\tau(H)} A(s,\xi)\,dW_s$$

But for $\tau(H)$ defined by (26) the process $\beta(H)=\int_0^{\tau(H)} A(s,\xi)\,dW_s$ is, as it is well known, a Wiener one, and hence the variable $\sqrt{H}\,(\tilde{\lambda}(H)-\lambda)$ is normal, $N(0,1)$.

Thus the following theorem can be stated.

Theorem 5. If $0<C\le|A(t,\xi)|\le C<\infty$ then the sequential maximum likelihood estimator $\tilde{\lambda}(H)$:

(i) is unbiased, $M_\lambda\tilde{\lambda}(H)\equiv\lambda$,

(ii) has a constant variance, $D_\lambda\tilde{\lambda}(H)\equiv\frac{1}{H}$

(iii) is normal, $N(\lambda,\frac{1}{H})$.

The properties of sequential maximum likelihood estimators just stated are undoubtedly their advantages. However a natural question arises whether these advantages are consequences of a rather long average observation time $M_\lambda\tau(H)$. This problem in case of an arbitrary function $A(t,x)$ is very difficult. A.A.Novikov has studied $M_\lambda\tau(H)$ in case $A(t,x)=-x_t$, i.e. for the process $\xi=(\xi t)$, $t\geq 0$, with the differential

(28) $$d\xi_t=-\lambda\xi_t\,dt+dW_t\ ,\quad \xi_0=0 .$$

He has obtained the following result [14], [15].

Theorem 6. If $H = const$, then

$$(29)\qquad M_\lambda \tau(H) \sim \begin{cases} 2\lambda H, & \lambda \to \infty; \\ B_1\sqrt{H}, & \lambda \to 0,\ B_1 \sim 2{,}09; \\ \dfrac{\ln 8\lambda^2 H}{2|\lambda|}, & \lambda \to -\infty. \end{cases}$$

Let us now compare mean square deviations $M_\lambda(\lambda_T(\xi)-\lambda)^2$ and $M_\lambda(\tilde{\lambda}(H)-\lambda)^2$ for ordinary $(\lambda_T(\xi))$ and sequential $(\tilde{\lambda}(H))$ maximum likelihood estimations. When comparing the ordinary estimation method with observation time T and the sequential method with stopping at a random time τ, it is naturally to assume $M_\lambda \tau \le T$, $-\infty < \lambda < \infty$. The time $\tau = \tau(H)$ defined above does not possess this property.

Nevertheless, let $H = H(\lambda, T)$ be such that

$$M_\lambda \tau(H(\lambda,T)) \equiv T$$

and

$$e(\lambda,T) = \frac{M_\lambda(\lambda_T(\xi)-\lambda)^2}{M_\lambda(\tilde{\lambda}(H(\lambda,T))-\lambda)^2}$$

The value $e(\lambda,T)$ characterizes efficiency of sequential methods as compared to ordinary ones and has a sense if it is a priori known that the parameter λ to be estimated lies in a neighborhood of a given point λ_0. In this case, when constructing the estimator $\tilde{\lambda}(H)$, it is naturally to take the "threshold" since, for λ close to λ_0,

$$\frac{M_\lambda(\lambda_T(\xi)-\lambda)^2}{M_\lambda(\tilde{\lambda}(H(\lambda_0,T))-\lambda)^2} \sim \frac{M_\lambda(\lambda_T(\xi)-\lambda)^2}{M_\lambda(\tilde{\lambda}(H(\lambda,T))-\lambda)^2} = e(\lambda,T)$$

The efficiency $e(\lambda,T)$ has been studied by A.A.Novikiv [15] for the process (28). He has got the following result.

Theorem 7. For the process (28)

$$(30)\qquad e(\lambda,T)=\begin{cases} 1-\dfrac{23}{4(\lambda T)}+O\left(\dfrac{1}{(\lambda T)^2}\right), & \lambda T\to\infty,\\ C_0-C_1(\lambda T)+O\left((\lambda T)^2\right), & \lambda T\to 0,\\ \dfrac{e^{|\lambda|T}}{4\sqrt{\pi}(|\lambda|T)^{1/2}}\left[1+O\left(\dfrac{1}{|\lambda|T}\right)\right], & \lambda T\to-\infty,\end{cases}$$

where $C_0\sim 3{,}04$; $C_1\sim 2{,}13$

From this theorem one sees that for T fixed the gain of the sequential maximum likelihood estimator is essential if $\lambda\leq 0$.

References

[1] А.Н.Ширяев, Исследования по статистическому последовательному анализу, Математические заметки, 3, №6 (1968), 739-754.

[2] Р.Ш.Липцер, А.Н.Ширяев, Нелинейная фильтрация марковских диффузионных процессов, Труды МИАН, 104 (1968), 135-179.

[3] G.Kallianpur, C.Striebel, Stochastic differential equations occuring in the estimation of continuous parameter stochastic processes, Теория вероят. и ее примен., XIУ (1969), 4, 597-622

[4] М.П.Ершов, Последовательное оценивание диффузионных процессов, Теория вероят. и ее примен., ХУ (1970), 4, 703-717.

[5] T.Kailath, An innovation approach to least-squeres estimation, Part I, II, IEEE Trans. on Autom. Control., Vol. Ac-13 (1968), N 6, 646-655, 655-660.

[6] P.Frost, T.Kailath, An innovation approach to least-squares estimation, Part III, IEEE Trans. on Autom. Control, Vol. AC-16

(1971), N 3, 217-226.

[7] M.Fujisaki, G.Kallianpur, H.Kunita, Stochastic differential equations for the non-linear filtering problem, Osaka J. Math.,1972

[8] Р.Ш.Липцер, А.Н.Ширяев, Статистика случайных процессов (нелинейная фильтрация и смежные вопросы), Изд-во "Наука", Москва.

[9] Р.Ш.Липцер, А.Н.Ширяев, Об абсолютной непрерывности мер, соответствующих процессам диффузионного типа, относительно винеровской, Изв. АН СССР, сер. матем. 1972, 36, № 4.

[10] R.Bucy, R.Kalman, New results in linear filtering and prediction theory, J. of Basic Eng. Trans. of the ASME (1961), 95-108.

[11] J.Schalkwijk, T.Kailath, A coding scheme for additive noise channels with feedback. Part I: no band-width constraint, IEEE Trans.on Inform. Theory, IT-12 (1966), 172-182.

[12] К.Зигангиров, Передача сообщений по двоичному гауссовскому каналу с обратной связью. Проблемы передачи информации, 3 (1967), 2, 98-101.

[13] Р.Хасьминский, Добавление (задача 72) в книге Дж.Турина "Лекции о цифровой связи", Изд-во "Мир", Москва, 1972.

[14] А.А.Новиков, Последовательное оценивание параметров диффузионных процессов, Теория вероят. и ее примен., XVI (1971), 2, 394-396.

[15] А.А.Новиков, Стохастические интегралы и последовательное оценивание, Канд. диссертация, МИАН, 1972.

Steklov Mathematical Institute
of the Academy of Sciences of the USSR
Moscow

ON BRANCHING MARKOV PROCESSES WITH DERIVATIVE

Tunekiti Sirao

1. Introduction. The purpose of this paper is to show that the following Cauchy problems (1) and its special case (2) can be interpreted through a kind of branching Markov processes.

$$(1)\qquad \begin{cases} \dfrac{\partial u}{\partial t} = \dfrac{1}{2}\dfrac{\partial^2 u}{\partial x^2} + P(u, \dfrac{\partial u}{\partial x}), & t > 0,\ x \in R^1, \\ u(0,x) = f(x), \end{cases}$$

where P is a polynomial of u and $\partial u/\partial x$ with constant coefficients and f is a bounded continuous function with bounded derivative.

$$(2)\qquad \begin{cases} \dfrac{\partial u}{\partial t} = \dfrac{1}{2}\dfrac{\partial^2 u}{\partial x^2} + P(u)\dfrac{\partial u}{\partial x}, & t > 0,\ x \in R^1, \\ u(0,x) = f(x), \end{cases}$$

where P is a polynomial of u with constant coefficients and f is a bounded continuous function. Of course (2) can be solved by means of Ito's stochastic integral in much more general cases.[1)] The probability measure of the process obtained in this case depends on the initial datum f. But if P in (1) is a polynomial of u alone, then (1) is understood through a branching Markov process whose probability measure does not depend on f. Moreover the coefficients in P determine the jumping law of the process at its splitting times.[2)] A similar interpretation for (1) where P is a polynomial of u and $\partial u/\partial x$ is given in 2-3.

In general, (1) does not have any global solution. But (2) has the bounded unique global solution $u(t,x;f)$ which is infinitely differentiable in t and x and inherits the monotonicity of the initial datum f. Furthermore $u(t,x;f)$ is a probability density if f is so. However a probability measure can not generally be an initial datum of (2). For example, $\partial u/\partial t = (1/2)\partial^2 u/\partial x^2 - u^k \partial u/\partial x$ is not solvable for the initial datum of

1) cf. [4], [7].
2) cf. [1], [6].

δ -measure when $k \geqq 3$. (See 4.)

2. A branching Markov process with derivative. The construction of a branching Markov process with derivative is essentially same to the one of the corresponding signed branching Markov process with age which was treated in [3], [5], [6]. The difference between them is the one of their state spaces, and we will not give here a rigorous discussion.

Let

(3) $$P(u, \frac{\partial u}{\partial x}) = \sum_{\substack{0 \leq p \leq N \\ 0 \leq q \leq M}} c_{pq} u^p (\frac{\partial u}{\partial x})^q,$$

and

(4) $$C = \sum_{\substack{0 \leq p \leq N \\ 0 \leq q \leq M}} |c_{pq}|,$$

where the coefficients c_{pq} are constants. Then we first construct a Brownian motion with age in the following way. Let

$$E_0 = R^1 \times N,$$

where $R^1 = (-\infty, \infty)$ and $N = \{0,1,2,\ldots\}$, and $\beta^0(t)$ be an $\exp(-Ct)$ subprocess of 1-dimensional Brownian motion $\beta(t)$. At the terminal time σ of $\beta^0(t)$, the process restart from $\beta^0(\sigma -)$ and behaves as $\beta^0(t)$ does. Repeating this procedure, we get the process $\bar{\beta}(t)$ which is equivalent to $\beta(t)$ and the integral valued process $n(t)$ which describes how many deaths of $\beta^0(t)$ occurred before the time t. ($n(t)$ is a Poisson process with parameter C in the present case.) Thus we have the motion $B(t) = (\bar{\beta}(t), n(t))$ on E_0 with its starting point $(x,0) \in E_0$. For another starting points $(x,k) \in E_0$, $k \geqq 1$, the first coordinate $\bar{\beta}(t)$ is defined as in the above and $n(t)$ is defined as the sum of k and the number of times of death of $\beta^0(t)$ before t. Thus we have a strong Markov process $B(t) = (\bar{\beta}(t), n(t))$ on E_0. $B(t)$, $\bar{\beta}(t)$, $n(t)$ are called a Brownian motion with age, a position and age of particle, respectively.

We next enlarge the state space of the process from E_0 to S as follows. Let

$$R = R^1 \cup \{\delta\},$$

where δ is an extra and isolated point, and

$E = R \times N.$

Then we introduce a symbol D to handle the term $\partial u/\partial x$ in (1) and S is defined as the smallest set satisfying the following conditions:

1° $E \subset S$,

2° $D\,\underline{\underline{z}} \in S$ for $\underline{\underline{z}} \in S$,

3° $(\underline{\underline{z}}_1, \underline{\underline{z}}_2) \in S$ for $\underline{\underline{z}}_1, \underline{\underline{z}}_2 \in S$,

4° $((\underline{\underline{z}}_1,\underline{\underline{z}}_2),\underline{\underline{z}}_3) = (\underline{\underline{z}}_1,(\underline{\underline{z}}_2,\underline{\underline{z}}_3))$ ($= (\underline{\underline{z}}_1,\underline{\underline{z}}_2,\underline{\underline{z}}_3)$) for $\underline{\underline{z}}_1,\underline{\underline{z}}_2,\underline{\underline{z}}_3 \in S$.

(As was stated above, $D\,\underline{\underline{z}}$ is an artificial state to treat the term $\partial u/\partial x$.) For the simplicity of the description, we will use occasionally the following notations to express points of S.

a) $\underline{\underline{z}} = \prod_{i=1}^{n} \underline{\underline{z}}_i$ for $\underline{\underline{z}} = (\underline{\underline{z}}_1,\underline{\underline{z}}_2,\dots,\underline{\underline{z}}_n)$,

b) $D^{\ell}\underline{\underline{z}} = \underbrace{D(D\cdots(D}_{\ell}\,\underline{\underline{z}})\cdots)$, and accordingly $D^0\underline{\underline{z}} = \underline{\underline{z}}$.

By these notations, any point $\underline{\underline{z}}$ of S is expressed in the form of

$$(5)\quad \underline{\underline{z}} = \prod_{i_1=1}^{k} D^{\varepsilon_{i_1}} \prod_{i_2=1}^{k_{i_1}} D^{\varepsilon_{i_1 i_2}} \cdots\cdots D^{\varepsilon_{i_1 i_2\cdots i_\ell}} \prod_{i_{\ell+1}=1}^{k_{i_1 i_2\cdots i_\ell}} (x_{i_1 i_2\cdots i_{\ell+1}}, n_{i_1 i_2\cdots i_{\ell+1}}),$$

$$(x_{i_1 i_2\cdots i_{\ell+1}}, n_{i_1 i_2\cdots i_{\ell+1}}) \in E, \qquad \varepsilon_{i_1 i_2\cdots i_j} = 0 \text{ or } 1,\ 1 \leqq j \leqq \ell,$$

or shortly

$$(6)\quad \underline{\underline{z}} = \prod_{\gamma\in\Gamma} D^{\varepsilon_\gamma} \prod_{i=1}^{k_\gamma} (x_{\gamma_i}, n_{\gamma_i}), \qquad \varepsilon_\gamma = 0 \text{ or } 1,\ (x_{\gamma_i}, n_{\gamma_i}) \in E,$$

where Γ is a finite set of γ. $\underline{\underline{z}}$ in (6) is denoted occasionally by

$$(7)\quad \underline{\underline{z}} = (\underline{\underline{x}}, \underline{\underline{n}}) \text{ with } \underline{\underline{x}} = \prod_{\gamma\in\Gamma} D^{\varepsilon_\gamma} \prod_{i=1}^{k_\gamma} x_{\gamma_i},\quad \underline{\underline{n}} = \prod_{\gamma\in\Gamma} \prod_{i=1}^{k_\gamma} n_{\gamma_i},$$

because D effects only on $x \in R$ in the sequel. When we ignore D, S is identical to $\mathbb{E} = \bigcup_{n=1}^{\infty} E^n$ and each point $\underline{\underline{z}}$ of S can be obtained from a point of $\mathbb{E}$ inserting D between components of it and making groups by parentheses For example $((x_1,n_1), D((x_2,n_2),(x_3,n_3))) \in S$ is obtainable from $((x_1,n_1), (x_2,n_2),(x_3,n_3)) \in \mathbb{E}$. We now introduce the topology in S. Let

$$\underline{\underline{z}}_j = \prod_{i_1=1}^{k^{(j)}} D^{\varepsilon^{(j)}_{i_1}} \prod_{i_2=1}^{k^{(j)}_{i_1}} \cdots D^{\varepsilon^{(j)}_{i_1 i_2\cdots i_{\ell(j)}}} \prod_{i_{\ell(j)+1}=1}^{k^{(j)}_{i_1 i_2\cdots i_{\ell(j)}}} (x^{(j)}_{i_1\cdots i_{\ell(j)+1}}, n^{(j)}_{i_1\cdots i_{\ell(j)+1}}),\ j \geqq 1,$$

$$\underset{=0}{z} = \prod_{i_1=1}^{k} D^{\varepsilon_{i_1}} \prod_{i_2=1}^{k_{i_1}} \cdots D^{\varepsilon_{i_1 i_2 \cdots i_\ell}} \prod_{i_{\ell+1}=1}^{k_{i_1 i_2 \cdots i_\ell}} (x_{i_1 \cdots i_{\ell+1}}, n_{i_1 \cdots i_{\ell+1}}).$$

Then we say that $\{\underset{=j}{z}\}$ converges to $\underset{=0}{z}$ when

$$(8)\quad \begin{cases} \lim_{j\to\infty} \ell(j) = \ell \quad , \lim_{j\to\infty} k^{(j)}_{i_1 i_2 \ldots i_s} = k_{i_1 i_2 \ldots i_s} , \\ \lim_{j\to\infty} \varepsilon^{(j)}_{i_1 i_2 \ldots i_s} = \varepsilon_{i_1 i_2 \ldots i_s}, \lim_{j\to\infty} x^{(j)}_{i_1 i_2 \ldots i_{\ell(j)+1}} = x_{i_1 i_2 \ldots i_{\ell+1}} \\ \lim_{j\to\infty} n^{(j)}_{i_1 i_2 \ldots i_{\ell(j)+1}} = n_{i_1 i_2 \ldots i_{\ell+1}} \end{cases}$$

hold for any $1 \leqq i_s \leqq k_{i_1 i_2 \ldots i_{s-1}}$, $1 \leqq s \leqq \ell$, where $k_{i_0} = k$. By this topology, S is a locally compact separable Hausdorff space with the second axiom of countability.

Let S_0, S_1, S_2, S_3 be four copies of S. Then the topological sum $\bigcup_{i=0}^{3} S_i$ is a locally compact separable Hausdorff space and satisfies the second axiom of countability. We denote it by $S \times J$, and its one point compactification by $\tilde{S}$, i.e.

$$S \times J = \bigcup_{i=0}^{3} S_i, \quad \tilde{S} = (S \times J) \cup \{\Delta\} \quad .$$

Now we are in a position to construct <u>a branching Markov process with derivative</u> corresponding to (1) where P is given by (3). A Brownian motion with age $B(t)$ on E_0 can be considered to be a process on E where any $(\delta, k) \in E - E_0$ is regarded as a trap of $B(t)$. Let $B_0(t)$ be an $\exp(-C(x,k)t)$ subprocess of $B(t)$ on E, where $C(x,k) = C$ on E_0 and $C(\delta, k) = 0$ on $E - E_0$. For

$$(\underset{=}{z}, j) = [\prod_{\gamma\in\Gamma} D^{\varepsilon_\gamma} \prod_{i=1}^{k_\gamma} (x_{\gamma i}, n_{\gamma i}), j] \in S \times J,$$

we take $(\sum_{\gamma\in\Gamma} k_\gamma)$ independent copies $B^0_{\gamma i}(t)$, $1 \leqq i \leqq k_\gamma$, of $B^0(t)$ starting from $(x_{\gamma i}, n_{\gamma i})$, respectively. Then define $Y^0(t)$ by

$$Y^0(t) = [\prod_{\gamma\in\Gamma} D^{\varepsilon_\gamma} \prod_{i=1}^{k_\gamma} B^0_{\gamma i}(t), j], \; t < \tau_1,$$

where τ_1 is the smallest life time σ_{γ_i} of $B^0_{\gamma_i}(t)$. $Y^0(t)$ is <u>the non-branching part</u> of the process we are constructing. At $\tau_1 = \sigma_{\gamma_0 i_0}$ (say), the $\gamma_0 i_0$-th particle located at $B^0_{\gamma_0 i_0}(\tau_1-) = (y,n)$ splits into $(p+q)$ particles with probability $|c_{pq}|/C$. Then it restart from

$$[\prod_{\gamma\in\Gamma} D^{\varepsilon_\gamma} \prod_{i=1}^{k_\gamma} \underline{z}_{\gamma i}, j'],$$

where

$$\underline{z}_{\gamma i} = \begin{cases} B^0_{\gamma i}(\tau_1-), & (\gamma,i) \neq (\gamma_0,i_0) \\ ((y,n),\underbrace{(y,0),\ldots,(y,0)}_{p-1}, \underbrace{D(y,0),D(y,0),\ldots,D(y,0)}_{q}), & (\gamma,i)=(\gamma_0,i_0) \\ & p+q > 0, \\ (\delta,n), & (\gamma,i)=(\gamma_0,i_0),\ p+q=0, \end{cases}$$

and

$$j' = \begin{cases} 1, & j=0 \text{ and } c_{pq} > 0 \text{ or } j=3 \text{ and } c_{pq} < 0, \\ 2, & j=0 \text{ and } c_{pq} < 0 \text{ or } j=3 \text{ and } c_{pq} > 0, \\ 3, & j=1 \text{ and } c_{pq} < 0 \text{ or } j=2 \text{ and } c_{pq} > 0, \\ 0, & j=1 \text{ and } c_{pq} > 0 \text{ or } j=2 \text{ and } c_{pq} < 0. \end{cases}$$

Repeating this procedure and setting

$$Y(t) = \Delta \quad , \quad t \geqq \tau_\infty = \lim_{n\to\infty} \tau_n,$$

where τ_n denotes the n-th splitting time in the above procedure, we can get the process $Y(t) = [X(t), j(t)]$, $X(t) \in S$, $j(t) \in J$. Since $Y(t)$ is obtained by path-stitching method, $Y(t)$ is strong Markov.[1)] An example of a trajectory of $X(t)$ starting from (x_0,n_0) is shown below.

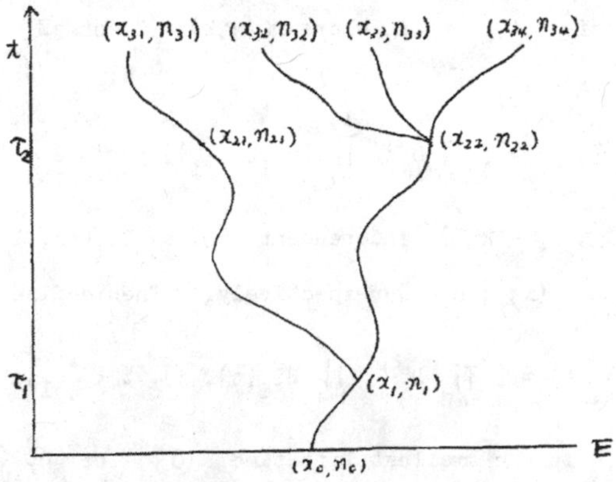

1) cf. [1], [3].

You can read the fiqure in the following way. A sample path started from $(x_0,n_0) \in E$ was governed by the probability law of $B^0(t)$ until its terminal time τ_1 and located at (x_1,n_1) at $t = \tau_1$. Then the split of probability $|c_{11}|/C$ occurred and the path jumped to $((x_1,n_1),D(x_1,0))$. The process restarted from the point $((x_1,n_1),D(x_1,0))$ and the second particle died at the time τ_2. The split of probability of $|c_{21}|/C$ occurred and the process restarted from $((x_{21},n_{21})D((x_{22},n_{22}),(x_{22},0),D(x_{22},0)))$. After τ_2, no jump occurred before t and $X(t) = ((x_{31},n_{31}),D((x_{32},n_{32}),(x_{33},n_{33}),D(x_{34},n_{34})))$.

Remark 1. If we ignore the symbol D from a trajectory of $Y(t)$, we will get the corresponding one of a signed branching Markov process with age. If we forget D and the second coordinate $j(t)$ from a trajectory of $Y(t)$, we will get the one of a branching Markov process with age and if we take away D, $j(t)$ and $n(t)$, then we will get the trajectory of a branching Markov process in the sense of [1].

<u>3. The local solution of (1).</u> For a bounded continuous function g on R^1, let us define the corresponding function $\tilde{g}$ on $\tilde{S}$ so that the expectation of $\tilde{g}(Y(t))$ gives the local solution of (1) with the initial datum g. We first assume that g is a $C_0^\infty(R^1)$-function.[1)] Then, putting $g(\delta) = 1$, we regard g as the function on $R^1 \cup \{\delta\}$ and set

$$g((x,n)) = 2^n g(x), \qquad (x,n) \in E,$$

$$\hat{g}((\underline{x},\underline{n})) = \prod_{i=1}^{k} g(x_i,n_i), \quad (\underline{x},\underline{n}) = \prod_{i=1}^{k} (x_i,n_i) \in E^k. \text{ 2)}$$

We have obviously

$$\hat{g}((\underline{x},\underline{n})) = 2^{|\underline{n}|} g((\underline{x},\underline{0})), \tag{9}$$

where

$$|\underline{n}| = \sum_{i=1}^{k} n_i, \quad (\underline{x},\underline{0}) = \prod_{i=1}^{k} (x_i,0).$$

Since $\hat{g}((\underline{x},\underline{n}))$, $(\underline{x},\underline{n}) \in E^k$, is differentiable in $x_1,x_2,\ldots,x_k$ with exception of

---

1) $C_0^\infty(R^1)$ is the collection of infinitely differentiable functions with compact supports.
2) See (7).

$x_i = \delta$, we can define $\hat{g}(D(\underline{x},\underline{n}))$ by

$$\hat{g}(D(\delta,n)) = 0,$$

and

$$\hat{g}(D(\underline{x},\underline{n})) = \sum_{i=1}^{k} \frac{\partial}{\partial x_i} \hat{g}((\underline{x},\underline{n})), \quad (\underline{x},\underline{n}) \in E^k.$$

If $\hat{g}(\underline{z}_1)$ and $\hat{g}(\underline{z}_2)$ are defined, we set

$$\hat{g}((\underline{z}_1, D\underline{z}_2)) = \hat{g}(z_1)\frac{\partial}{\partial \underline{x}_2} \hat{g}(\underline{z}_2), \tag{10}$$

where $\partial/\partial \underline{x}$ denotes

$$\sum_{\gamma i} \frac{\partial}{\partial x_{\gamma_i}} \quad \text{for } \underline{z} = \prod_{\gamma \in T} D^{\varepsilon_\gamma} \prod_{i=1}^{k_\gamma} (x_{\gamma_i}, n_{\gamma_i}) = (\underline{x},\underline{n}).$$

Now $\tilde{g}$ is defined on $\tilde{S}$ by

$$\tilde{g}([\underline{z},j]) = \begin{cases} (-1)^{[\frac{j}{2}]}\hat{g}(\underline{z}), & [\underline{z},j] \neq \Delta \\ 0 & , [\underline{z},j] = \Delta \end{cases}, \tag{11}$$

where [ ] denotes Gauss' symbol. Then it follows from (9), (10) and (11) that

$$\tilde{g}([(\underline{x},\underline{n}),j]) = (-1)^{[\frac{j}{2}]} 2^{|\underline{n}|} \tilde{g}([(\underline{x},\underline{0}),0]), \tag{12}$$

and

$$\tilde{g}([(\underline{z}_1, D\underline{z}_2),j]) = \tilde{g}([\underline{z}_1,j])\frac{\partial}{\partial \underline{x}_2} \tilde{g}([\underline{z}_2,0]), \quad \underline{z}_2 = (\underline{x}_2,\underline{n}_2). \tag{13}$$

We next consider $\tilde{g}(Y(t))$. Setting $\tau_0 = 0$, we define U_ℓ by

$$U_\ell(t,[\underline{z},j];g) = E_{[\underline{z},j]}[\tilde{g}(Y(t)); \quad \tau_\ell \leq t < \tau_{\ell+1}], \quad \ell \geq 0, \text{ 1)}$$

where τ_ℓ is the ℓ-th splitting time of $Y(t)$ and $E_{[\underline{z},j]}$ denotes the integration by the probability measure of $Y(t)$ starting from $[\underline{z},j] \in S$. Then, by the independence of particles behavior, we have

$$U_\ell(t,[(t,[(\underline{x},\underline{n}),j];g) = (-1)^{[\frac{j}{2}]} 2^{|\underline{n}|} U_\ell(t,[(\underline{x},\underline{0}),0];g), \text{ 2)} \tag{14}$$

and

---

1) U_ℓ is always finite because $g \in C_0^\infty(R^1)$.
2) $(\underline{x},\underline{0})$ denotes the element $(\underline{x},\underline{n})$ where all components of $\underline{n}$ are zero.

$$(15)\qquad U_\ell(t,[(\underline{\underline{z}}_1,\underline{\underline{z}}_2),j];\,g) = (-1)^{[\frac{j}{2}]}\sum_{r=0}^{\ell} U_r(t,[\underline{\underline{z}}_1,0];\,g)U_{\ell-r}(t,[\underline{\underline{z}}_2,0];\,g),\quad \ell \geqq 0$$

Moreover the effects of killing and age cancel each other.[1] So we get

$$(16)\qquad U_0(t,[(\underline{\underline{x}},\underline{\underline{0}}),0];g) = \underbrace{\int_{-\infty}^{\infty}\cdots\int_{-\infty}^{\infty}}_{k}\prod_{i=1}^{k} p(t,x_i,y_i)\hat{g}((\underline{\underline{y}},\underline{\underline{0}}))dy_1dy_2\cdots dy_k,$$ [2]

where $p(t,x,y)$ denotes the transition density of Brownian motion $\beta(t)$ and

$$(\underline{\underline{x}},\underline{\underline{0}}) = \prod_{i=1}^{k}(x_i,0),\qquad (\underline{\underline{y}},\underline{\underline{0}}) = \prod_{i=1}^{k}(y_i,0).$$

Accordingly, by (10),

$$(17)\qquad U_0(t,[(\underline{\underline{z}}_1,D\underline{\underline{z}}_2),0];g) = U_0(t,[\underline{\underline{z}}_1,0];g)\frac{\partial}{\partial \underline{\underline{x}}_2}U_0(t,[\underline{\underline{z}}_2,0];g).$$

Since τ_ℓ is a Markov time of $Y(t)$ for any $\ell \geqq 0$, the strong Markov property of $Y(t)$ implies

$$U_{\ell+1}(t,[\underline{\underline{z}},0];g) = E_{[\underline{\underline{z}},0]}[E_{Y(\tau_1)}[\tilde{g}(Y(t-s);\ \tau_\ell \leqq t-s \leqq \tau_{\ell+1}]_{s=\tau_1};\ \tau_1 <$$

Then, taking the particle which splitted at τ_1 into account and using (12), we obtain for $\underline{\underline{z}} = \prod D^{\varepsilon_\gamma}\prod(x_{\gamma_i},0)$

$$(18)\qquad U_{\ell+1}(t,[\underline{\underline{z}},0];g) = \sum_{\gamma\in\Gamma}\sum_{i=1}^{k_\gamma}\int_0^t ds\underbrace{\int_{-\infty}^{\infty}\cdots\int_{-\infty}^{\infty}}_{\sum k_\gamma}\prod_{\gamma'\in\Gamma}\prod_{j=1}^{k_{\gamma'}}p(s,x_{\gamma'_j},y_{\gamma'_j})$$

$$\sum_{\substack{0\leqq p\leqq N\\ 0\leqq q\leqq M}} c_{pq}U_\ell(t-s,[\underline{\underline{y}}_{\gamma ipq},0];g)\bigotimes_{\gamma',j}dy_{\gamma'_j},\quad \ell\geqq 0,$$ [3]

where $\otimes\, dy_{\gamma'_j}$ denotes the product measure of $dy_{\gamma'_j}$ and

$$\underline{\underline{y}}_{\gamma ipq} = \prod_{\gamma\in\Gamma} D^{\varepsilon_{\gamma'}}\underline{\underline{y}}_{\gamma'_i},$$

---

1),2) cf. [5] or [6].
3) cf. Lem.5.2 and 5.3 in [6].

$$\underline{\underline{y}}_{\gamma' i} = \begin{cases} \prod_{j=1}^{k_{\gamma'}} (y_{\gamma' j},0), & \gamma' \neq \gamma, \\ (\prod_{j=1}^{i-1}(y_{\gamma j},0),\underbrace{(y_{\gamma i},0),\ldots,(y_{\gamma i},0)}_{p},\underbrace{D(y_{\gamma i},0),\ldots,D(y_{\gamma i},0)}_{q},\prod_{j=1}^{k_\gamma}(y_{\gamma j},0)), & \gamma' = \gamma,\ p+q > 0, \\ (\prod_{j=1}^{i-1}(y_{\gamma j},0),(\delta,0),\prod_{j=i+1}^{k_\gamma}(y_{\gamma j},0), & \gamma' = \gamma,\ p+q = 0. \end{cases}$$

Hence, (17) and the relation $\partial p/\partial x = -\partial p/\partial y$ imply inductively

(19) $$U_\ell(t,[(\underline{\underline{z}}_1,D\underline{\underline{z}}_2),0];g) = \frac{\partial}{\partial \underline{\underline{x}}_2} U_0(t,[(\underline{\underline{z}}_1,\underline{\underline{z}}_2),0];g), \quad \ell \geqq 0.$$

Now let f be a bounded continuous function on R^1. According to (16), we have for $\underline{\underline{z}} = \prod_{i=1}^k (x_i,n_i) \in E^k$

$$E_{[\underline{\underline{z}},0]}[\tilde{f}(Y(t));t < \tau_1] = 2^{|\underline{\underline{n}}|} \prod_{i=1}^k \int_{-\infty}^{\infty} p(t,x_i,y_i)f(y_i)dy_i .$$

Accordingly, if we take $\partial f/\partial x$ in the sense of distribution of L. Schwartz, it follows that

$$E_{[D^\ell \underline{\underline{z}},0]}[\tilde{f}(Y(t));t < \tau_1] = 2^{|\underline{\underline{n}}|}\int_{-\infty}^{\infty} \cdots \int_{-\infty}^{\infty} (-1)^\ell \sum_{i=1}^k \frac{\partial^\ell}{\partial y_i^\ell} \prod_{i=1}^k p(t,x_i,y_i)$$

$$\cdot \prod_{i=1}^k f(y_i)dy_1 \cdots dy_k$$

$$= \frac{\partial^\ell}{\partial \underline{\underline{x}}^\ell} E_{[\underline{\underline{z}},0]}[\tilde{f}(Y(t));\ t < \tau_1], \quad \ell \geqq 0.$$

In this sense, we can define $U_\ell(t,[\underline{\underline{z}},j];\ f)$ for all $\ell \geqq 0$, $t > 0$, $[\underline{\underline{z}},j] \in S$ and bounded continuous functions f. Furthermore all relations from (14) to (19) still hold when we replace g with f.

<u>Definition 1</u>. <u>If</u>

$$\sum_{\ell=0}^{\infty} |U_\ell(t,[\underline{\underline{z}},j];f)| < \infty ,$$

then

$$U(t,[\underline{\underline{z}},j];f) = \sum_{\ell=0}^{\infty} U_\ell(t,[\underline{\underline{z}},j];f)$$

is called the expectation of $\widetilde{f}(Y(t))$ and is denoted by $E_{[\underline{\underline{z}},j]}[\widetilde{f}(Y(t))]$.

Now assume that $E_{[\underline{\underline{z}}_1,j]}[\widetilde{f}(Y(t))]$, $E_{[\underline{\underline{z}}_2,0]}[\widetilde{f}(Y(t))]$ and $E_{[D\underline{\underline{z}}_2,0]}[\widetilde{f}(Y(t))]$ exist and are finite. Then we have by (15) and (19)

$$E_{[(\underline{\underline{z}}_1,D\underline{\underline{z}}_2),j]}[\widetilde{f}(Y(t))] = \sum_{\ell=0}^{\infty} U_\ell(t,[(\underline{\underline{z}}_1,D\underline{\underline{z}}_2),j];f)$$

$$= \sum_{\ell=0}^{\infty} \sum_{r=0}^{\ell} U_r(t,[(\underline{\underline{z}}_1,j)];f)U_{\ell-r}(t,[(D\underline{\underline{z}}_2,0)];f)$$

$$= U(t,[\underline{\underline{z}}_1,j];f) \frac{\partial}{\partial \underline{\underline{x}}_2} U(t,[\underline{\underline{z}}_2,0];f).$$

Theorem 1. *Let f be a bounded continuous function with bounded derivative and $Y(t)$ be a branching Markov process with derivative corresponding to*

$$P(u, \frac{\partial u}{\partial x}) = \sum_{\substack{0 \le p \le N \\ 0 \le q \le M}} c_{pq} u^p (\frac{\partial u}{\partial x})^q.$$

Then there exists $T_0 > 0$ such that $u(t,x;f) = E_{[(x,0),0]}[\check{f}(Y(t))]$ is the unique solution of

$$(1) \quad \begin{cases} \dfrac{\partial u}{\partial t} = \dfrac{1}{2} \dfrac{\partial^2 u}{\partial x^2} + P(u, \dfrac{\partial u}{\partial x}), \quad 0 < t < T_0 \\ u(0,x) = f(x). \end{cases}$$

Proof. If both of

$$\sum_{\ell=0}^{\infty} |U_\ell(s,[(x,0),0];f)|, \qquad \sum_{\ell=0}^{\infty} |U_\ell(s,[D(x,0),0];f)|$$

are bounded uniformly in $(t,x) \in [0,T_0) \times R^1$, then it follows from (4), (18) and (20) that

$$u(t,x;f) = \int_{-\infty}^{\infty} p(t,x,y)f(y)dy + \int_0^t ds \int_{-\infty}^{\infty} p(s,x,y)$$

$$\cdot \sum_{\substack{0 \le p \le N \\ 0 \le q \le M}} c_{pq}\, u(t-s,y;f)^p (\frac{\partial u(t-s,y;f)}{\partial y})^q dy,$$

which proves the theorem. Hence it is enough to prove the existence of $T_0 > 0$.

Let

$$\|f\|_1 = \|f\| \vee \|f'\| \vee 1 = \max(\|f\|, \|f'\|, 1),$$

where $\|f\| = \sup|f(x)|$, and

$$K = \left\{ \sup_{1\leq \ell < \infty} \frac{2^{N+M+1}(1+\log(\ell+1))^{N+M}}{\sqrt{\ell+1}} \frac{\pi}{\sqrt{\ell+1}} \frac{((\ell+2)/2)}{\Gamma((\ell+1)/2)} \right\} \vee \|f\|_1 .$$

In addition, we assume that

$$(21) \qquad |U_r(t,[(x,0),0];f)| \ , \ |U_r(t,[D(x,0),0];f)| \leqq \frac{K(K^{N+M}C)^r}{r+1} t^{\frac{r}{2}},$$

$$0 \leqq r \leqq \ell \ , \ 0 \leqq t \leqq 1,$$

which holds for $r = 0$. Then (15) and (18) imply

$$|U_{\ell+1}(t,[(x,0),0];f)|$$

$$\leqq \int_0^t ds \int_{-\infty}^{\infty} p(s,x,y) \sum_{\substack{0 \leqq p \leqq N \\ 0 \leqq q \leqq M}} |c_{pq}|$$

$$\cdot |U_\ell(t-s,[(\underbrace{(y,0),\ldots,(y,0)}_{p},\underbrace{D(y,0),\ldots,D(y,0)}_{q}),0];f)|\,dy$$

$$= \int_0^t ds \int_{-\infty}^{\infty} p(s,x,y) \sum_{p,q} |c_{pq}| \sum_{r_1+r_2+\ldots+r_{p+q}=\ell} \left| \prod_{i=1}^{p} U_{r_i}(t-s,[(y,0),0];f) \right.$$

$$\left. \cdot \prod_{j=p+1}^{r+q} U_{r_j}(t-s,[D(y,0),0];f) \right| dy.$$

Applying (21) to the above inequality, we have

$$|U_{\ell+1}(t,[(x,0),0];f)| \leqq (K^{N+M}C)^{\ell+1} \int_0^t (t-s)^{\frac{\ell}{2}} ds$$

$$\cdot \sum_{r_1+r_2+\ldots+r_{p+q}=} \prod_{i=1}^{p+\ell} \frac{1}{r_i+1} .$$

Since

$$\sum_{r_1+\dots+r_p=\ell} \prod_{i=1}^{p} \frac{1}{r_i+1} \leqq \frac{2^{p-1}\{1+\log(\ell+1)\}^p}{\ell+1}$$

holds for all $\ell \geqq 0$ and $p \geqq 1$, it follows that

$$|U_{\ell+1}(t,[(x,0),0];f)| < \frac{K(K^{N+M}C)^{\frac{\ell+1}{2}}}{\ell+2} t^{\frac{\ell+1}{2}}, \quad 0 \leqq t \leqq 1.$$

Similarly we can get

$$|U_{\ell+1}(t,[D(x,0),0];f)| < (K^{N+M}C)^{\ell+1} \int_0^t \frac{(t-s)^{\frac{\ell}{2}}}{\sqrt{s}} ds \cdot \frac{2^{N+M+1}\{1+\log(\ell+1)\}^{N+M}}{\ell+1}$$

$$\leqq \frac{K(K^{N+M}C)^{\ell+1}}{2(\ell+1)} t^{\frac{\ell+1}{2}}$$

$$\leqq \frac{K(K^{N+M}C)^{\ell+1}}{\ell+2} t^{\frac{\ell+1}{2}}, \quad 0 \leqq t \leqq 1,$$

and accordingly (21) holds for all $r \geqq 0$. Therefore, if we take $(K^{N+M}C)^{-2}$ for T_0, both of

$$\sum_{\ell=0}^{\infty} |U_\ell(t,[D^{\varepsilon}(x,0),0];f)|, \qquad \varepsilon = 0,1,$$

are bounded in $(t,x) \in [0,T_0) \times R^1$, as was to be proved.

Theorem 2. Let f be a bounded continuous function on R^1 and $Y(t)$ be a branching Markov process with derivative corresponding to

$$P(u)\frac{\partial u}{\partial x} = \Big(\sum_{p=0}^{N} c_p u^p\Big)\frac{\partial u}{\partial x}.$$

Then $u(t,x;f) = E_{[(x,0),0]}[\tilde{f}(Y(t))]$ is the local solution of

$$(2) \qquad \begin{cases} \dfrac{\partial u}{\partial t} = \dfrac{1}{2}\dfrac{\partial^2 u}{\partial x^2} + \displaystyle\sum_{p=0}^{N} c_p u^p \frac{\partial u}{\partial x}, \\ u(0,x) = f(x) \end{cases}$$

and $u(t,x;f)$ can be extended to $[0,\infty) \times R^1$ so that u satisfies (2).

Proof. A similar computation as in the proof of Theorem 1 shows that $u(t,x;f)$ exists in $[0,T_0] \times R^1$ and satisfies (2) where T_0 depends on

$\|f\|$.[1] Then, regarding $P(u) = \sum c_p u^p$ as the drift term and using Ito's stochastic integral, we have

$$u(t,x;f) = E_x[\exp\left\{\int_0^t P(u(t-s,\beta(s);f))d\beta(s)\right.$$

$$\left. - \frac{1}{2}\int_0^t P^2(u(t-s,\beta(s);f))ds\right\} f(\beta(t))], \quad 0 \leqq t \leqq T_0,$$

because u is the solution of (2). Therefore the inequality

$$\|u(t,\cdot;f)\| \leqq \|f\|$$

holds for all $t \leqq T_0$. So we can define $u(t+s,x;f)$

$$u(t+s,x;f) = E_{[(x,0),0]}[\widetilde{u(t,\cdot;f)(Y(s))}], \quad 0 \leqq s,\ t \leqq T_0.$$

Of course $\tilde{u}(t+s,x;f)$ is identical to $u(t_1+s_1,x;f)$ if $t+s = t_1+s_1$ and $0 \leqq s,t,s_1,t_1 \leqq T_0$. Thus $u(t,x;f)$ is extended to $[0,2T_0]\times R^1$ and repeating the procedure we have the global solution of (2).

4. Solvability of (2) where the initial datum is a probability distribution. In the sequel, $u(t,x;f)$ denotes the solution of (2) and $u_\ell(t,x;f)$ does $U_\ell(t,[(x,0),0];f)$, $\ell \geqq 0$.

Proposition 1. $f \geqq g$ implies $u(t,x;f) \geqq u(t,x;g)$.

Proof. Let us set

$$w(t,x) = u(t,x;f) - u(t,x;g),$$

$$q(t,x)w(t,x) = \left\{P(u(t,x;f)) - P(u(t,x;g))\right\} \frac{\partial u(t,x;f)}{x}.$$

Then $w(t,x)$ satisfies

$$w(t,x) = \int_{-\infty}^{\infty} p(t,x,y)\left\{f(y) - g(y)\right\} dy$$

$$+ \int_0^t ds \int_{-\infty}^{\infty} p(s,x,y)\left\{Q(t-s,y)+P(u(t-s,y;g)) \frac{\partial w(t-s,y)}{y}\right\} dy,$$

and accordingly

---

1) In this case, we can take $\|f\|$ instead of $\|f\|_1$ to determine K in (21).

$$w(t,x) = E_x[\exp\{\int_0^t [Q(t-s,\beta(s)) - \frac{1}{2}P^2(u(t-s,\beta(s);g))]ds$$

$$+ \int_0^t P(u(t-s,\beta(s);g))d\beta(s)\}(f(\beta(t))-g(\beta(t)))]$$

$$\geqq 0,$$

as was to be proved.

Proposition 2. $u(t,x;f)$ inherits the monotonicity of f.

Proof. Let f be a monotone increasing function and set $f_h(x) = f(x+h)$. Then it follows from the homogeneity of $\beta(t)$ that

$$u(t,x;f_h) = E_{[(x,0),0]}[\tilde{f}_h(Y(t))]$$

$$= E_{[(x+h,0),0]}[\tilde{f}(Y(t))]$$

$$= u(t,x+h;f).$$

Hence, by Proposition 1,

$$u(t,x+h;f) \geqq u(t,x;f), \qquad t \geqq 0,\ h \geqq 0,\ x \in R^1.$$

Proposition 3. If $f(x)$ converges to a as x tends to infinity, then

$$\lim_{x\to\infty} u(t,x;f) = \lim_{x\to\infty} u_0(t,x;f) = \lim_{x\to\infty} f(x)\ (=a).$$

Proof. We have

$$\lim_{x\to\infty} u_0(t,x;f) = a\ , \quad \lim_{x\to\infty} \frac{\partial u_0(t,x;f)}{\partial x} = 0,$$

and accordingly

$$u_1(t,x;f) = \int_0^t ds \int_{-\infty}^{\infty} p(s,x,y) \sum_{0\leqq p\leqq N} c_p\, u_0(t-s,y;f)^p \frac{\partial u_0(t-s,y;f)}{\partial y}\, dy$$

$$\longrightarrow 0 \quad \text{as} \quad x \longrightarrow \infty .$$

Similarly, we have

$$u_\ell(t,x;f) \longrightarrow 0 \quad \text{as} \quad x \longrightarrow \infty \quad , \qquad \ell \geqq 1.$$

Since $\sum_\ell |u_\ell(t,x;f)|$ converges uniformly in $(t,x) \in [0,T_0] \times R^1$, these relations prove Proposition 3.

Proposition 4.

$$\int_{-\infty}^{\infty} f(x)dx = \int_{-\infty}^{\infty} u(t,x;f)dx, \quad t \geqq 0.$$

Proof. We first remark that

$$\int_{-\infty}^{\infty} f(x)dx = \int_{-\infty}^{\infty} u_0(t,x;f)dx,$$

and

$$\lim_{x \to \pm\infty} u_0(t,x;f) = 0.$$

Hence, by the same consideration as in Proposition 3, we have

$$\lim_{x \to \pm\infty} u(t,x;f) = 0.$$

Then it follows that

$$\int_a^b \{u(t,x;f) - u_0(t,x;f)\}\, dx$$

$$= \int_a^b dx \int_0^t ds \int_{-\infty}^{\infty} p(s,x,y)\frac{\partial}{\partial y} \sum_{0 \leqq p \leqq N} \frac{c_p}{p+1} u(t-s,y;f)^{p+1} dy$$

$$= \int_0^t ds \sum_{0 \leqq p \leqq N} \frac{c_p}{p+1} \int_{-\infty}^{\infty} \{p(s,b,y) - p(s,a,y)\} u(t-s,y;f)^{p+1} dy$$

$$\longrightarrow 0 \quad \text{as} \quad a \to -\infty,\ b \to \infty,$$

i.e.

(22) $$\lim_{\substack{a \to -\infty \\ b \to \infty}} \int_a^b u(t,x;f)dx = \int_{-\infty}^{\infty} f(x)dx.$$

Since Proposition 1 and (22)imply that

$$\int_{-\infty}^{\infty} |u(t,x;f)|\, dx \leqq \int_{-\infty}^{\infty} u(t,x;|f|)dx - \int_{-\infty}^{\infty} u(t,x;-|f|)dx$$

$$= 2 \int_{-\infty}^{\infty} |f(x)|\, dx \qquad ,$$

(22) proves the proposition.

Proposition 5. If $\{f_n(x)\}$ converges to $f(x)$ boundedly

$$\lim_{n\to\infty} u(t,x;f_n) = u(t,x;f).$$

Proof. Since the result holds when we replace u with u_ℓ, $\ell \geq 0$, we obtain the proposition.

Proposition 6.[1] The solution of Burger's equation

$$\begin{cases} \dfrac{\partial u}{\partial t} = \dfrac{\varepsilon}{2}\dfrac{\partial^2 u}{\partial x^2} - u\dfrac{\partial u}{\partial x} \\ u(0,x) = f(x) \end{cases}$$

is given by

$$u(t,x) = \frac{E_x[\exp\{-\frac{1}{\varepsilon}\int_0^{\gamma(t)} f(y)dy\} f(\gamma(t))]}{E_x[\exp\{-\frac{1}{\varepsilon}\int_0^{\gamma(t)} f(y)dy\}]},$$

where $\gamma(t) = \beta(0) + \sqrt{\varepsilon}(\beta(t) - \beta(0))$.

Proof. We will consider the case where $\varepsilon = 1$. Let us put

$$F(x) = -\int_0^x f(y)dy ,$$

and

$$w_\ell(t,x) = \sum_{r=0}^{\ell} u_r(t,x;f)\frac{E_x[F(\beta_t)^{\ell-r}]}{(\ell-r)!},^{2)} \qquad \ell \geq 0.$$

Then we assume that

$$(23) \qquad w_r(t,x) = -\frac{\partial}{\partial x}\frac{E_x[F(\beta_t)^{r+1}]}{(r+1)!}, \qquad r \leq \ell ,$$

which holds for $\ell = 0$. Since it follows from (18) that

$$\frac{\partial w_{\ell+1}}{\partial t} = \frac{1}{2}\sum_{r=0}^{\ell+1}\left\{\frac{\partial^2 u_r}{\partial x^2}\frac{E_x[F(\beta_t)^{\ell+1-r}]}{(\ell+1-r)!} + u_r\frac{\partial^2}{\partial x^2}\frac{E_x[F(\beta_t)^{\ell+1-r}]}{(\ell+1-r)!}\right.$$

$$\left. - 2\sum_{s=0}^{r-1} u_{r-1-s}\frac{\partial u_s}{\partial x}\frac{E_x[F(\beta_t)^{\ell+1-r}]}{(\ell+1-r)!}\right\}$$

---

1) For a precise discussion, see [2]. 2) β_t denotes $\beta(t)$.

$$= \frac{1}{2}\sum_{r=0}^{\ell+1}\left\{ \frac{\partial^2 u_r}{\partial x^2}\,\frac{E_x[F(\beta_t)^{\ell+1-r}]}{(\ell+1-r)!} + u_r\frac{\partial^2}{\partial x^2}\,\frac{E_x[F(\beta_t)^{\ell+1-r}]}{(\ell+1-r)} \right\}$$

$$- \sum_{s=0}^{\ell} \frac{\partial u_s}{\partial x} w_{\ell-s}$$

$$= \frac{1}{2}\,\frac{\partial^2 w_{\ell+1}}{\partial x^2} \quad ,$$

we have

$$w_{\ell+1}(t,x) = E_x[w_{\ell+1}(0,\beta_t)] = -\frac{\partial}{\partial x}\,\frac{E_x[F(\beta_t)^{\ell+2}]}{(\ell+2)!} .$$

Therefore (23) holds for all $r \geqq 0$ and accordingly

$$E_x[\exp(F(\beta_t))f(\beta_t)] = \sum_{\ell=0}^{\infty} \frac{E_x[F(\beta_t)^{\ell} f(\beta_t)]}{\ell!}$$

$$= \sum_{\ell=0}^{\infty} w_\ell(t,x)$$

$$= u(t,x;f)E_x[\exp(F(\beta_t))],$$

as was to be proved.

Propositions 1 - 4 show that $u(t,x;f)$ is a probability density or a cumulative distribution function according as f is a probability density or a cumulative distribution function. Proposition 5 indicates that a sequence of cumulative distribution functions converges when their densities converge boundedly. So the question that a probability measure can be generally an initial datum of (2) or not arises there. Let $\Phi(dx)$ be a probability measure on R^1. We define $u(t,x;d\Phi) = U(t,[(x,0),0];d\Phi)$ in the following way. Set

$$F(x) = \Phi((-\infty,x]),$$

$$u_0(t,x;d\Phi) = U_0(t,[(x,0)];d\Phi) = \frac{\partial}{\partial x} U_0(t,[(x,0),0];F),$$

$$U_0(t,[(\underline{\underline{x}},\underline{\underline{n}}),j];d\Phi) = (-1)^{[\frac{j}{2}]}\widehat{u_0}(t,\cdot;d\Phi)(\underline{\underline{x}},\underline{\underline{n}}), \quad (\underline{\underline{x}},\underline{\underline{n}}) \in E^k,$$

$$U_0(t,[D(\underline{\underline{x}},\underline{\underline{n}}),j];d\Phi) = \frac{\partial}{\partial \underline{\underline{x}}} U_0(t,[(\underline{\underline{x}},\underline{\underline{n}}),j];d\Phi),$$

$$U_0(t,[(\underline{z}_1,D\underline{z}_2),j];d\Phi) = U_0(t,[\underline{z}_1,j];d\Phi)\ \frac{\partial}{\partial \underline{x}_2} U_0(t,[\underline{z}_2,0];d\Phi),$$

where $\underline{z}_2 = (\underline{x}_2,\underline{n}_2)$, and

$$U_{\ell+1}(t,[\underline{z},j];d\Phi) = E_{[\underline{z},j]}[U_\ell(t-s,Y(s);d\Phi)|_{s=\tau_1};\ \tau_1 < t], \qquad \ell \geqq 0.$$

Then U is defined by

$$U(t,[\underline{z},j];d\Phi) = \sum_{\ell=0}^{\infty} U_\ell(t,[\underline{z},j];d\Phi),$$

where $\sum_{\ell=0}^{\infty}|U_\ell(t,[\underline{z},j];d\Phi)|$ is assumed to be finite.

Definition 2. *Let* $u(t,x;d\Phi)$ *be a probability density function of* $\Phi(t,dx)$. *If* $u(t,x;d\Phi)$ *satisfies*

(24) $$u(t,x;d\Phi) = u_0(t,x;d\Phi) + \int_0^t ds \int_{-\infty}^{\infty} p(s,x,y) \sum_{0\leqq p\leqq N} c_p u(t-s,y;d\Phi)^p \frac{\partial u(t-s,y;d\Phi)}{\partial y}\, dy,$$

and $\Phi(t,dx)$ *converges to* $\Phi(dx)$ *weakly as* t *tends to zero*, i.e.

$$\lim_{t\to 0} \int_{-\infty}^{\infty} f(x)\,\Phi(t,dx) = \int_{-\infty}^{\infty} f(x)\,\Phi(dx)$$

holds for any bounded continuous function f, *then* $u(t,x;d\Phi)$ *is said to be a solution of* (2) *with the initial datum* $\Phi(dx)$.

Let us now consider the equation

(25) $$\frac{\partial u}{\partial t} = \frac{1}{2}\frac{\partial^2 u}{\partial x^2} - u^p \frac{\partial u}{\partial x}, \quad p \geqq 2.$$

Theorem 3. [1] *For* $p = 2$, (25) *can be solved for any initial datum of probability measure* $\Phi(dx)$. *But, for* $p \geqq 3$, (25) *can not be solved for an initial datum of* δ *-measure*.

Proof. 1°. The case of $p = 2$. Let us put

$$F(x) = -\Phi((-\infty,x]),$$

$$W_\ell(t,x) = \frac{E_x[F(\beta(t))^\ell]}{\ell!}, \qquad \ell \geqq 0,$$

1) This is a question given by H. P. McKean.

and

$$u_\ell(t,x;d\Phi) = U_\ell(t,[(x,0),0];d\Phi), \qquad \ell \geqq 0.$$

Then

$$u_0(t,x;d\Phi) = -\frac{\partial}{\partial x} E_x[F(\beta(t))] = -\frac{\partial W_1(t,x)}{\partial x}$$

is a $C^\infty((0,\infty)\times R^1)$ probability density function.

Now we assume that

$$(26) \qquad -\frac{\partial W_{r+1}(t,x)}{\partial x} = \sum_{k=0}^{r} u_k(t,x;d\Phi)W_{r-k}(t,x), \quad 0 \leqq r \leqq \ell,$$

$$u_r(t,x;d\Phi) = O(t^{-\frac{1}{2}}) \quad \text{as} \quad t \longrightarrow 0, \quad 0 \leqq r \leqq \ell,$$

which hold for $\ell = 0$. Since

$$\frac{\partial u_r}{\partial t} = \frac{1}{2}\frac{\partial^2 u_r}{\partial x^2} - \sum_{k=0}^{r-1} u_k \frac{\partial u_{r-1-k}}{\partial x}, \quad 0 \leqq r \leqq \ell,$$

holds, by (26), we have

$$\sum_{r=0}^{\ell} u_r(t,x;d\Phi) \frac{\partial u_{\ell-r}(t,x;d\Phi)}{\partial x}$$

$$= -\sum_{r=0}^{\ell}\left(\frac{\partial W_{r+1}(t,x)}{\partial x} + \sum_{k=0}^{r-1} u_k(t,x;d\Phi)W_{r-k}(t,x)\right)\frac{\partial u_{\ell-r}(t,x;d\Phi)}{\partial x}$$

$$= -\frac{1}{2}\frac{\partial^2}{\partial x^2}\sum_{r=0}^{\ell} W_{r+1}(t,x)u_{\ell-r}(t,x;d\Phi) + \frac{1}{2}\sum_{r=0}^{\ell}\frac{\partial^2 W_{r+1}(t,x)}{\partial x^2}u_{\ell-r}(t,x;d\Phi)$$

$$+ \frac{1}{2}\sum_{r=0}^{\ell} W_{r+1}(t,x)\frac{\partial^2 u_{\ell-r}(t,x;d\ )}{\partial x^2}$$

$$- \sum_{r=1}^{\ell}\sum_{k=0}^{r-1} u_k(t,x;d\Phi)W_{r-k}(t,x)\frac{\partial u_{\ell-r}(t,x;d\Phi)}{\partial x}$$

$$= -\frac{1}{2}\frac{\partial^2}{\partial x^2}\sum_{r=0}^{\ell} W_{r+1}(t,x)\,u_{\ell-r}(t,x;d\Phi) + \sum_{r=0}^{\ell}\frac{\partial W_{r+1}(t,x)}{\partial t}u_{\ell-r}(t,x;d\Phi)$$

$$+ \sum_{r=0}^{\ell} W_{r+1}(t,x)\frac{\partial u_{\ell-r}(t,x;d\Phi)}{\partial t}$$

$$+\sum_{r=1}^{\ell} W_{r+1}(t,x)\sum_{k=0}^{\ell-r-1} u_k(t,x;d\Phi)\,\frac{\partial u_{\ell-r-1-k}(t,x;d\Phi)}{\partial x}$$

$$-\sum_{r=1}^{\ell}\sum_{k=0}^{r-1} u_k(t,x;d\Phi)W_{r-k}(t,x)\,\frac{u_{\ell-r}(t,x;d\Phi)}{x}$$

Since the last two summations cancel each other, it follows from the definition of $u_{\ell+1}(t,x;d\Phi)$ that

$$u_{\ell+1}(t,x;d\Phi) = -\int_0^t ds \int_{-\infty}^{\infty} p(s,x,y)\Big\{\sum_{r=0}^{\ell} u_r(t-s,y;d\Phi)\frac{\partial u_{\ell-r}(t-s,y;d\Phi)}{\partial y}\Big\}\,dy$$

$$= -\int_0^t ds\,\frac{\partial}{\partial s}\int_{-\infty}^{\infty} p(s,x,y)\sum_{r=0}^{\ell} W_{r+1}(t-s,y)u_{\ell-r}(t-s,y;d\Phi)\,dy$$

$$= -\frac{\partial W_{\ell+2}}{\partial x} - \sum_{r=0}^{\ell} W_{r+1}(t,x)u_{\ell-r}(t,x;d\Phi),$$

which shows that (26) holds for $r = \ell + 1$. Thus, as in the proof of Proposition 6, we have

$$\frac{-\frac{\partial}{\partial x} E_x[\exp F(\beta(t))]}{E_x[\exp F(\beta(t))]} = u(t,x;d\Phi),\ ^{1)}$$

as was to be proved, because the weak convergence is evident.

2°. The case of $p \geq 3$. Let $\Phi(dx)$ be the δ-measure with the unit mass at $x = a$. Then we have

$$u_0(t,x;d\Phi) = p(t,x,a)$$

and accordingly

$$\int_{-\infty}^{\infty} p(s,x,y)u_0(t-s,y;d\Phi)^p\,\frac{\partial u_0(t-s,y;d\Phi)}{\partial y}\,dy$$

is not integrable over $(0,t)$, $t > 0$, with respect to Lebesgue measure. Since $u(t,x;d\Phi) - u_0(t,x;d\Phi)$ tends to zero with t for any solution $u(t,x;d\Phi)$, the above computation shows the last half of the theorem.

1) $\varphi(t,z) = E_x[\exp\{zF(\beta(t))\}]$ is continuous in (t,z) and analytic in z. Hence $\varphi(0,z)\neq 0$ implies the existence of $u(t,x;d\Phi)$ for small $t > 0$.

References

[1] N. Ikeda, M. Nagasawa and S. Watanabe: Branching Markov Processes I, II, III Jour. Math. Kyoto Univ., vol.8(1968), 233-278, 365-410, vol.9(1969), 97-162.

[2] E. Hopf: The partial differential equation $u_t + uu_x = \mu u_{xx}$. Comm. Pure Appl. Math., vol.3(1950), 201-230.

[3] M. Nagasawa: Construction of branching Markov processes with age and sign. Kodai Math. Sem. Rep., vol.20(1968), 469-508.

[4] M. Nisio: On stochastic differential equations associated with certain quasilinear parabolic equation. Jour. Math. Soc. Japan, vol.22, No.3(1970), 278-292.

[5] T. Sirao: A probabilistic treatment of semi-linear parabolic equations. Proc. Japan Acad., vol.42(1966), 885-890.

[6] ______ : On signed branching Markov processes with age. Nagoya Math. Jour., vol.32(1968), 155-225.

[7] H. Tanaka: Local solutions of stochastic differential equations associated with certain quasilinear parabolic equations. Jour. Fac. Sci. Univ. Tokyo, Sect. I, 14(1967), 313-326.

Tokyo Metropolitan University

ON LIMIT THEOREMS FOR RANDOM PROCESSES

V. Statulevičius

1. Let $\xi(t)$ be a Markov process on a probability space $(\mathcal{U}, \mathcal{F}, P)$ with values in measurable spaces $(\Omega_t, \mathcal{F}_t)$, $t=0,1,\dots,n$, $P_t(\omega, A)$ be its transition probability function (from a state $\omega \in \Omega_{t-1}$ at time $t-1$ into a set $A \in \mathcal{F}_t$ at time) and $P_0(A)$, $A \in \mathcal{F}_0$, be an initial probability distribution.

In limit theorems (see [1]) for the sums

$$S_n = X_1 + X_2 + \dots + X_n$$

of random variables X_K related to the Markov chain $\xi(t)$ (i.e. $X_K = g_K(\xi(K))$, where $g_K(\omega)$ is an $\mathcal{F}_K$ -measurable function defined on Ω_K, $K = 1, 2, \dots, n$), the ergodicity co-efficient in the first n steps of the chain:

$$\alpha^{(n)} = \min_{1 < K \leq n} \alpha_K$$

is usually used, where $\alpha_K = \alpha(P_K)$ is the ergodicity co-efficient of the transition function $P_K(\omega, A)$:

$$1 - \alpha_K = \sup_{\substack{\omega, \tilde{\omega} \in \Omega_{K-1} \\ A \in \mathcal{F}_K}} |P_K(\omega, A) - P_K(\tilde{\omega}, A)|.$$

For instance, if $|X_K| \leq C^{(n)}$, $K = 1, 2, \dots, n$, with probability 1, then there exists ([1], theorem 1) an absolute constant C, such that

$$\sup_x \left| P\{Z_n < x\} - \Phi(x) \right| \leq C \frac{C^{(n)}}{\alpha^{(n)} B_n}, \tag{1}$$

where $Z_n = \dfrac{S_n - ES_n}{B_n}$, $B_n^2 = DS_n$, $\Phi(x) = \dfrac{1}{\sqrt{2\pi}} \displaystyle\int_{-\infty}^{x} e^{-\frac{y^2}{2}} dy$.

Moreover, there exist examples where estimate (1) is the best possible. It turns out that, in many limit theorems for distributions P_{Z_n}, the quantity $\frac{1}{\alpha^{(n)}}$ can be replaced by

$$\Lambda_n = \max_{1<K\leq n} \sum_{\ell=K}^{n} (1-\alpha_{K\ell}) =$$

$$= \max_{1<K\leq n} \sum_{\ell=K}^{n} \sup_{\substack{\omega,\tilde{\omega}\in\Omega_K \\ A\in\mathcal{F}_\ell}} \left| P_{K\ell}(\omega, A) - P_{K\ell}(\tilde{\omega}, A) \right|,$$

where $\alpha_{K\ell} = \alpha(P_{K\ell})$ is the ergodicity co-efficient of the transition function $P_{K\ell}(\omega, A)$

Since

$$1-\alpha_{K\ell} \leq \prod_{j=K+1}^{\ell} (1-\alpha_j) \leq (1-\alpha^{(n)})^{\ell-K},$$

we always have

$$\Lambda_n \leq 1 + \frac{1-\alpha^{(n)}}{\alpha^{(n)}}.$$

Using $\alpha^{(n)}$ or Λ_n, it is possible to generalize most of the well-known results for distributions of sums of independent random variables on P_{Z_n}.

<u>Theorem 1</u>. <u>If</u> $|X_K| \leq C^{(n)}$, $K = 1, 2, \ldots, n$, <u>with probability</u> 1, <u>there exists an absolute constant</u> C_1 <u>such that</u>

$$\sup_x \left| P\{Z_n - x\} - \Phi(x) \right| \leq C_1 \frac{C^{(n)}\Lambda_n}{B_n}.$$

As a corollary we have

<u>Theorem 2</u>. <u>If</u> $DX_K < \infty$, $K = 1, 2, \ldots, n$,

$$\frac{\Lambda_n}{B_n^2} \sum_{K=1}^{n} \int_{|x|>\varepsilon\frac{B_n}{\Lambda_n}} x^2 dF_{X_K}(x + EX_K) \longrightarrow 0 \quad (n \to \infty)$$

<u>for any</u> $\varepsilon > 0$, <u>then</u>

$$\sup_x \left| P\{Z_n < x\} - \Phi(x) \right| \longrightarrow 0 \quad (n \to \infty).$$

In the sequel, we assume $EX_K = 0$ for simplicity and everywhere $DX_K < \infty$, $K = 1, 2, \ldots, n$.

Theorem 3. There exists an absolute constant C_2, such that

$$|P\{Z_n < x\} - \Phi(x)| \leq$$

$$\leq \frac{C_2 \Lambda_n^2}{(1+|x|)^3 B_n} \sum_{k=1}^{n} \Big(\sup_{\omega} \sup_{0<z\leq(1+|x|)B_n\Lambda_n^{-1}} z \int_{\{\bar\omega : |X_k(\bar\omega)|>z\}} X_k^2(\bar\omega) P_k(\omega, d\bar\omega) +$$

$$+ \sup_{\omega} \sup_{0<z\leq(1+|x|)B_n\Lambda_n^{-1}} \Big| \int_{\{\bar\omega : |X_k(\bar\omega)|\leq z\}} X_k^3(\bar\omega) P_k(\omega, d\bar\omega) \Big| .$$

Hence it follows, for example, that, for any δ, $0<\delta\leq 1$, the estimate

$$|P\{Z_n < x\} - \Phi(x)| \leq$$

$$\leq C_3 \frac{\Lambda_n^{1+\delta}}{(\Lambda+|x|)^{1+\delta} B_n^{2+\delta}} \sum_{k=1}^{n} \sup_{\omega} \int_{\Omega_k} |X_k(\bar\omega)|^{2+\delta} P_k(\omega, d\bar\omega)$$

holds true.

If we make use only of the unconditional moments of the variables X_k to estimate $F_{Z_n}(x) - \Phi(x)$, we obtain the following result:

Theorem 4. For any integer $s \geq 3$,

$$|P\{Z_n < x\} - \Phi(x)| \leq$$

$$\leq C_4 \frac{L_{sn}^{\frac{1}{s-2}} + L_{sn} \ln^{\frac{s}{2}} \left(1 + L_{sn}^{-\frac{1}{s-2}}\right)}{(1+|x|)^3},$$

where

$$L_{sn} = \Lambda_n^{s-1} B_n^{-s} \sum_{k=1}^{n} E|X_k|^s$$

is the Liapunov ratio of the s th order, C_4 is an absolute constant.

Let us discuss the problem of estimation for probabilities of large deviations.

Theorem 5. If there exist positive constants $A, C_1, C_2, C_3, \ldots, C_n$ such that

$$\left| \frac{1}{z^2} \ln E\, e^{zX_k} \right| \le C_k, \qquad k = 1, 2, \ldots, n$$

in the circle $|z| < A$ and

$$\frac{1}{B_n^2} \sum_{k=1}^{n} C_k \le H,$$

then there exists a δ, $0 < \delta_H < 1$, dependent only on H such that in the interval

$$0 \le x \le \delta \Delta_n, \quad \delta < \delta_H, \quad \Delta_n = \frac{A B_n}{\Lambda_n}$$

the following relations hold true:

$$\frac{P\{Z_n > x\}}{1 - \Phi(x)} = \exp\left\{\frac{x^3}{\Delta_n} \lambda_n\left(\frac{x}{\Delta_n}\right)\right\} \left(1 + \theta_{H,\delta} \frac{x+1}{\Delta_n}\right),$$

(2) $$\frac{P\{Z_n < -x\}}{\Phi(-x)} = \exp\left\{-\frac{x^3}{\Delta_n} \lambda_n\left(-\frac{x}{\Delta_n}\right)\right\} \left(1 + \theta'_{H,\delta} \frac{x+1}{\Delta_n}\right),$$

where $\theta_{H,\delta}$ and $\theta'_{H,\delta}$ are finite for all H, $0 < H < \infty$ and $\delta < \delta_H$, $\lambda_n(t) = \sum_{k=0}^{\infty} \lambda_{kn} t^k$ is the power series of Cramér with

$$|\lambda_{kn}| \le \frac{1}{(k+3)\, \delta_H^{k+2}}, \qquad k = 0, 1, 2, \ldots$$

(an equation for δ_H and estimates for $\theta_{H,\delta}$, $\theta'_{H,\delta}$ could be found in [2] or [3]).

Theorem 6. Let

$$E e^{hX_k} \le e^{d_k h^2}$$

for $0 \le h \le A$ and some positive numbers d_k, $k = 1, 2, \ldots, n$. Then there exist absolute constants $\alpha > 0$ and $\beta > 0$ such that

$$P\{S_n > x\} \le \exp\left\{-\frac{\beta x^2}{D_n}\right\}$$

when $0 \leq x \leq \alpha \frac{D_n}{\Lambda_n}$ and

$$P\{S_n > x\} \leq \exp\left\{ \frac{\alpha\beta x}{\Lambda_n} \right\}$$

when $x \geq \alpha \frac{D_n}{\Lambda_n}$; $D_n = \sum_{K=1}^{n} d_K$.

Theorem 7. (together with R. Lapinskas). Let the variables X_K, $K = 1, 2, \ldots, n$, take values in the m-dimensional space R^m and $\mathfrak{N}$ be a class of all measurable convex sets in R^m. Then there exists a constant $C(m)$ dependent only on m such that

$$\sup_{A \in \mathfrak{N}} \left| P\{S_n \in A\} - \Phi_{S_n}(A) \right| \leq$$

$$\leq C(m) \sup_{|t|=1} \frac{\sum_{K=1}^{n} \sup_{\omega} \int_{\Omega_K} \left|(X_K(\bar{\omega}), t)\right|^3 P_K(\omega, d\bar{\omega})}{\alpha^{(n)2} (D(S_n, t))^{3/2}}, \tag{3}$$

where $\Phi_{S_n}(A)$ is the normal law corresponding to S_n, (x, t) is the inner product of the vectors $x \in R^m$ and $t \in R^m$.

Taking into account that

$$D(S_n, t) \geq \frac{\alpha^{(n)}}{32} \sum_{K=1}^{n} D(X_K, t),$$

one can make use of the numerous results for sums of independent m-dimensional random variables.

Let random variables $\xi_1, \xi_2, \ldots, \xi_n$ be independent and let functions $g_i(t)$, $t \geq 0$, $i = 1, 2$, satisfy the conditions

$$g_1(t) < g_2(t), \quad g_1(0) < 0 < g_2(0),$$

$$|g_i(t+s) - g_i(t)| < Ks, \quad s > 0, \quad i = 1, 2,$$

where K is a constant. Put $S_K = \sum_{j=1}^{K} \xi_j$, $B_K^2 = D S_K$, $E\xi_K = 0$, $K = 1, 2, \ldots, n$ and let $W(t)$ be a Wiener process.

S.V. Nagayev's theorem [4] is generalized as follows.

<u>Theorem 8. If</u> $\sigma^2 \le \mathcal{D}\xi_K$, $E|\xi_K|^3 \le \beta$, $K = 1,\dots,n$ <u>then there exists an absolute constant</u> C_6 <u>such that</u>

$$\left| P\left\{ g_1\left(\frac{B_K^2}{B_n^2}\right) < \frac{S_K}{B_n} < g_2\left(\frac{B_K^2}{B_n^2}\right),\ K = 1,\dots,n \right\} - \right.$$
$$\left. - P\{ g_1(t) < w(t) < g_2(t),\ 0 \le t \le 1 \} \right| \le$$
$$\le C_6 \frac{\beta^2 (1+K)}{\sigma^6 \sqrt{n}} .$$

For most chains, for instance, for the chain $\xi(k) = S_n$, $K = 1,\dots,n$ we have $\alpha^{(n)} = 0$ or Λ_n is of the same order as n. In such cases the above statements are trivial.

Let the symbol "$\frown$" above a random variable denote that it is centred:

$$\widehat{\xi} = \xi - E\xi .$$

Introduce a new notation

$$\widehat{E} XY \cdots ZW = E X \widehat{Y \cdots \widehat{Z \widehat{W}}}$$

if the right-hand side (for random variables $X,\dots,W$) has sense.

If $X_1,\dots,X_n$ are related to a Markov chain, then the moments centred in such a way have a comparatively simple expression:

$$\widehat{E} X_{l_1}^{\nu_1} X_{l_2}^{\nu_2} \cdots X_{l_r}^{\nu_r} =$$

$$= \int_{\Omega_{l_1}} \cdots \int_{\Omega_{l_r}} X_{l_1}^{\nu_1}(\omega_1) P_{l_1}(d\omega_1) \prod_{\rho=2}^{r} X_{l_\rho}^{\nu_\rho}(\omega_\rho)\left(P_{l_{\rho-1}\, l_\rho}(\omega_{\rho-1}, d\omega_\rho) - P_{l_\rho}(d\omega_\rho)\right),$$

for any $1 \le l_1 < l_2 < \dots < l_r \le n$. Here $P_l(A) = P\{\xi(l) \in A\}$.

<u>Theorem 9. Let there exist positive constants</u> L <u>and</u> K <u>such that, for any integers</u> $1 \le l_1 < l_2 < \dots < l_r \le n$ <u>and</u> $0 \le \nu_i < \infty$, $i = 1,\dots,r$, <u>with</u> $\nu_1 + \nu_2 + \dots + \nu_r = \nu$, <u>and some</u> β_n, $0 < \beta_n < 1$,

$$\left|\widehat{E} X_{l_1}^{\nu_1} \cdots X_{l_2}^{\nu_2}\right| \leq \nu!\, L K^{\nu-2} e^{-\beta_n (l_2 - l_1)} \sqrt{\mathcal{D} X_{l_1} \mathcal{D} X_{l_2}},$$

$$\left| E X_l^{\nu} \right| \leq \nu! \quad L K^{\nu-2} \mathcal{D} X_l, \quad l = 1, \ldots, n.$$

Then, for $P\{Z_n > x\}$, the relations (2) are true with

$$\Delta_n = \frac{c_{L,K} \beta_n B_n}{\max\limits_{1 \leq K \leq n} \sqrt{\mathcal{D} X_K}} \qquad H = C_{L,K} \frac{\sum_{K=1}^{n} \mathcal{D} X_K}{\beta_n B_n}.$$

Positive constants $c_{L,K}$ and $C_{L,K}$ depend only on L and K

2. If we have random variables $X_1, X_2, \ldots, X_n$ related by a more general dependence then Markov, for instance, if only the strong mixing condition is satisfied

$$\sup_{\substack{A \in \mathcal{F}_{0,K} \\ B \in \mathcal{F}_{l,n}}} \left| P(AB) - P(A) P(B) \right| \leq \gamma_n (l - K),$$

or

$$\sup_{\substack{A \in \mathcal{F}_{0,K} \\ B \in \mathcal{F}_{l,n}}} \left| P(B/A) - P(B) \right| \leq \delta_n (l - K),$$

where $\gamma_n(\tau)$ and $\delta_n(\tau)$ tend to 0 at a definite rate when $\tau \to \infty$ and $\mathcal{F}_{Kl} = \sigma\{X_u,\ K < u \leq l\}$ then it is rather difficult to expect a good rate of convergence of P_{Z_n} to the limit distribution as well as to obtain more or less strong results for the probabilities of large deviations or asymptotic expansions of the distribution P_{Z_n} . An exception is the case of the so-called regularity of Markov's type when

$$\left| P(B/CA) - P(B|C) \right| \leq \alpha_n (l - K)$$

for any $A \in \mathcal{F}_{0K}$, $C \in \mathcal{F}_{K,l}$, $B \in \mathcal{F}_{l,n}$ and $\alpha_n(\tau) \downarrow 0$, when τ tend to ∞ at a definite rate. In this case, it is possible to generalize all the theorems of section 1 for P_{Z_n} . Here is only one of them.

<u>Theorem 10</u>. <u>If</u> $\alpha_n(\tau) \le \exp\{-\tilde{\mathcal{Z}}^{(n)}\tau\}$ <u>then</u>

$$\left|P\{Z_n < x\} - \Phi(x)\right| \le C_7 \frac{\sum_{k=1}^{n} \operatorname{vraisup} E\{|X_k|^3 | \mathcal{F}_{0,k-1}\}}{\mathcal{Z}^{(n)2} B_n^3 (1+|x|^3)} .$$

3. Let $X(t) = \{X_a(t)\}_{a=\overline{1,r}}$, $t = \dots, -1, 0, 1, \dots$, be an r -dimensional stationary in the wide sense random sequence with mean 0 and real components. Let $c_{a_1 \dots a_n}(t_1, \dots, t_n) = \Gamma\{X_{a_1}(t_1) \dots, X_{a_n}(t_n)\}$ be a simple semi-invariant of the variables $X_{a_1}(t_1), \dots, X_{a_n}(t_n)$ (if it exists), and

$$F(\lambda) = \{F_{ab}(\lambda)\}_{a=\overline{1,r}}^{b=\overline{1,r}}$$

and

$$f(\lambda) = \{f_{ab}(\lambda)\}_{a=\overline{1,r}}^{b=\overline{1,r}}$$

the spectral function (s.f.) and the spectral density (s.d.) of the sequence $X(t)$, respectively.

Assume that there exist the s.d.'s of the first and higher orders. Conditions under which, for a fixed set $(a_1, \dots, a_q)$, $1 \le a_j \le r$, the s.d. of the $(q-1)$ -th order $f_{a_1 \dots a_q}$ exists, are:

1) $E|X_k(t)|^q < \infty$ for all $k \in \{a_1, \dots, a_q\}$ and $t = \dots, -1, 0, 1, \dots$,

2) for any t_j, $t = \dots, -1, 0, 1, \dots$,

$$c_{a_1 \dots a_q}(t_1 + t, \dots, t_q + t) = c_{a_1 \dots a_q}(t_1, \dots, t_q),$$

3) there exists a complex-valued function $f_{a_1 \dots a_q}(\lambda_1, \dots, \lambda_{q-1})$ such that, for any $t_j = \dots, -1, 0, 1, \dots$,

$$c_{a_1 \dots a_q}(t_1, \dots, t_{q-1}, 0) = \int_{-\pi}^{\pi} \dots \int_{-\pi}^{\pi} f_{a_1 \dots a_q}(\lambda_1, \dots, \lambda_{q-1}) \times$$

$$\times \exp\left\{i \sum_{1}^{n-1} t_j \lambda_j\right\} d\lambda_1 \dots d\lambda_{n-1}.$$

Let

$$\mathcal{I}_N(\lambda) = \left\{ \frac{1}{2\pi N} \sum_{s=1}^{N} e^{-is\lambda} X_a(s) \sum_{t=1}^{N} e^{it\lambda} X_b(t) \right\}_{a=\overline{1,r}}^{b=\overline{1,r}}$$

be the matrix of the periodogram of the second order for the sample $\{X(t),\ 0 < t \leq N\}$ of size N. Consider the asymptotic behaviour of the matrix-valued random processes

$$(4) \qquad \zeta_N(\lambda) = \sqrt{N}\left[\int_0^\lambda \mathcal{I}_N(\alpha)\,d\alpha - F(\lambda)\right], \quad 0 \leq \lambda \leq \pi,$$

and

$$(5) \qquad \xi_N(\lambda) = \sqrt{N}\left[\int_0^\lambda \mathcal{I}_N(\alpha)\,d\alpha - E\int_0^\lambda \mathcal{I}_N(\alpha)\,d\alpha\right], \quad 0 \leq \lambda \leq \pi$$

the components of which take complex values. Denote by $C^{r\times r}[0,\pi]$ the space of all $r\times r$ -matrix-valued continuous functions $x(\lambda) = \{x_{ab}(\lambda)\}_{a=\overline{1,r}}^{b=\overline{1,r}},\ 0 \leq \lambda \leq \pi$ with the distance

$$d(x,y) = \sup_{0\leq\lambda\leq\pi}\ \max_{1\leq a,\, b\leq r} |x_{ab}(\lambda) - y_{ab}(\lambda)|.$$

Random processes (4) and (5) generate probability measures on the Borel σ -algebra of the space $C^{r\times r}[0,\pi]$, and we can consider them as random elements ζ_N and ξ_N with values in $C^{r\times r}[0,\pi]$. Together with R. Bentkus we found conditions under which the random elements ζ_N and ξ_N converge in distribuion to a random element ζ, which is defined by an $r\times r$ -matrix-valued Gaussian process

$$\zeta(\lambda) = \{\zeta_{ab}(\lambda)\}_{a=\overline{1,r}}^{b=\overline{1,r}}, \quad 0 \leq \lambda \leq \pi,$$

with $E\zeta(\lambda) = 0,$

$$E\zeta_{a_1b_1}(\lambda)\zeta_{a_2b_2}(\mu) =$$

$$= 2\pi\int_0^\lambda\int_0^\mu f_{a_1b_1a_2b_2}(\alpha,-\alpha,\beta)\,d\alpha\,d\beta + 2\pi\int_0^{\min(\lambda,\mu)} f_{a_1b_2}(\alpha)\, f_{b_1a_2}(-\alpha)\,d\alpha$$

and

$$E\zeta_{a_1b_1}(\lambda)\overline{\zeta_{a_2b_2}(\mu)} = E\zeta_{a_1b_1}(\lambda)\zeta_{b_2a_2}(\mu).$$

The rate of convergence, theorems of large deviations, the problem related to the existence only of some moments have been investigated. Here is a pair of more evident results.

Theorem 11. (together with R. Bentkus). Let a random sequence $\{X(t),\ t=\dots,-1,0,1,\dots\}$ be such that:

1) for any $(a_1,\dots,a_q)$, $1\le a_j\le r$, $q=2,3,\dots$, there exists the s.d. of the $(q-1)$ -th order $f_{a_1\dots a_q}$ and it is bounded;

2) for any (a_1,a_2,a_3,a_4), $1\le a_j\le r$,

$$\lim_{h\to 0}\int_{-\pi}^{\pi}\int_{-\pi}^{\pi}\left|f_{a_1a_2a_3a_4}(\alpha,h-\alpha,\beta)-f_{a_1a_2a_3a_4}(\alpha,-\alpha,\beta)\right|d\alpha\,d\beta=0.$$

Then $\zeta_N\xrightarrow{\mathcal{D}}\zeta$ and $\xi_N\xrightarrow{\mathcal{D}}\xi$ as $N\to\infty$.

Theorem 12. (together with R. Bentkus). Let a Gaussian sequence $\{X(t),\quad t=\dots,-1,0,1,\dots\}$ be such that for $a=\overline{1,r}$ the s.d. f_{aa} exists and $\int_{-\pi}^{\pi}f_{aa}^2(\lambda)\,d\lambda<\infty$. Then $\zeta_N\xrightarrow{\mathcal{D}}\zeta$ and $\xi_N\xrightarrow{\mathcal{D}}\xi$ when $N\to\infty$.

References

1. В.А.Статулявичус, Предельные теоремы для случайных величин, связанных в цепь Маркова, I, II, Лит.матем.сб., IX, № 2 (I969), 346-362, № 3 (I969), 635-672.
2. В.А.Статулявичус, О предельных теоремах для случайных функций, I, Лит.матем.сб., X, № 3 (I970), 583-592.
3. V.A. Statulevicius, On large deviations, Z.Wahrscheinlichkeitstheorie verw. Gebiete, 6, 2(1966), 133-144.

4. С.В.Нагаев, О скорости сходимости в одной граничной задаче, I, Теория вероят. и ее примен., XУ, 2(I970), I79-I99.

Institute of Physics and Mathematics
of the Academy of Sciences of the Lithuanian SSR
Vilnius

A REMARK ON THE CRITERION OF CONTINUITY OF GAUSSIAN SAMPLE FUNCTION

V.N.Sudakov

1. Let H be the Hilbert space of Gaussian random variables $x(\omega)$, $\omega \in (E, \gamma)$, with zero means. The space (E, γ) can be considered as a linear space with Gaussian measure, H being the set of all linear measurable functionals on (E, γ). It is known that, for some classes of stochastic processes, the boundedness of their sample functions implies the continuity of the latter (for example, this is true for Gaussian stationary processes). At the same time, there exist Gaussian random processes whose sample functions $x(\omega)$ are bounded with probability 1 but not continuous in $x \in K$ in relative topology on a set $K \subset H$. The investigation of the oscilation function of a Gaussian process is related to this phenomenon [4]. If the sample functions of a process K, $K \subset H$, are bounded with probability 1, then the process K is said to have GB -property $(K \in GB)$. If in addition the realizations are continuous, the process K is said to have GC -property $(K \in GC)$.

Perhaps the most convenient criteria for verifying GB -property were formulated in terms of ε -entropy of the set K (see [2], [7], [8]). As it was pointed out in [8], all the ε-entropy type conditions for verifying GB -property are the consequences of the monotonicity property (in a certain sense) of the functional $h_1(K)$, which is a generalization to the infinite-dimensional case of H.Minkowski's "mixed volume" of the first degree of homogeneity. If regarded from the point of view of probability theory, the value $h_1(K)$, originally defined by inner geometry of the set $K \subset H$, appears proportional to the mathematical expectation of the supremum of the set of Gaussian random variables K . The finiteness of $h_1(K)$ is both a necessary and sufficient condition for a process to have the GB -property. Some upper bound will be given below for $h_1(K)$ in terms of the probability for a process to cross the unit level. This upper bound, which may be obtained with the help of the solution of the isoperimetric problem on the surface of an n -dimensional sphere [5] , is essentially more precise than that which can be obtained with the help of a witty idea of X. Fernique from his paper [3] . For the case of a centrally symmetric K a bound similar to that of mine could be obtained if one had been able to prove a lemma of [6] (it should be mentioned at this point that the solution of the isoperimetric problem on the surface of the n -dimensional sphere within the class of the centrally symmetric sets is nontrivial). We note that all the arguments of L.Shepp can be applied in case when K is an arbitrary convex subset containing the origin, and an assertion on page 5 of [6] can be justified if [5] is taken into account.

Here we shall prove a general criterion in terms of $h_1(K)$ to verify the GC -property. As a corollary we obtain a new necessary condition of continuity in terms of the ε -entropy.

2. <u>Definition</u>. Let $K \subset H$. Then

$$h_1(K) = \sqrt{2\pi}\; E \sup\{x(\omega) : x \in K\}.$$

The factor $\sqrt{2\pi}$ arises from the desire to have the value of $h_1(I)$ equal to the length of I for the intervals I . (The geometric sense of this characteristic is discussed in [8] .)

<u>Definition</u>. $\delta(K) = \sup\{d : \gamma(dK^0) = 0\}$ (K^0 is a polar of K , i.e. the set of all linear forms whose modulus is less than 1 on K ; γ is the Gaussian measure).

Proposition 1. Let $K \subset H$ be a convex symmetric subset. Then $K \in GC \Longleftrightarrow \delta(K)=0$.

Proof. See [2], Theorem 4.6. For another proof see [7], Proposition 6, where the assertion is a consequence of the following representation of all linear forms on $F=\mathcal{L}(K)$ ($\mathcal{L}(K)$ denotes the linear span of K) bounded on K and continuous on K in relative topology:

$$F = \bigcap_{\lambda>0} \lambda \bigcup_{j=0}^{\infty} (K \cap L_j)^0, \tag{*}$$

where the polar is considered in the space $(F, \|\cdot\|_K)^*$ and the sequence of closed subspaces $L_j \subset H,\ j=0,1,\dots,$ has the properties:

(1) $\quad H=L_0 \supset L_1 \supset \dots, \quad \bigcap_j L_j = \{0\}; \quad \dim L_j / L_{j+1} < \infty.$

Proposition 2. Let $K \subset H$ be a convex bounded symmetric subset and $L \subset H$ be a closed subspace with finite defect. If, for a certain $\varepsilon>0$, the condition $\gamma((1+\varepsilon)K^0)=0$ is satisfied, then $\gamma(K \cap L)^0=0$, and if $\gamma K^0=0$, then $\gamma(Pr_L K)^0=0$. (Pr_L denotes the ortogonal projection operator into the subspace $L \subset H$.)

Proof. The second assertion follows from the fact that K^0 is a null-set for all conditional Gaussian measures on the quotient classes relative to L^0. Now we shall prove the first assertion. The defect of the subspace $L \subset H$ may be always assumed to equal 1. If $y_L \in H^* \subset (E,\gamma)$, $\|y_L\|_{H^*}=1$ is the functional with the kernel L, then

$$(K \cap L)^0 = K^0 + \{\lambda y_L : \lambda \in \mathbb{R}\}.$$

Let us consider a measurable decomposition ξ_L of the space (E,γ) into straight lines parallel to $\{\lambda y_L\}$. The space (E,γ) may be taken as a product of the straight line $\{\lambda y_L\}$ with the standard Gaussian measure and some space (E_1, γ_1) with a Gaussian measure. The space E_1 can be identified with the subspace of the space E where the variable $x_L \in H$, orthogonal in H to L, vanishes. In view of the boundedness of the set $K \subset H$, the intersection of the set K^0 with the straight line $\{\lambda y_L\}$ contains a segment and hence has a positive conditional (one-dimensional) Gaussian measure. Since the set K^0 is convex and K is bounded, every straight line parallel to $\{\lambda y_L\}$, whose intersection with the set K^0 is nonempty, intersects essentially the set $(1+\varepsilon)K^0$, i.e. the conditional measure of this intersection is positive. But the condition $\gamma[(1+\varepsilon)K^0]=0$ implies in this case that the set of straight lines intersecting the set K^0 has the marginal (γ/ξ_L-) measure zero. So we get the first assertion. More precise arguments would show that the assertion is true even if ε is equal to zero.

<u>Proposition 3</u>. <u>Let</u> $K \subset H$ <u>be a convex bounded symmetric set</u>, $L \subset H$ <u>be a closed subspace with finite defect</u>. <u>Then</u> $\delta(K) = \delta(K \cap L) = \delta(Pr_L K)$.

The proof follows immediately from the previous proposition.

<u>Proposition 4</u>. <u>Let</u> A_λ <u>be a linear operator in space</u> (E, γ), <u>stretching the space</u> E $\lambda > 1$ <u>times in a certain direction</u>. <u>Then, for every convex symmetrical set</u> $V \subset E$, <u>the following inequality holds true</u>: $\gamma(A_\lambda V) \geqslant \gamma V$.

<u>Proof</u>. We may restrict ourselves by the finite-dimensional case. Let $\gamma = \gamma_n$ be the standard Gaussian measure in the space $E = \mathbb{R}^n$. An operator A_λ is now an arbitrary self-adjoint operator with the spectrum $(1, 1, \dots, \lambda)$, $\lambda > 0$. Let $L \subset \mathbb{R}^n$ be the eigen-subspace corresponding to the eigen-value 1, and let $L_t = L + t e_n$ where $\|e_n\| = 1$ and $A_\lambda e_n = \lambda e_n$. Let γ_{n-1} be the standard (conditional) Gaussian measure on L $(\gamma_{n-1} = \gamma_n Pr_L^{-1})$. Then

$$Pr_L(A_\lambda V \cap L_t) = Pr_L(V \cap L_{\frac{t}{\lambda}}) \tag{2}$$

Taking into account the convexity of V , we conclude that

$$\frac{\lambda+1}{2\lambda}(V\cap L_t)+\frac{\lambda-1}{2\lambda}(V\cap L_{-t})\subset V\subset L_{\frac{t}{\lambda}}$$

or, in an equivalent form,

$$(3)\qquad \frac{\lambda+1}{2\lambda}Pr_L(V\cap L_t)+\frac{\lambda-1}{2\lambda}Pr_L(V\cap L_{-t})\subset Pr_L(V\cap L_{\frac{t}{\lambda}}).$$

The sets $Pr_L(V\cap L_t)\subset L$ and $Pr_L(V\cap L_{-t})\subset L$ are mutually symmetric with respect to the origin. V.A.Zalgaller's theorem [10] (based on the application to the left-hand side of (3) the Brunn-Minkowski inequality for mixed volumes) states that, for any two convex mutually symmetric (with respect to a certain point) subsets A and A' of a finite-dimensional space, the inequality $\gamma(\alpha A+(1-\alpha)A')\geq A$ holds for an arbitrary α, $0\leq\alpha\leq 1$. In our case $A=Pr_L(V\cap L_t)$ and Zalgaller's theorem gives:

$$\gamma_{n-1}\Big(\frac{\lambda+1}{2\lambda}Pr_L(V\cap L_t)+\frac{\lambda-1}{2\lambda}Pr_L(V\cap L_{-t})\Big)\geq\gamma_{n-1}Pr_L(V\cap L_t).$$

Taking into account (2) and (3) we can write:

$$(4)\qquad \gamma_{n-1}Pr_L(A_\lambda V\cap L_t)\geq\gamma_{n-1}Pr_L(V\cap L_t).$$

But

$$\gamma_n V=\int_{-\infty}^{\infty}\gamma_{n-1}Pr_L(V\cap L_t)\gamma_1(dt),$$

$$\gamma_n(A_\lambda V)=\int_{-\infty}^{\infty}\gamma_{n-1}Pr_L(A_\lambda V\cap L_t)\gamma_1(dt).$$

The inequality (4), which holds for every t , now implies the desirable conclusion.

Remark. Proposition 4 remains true in case $\lambda=\infty$.

If $K\in GB$ is a convex set containing the origin then $h_1(K)<\infty$, and the following inequality can be proved:

$$h_1(K)\leq\frac{1}{\Phi^{-1}(\gamma K^0)}e^{-\frac{1}{2}(\Phi^{-1}(\gamma K^0))^2}+\sqrt{2\pi}\,\gamma K^0 \quad\text{if}\quad \gamma K^0>\frac{1}{2}$$

where

$$\Phi(x)=\frac{1}{\sqrt{2\pi}}\int_{-\infty}^{x} e^{-\frac{1}{2}u^2}du .$$

This inequality will not be referred to below. The finiteness of all moments of random variable sup K follows from a result of X.Fernique [3] or L.Shepp [6]with the said in the beginning of the paper taken into account.

Proposition 5. Let $K\subset H$, $K\in GB$, be a convex symmetrical subset of a Hilbert space H, $\{L_j, j=0,1,\dots\}$ be a sequence of closed subspaces satisfying the condition (1). Then $h_1(Pr_j K)\searrow\sqrt{2\pi}\,\delta(K)$ and $h_1(K\cap L_j)\searrow\sqrt{2\pi}\,\delta(K)$.

Proof. Any Gaussian subspace L_j can be considered as the space of (Gaussian) linear functionals, which are measurable with respect to a measurable decomposition η_j of the measurable space (E,γ) into subspaces parallel to the finite-dimensional subspace $L_j^0\subset H^*\subset E$. Let $\tilde{L}_j\subset L^2(E,\gamma)$ denote the subspace consisting of all (not necessarily linear) functions from the space $L^2(E,\gamma)$ which are measurable with respect to the decomposition η_j, so that $L_j=H\cap\tilde{L}_j$. The operator Pr_j is the operator of the conditional (in the strict sense) mathematical expectation.

Now we can go over to the proof. The sequence of the sets $\{K\cap L_j, j=0,1,\dots\}$ is decreasing in the set-theoretic sense, hence the sequence of nonnegative measurable functions sup $(K\cap L_j)$ is monotonically decreasing and thus tends in the mean to a certain limit function. This limit is measurable with respect to every decomposition η_j and therefore is equivalent to a constant. It follows from Proposition 3 that this constant is equal to $\delta(K)$. The definition of $h_1(K)$ implies now the second assertion of the Proposition 5.

Now we shall prove the first assertion. If Pr is an operator of the conditional (in the strict sense) mathematical expectation,

then

(5) $$\sup Pr\, K \leq Pr \sup K .$$

(This inequality is a generalization of the evident inequality $\sup(x_K + y_K) \leq \sup x_K + \sup y_K$. For any $K \geq 0$, the sequence $Pr_j \sup Pr_K K$ $(j = K, K+1, \ldots)$ is a martingale and it follows from the Doob theorem [1] on the convergence of martingals that this sequence tends in the mean to a constant $C_K \geq 0$. It is easy to see that $C_0 \geq C_1 \geq \ldots$. In fact $C_K = C_K = E Pr_j \sup Pr_K K$ for any $j = K, K+1, \ldots$; in particular $C_K = E \sup Pr_K K$ and taking into account the monotonicity of the functional h_1 (see [8]) we conclude that $h_1(K) \geq h_1(Pr_1 K) \geq$ $\geq h_1(Pr_2 K) \geq \ldots$ i.e. $C_0 \geq C_1 \geq \ldots$; let $C = \lim C_K$. Now for $j \geq K$ we can write

$$E\left|\sup Pr_j K - c\right| \leq E\left|\sup Pr_j K - Pr_j \sup Pr_K K\right| +$$
$$+ E\left|Pr_j \sup Pr_K K - C_K\right| + \left|C_K - C\right| .$$

The right-hand side being arbitrarily small if K and j are sufficiently large, from inequality (5) we have

$$E\left|Pr_j \sup Pr_K K - \sup Pr_j K\right| = E\left(Pr_j \sup Pr_K K - \sup Pr_j K\right) = C_K - C_j ,$$

and, for a fixed K, the second term is arbitrarily small, if j is large enough, since $Pr_j \sup Pr_K K \to C_K$.

This proves that $\sup Pr_j K$ tends to a constant C in the mean (in fact it converges also in the quadratic mean). It follows from Proposition 4 that $C = \delta(K)$. Taking into account the definition of $h_1(K)$ we get as above the desirable conclusion.

Corollary. In the evident inequality: $h_1(K) \geq \sqrt{2\pi}$ if $\gamma K^0 = 0$, the constant $\sqrt{2\pi}$ is the best possible.

Proposition 6. For a convex symmetrical subset K of a Hilbert space H to belong to the class GC , any of the following conditions is necessary and sufficient:

1. There exists a sequence $\{L_j, j=0,1,\dots\}$ of closed subspaces of the space H, for which the conditions (1) hold, and $h_1(Pr_j K)\to 0$.

2. There exists a sequence $\{L_j, j=0,1,\dots\}$ satisfying the conditions (1) for which $h_1(K\cap L_j)\to 0$.

3. For every sequence of closed subspaces $\{L_j, j=0,1,\dots\}$, under the conditions (1), $h_1(Pr_j K)\to 0$.

4. For every sequence of closed subspaces $\{L_j, j=0,1,\dots\}$, under the conditions (1), $h_1(K\cap L_j)\to 0$.

The proof follows from the previous Proposition.

3. Now we are going to formulate and to prove a new necessary condition to verify the GC -property in terms of the ε -entropy. $N(K,\varepsilon)$ being the minimal power of the ε -net of a precompact set $K\subset H$, the following inequality can be proved:

$$(6)\qquad h_1(K)\le 22\sum_{K=-\infty}^{\infty} 2^{-K}\sqrt{\log_2 N(K,2^{-K})}.$$

We are interested in the lower bound for the value $h_1(K)$. Let $M(K,\varepsilon)$ be the maximal power of an ε -grating, i.e. of a subset of K consisting of the elements whose mutual pairwise distances are greater than ε. It is well known that $M(K,\varepsilon)\ge N(K,\varepsilon)\ge M(K,2\varepsilon)$.

Proposition 7. Let $M=M(K,\varepsilon)\ge 10$. Then, for an arbitrary $\varepsilon>0$ the following inequality holds true

$$h_1(K)\ge\left(1-\frac{1}{e}\right)\sqrt{\frac{\pi}{2}}\,\varepsilon\sqrt{\ln M}.$$

Proof. The Chebyshev's inequality gives

$$h_1(K)\ge 6^{-1}\sqrt{2\pi}\,(1-\gamma(6K)^0).$$

Denote by B_n the simplex with vertices $\frac{1}{\sqrt{2}}e_1,\dots,\frac{1}{\sqrt{2}}e_n$, where $e_1,\dots,e_n$ is a sequence of ortonormal vectors. Taking into account the monotonicity of the functional h_1 (see [8]) we obtain the inequality:

$$h_1(K) \geq h_1(\varepsilon B_{M(K,\varepsilon)}) ;$$

hence $(0 < \alpha(z) < \frac{1}{z^2}, \quad \sigma\varepsilon < 1)$

$$h_1(K) \geq \frac{\sqrt{2\pi}}{\sigma}(1 - \gamma(\sigma\varepsilon B_M)^0) =$$

$$= \frac{\sqrt{2\pi}}{\sigma}\left(1 - \left[\frac{1}{\sqrt{2\pi}}\int_{-\infty}^{\frac{\sqrt{2}}{\varepsilon\sigma}} \exp(-\tfrac{1}{2}u^2)\,du\right]^M\right) =$$

$$= \frac{\sqrt{2\pi}}{\sigma}\left(1 - \left[1 - \frac{\sigma\varepsilon}{2\sqrt{\pi}}\exp\left(-\frac{1}{\sigma^2\varepsilon^2}\right)\left(1 - \theta\left(\frac{\sqrt{2}}{\sigma\varepsilon}\right)\right)\right]^M\right) \geq$$

$$\geq \frac{\sqrt{2\pi}}{\sigma}\left(1 - \exp\left\{-\exp\left[-\frac{1}{\sigma^2\varepsilon^2} + \ln M + \ln\frac{\sigma\varepsilon}{4\sqrt{\pi}}\right]\right\}\right).$$

Because of monotonicity in z of the function $z^2 \ln \frac{zM}{4\sqrt{\pi}}$, the condition $M \geq 10$ and the inequality $z^2 \ln \frac{zM}{4\sqrt{\pi}} > 1$ true for $z = \frac{2}{\log M}$,. we get the inequalities $\sigma\varepsilon < \frac{2}{\sqrt{\ln M}} < 1$ and $\sigma < \frac{2}{\varepsilon\sqrt{\ln M}}$ in case σ and ε are subject to the relation

$$-\frac{1}{\sigma^2\varepsilon^2} + \ln M + \ln\frac{\sigma\varepsilon}{4\sqrt{\pi}} = 0 .$$

Thus the inequality of Proposition 7 is proved.

Proposition 8.

$$\delta(K) \geq \tfrac{1}{2}\left(1 - \tfrac{1}{e}\right) \limsup_{\varepsilon\to 0} \varepsilon\sqrt{\ln M(K,\varepsilon)} .$$

Proof. For an arbitrary closed linear subspace $L \subset H$ with finite defect $d = \dim H/L$, the orthogonal projection $Pr_L K$ of any precompact set K into this subspace has the same order of growth of the ε-entropy as K itself: $\limsup \varepsilon^2 \ln M(Pr_L K, \varepsilon) = \limsup \varepsilon^2 \ln M(K,\varepsilon)$. Now note that for a certain constant C dependent upon the set K, the following inequality holds

$$M(Pr_L K, \varepsilon) \geq M(K,\varepsilon)\cdot\frac{c(K)}{\varepsilon^d}$$

since the set K is contained in the orthogonal sum of the set $Pr_L K$ and a d -dimensional bounded set. From this inequality it follows that

$$\lim\sup \varepsilon\sqrt{\ln M(K,\varepsilon)} \leq \lim\sup \varepsilon\sqrt{\ln M(Pr_L K,\varepsilon)}.$$

If $\varepsilon \to 0$, then from Proposition 7 we get

$$h_1(K) \geq \lim\sup\left(1-\frac{1}{e}\right)\sqrt{\frac{\pi}{2}}\, \varepsilon\sqrt{\ln M(K,\varepsilon)}.$$

Therefore, for any subspace L with $\dim H/L < \infty$, we have the inequality

$$h_1(Pr_L K) \geq \lim\sup\left(1-\frac{1}{e}\right)\sqrt{\frac{\pi}{2}}\, \varepsilon\sqrt{\ln M(K,\varepsilon)}.$$

Now let $\{L_j, j=0,1,\ldots\}$ be a sequence of closed subspaces of the space H satisfying the conditions (1). According to Proposition 5 we get finally

$$\delta(K)=\frac{1}{\sqrt{2\pi}}\lim_j h_1(Pr_{L_j}K) \geq \frac{1}{2}\left(1-\frac{1}{e}\right)\lim_{\varepsilon\to 0}\sup \varepsilon\sqrt{\ln M(K,\varepsilon)}.$$

Note that Dudley's theorem ([2]) asserting that

$$\sum_{k=-\infty}^{\infty} 2^{-k}\sqrt{\log_2 N(K,2^{-k})} < \infty \Longrightarrow K\in GC$$

can be now obtained as a consequence of the inequality (6) and Proposition 6. In fact, if a sequence of closed subspaces $\{L_j, j=0,1,\ldots\}$ satisfies the conditions (1), only a finite number of terms of the decreasing sequence $\log_2 N(K\cap L_j,\varepsilon),\ j=0,1,\ldots$, differ from zero.

The author is thankful to Prof. R.Dudley who noticed that in formulation of theorem 5 of [8] the condition $\int_0^1 \varepsilon^2 dH = -\infty$ should be replaced by the condition

$$\lim\sup \varepsilon^2 \ln M(K,\varepsilon)=\infty.$$

References

[1] J. Doob: Stochastic Processes, New York - London, 1953.

[2] R.M. Dudley: The sizes of compact subsets of Hilbert space and continuity of Gaussian processes, J. Functional Analysis, I(1967), 1290-330.

[3] X. Fernique: Intérgralite des vecterns gaussiens, C.R.Acad.Sci. Paris, 270(1970), 1698-1699.

[4] K. Ito, M. Nisio: On the oscillation function of Gaussian processes, Math.Scand., 22(1968).

[5] E. Schmidt: Beweis der isoperimetrischen Eigenschaft der Kugel im hyperbolischen und sphärischen Raum jeder Dimensionszahl, Math.Z., 49(1943), 1-109.

[6] L. Shepp: On the Supremum of Gaussian Sequences, Bell Telephone Laboratories, preprint, (1969).

[7] V.N. Sudakov: Gaussian and Cauchy measures and ε-entropy, Sov.Math.Dokl., 10(1969), 310-313. (Russian)

[8] V.N. Sudakov: Gaussian random processes and measures of solid angles in Hilbert space, Sov.Math.Dokl., 12(1971), 412-415. (Russian)

[9] A.M. Vershik, V.N. Sudakov: Probability measures in infinite-dimensional spaces, Seminars in mathematics, V.A.Steklov Mathematical Institute, Leningrad, vol. 12.(Russian)

[10] В.А.Залгаллер, Смешанные объемы и вероятность попадания в выпуклые области при многомерном нормальном распределении, Матем. заметки, 2, № I, (1967), 97-I04.

Steklov Mathematical Institute
of the Academy of Sciences of the USSR
Leningrad

β-TRANSFORMATIONS AND SYMBOLIC DYNAMICS

Yōichirō Takahashi

§1. β-transformations

1.1. The β-transformations are rich not only in the ergodic structures but also in the structures as topological or symbolic dynamics, and offer a nice class of examples. They originate in the number-theoretical expansions of a number $t \in [0,1)$ with respect to a real $\beta > 1$;

$$(1) \quad t = \sum_{n \geq 0} a_n \beta^{-n-1} \qquad (a_n : \text{ integers}).$$

Let $\beta = (\beta_0, \cdots, \beta_{s-1})$, $\beta_i > 1$, $\sum_{t=0}^{s-2} \beta_i^{-1} < 1 \leq \sum_{i=0}^{s-1} \beta_i^{-1}$.

Definition. A *β-transformation* is a map T_β defined on the unit interval $[0,1)$ by the relation

$$(2) \qquad T_\beta t = \beta_k t - k \quad \text{if} \quad \sum_{i=0}^{k-1} \beta_i^{-1} \leq t < \sum_{i=0}^{k} \beta_i^{-1}$$

This transformation has been studied by A. Renyi, W. Parry, Sh. Ito, the author et al in the case of $\beta_0 = \cdots = \beta_{s-1}$, and by Shiokawa, Sh. Ito, M. Mori in the general cases. Let $\pi_\beta(t)$ be a sequence whose n-th coordinate is

$$(3) \qquad \pi_\beta(t)(n) = k \quad \text{if} \quad \sum_{i=0}^{k-1} \beta_i^{-1} \leq t < \sum_{i=0}^{k} \beta_i^{-1} \quad .$$

Then, π_β defines a Borel homomorphisms of $([0,1), T_\beta)$ into $(A^{\mathbb{N}}, \sigma)$ where $A = \{0, 1, \cdots, s-1\}$. Conversely, $t \in [0,1)$ is obtained from $\omega = \pi_\beta(t)$ by the formula :

$$(4) \qquad t = \rho_\beta(\omega) = \sum_{n\geq o} \prod_{k=o}^{n-1} \beta\omega(k)^{-1} \sum_{i=o}^{\omega(n)-1} \beta_i^{-1}$$

If $\beta_o = \cdots = \beta_{s-1}$, then this formula gives (1) with $a_n = \pi_\beta(t)(n)$.

Let X_β be the closure of the image $\pi_\beta([0,1))$ by the map π_β with respect to the product topology in A^N, and denote by σ the one-step shift transformation to the left. Then the (topological) subshift (X_β, σ) is a *realization* of the Borel dynamical system $([0,1), T_\beta)$, and will be called β-*shift*.

1.2. The so-called β-*expansion of one* is the sequence ω_β defined by the relation :

$$(5) \qquad \omega_\beta = \max X_\beta \quad \text{(lexicographical order)}.$$

It is the unique element of X_β for which $\rho_\beta(\omega)=1$. Let $(\overline{X}_\beta, \overline{\sigma})$ be the (topological) natural extension to the bilateral subshift of one-sided (X_β, σ) and we will also call it a (bilateral) β-shift. Then

$$(6) \qquad \overline{X}_\beta = \{\omega \in A^Z \mid \sigma^n \omega \leq \omega_\beta \quad \text{for all} \quad n \in Z\}.$$

Here, the order $\leq$ is lexicographical, too. Thus, the sequence ω_β completely determines the topological, and therefore measure-theoretical properties of the subshift (X_β, σ), and is also a key to construct isomorphisms of β-automorphisms to mixing Markov automorphisms in §5.

1.3. Most of the results in §§1-3 will be found in [10] and those in §5 will be found in [21] (simple case) and the recent work [8] of Sh. Ito and M. Mori (general case).

§2. Markov property

2.1. The following concept is an essential tool of our study.

Definition. A subshift (X,σ) over a symbol set A is called *Markov* if there is a subset W of A^{p+1} for some $p\geq o$ such that

$$(7) \qquad X=m(W)\equiv\{\omega\,|\,(\omega(n),\cdots,\omega(n+p))\in W(\forall n)\}$$

This class of subshift is characterized by the following :

Proposition 1. For any bilateral subshift $(\overline{X},\overline{\sigma})$ the following statesments are equivalent:

(a) *The bilateral subshift $(\overline{X},\overline{\sigma})$ is Markov.*

(b) *The one-sided subshift (X,σ) is Markov where* $X=\{(\omega(n),\ n\geq 0)\,|\ \omega\in\overline{X}\}$.

(c) *The projection $\pi:\overline{X}\to X$ is an open map.*

(d) *The shift $\sigma:X\to X$ is an open map.*

Some of the properties of Markov subshifts will be found in 3.3 and 3.4.

2.2. The following theorem shows that $(X_\beta,\ \sigma)$ is not Markov for almost all β's when $\beta_o=\cdots=\beta_{s-1}$.

Theorem. 2. A β-shift $(X_\beta,\ \sigma)$ is Markov if and only if β is a root of the algebraic equation

$$(8) \qquad 1-t^{-p-1}=\sum_{k=o}^{p}\omega_\beta(k)t^{-k-1}$$

$$\text{and} \quad 1-t^{-p-1}>\sum_{k=o}^{p}\omega_\beta(k+q)t^{-k-1} \quad \text{for } q=1,2,\cdots,p.$$

A similar result for general β is obtained by I. Shiokawa.

2.3. Nevertheless, any X_β is closely approximated both from above and below by Markov X_β's. In fact, the family $\{X_\beta | \beta>1\}$ satisfies the following properties:

(a) monotone: $X_\beta \subset X_\alpha$ if $\beta<\alpha$

(b) continuous: $X_\beta = \bigcap_{\alpha>\beta} X_\alpha$,

$$\pi_\beta([0,1)) \subset \bigcup_{\alpha<\beta} X_\alpha \quad X_\beta = \overline{\bigcup_{\alpha<\beta} X_\alpha}$$

(c) the topological entropy $e(X_\beta, \sigma) = \log \beta_o$ if $\beta_o = \cdots = \beta_{s-1}$.

§3. Ergodic and dynamical properties

3.1. *Theorem.3. The β-automorphisms are Bernoulli.*

Although Smordinsky proved this theorem independently, we will outline a proof which is symbolic and interesting in itself. Let $[u]$ be the cylinder set of ω's such that u is the initial word of ω. It is enough to verify Ornstein's weak Bernoulli condition

$$(9) \quad \sup_n \sum_{u\in A^n} \sum_{v\in A^n} |\mu([u]\cap\sigma^{-k+n}[v]) - \mu[u]\cdot\mu[v]| \to 0 \quad (k\to\infty)$$

for the invariant measure μ_β obtained by W. Parry :

$$(10) \quad \mu_\beta\{\omega' | \omega' \leqq \omega\} = C_\beta \sum_{n\geqq o} \beta^{-n-1} \min\{\rho_\beta(\omega), \rho_\beta(\sigma^n \omega_\beta)\}.$$

The condition (9) can be reduced to an estimate of the number of words in X_β, which is done by modifying the method in the following 3.2 slightly.

3.2. *Classification of words and topological entropy:*
Let $W_p(X_\beta)$ be the set of words $u=(a_o,\cdots,a_{p-1})$ appearing in X_β and let $W_p^o(X_\beta)$ be the subset of those u's for which $u^+=(a_o,\cdots,a_{p-2},a_{p-1}+1)\in W_p(X_\beta)$. Then, it can be shown that

$$(11) \qquad W_p(X_\beta)=\bigcup_{k=o}^{p} W_{p-k}^o(X_\beta)\cdot(\omega_\beta(0),\cdots,\omega_\beta(k-1))$$

from which we obtain

$$\lim_{n\to\infty} \beta^{-n}\ \mathrm{card}(W_n^o(X_\beta))={}^1/M_\beta\ ,$$

$$M_\beta = \sum_{n\geq o} (n+1)\,\omega_\beta(n)\,\beta^{-n-1}\ .$$

and

$$(12) \qquad \lim_{n\to\infty} \beta^{-n}\ \mathrm{card}(W_n(X_\beta)) = {}^1/M_\beta(1-\beta^{-1})\ .$$

In particular, we obtain (c) in 2.3.

We note that the convergence to topological entropy of type (12) is known only for aperiodic Markov subshifts except for our case.

3.3. *Periodic points and topological entropy:* the following proposition gives some information on the character of topological entropy and can be extended to expansive dynamical systems (see [3]) :

Proposition 4. *For aperiodic Markov subshifts* (X, σ),

$$(13) \qquad \lim_{n\to\infty} \frac{1}{n} \log\ \mathrm{card}\ \{\omega\in X|\sigma^n\omega=\omega\}=e(X,\sigma).$$

Furthermore, (13) *holds for any Markov subshift if we replace* lim *by* lim sup, *and for any subshift replacing* $=$ *by* $\leqq$*, too.*

In the proof of this proposition useful is the decomposition of state space into irreducible components ([11])

3.4. Here, we must mention on the maximality of entropy and a generalized notion.

Definition. An invariant probability measure for a topological dynamics is called *maximal* if the metrical entropy coincides with the topological entropy.

Such a measure does exist for expansive systems, while the uniqueness theorem in general is obtained only for irreducible Markov subshifts. There is a more general notion of maximality of entropy (see [20]), by which Sh. Ito and M. Mori succeeded to prove that $(\beta_o,\beta_1,\cdots,\beta_{s-1})$-automorphisms are Bernoulli.

Let (X, ϕ) be a fixed topological dynamics, U a function defined on X with values in $(-\infty, +\infty]$ which is upper semi-continuous. Consider the *minimizing problem* of the function

$$(16) \qquad f_U(\mu) = \int U d\mu - h(\mu, \phi).$$

Definition. An invariant probability measure μ for (X, ϕ) is called *minimal* with respect to the *free energy* f_U corresponding to the *potential* U if the function f_U attains its minimum at μ.

3.5. Remark. 1) If the transformation ϕ is expansive, then there exists a solution of minimizing problem (16). But the uniqueness is known (at least to the author) merely for shift transformations with those U which depend only on finite number of coordinates and are irreducible ([20]).

2) Let $(X,\phi)=(A^Z,\sigma)$ be a shift. If U depends only on $(\omega(k))$ $k\geq 0$, and satisfies $\sum_{n\geq 0} \sup\{|U(\omega)-U(\omega')| : \omega(k)=\omega'(k), k\leq n\}<\infty$, then the corresponding "Gibbsian" measure μ_U *) is a solution of (16). (The proof can be given by modifying the proof of [6], and the system (X,σ,M_U) is weak Bernoulli.)

3) Furthermore, $\log J-U(\omega) = h(\omega)-h(\sigma(\omega))+c$ for some constant c and a continuous function h, where $J(\omega)=\mu(\omega|\sigma\omega)$ (Jacobian). It seems to be open whether $\log J - U = h - h\circ\sigma + c$ for some constant c, a function h, a Jacobian J of for some generator, whenever μ is a solution of (16); in other words, the class of potentials $\{\log J - h + h\circ\sigma + c \mid c,h\}$ is invariant under isomorphisms.

§4. Isomorphisms

4.1. In view of the Ornstein's isomorphism theorem, it is of importance to construct isomorphism of β-automorphisms to other Bernoulli automorphisms. We can do this making a full use of symbolic properties of β-shifts, i.e., of the sequences ω_β's, and obtain the uniqueness of maximal invariant measures as a corollary.

We know two examples of concrete isomorphisms, Meškin's [14] and Adler-Weiss' [2]. We may say that both of them are based essentially on the notion of entropical maximality in somewhat generalized sense, upon which we will also depend. The extensions will be possible, for example, to f-expansion transformations.

4.2. Let $\tau(\overline{\omega})$ be the hitting time to the set

$$\{\overline{\omega}\in \overline{X}_\beta \mid (\overline{\omega}(-n),\cdots,\overline{\omega}(0)=(\omega_\beta(0),\cdots,\omega_\beta(n)) \text{ for no } n\geq 0\}$$

---

*) Recently the author found a paper of Sinai (Uspehi Mat. Nauk. 166 (1972)), where he discusses those U depending also on $(\omega(-k))$ $k\geq 0$ under stronger conditions.

and

$$\psi_\beta(\overline{\omega}) = \begin{cases} \tau(\overline{\omega}) & \text{if} \quad \tau(\overline{\omega})>0, \\ -\overline{\omega}(0) & \text{if} \quad \tau(\overline{\omega})=0. \end{cases}$$

Then, it can be shown that the map ψ_β is a Borel injection of X_β, anti-commuting with the shift transformation, to the Markov subshift (Z_β,σ) over the countable symbol set $I=\{-(s-1),\cdots, 0, 1,\cdots,\infty\}$ where

$$Z_\beta = m(M_\beta) \equiv \{\eta\in I^Z \mid M_\beta(\eta(k),\eta(k+1))=1 \ \ \forall k\in Z\}$$

and

$$M_\beta(i,j)= \begin{cases} 1 \ \text{ if } \ i\leqq 1 \ \text{ and } \ j\leqq 0, \ \text{ if } \ i=j+1\geqq 1 \\ \quad \text{or if } \ i\leqq 0, \ j\leqq 1 \ \text{ and } \ \omega_\beta(j-1)>|i|, \\ 0 \ \text{ otherwise.} \end{cases}$$

Let

$$U_\beta(i,j)= \begin{cases} \log \beta_{|i|} & \text{if} \ \ i\leqq 0 \ \text{ and } \ M_\beta(i,j)=1 \\ \log \beta_{s-1} & \text{if} \ \ i=1 \ \text{ and } \ j\leqq 0 \\ \log \beta_{\omega(i)} & \text{if} \ \ i=j+1\geqq 1 \\ +\infty & \text{if} \ \ M_\beta(i,j)=0 \end{cases}$$

4.3. For the potential U_β it can be shown that there exists a unique invariant measure ν_β for which the free energy $f_{U_\beta}(\nu_\beta)$ is minimal. Furthermore, $(Z_\beta,\nu_\beta,\sigma)$ is mixing Markov. On the other hand, it is not difficult to deduce the equality $f_{U_\beta}(\psi_\beta(\mu_\beta))=f_{U_\beta}(\nu_\beta)$ by a direct computation. (if $\beta_o=\cdots=\beta_{s-1}$, then ν_β has maximal entropy on Z_β and this equality means that $h(\psi_\beta(\mu_\beta))=h(\nu_\beta)$ Consequently, $\psi_\beta(\mu_\beta)=\nu_\beta$. Thus, we obtain

Theorem 5 ([21],[8]). *The map* ψ_β *is an isomorphism* (mod 0) *of* β*-automorphism* $(X_\beta,\sigma,\mu_\beta)$ *to a mixing Markov automorphism* $(Z_\beta,\sigma^{-1},\nu_\beta)$.

Corollary. A *maximal invariant measure for* β*-transformation is unique.*

4.4. We note that our Theorem 5 asserts that the *dual* β-endomorphism $(X^*_\beta,\sigma,\mu_\beta)$ is isomorphic to a mixing Markov endomorphism, where $X^*_\beta=\{(\overline{\omega}(-n))\ n\geq 0\,|\,\overline{\omega}\in\overline{X}_\beta\}$. However, β-endomorphisms do not seem to be isomorphic to mixing Markov endomorphisms except those β's for which Markov property is verified. (cf. the lecture of Kubo-Murata-Totoki).

REFERENCE

[1] R. L. Adler, A. Konheim and M. H. McAndrew, Topological entropy. Trans. AMS. 114, 309-319(1965)

[2] R. L. Adler and B. Weiss, Similarity of automorphisms of the torus. Mem. AMS. 98.

[3] R. Bowen, Periodic points and measures for Axiom A diffeomorphisms, Trans. AMS 154, 377-397(1971)

[4] E. I. Dinaburg, Relation between various entropical characteristics of dynamical systems. Izv. Akad. Nauk. SSSR Ser. Mat. 35, 324-366 (1971) (in Russian)

[5] N. A. Friedman and D. S. Ornstein, On isomorphism of weak Bernoulli transformations. Advances in Math. 5, 365-394(1971)

[6] G. Gallavotti and S. Miracle-Sole, Absence of phase transition in hard-core one-dimensional systems with long-range interactions, J. Math. Phys. 11, 147-154(1970)

[7] Sh. Ito, An estimate from above for the entropy and the topological entropy of a C^1-diffeomorphism. Proc. Japan Acad. 46, 226-230(1970)

[8] Sh. Ito, and M. Mori , Isomorphisms of piecewise linear transformations to Markov automorphisms, to appear

[9] Sh. Ito, H. Murata, and H. Totoki, Remarks on the isomorphism theorem for weak Bernoulli transformations in general case, Publ. RIMS, Kyoto Univ., 7, 541-580(1972)

[10] Sh. Ito, and Y. Takahashi, Markov subshifts and β-transformations, to appear.

[11] N. Iwahori, On stochastic matrices of a given Frobenius type. J. Fac. Sci. Univ. Tokyo, Sect I, 13, 139-161(1966)

[12] H. B. Keynes and J. B. Robertson, Generators for topological entropy and expansiveness, Math. Sys. Theory, 3, 51-59(1969)

[13] G. Maruyama, Some aspects of Ornstein's theory of isomorphism problems in ergodic theory, Publ. RIMS, Kyoto Univ., 7, 511-540(1972)

[14] L. D. Mešalkin, A case of isomorphism of Bernoulli schemes, DAN SSSR 128, 41-44(1959) (Russian)

[15] D. S. Ornstein, Bernoulli shifts with the same entropy are isomorphic. Advances in Math, 4, 337- 348(1970)

[16] W. Parry, On the β-expansions of real numbers. Acta Math. Acad. Sci. Hung. 11, 401-416(1970)

[17] W. Parry, Instrinsic Markov chains Trans. AMS. 112, 55-66(1964)

[18] A. Renyi, Representations for real numbers, Acta Math. Acad. Sci. Hung. 8, 477-493(1957)

[19] M. Smorodinsky, β-automorphisms are weak Bernoulli, preprint

[20] F. Spitzer, Interaction of Markov processes, Advance in Math. 5, 246-290(1970)

[21] Y. Takahashi, Isomorphisms of β-automorphisms to Markov automorphisms, to appear in Osaka J. Math.

Hiroshima University

ON LOCATION PARAMETER FAMILY OF DISTRIBUTIONS WITH UNIFORMLY MINIMUM VARIANCE UNBIASED ESTIMATOR OF LOCATION

Kei Takeuchi

Suppose that $X_1, \ldots, X_n$ are real random variables distributed according to a continuous distribution with a location parameter. Let the density functions of X_i be denoted by $f(x-\theta)$. The purpose of this paper is to prove that under some set of regularity conditions, uniformly minimum variance unbiased (UMVU) estimator exists if and only if the distribution is either normal or expo-gamma type in one case and exponential in another. The proof of the main theorems is done in three steps. First it is shown that if UMVU estimator exists, it is equivalent to the Pitman or the best location invariant estimator. Secondly, it will be shown that the Pitman estimator is UMVU only if it is a sufficient statistic. Finally, under a set of regularity conditions it is shown that the location parameter family admits one-dimensional sufficient statistic if and only if the distribution is either normal or expo-gamma type, or in another situation, exponential.

§ 1. An estimator $\hat{\theta}(X_1, \ldots, X_n)$ is called to be locaiton invariant if for almost all $(x_1, \ldots, x_n)$

$$\hat{\theta}(x_1+a, \ldots, x_n+a) = \hat{\theta}(x_1, \ldots, x_n) + a \qquad \text{for all } a$$

and is called to be almost location invariant if for each a the above holds almost everywhere in $x_1, \ldots, x_n$.

<u>Lemma 1.</u> If $\hat{\theta}$ is an almost location invariant estimator, then there exists a location invariant eastimator $\hat{\theta}^*$ which is almost everywhere equal to $\hat{\theta}$.

<u>Proof.</u> Let us define a set of estimators,

$$\hat{\theta}_{A,B} = \frac{1}{A+B} \int_{-A}^{B} \{\hat{\theta}(x_1+t, \ldots, x_n+t) - t\}dt$$

where A and B are positive numbers. Then

$$\int_S |\hat{\theta}_{A,B} - \hat{\theta}| \Pi dx_i$$

$$= \frac{1}{A+B} \int_S \left| \int_{-A}^{B} \{\hat{\theta}(x_1+t, \ldots, x_n+t) - \hat{\theta}(x_1, \ldots, x_n)\}dt \right| \Pi dx_i$$

$$\leq \frac{1}{A+B} \int_S \int_{-A}^{B} |\hat{\theta}(x_1+t, \ldots, x_n+t) - \hat{\theta}(x_1, \ldots, x_n)| \ dt \Pi dx_i$$

$$= \frac{1}{A+B} \int_{-A}^{B} \int_S | \hat{\theta}(x_1+t, \ldots, x_n+t) - \hat{\theta}(x_1, \ldots, x_n)| \Pi dx_i dt$$

$$= 0$$

for any bounded interval S in R^n. Therefore

$$\hat{\theta}_{A,B} = \hat{\theta} \qquad \text{almost everywhere.}$$

Now consider the double sequence of estimates $\hat{\theta}_{m+\alpha,\, n+\alpha}$ where m, n = 1,2,... and α is a const. Then obviously it holds that

$$\hat{\theta}_{m+\alpha,\, n+\alpha} = \hat{\theta} \qquad \text{for all } m, n$$

almost everywhere in $(x_1, \ldots, x_n)$. Therefore for almost all $(x_1, \ldots, x_n)$ the limit

$$\lim_{\substack{m\to\infty \\ n\to\infty}} \hat{\theta}_{m+\alpha,\, n+\alpha} = \hat{\theta}^*_\alpha$$

exists and is equal to $\hat{\theta}$. From the construction of $\hat{\theta}^*_\alpha$ it can be easily seen that

$$\hat{\theta}^*_\alpha(x_1+a, \ldots, x_n+a) - a \equiv \hat{\theta}^*_{\alpha+a}(x_1, \ldots, x_n)$$

and

$$\hat{\theta}^*_\alpha = \hat{\theta}^*_\beta \qquad \text{if} \quad \alpha-\beta \text{ is integer}$$

when either side exists. We may define $\hat{\theta}^*_\alpha$ for all values of x_i so that the above two equations hold for all values of x_i's , and it is almost everywhere equal to $\hat{\theta}$.

Define

$$\hat{\theta}^* = \int_0^1 \hat{\theta}^*_\alpha \, d\alpha ,$$

then it is seen that $\hat{\theta}^*$ is again almost everywhere equal to $\hat{\theta}$. $\hat{\theta}^*$ is easily shown to be location invariant. (Q.E.D.)

Lemma 2. (Pitman's theorem) The best invariant estimator, i.e. the estimator which minimizes the mean square error among all location invariant estimators is given by

$$\hat{\theta}_p = \int_{-\infty}^{\infty} t\Pi f(x_i-t)dt / \int_{-\infty}^{\infty} \Pi f(x_i-t)dt$$

provided that its variance is finite.

Lemma 3. (Girshick-Savage) Pitman's estimator $\hat{\theta}_p$ is expressed alternatively as

$$\hat{\theta}_p = X_1 - E_0\{X_1 \mid D\}$$

where D denotes the set of differences $X_2-X_1, \ldots, X_n-X_1$ and $E_0\{\ \mid D\}$ the conditional expectation given D when $\theta = 0$. X_1 in the above can be raplaced by any location invariant statistic.

Lemma 4. $E_\theta\{\hat{\theta}_p \mid D\} \equiv \theta$

Theorem 1. If UMVU estimator exists, then $\hat{\theta}_p$ is UMVU.

Proof. Suppose that $\hat{\theta}^*$ is the UMVU estimator. Define $\hat{\theta}^*_\alpha$ by

$$\hat{\theta}^*_\alpha (x_1, \ldots, x_n) = \hat{\theta}^*_\alpha(x_1+\alpha, \ldots, x_n+\alpha) - \alpha$$

Then we have

$$E_\theta(\hat{\theta}^*_\alpha) = E_{\theta+\alpha}(\hat{\theta}^*) - \alpha = \theta$$

therefore it is also unbiased for θ. Since $\hat{\theta}^*$ is UMV, we have

$$V_\theta(\hat{\theta}^*) \leqq V_\theta(\hat{\theta}^*_\alpha) = V_{\theta+\alpha}(\hat{\theta}^*)$$

But α can be arbitrary, hence $V_\theta(\hat{\theta}^*) = V_\theta(\hat{\theta}^*_\alpha)$. Therefore $\hat{\theta}^*_\alpha$ is also UMV. Since UMVU is essentially unique, we have

$$\hat{\theta}^*_\alpha \equiv \hat{\theta}^* \qquad \text{a.e.}$$

which implies that $\hat{\theta}^*$ is almost location invariant. By Lemma 1 there exists $\hat{\theta}^{**}$ which is location invariant and is almost everywhere equal to $\hat{\theta}^*$, and is also UMV. $\hat{\theta}_p$ is unbiased and therefore

$$V_\theta(\hat{\theta}_p) \geqq V_\theta(\hat{\theta}^{**})$$

but $\hat{\theta}_p$ is best among the invariant estimators so we have

$$V_\theta(\hat{\theta}_p) \leqq V_\theta(\hat{\theta}^{**})$$

therefore $V_\theta(\hat{\theta}_p) \equiv V_\theta(\hat{\theta}^{**})$, and essential uniqueness of UMVU estimator implies

$$\hat{\theta}_p \equiv \hat{\theta}^{**}$$

which establishes the theorem. (Q.E.D.)

§ 2. Now let us consider the conditional probability of $\hat{\theta}_p$ given D. Since $\hat{\theta}_p$ is real valued, we can define a conditional probability distribution $F_\theta(\ \mid D)$ almost everywhere in D, such that for any integrable function $\phi(\hat{\theta}_p)$ of $\hat{\theta}_p$ we have,

$$E_\theta\{\phi(\hat{\theta}_p) \mid D\} \equiv \int \phi(y)\, dF_\theta(y|D)$$

For all D, the class of conditional distributions $F_\theta(y|D)$, $-\infty<\theta<\infty$ also consists a location parameter family, i.e. we have

$$F_\theta(y|D) = F_0(y-\theta|D) \qquad \text{for all } \theta$$

for almost all D. We may define $F_\theta(\ |\ D)$ so that the above relation holds for all θ and for all D.

Lemma 5. Let $\phi_\theta(t|D)$ be defined by

$$\phi_\theta(t|D) = E_\theta\{\exp it\hat{\theta}_p|D\},$$

then $\phi_\theta(t|D)$ is the characteristic function of the conditional distribution of $\hat{\theta}_p$ given D, for almost all D.

Proof. For any countable set of real values t_j, $j = 1, 2, \ldots,$ we have

$$\phi_\theta(t_j|D) \equiv \int e^{it_j y}\, dF_\theta(y|D) \qquad \text{for almost all } D.$$

Since the characteristic funciton is continuous, we have the Lemma.

Lemma 6. $\phi_\theta(t|D) = e^{i\theta t}\phi_0(t|D)$ for almost all D.

Lemma 7. If $\phi_0(t\ |D) \equiv \phi_0(t)$ for almost all D, then $\hat{\theta}_p$ is a sufficient statistic.

Proof. By Lemma 6, the conditional distribution $F_0(\ |\ D)$ can be defined independently of D, therefore $\hat{\theta}_p$ and D are stochastically independent. Therefore the conditional distribution of D given $\hat{\theta}_p$ is independent of D. By Lemma 6, it is true for all θ. From Lemma 3, any statistic $T(X_1, \ldots, X_n)$ can be expressed as a function of $\hat{\theta}_p$ and D, therefore given $\hat{\theta}_p$, the conditional distribution of T depends only on the conditional distribution of D, which is equal to its marginal distribution since it is independent of $\hat{\theta}_p$ and is independent of θ. Thus the lemma is established.

Lemma 8. Let $\psi(D)$ be a (complex valued) bounded measurable function of D, and suppose that $E\{|\hat{\theta}_p|\}<\infty$, then

$$E_0\{\hat{\theta}_p \exp it\hat{\theta}_p\psi(D)\} = -iE\{\frac{d}{dt}\phi_0(t|D)\psi(D)\}$$

Proof. The differentiation within the parentheses is permitted.

Lemma 9. Suppose that $h_1(D)$ and $h_2(D)$ are two complex valued functions and that $E|h_1(D)|<\infty$, $E|h_2(D)|<\infty$. There exists a bounded complex valued function $\psi(D)$ such that

$$E\{h_1(D)\psi(D)\} = 0 \quad \text{and} \quad E\{h_2(D)\psi(D)\} \neq 0$$

unless $h_2(D)/h_1(D)$ = constant.

This lemma can be derived immediately from the following

Lemma 9'. Suppose that $h_{ij}(D)$ $i = 1, 2, 3$ $j = 1, 2$ are real valued functions with finite expectations. Then there exists a pair of bounded functions $\psi_1(D)$ and $\psi_2(D)$ such that

$$E\{h_{i1}\psi_1 + h_{i2}\psi_2\} = 0 \quad \text{for} \quad i = 1, 2$$

and

$$E\{h_{31}\psi_1 + h_{32}\psi_2\} \neq 0$$

unless there exists a pair of constants C_1, C_2 such that

$$h_{3j}(D) \equiv C_1 h_{1j}(D)\psi_j(D) + C_2 h_{2j}(D)\psi_j(D) \quad \text{for} \quad j = 1, 2.$$

Outline of the proof. We may define

$$\psi_1(D) = 1 \qquad \text{if } h_{31}(D) > \lambda_1 h_{11}(D) + \lambda_2 h_{21}(D)$$

$$= -C_1 \qquad \text{otherwise}$$

$$\psi_2(D) = 1 \qquad \text{if } h_{32}(D) > \lambda_1 h_{12}(D) + \lambda_2 h_{22}(D)$$

$$= -C_2 \qquad \text{otherwise}$$

and with appropriate choice of constants λ_1, λ_2 and C_1, C_2 we get ψ_1 and ψ_2 satisfying the condition.

Lemma 10. If for all t

$$\frac{d}{dt}\ \phi_0(t|D) \equiv C(t)\phi_0(t|D) \qquad \text{for almost all } D,$$

then $\phi_0(t|D) = \phi_0(t)$, that is, it can be determined independedtly of D.

Proof. Since $\phi_0(t|D)$ and $\frac{\partial}{\partial t}\ \phi_0(t|D)$ are continuous, we have for almost all D,

$$\frac{d}{dt}\ \phi_0(t|D) \equiv C(t)\phi_0(t|D)$$

for all t. Therefore we have for almost all D,

$$\phi_0(t|D) = \exp[\textstyle\int_0^t C(t)dt + A(D)]$$

Since $\phi_0(0|D) = 1$ for all D, we have $A(D) = 0$, and

$$\phi_0(t) = \exp[\textstyle\int_0^t C(t)dt]$$

Lemma 11. If $\hat{\theta}_p$ is UMV, then for any complex valued $\psi(\hat{\theta}_p, D)$ such that $|\psi(\hat{\theta}_p, D)|$ is bounded and

$$E_\theta\{\psi(\hat{\theta}_p, D)\} = 0 \qquad \text{for all } \theta,$$

we have

$$E_\theta\{\ \hat{\theta}_p\psi(\hat{\theta}_p, D)\} = 0$$

Proof. The lemma is well known for real valued ψ, but it can be immediately extended to complex valued case.

Now the second main theorem is easily established :

<u>Theorem 2</u>. $\hat{\theta}_p$ <u>is UMV only if it is a sufficient statistic</u>.

<u>Proof</u>. Suppose $\hat{\theta}_p$ is not sufficient. Then from Lemmas 7 and 10, there is a value t_0, such that

$$\frac{d}{dt}\ \phi_0(t_0|D) \not\equiv C\phi_0(t_0|D)$$

for any constant C. By Lemma 9' there exists a bounded function $\psi(D)$ such that

$$E\{\phi_0(t_0|D)\psi(D)\} = 0$$

and

$$E\{\frac{d}{dt}\ \phi_0(t_0|D)\psi(D)\} \not\equiv 0$$

Define

$$\psi(\hat{\theta}_p, D) = \exp\{it_0\hat{\theta}_p\}\psi(D) ,$$

then we have

$$E_\theta\{\psi(\hat{\theta}_p, D)\} = E\{\phi_\theta(t_0|D)\psi(D)\}$$

$$= e^{it_0\theta}\ E\{\phi_0(t_0|D)\psi(D)\} = 0$$

for all θ, and

$$E_0\{\hat{\theta}_p\psi(\hat{\theta}_p, D)\} = -iE\{\frac{d}{dt}\ \phi_0(t_0|D)\psi(D)\} \not\equiv 0$$

which contradicts the assumption that $\hat{\theta}_p$ is UMV. (Q.E.D.)

§ 3. Now we assume that

A 1. $f(x) > 0$ for all x,

A 2. f is twice differentiable.

<u>Lemma 12</u>. Under the assumptions A1 ~ 2 , $\hat{\theta}_p$ is sufficient only if $\log f(x-\theta) = s(\theta)t(x) + c(\theta) + h(x)$ and $\hat{\theta}_p$ is a function of $\sum_{i=1}^{n} t(X_i)$.

Proof. This is a straightforward application of Dynkin's theorem.

Theorem 3. Under the assumptions A $1\sim 2$, UMVU estimator exists if and only if $f(x)$ is either of the following type :

$$f(x) = C \exp - \frac{(x-a)^2}{2\sigma^2} \qquad \text{(normal)}$$

$$C \exp \{-\exp \frac{(x-a)}{b} + n\frac{(x-a)}{b}\} \qquad \text{(exponential-gamma)}$$

where a, b, C, n are constants.

Proof. UMVU estimator exists only if

$$\log f(x-\theta) = s(\theta)t(x) + c(\theta) + h(x)$$

Differentiating in x and then in θ we have,

$$\frac{f'(x-\theta)}{f(x-\theta)} = s(\theta)t'(x) + h'(x)$$

$$-\frac{f'(x-\theta)}{f(x-\theta)} = s'(\theta)t(x) + c'(\theta)$$

hence

$$s(\theta)t'(x) + s'(\theta)t(x) + h'(x) + c'(\theta) = 0 \qquad (*)$$

Differentiating the above first by x and then by θ, we have

$$s'(\theta)t''(x) + s''(\theta)t'(x) \equiv 0$$

or

$$\frac{t''(x)}{t'(x)} \equiv - \frac{s''(\theta)}{s'(\theta)}$$

both sides of which must be constant. Therfore we have

$$\log t'(x) = Ax+B$$

$$\log s'(\theta) = -A\theta+C$$

where A, B, C are constants.

When $A = 0$, we have

$$t'(x) = e^B = B'$$

$$t(x) = B'x + D$$

$$s'(\theta) = e^C = C'$$

$$s(\theta) = C'\theta + D'$$

Substituting in (*) we have

$$B'C'(x+\theta) + h'(x) + c'(\theta) = 0$$

$$h'(x) = - B'C'x + E , \quad c'(\theta) = -B'C'\theta + E'$$

And finally we have

$$- 2 \log f(x-\theta) = B'C'(x-\theta)^2 + D$$

which corresponds to the first case.

When $A \neq 0$,

$$t'(x) = e^{Ax+B}$$

$$t(x) = C'e^{Ax} + D'$$

$$s'(\theta) = e^{-A\theta+D}$$

$$s(\theta) = - C''e^{-A\theta} + D''$$

where C' and C'' are positive constants. From (*) we have

$$h(x) = Ee^{Ax} + Fx$$

$$c(\theta) = E'e^{-A\theta} - F\theta$$

and we have the second case.

For both cases, the sufficient statistic

$$\overline{X} = \frac{1}{n} \sum_i X_i \qquad \text{for the normal}$$

$$\frac{1}{n} \sum_i e^{-X_i/b} \qquad \text{for the exponential-gamma type}$$

is complete, therefore for any parameter which admits an unbiased estmator, there exists UMVU estimator. This UMVU estimator of θ is explicitly given by

$$\hat{\theta} = \overline{X} - a$$

or
$$\hat{\theta} = \log \left(\frac{1}{n} \sum_i e^{-X_i/b}\right) - a + C_m b$$

where C_m is a constant depending on m. (Q.E.D.)

For the second case we consider we assume the following :

A 3. $f(x) > 0$ for $x > a$ and

$= 0$ for $x \leq a$

and $f(x)$ is continuous.

<u>Lemma 13</u>. Under the assumption A 3, $T = \min X_i$ is a function of sufficient statistic for the location parameter family.

<u>Proof</u>. The joint density function is given by

$$\Pi f(x_i - \theta) \quad \text{for} \quad \min x_i > \theta + a$$

and equal to zero otherwise

or equivalenty,

$$\chi(t-\theta) \Pi f(x_i - \theta)$$

where $t = \min x_i$ and $\chi(t) = 1$ if $t > a$, $= 0$ if $t \leq a$. Suppose that $y = y(X_1, \ldots, X_n)$ is a sufficient statistic, then we have

$$\chi(t-\theta) \Pi f(x_i - \theta) \equiv g(x) h(y, \theta) \qquad \text{for all} \quad \theta$$

almost everywhere in x. Since the left side is positive if and only if $t - \theta > a$, so is $h(y, \theta)$, which implies that t is a function of y.

<u>Lemma 14</u>. Under the assumption A 3, $\min X_i$ is independent of D if and only if $f(x)$ is exponential , i.e.

$$f(x) = c \exp - c(x-a) \qquad \text{for} \quad x > a$$

$$= 0 \qquad \text{for} \quad x \leq a$$

Proof. Without loss of generality we may assume that $a = 0$. Let $X_{(1)} < \cdots < X_{(n)}$ be the order statistic obtained from $X_1, \cdots, X_n$, and put $T = \min X_i = X_{(1)}$, and $Y_1 = X_{(2)} - X_{(1)}, \cdots, Y_{n-1} = X_{(n)} - X_{(1)}$. Then the joint density is given by

$$f(t) \prod_{i=1}^{n-1} f(t+y_i) \quad \text{for} \quad t > 0 \quad \text{and} \quad 0 \leqq y_1 \leqq \cdots \leqq y_{n-1}$$

and 0 otherwise.

If T is independent of D, it is also independent of $Y_1, \cdots, Y_{n-1}$. Therefore we have

$$f(t) \prod_{i=1}^{n-1} f(t+y_i) \equiv g(t)h(y_1, \cdots, y_{n-1})$$

or

$$\log f(t) + \sum_{i=1}^{n-1} \log f(t+y_i) \equiv \log g(t) + \log h(y_1, \cdots, y_{n-1}),$$

for almost all t and $y_1, \cdots, y_{n-1}$. By putting $y = y_1 = \cdots = y_{n-1}$ we have

$$\log f(t+y) = \{\log g(t) - \log f(t) + \log h(y, \cdots, y)\}/(n-1)$$

which is possible only if $\log f(t)$ is a linear function.

Theorem 4. Under the assumption A 3, UMVU estimator of θ exists if and only if $f(x)$ is exponential.

Proof. UMVU estimator exists only if $\hat{\theta}_p$ is sufficient and is independent of D. Since $\min X_i$ is a function of any sufficient statistic, it is independent of D, therefore the distribution $f(x)$ must be exponential. (Q.E.D.)

The last case we discuss is :

A 4. $f(x) > 0$ for $a < x < b$ and
$= 0$ for $x \leqq a$ and $x \geqq b$

Lemma 15. Under the assumption A 4, $T_1 = \min X_i$, $T_2 = \max X_i$ are functions of sufficient statistic.

Theorem 5. Under the assumption A 4, no UMVU estimator for the location parameter exists.

Proof. If UMVU exists, $\hat{\theta}_p$ is sufficient, hence $R = \max X_i - \min X_i$ is a function of $\hat{\theta}_p$, which is independent of D. On the other hand R is obviously a function of D, which is a contradiction. (Q.E.D.)

Remark. The condition for the first case of this section can be weakened a little. Pfanzagle [3] gave some results about the problem.

References

[1] E.B. Dynkin : On sufficient and necessary statistics for a family of probability distributions, Selected Translations in Mathematical Statistics and Probability Vol. 1 (1961), 17-40.

[2] E. J. G. Pitman : The estimation of the location and scale parameters of a continuous distributions of any given form, Biometrika 30 (1938), 390-421.

[3] J. Pfanzagle : Transformation groups and sufficient statistics, Ann. Math. Statist. 43 (1972), 553-568.

Faculty of Economics

University of Tokyo

ON MARKOV PROCESS CORRESPONDING TO BOLTZMANN'S EQUATION OF MAXWELLIAN GAS

Hiroshi Tanaka

§1. Introduction. The basic equation in the kinetic theory of dilute monoatomic gases is the famous Boltzmann's equation. In the spatially homogeneous case, the initial value problem of this equation was solved for a gas of hard balls by Carleman [1], for Maxwellian gas with cutoff by Wild [14], and for bounded total collision cross-section by Povzner [8] (in modified spatially inhomogeneous case), but it seems that no results (for existence and uniqueness) have been obtained for Maxwellian gas without cutoff. On the other hand, H. P. McKean [5] introduced a class of Markov processes associated with certain nonlinear (parabolic) equations such as Boltzmann's equation, and brought a new light in the field of investigation of such equations by probabilistic methods (see also [6]). Then, there appeared works by D. P. Johnson [3], T. Ueno [11] [12], Y. Takahashi [9] and H. Tanaka [10], mostly concerned with Boltzmann's equation of cutoff type and certain nonlinear equations with similar structure. Especially, Ueno [12] constructed Markov processes which describe motions of infinitely many interacting particles, while Takahashi [9] introduced interaction semigroups and discussed their relationship to branching semigroups. In this paper we are exclusively concerned with non-cutoff Maxwellian gas ; our purpose is to construct a Markov process in the sense of McKean [5] corresponding to the 3-dimensional Maxwellian gas without cutoff by solving appropriate stochastic differential equation (the equation (2.10) in §2). The theory of stochastic differential equations was initiated by K. Itô [2] and, in the case of diffusions, equations similar to (2.10) were considered by McKean [7] in connection with certain nonlinear parabolic equations. The results are only summarized ; full proofs will be published elsewhere.

We consider a monoatomic dilute gas composed of a large number of molecules moving in the space and assume that there are no outside forces. Let $Nu(t,x)dx$ be the number of molecules with velocities x within the differential element dx at time t, where N is the total number of molecules. Then under the assumption of spatial homogeneity, $u(t,x)$ satisfies the following Boltzmann's equation :

$$(1.1) \qquad \frac{\partial u(t,x)}{\partial t} = \int_{S_o \times R^3} \{u(t,x^*)u(t,y^*)-u(t,x)u(t,y)\}|x-y|Q(|x-y|,\theta)\sin\theta d\theta d\psi dy,$$

where $S_o=(0,\pi)\times[0,2\pi)$ and θ,ψ are points in $(0,\pi)$ and $[0,2\pi)$ respectively. Denote by $S_{x,y}$ the sphere with center $\frac{x+y}{2}$ and diameter $|x-y|$, and on this sphere we consider a spherical coordinate system with polar axis defined by the relative velocity $x-y$. x^* and y^* are the post-collisional velocities. Let θ and ψ be the colatitude of x^*(the angle between two vectors $x-y$ and x^*-y^*) and the logitude of x^*, respectively. By the conservation laws of momentum and energy x^* and y^* are always situated on $S_{x,y}$ and constitute a diameter of $S_{x,y}$, and so y^* is also determined by θ and ψ. For each x and y the origin of the longitude ψ may be arbitrary chosen within the requirement that x^* and y^* as functions of (x,y,θ,ψ) should be Borel measurable. A nonnegative function Q is determined by the intermolecular force and is called the differential collision cross-section. In the model of gas of hard balls Q is a positive constant, while in the Maxwellian model in which molecules repel each other with a force inversely proportional to the fifth power of their distance, $|x-y|Q(|x-y|,\theta)$ turns out to be a function $Q_M(\theta)$ of θ alone ; in the latter case $Q_M(\theta)$ is a decreasing function of θ with $Q_M(\theta) \sim \text{const.}\ \theta^{-\frac{5}{2}}$, $\theta\downarrow 0$, and so the total collision cross-section is infinite (non cutoff) (see [13]). This is the case we consider in this paper.

§2. Markov processes and stochastic differential equation

In order to indicate our problem clearly, we first explain how a Markov process in the sense of McKean [5] is associated with Boltzmann's equation, taking gas of hard balls by example. The equation (1.1) for gas of hard balls is usually treated in the following form :

$$(2.1)\qquad \frac{\partial u(t,x)}{\partial t} = \int_{S^2\times R^3} \{u(t,x^*)u(t,y^*)-u(t,x)u(t,y)\}\,|(y-x),\ell)|\,d\ell\,dy,$$

where $x^*=x+(y-x,\ell)\ell$, $y^*=y-(y-x,\ell)\ell$, $\ell\in S^2$, and $d\ell$ is the uniform distribution on S^2. We set

$$(2.2)\qquad \begin{cases} u(t,\Gamma) = \displaystyle\int_\Gamma u(t,x)\,dx, & \Gamma\in\mathcal{B}(R^3) \\ u(t,\varphi) = \displaystyle\int \varphi(x)\,u(t,dx), & \varphi\in C_b(R^3) \end{cases}$$

where $C_b(R^3)$ denotes the space of real valued bounded continuous functions on R^3 (the notation (2.2) will be used throughout in this paper). Then, from (2.1) we have

$$(2.3)\qquad \frac{\partial u(t,\varphi)}{\partial t} = \int_{S^2\times R^3\times R^3} \{\varphi(x^*)-\varphi(x)\}\,|(y-x,\ell)|\,d\ell\,u(t,dx)\,u(t,dy),\quad \varphi\in C_b(R^3).$$

Povzner's result [8] may be stated as follows : given an initial data f (probability measure on R^3) such that $\int|x|^4 f(dx)<\infty$, there exists a unique solution $u(t,\cdot)$ (probability measure) of (2.3) such that $\mu(t)=\int|x|^4 u(t,dx)$ is locally bounded. Now, keeping f and $u(t,\cdot)$ as above, we consider the following equation for $v(t,\cdot)$:

$$(2.4)\quad \begin{cases} \dfrac{\partial v(t,\varphi)}{\partial t} = \displaystyle\int_{S^2\times R^3\times R^3} \{\varphi(x^*)-\varphi(x)\}\,|(y-x,\ell)|\,d\ell\, v(t,dx)u(t,dy) \\ v(0,\cdot) = \delta(z,\cdot), \qquad \varphi\in C_b(R^3) , \end{cases}$$

where $\delta(z,\cdot)$ denotes the unit distribution with mass at z. Then, using the result of [8] it is not hard to prove that (2.4) has a unique solution, which we denote by $P_f(t,z,\cdot)$. We can also prove that

(2.5a) $P_f(t,z,\cdot)$ is a probability measure on R^3,

$$(2.5b)\quad u(t,\Gamma) = \int_{R^3} P_f(t,z,\Gamma)f(dz), \quad \Gamma\in\mathcal{B}(R^3) ,$$

$$(2.5c)\quad P_f(t+s,z,\Gamma) = \int_{R^3} P_f(t,z,dy)P_{u(t)}(s,y,\Gamma).$$

The above properties of $P_f(t,z,\cdot)$ enable us to construct a Markov process as follows. Let Ω be the space of step functions $z(t)$ defined on $[0,\infty)$ and taking values in R^3, and $\mathcal{B}$ the σ-field generated by (measurable) cylinder sets in Ω. For each probability measure f on R^3 such that $\int |x|^4 f(dx)<\infty$, we can construct a probability measure $P_f(\cdot)$ on $(\Omega,\mathcal{B})$ so that $\{\Omega,z(t),P_f\}$ is a Markov process in the following sense.

$$(2.6a)\quad P_f\{z(0)\in dx\} = f(dx) ,$$

$$(2.6b)\quad P_f\{z(t+s)\in\Gamma\,|\,z(\tau):0\leq\tau\leq t\}=P_{u(t)}(s,z(t),\Gamma), \text{ a.s. } (P_f).$$

This is the Markov process in the sense of [5] associated with the gas of hard balls (2.1).

Now, our problem can be stated as follows : construct a Markov process $x(t)$ in the sense of [5] which is related to the Maxwellian gas without cutoff as $z(t)$ is to (2.1). So we take up the equation (1.1) with Q_M specified as in §1 and rewrite it as in the form (2.3). A formal but careful calculation yields

$$(2.7) \qquad \frac{\partial u(t,\varphi)}{\partial t} = \int_{S_o \times R^3 \times R^3} \{\varphi(x^*)-\varphi(x)\} Q_M(\theta)\sin\theta d\theta d\psi u(t,dx)u(t,dy).$$

Considering the singularity of $Q_M(\theta)$ at $\theta=0$, it may be natural to take φ (test function) from the space $C_o^1(R^3)$ of C^1-functions with compact supports. However, we do not treat (2.7) directly ; instead we consider a suitable stochastic differential equation as will be described below.

Let $S=(0,1)\times(0,\pi)\times[0,2\pi)$ and λ the measure on S defined by $d\lambda = d\alpha \otimes Q(d\theta)\otimes d\psi$, where $Q(d\theta)=Q_M(\theta)\sin\theta d\theta$. Although we have thus specified $Q(d\theta)$, we remark that the only property of $Q(d\theta)$ we need later is

$$(2.8) \qquad \int_{(0,\pi)} \theta Q(d\theta) < \infty,$$

and so our results remain valid for arbitrary measure $Q(d\theta)$ subject to the condition (2.8). Let $\tilde{\lambda}$ be the product measure $dt\otimes d\lambda$ on the space $(0,\infty)\times S$ and denote by $\mathfrak{F}$ the class of Borel sets in $(0,\infty)\times S$ which have finite $\tilde{\lambda}$-measure. A family of random variables $\{p(A,\omega)\}_{A\in\mathfrak{F}}$ defined on a probability space $(\Omega,\mathfrak{B},P)$ is called a <u>Poisson random measure</u> on $(0,\infty)\times S$ associated with the measure $\tilde{\lambda}$, if the following two conditions are satisfied.

(i) Each $p(A,\omega)$ is distributed according to the Poisson distribution with mean $\tilde{\lambda}(A)$.

(ii) If $A_1, A_2, \cdots \in \mathcal{F}$, $A_j \cap A_K = \phi\ (j \neq k)$ and $A = \bigcup_{k=1}^{\infty} A_k \in \mathcal{F}$,

then the family $\{p(A_k,\omega)\}_{k=1,2,\cdots}$ is independent and

(2.9) $$p(A,\omega) = \sum_{k=1}^{\infty} p(A_k,\omega) \quad \text{(a.s.)}.$$

Since a Poisson random measure always admits a suitable modification for which (2.9) holds for <u>all</u> ω, we may suppose, if necessary, that $\{p(A,\omega)\}$ satisfies this stronger version of (2.9). For a Poisson random measure $\{p(A,\omega)\}_{A\in\mathcal{F}}$ we set

$$\mathcal{B}_{\infty}^{t} = \sigma\{p(A,\omega) : A \subset (t,\infty)\times S\},$$

where the notation $\sigma\{$ —— $\}$ stands for the smallest σ-field that makes all the random variables in $\{$ —— $\}$ measurable. Next, we regard the unit interval (0,1) as a probability space by considering the Lebesgue measure (precisely, its restriction) on the Borel field of (0,1), and on this probability space we sometimes consider an R^3-valued stochastic process $\{y(t,\alpha),\ 0\leq t<\infty\}$ with path functions which are right continuous and have left hand limits. Such a process is called an <u>α-process</u> for simplicity ; similarly a random variable defined on the probability space (0,1) is called an α-random variable. Now our stochastic differential equation associated with (2.7) can be written as follows :

(2.10a) $$x(t,\omega) = x(0,\omega) + \int_{(0,t]\times S} a(x(s,\omega), y(s,\alpha), \theta, \psi)\, p(dsd\sigma,\omega).$$

(2.10b) $\{y(t,\alpha),\ 0\leq t<\infty\}$ is an α-process which is equivalent in law to $\{x(t,\omega),\ 0\leq t<\infty\}$.

(2.10c) For each $t\geq 0$, the σ-field

$$\sigma\{x(s,\omega),\ p(A,\omega) : 0\leq s\leq t,\ A\subset(0,t]\times S\}$$

is independent of $\mathcal{B}_{\infty}^{t}$.

Here $a(x,y,\theta,\psi)=x^*(x,y,\theta,\psi)-x$, $d\sigma=d\alpha d\theta d\psi$, and of course $\{p(A,\omega)\}$ is a Poisson random measure associated with $\tilde{\lambda}$ defined over a (basic) probability space $(\Omega,\mathcal{B},P)$. $\{x(t,\omega),\ 0\leq t\leq\infty\}$ is unknown process to be determined by the equation (2.10a) under the additional conditions (2.10b) and (2.10c). Solving (2.10) gives rise to an answer to the problem of finding Markov process associated with the Maxwellian gas, as will be seen in the next section.

§3. Solving the stochastic differential equation. Main results

Let f be a probability measure on R^3. By a solution of (2.10) with initial distribution f, we mean an R^3-valued stochastic process $\{x(t,\omega),\ 0\leq t\leq\infty\}$ defined over a suitable probability space $(\Omega,\mathcal{B},P)$ and satisfying the following conditions

(i) $x(t,\omega)$ is right continuous and has left hand limits with probability 1,

(ii) $P\{x(0,\omega)\in dx\} = f(dx)$,

(iii) the relations (2.10a), (2.10b) and (2.10c) hold for some Poisson random measure $\{p(A,\omega)\}$ associated with $\tilde{\lambda}$ and defined over the probability space $(\Omega,\mathcal{B},P)$.

In solving the existence problem, the usual method of successive approximation can not be applied directly, since the function $a(x,y,\theta,\psi)$ is not smooth. However, it can be shown that $a(x,y,\theta,\psi)$ has a nice property similar to the Lipschitz continuity (Lemma 1). Owing to this property a kind of successive approximation method works, but in the results the usual pathwise uniqueness will be replaced by weaker one, the uniqueness in law. Here, we say that the uniqueness in law holds for initial distribution f, if any two solutions of (2.10) with initial distribution f (which may be defined on different probability spaces) are identical in law as stochastic processes.

In what follows, $a(x,y,\theta,\psi)$ viewed as function of ψ alone is regarded as a periodic function on R^1 with period 2π.

<u>Lemma 1. There exists a Borel function</u> $\psi_o(x,y,\tilde{x},\tilde{y},)$ of $x,y,\tilde{x},\tilde{y} \in R^3$ <u>such that</u>

$$|a(x,y,\theta,\psi)-a(\tilde{x},\tilde{y},\theta,\psi+\psi_o(x,y,\tilde{x},\tilde{y}))| \leq K\{|x-\tilde{x}|+|y-\tilde{y}|\}\cdot\theta,$$

<u>where</u> K <u>is an absolute constant</u>.

In fact, ψ_o can be defined as follows. If $x=y$ or $\tilde{x}=\tilde{y}$, we set $\psi_o(x,y,\tilde{x},\tilde{y}) =0$. If $x \neq y$ and $\tilde{x}\neq\tilde{y}$, we set

$$x^o = \frac{|x-y|}{|\tilde{x}-\tilde{y}|} \frac{\tilde{x}-\tilde{y}}{2} + \frac{x+y}{2}$$

and denote by ρ the rotation around ℓ such that $\rho x = x^o$, where ℓ is the straight line passing through the point $\frac{x+y}{2}$ and perpendicular to the plane determined by the three points $\frac{x+y}{2}$, x, x^o (when $x^o=y$, we define ρ by $\rho z = x+y-z$). Now we set

$$x^{\#}(x,y,\theta,0)= \frac{|\tilde{x}-\tilde{y}|}{|x-y|} (\rho x^*(x,y,\theta,0)- \frac{x+y}{2}) + \frac{\tilde{x}+\tilde{y}}{2} .$$

Then $x^{\#}$ lies on the sphere $S_{\tilde{x},\tilde{y}}$ and hence we can define ψ_o $(0\leq\psi_o<2\pi)$ by the formula

$$x^{\#}(x,y,\theta,0) = x^*(\tilde{x},\tilde{y},\theta,\psi_o(x,y,\tilde{x},\tilde{y})).$$

One of our main results is

<u>Theorem A. Assume that</u> $\int |x| f(dx)<\infty$. <u>Then</u>,

(i) <u>there exists a solution</u> $\{x(t,\omega),\ 0\leq t<\infty\}$ <u>of</u> (2.10) <u>with initial distribution</u> f,

(ii) <u>the uniqueness in law holds for initial distribution</u> f,

(iii) <u>the probability distribution</u> $u(t,\cdot)$ <u>of</u> $x(t,\omega)$ <u>satisfies</u> (2.7) <u>for any</u> $\varphi \in C_o^1(R^3)$,

(iv) (a) <u>(conservation of momentum)</u> $E\{x(t,\omega)\}$ <u>is independent of</u> t,

(b) <u>(conservation of energy)</u> $E\{|x(t,\omega)|^2\}$ <u>is independent of</u> t, <u>provided</u> $\int |x|^2 f(dx)<\infty$.

Here is an outline of the proof of the existence part (i). Choose a sequence $\{\xi_n(\alpha)\}_{n=o,1,\cdots}$ of independent α-random variables with the uniform distribution on (0,1), and set $\mathcal{F}_n = \sigma\{\xi_k(\alpha),\ 0\leq k\leq n\}$. Over a suitable probability space $(\Omega,\mathcal{B},P)$ we take an R^3-valued random variable $x(0,\omega)$ with distribution f and a Poisson random measure $\{\tilde{p}(A,\omega)\}$ associated with $\tilde{\lambda}$ so that $x(0,\omega)$ is independent of $\{\tilde{p}(A,\omega)\}$, and set $x_o(t,\omega)=x(0,\omega)$ for all $t\geq 0$. Then, taking arbitrary $\mathcal{F}_o$--measurable α-process $\{y_o(t,\alpha),\ 0\leq t<\infty\}$ which is equivalent in law to the process $\{x_o(t,\omega),\ 0\leq t<\infty\}$, we put

$$x_1(t,\omega)=x(0,\omega)+\int_{(0,t]\times S} a(x_o(s,\omega),y_o(s,\alpha),\theta,\psi)\tilde{p}(dsd\sigma,\omega).$$

In general, for $n\geq 1$ we put

$$x_{n+1}(t,\omega)=x(0,\omega)+\int_{(0,t]\times S} a(x_n(s,\omega),y_n(s,\omega),\theta,\psi+\tilde{\psi}_n)\tilde{p}(dsd\sigma,\omega),$$

where $\{y_n(t,\alpha),\ 0\leq t<\infty\}$ is an $\mathcal{F}_n$-measurable *) α-process such that the process $\{(y_{n-1}(t,\alpha),y_n(t,\alpha)),\ 0\leq t<\infty\}$ is equivalent in law to the process $\{(x_{n-1}(t,\omega), x_n(t,\omega)),\ 0\leq t<\infty\}$, and $\tilde{\psi}_n=\tilde{\psi}_{n-1}+\psi_o(x_{n-1}(s,\omega),y_{n-1}(s,\alpha),x_n(s,\omega),y_n(s,\alpha))$, $\tilde{\psi}_o=0$.

Then by Lemma 1

$$\text{(3.1)}\qquad \begin{aligned} &|a(x_{n-1}(s),y_{n-1}(s),\theta,\psi+\tilde{\psi}_{n-1})-a(x_n(s),y_n(s),\theta,\psi+\tilde{\psi}_n)| \\ &\leq K\{|x_n(s)-x_{n-1}(s)|+|y_n(s)-y_{n-1}(s)|\}\theta, \end{aligned}$$

and hence

$$E|x_{n+1}(t)-x_n(t)|\leq \text{const.}\int_0^t E|x_n(s)-x_{n-1}(s)|ds,$$

which implies

$$\sum_{n=1}^{\infty} E|x_n(t)-x_{n-1}(t)|<\infty.$$

---

*) $\mathcal{F}_n$-measurability is imposed only to make the construction of $\{y_n(t,\alpha)\}$ easier.

Therefore, $x(t)=\lim_{n\to\infty} x_n(t)$ exists (a.s.), and hence the same for $y(t)=\lim_{n\to\infty} y_n(t)$. Again using (3.1), it can be shown that $\lim_{n\to\infty} a(x_n(t),y_n(t),\theta,\psi+\tilde{\psi}_n)$ exists and is equal to $a(x(t),y(t),\theta,\psi+\tilde{\psi}_\infty)$ with some $\tilde{\psi}_\infty=\tilde{\psi}_\infty(t,\alpha,\omega)$ except on a negligible set. Now setting

$$p(A,\omega)=\int_{(0,\infty)\times S} \chi_A(t,\alpha,\theta,\psi+\tilde{\psi}_\infty)\tilde{p}(dtd\sigma,\omega),$$

we obtain (2.10).

The uniqueness part (ii) follows from the following

Lemma 2. *Let f be the same as in the theorem and $\{x(t,\omega),\ 0\leqq t<\infty\}$ any solution of (2.10) with initial distribution f. Take arbitrary finite interval $[0,T]$ and let Δ be a partition of $[0,T]$: $0=t_0<t_1<\cdots<t_n=T$. Choosing an R^3-valued random variable $X(0)$ independent of $\{p(A)\}$ and with distribution f, we define a process $\{X_\Delta(t),\ 0\leqq t\leqq T\}$ as follows :*

$$\begin{cases} X_\Delta(t)=X(0)+\displaystyle\int_{(0,t]\times S} a(X(0),Y_0(\alpha),\theta,\psi)p(dsd\sigma), & 0\leqq t\leqq t_1 \\ X_\Delta(t)=X_\Delta(t_1)+\displaystyle\int_{(t_1,t)\times S} a(X_\Delta(t_1),Y_1(\alpha),\theta,\psi)p(dsd\sigma), & t_1<t\leqq t_2 \\ \quad , & \\ \quad , & \\ \quad , & \\ X_\Delta(t)=X_\Delta(t_{n-1})+\displaystyle\int_{(t_{n-1},t]\times S} a(X_\Delta(t_{n-1}),Y_{n-1}(\alpha),\theta,\psi)p(dsd\sigma), & t_{n-1}<t\leqq T, \end{cases}$$

where $Y_0(\alpha),\cdots,Y_{n-1}(\alpha)$ are α-random variables defined successively so that each $Y_k(\alpha)$ has the same distribution as $X_\Delta(t_k)$. Then, any finite dimensional distribution of $\{X_\Delta(t),\ 0\leqq t\leqq T\}$ converges to the corresponding finite dimensional distribution of $\{x(t),\ 0\leqq t\leqq T\}$ as $|\Delta|=\max_{1\leqq k\leqq n}|t_k-t_{k-1}|$ tends to 0.

The following theorem shows that the process $\{x(t,\omega),\ 0\leq t<\infty\}$ is the Markov process we were looking for. The proof is almost the same as that of Theorem A.

Theorem B. (i) Let $\{y(t,\alpha),\ 0\leq t<\infty\}$ be any α-process equivalent in law to the process $\{x(t,\omega),\ 0\leq t<\infty\}$ constructed in Theorem A. Then, for each $x\in R^3$ there exists a solution $z(t,\omega)$ for the stochastic differential equation

$$z(t,\omega)=x+\int_{(0,t]\times S} a(z(s,\omega),y(s,\alpha)\theta,\psi)p(dsd\sigma,\omega),$$

and the uniqueness in law holds for this equation.

(ii) Set

$$P_f(t,x,\Gamma)=P\{z(t,\omega)\in\Gamma\}\ ,\quad \Gamma\in\mathcal{B}(R^3)\ ,$$

and let $\{x(t,\omega),\ 0\leq t<\infty\}$ be the solution of (2.10). Then

$$P\{x(t+s,\omega)\in\Gamma|\mathcal{B}_t\}=P_{u(t)}(s,x(t,\omega),\Gamma)\quad \text{a.s.},$$

where $\mathcal{B}_t=\sigma\{x(s,\omega),p(A,\omega)\ :\ 0\leq s\leq t,\ \ A\subset(0,t]\times S\}.$

References

[1] T. Carleman, Problèmes Mathematiques dans la Théorie Cinétique des Gaz, Uppsala, 1957.

[2] K. Itô, On stochastic differential equations, Mem. Amer. Math. Soc. No. 4 (1951).

[3] D. P. Johnson, On a class of stochastic processes and its relationship to infinite particle gases, Trans. Amer. Math. Soc. 132(1968),275-295.

[4] M. Kac, Foundation of Kinetic theory, Proc. Third Berkeley Symp. on Math. Stat. and Prob. 3, 171-197.

[5] H. P. McKean, Jr., A class of Markov processes associated with nonlinear parabolic equations, Proc. Nat. Acad. Sci. 56(1966), 1907-1911.

[6] H. P. McKean, Jr., An exponential formula for solving Boltzmann's equation for a Maxwellian gas, J. Combinatorial Theory 2(1967), 358-382.

[7] H. P. McKean, Jr., Propagation of chaos for a class of non-linear parabolic equations, Lecture series in Differential Equations, session 7, Catholic Univ. (1967).

[8] A. Ya. Povzner, On Boltzmann's equation in the kinetic theory of gases, Mat. Sb. 58(1962), 63-86.

[9] Y. Takahashi, Markov semigroups with simplest interactions I, II (to appear in Proc. Japan Acad.).

[10] H. Tanaka, Propagation of chaos for certain purely discontinuous Markov processes with interactions, J. Fac. Sci., Univ. of Tokyo, Sec. I 17(1970), 259-272.

[11] T. Ueno, A class of Markov processes with interactions I, Proc. Japan Acad. 45(1969), 641-646 ; II, ibd., 995-1000.

[12] T. Ueno, A path space and the propagation of chaos for a Boltzmann's gas model, Proc. Japan Acad., 47(1971), 529-533.

[13] G. E. Uhlenbeck and G. W. Ford, Lectures in Statistical Mechanics, Amer. Math. Soc. Providence 1963.

[14] E. Wild, On Boltzmann's equation in the kinetic theory of gases, Proc. Camb. Phil. Soc. 47(1951), 602-609.

Department of Mathematics
Hiroshima University.

ON SOME PERTURBATIONS OF STABLE PROCESSES

Masaaki Tsuchiya

In the previous papers [4],[5], we have investigated drift-type perturbation of one-dimensional symmetric stable processes. The problem has close connection with the construction of the Markov process on the boundary induced by the process on the upper-half plane with generator

$$\frac{d^2}{dx^2} + \frac{d^2}{dy^2} + \frac{1-\alpha}{y}\frac{d}{dy} \quad (\alpha \geq 1) \text{ and an oblique reflecting barrier.}$$

In this report, we shall study more general perturbation of multi-dimensional symmetric stable processes with index $\alpha \geq 1$.

Let us consider the following equation:

$$\frac{\partial v(t,x)}{\partial t} = Av(t,x)$$

$$= b(x)\cdot\nabla v(t,x) + \int_{|u|\leq 1}\{v(t,x+u)-v(t,x)-\nabla v(t,x)\cdot u\} \times \frac{\phi(x,u)}{|u|^{d+\alpha}}du, \quad ^{1)}$$

where b is a R^d-valued function on R^d and ϕ is a non-negative function on $R^d \times S(1)$ [2]. The perturbation of a stable process with index α means that $\phi(x,u) = 1 + \psi(x,u)|u|^{\alpha-\beta}$ for some non-negative function $\psi(x,u)$ and $\alpha > \beta$. Then, the operator A is rewritten as

$$Av(x) = b(x)\cdot\nabla v(x) + \int_{|u|\leq 1}\{v(x+u)-v(x)-\nabla v(x)\cdot u\}\frac{\psi(x,u)}{|u|^{d+\beta}}du$$

$$+ \int_{|u|\leq 1}\{v(x+u)-v(x)-\nabla v(x)\cdot u\}\frac{du}{|u|^{d+\alpha}} .$$

The first term of the right side denotes the drift perturbation, the second denotes the jump perturbation and the last is the main term.

The stochastic equation associated with the above integro-differen-

1) ∇ is the gradient operator and we denote by $\cdot$ the inner product in R^d.

2) $S(1)$ is the unit ball in R^d.

tial equation is given by the following form:

$$(0.1) \quad dx(t) = b(x(t))dt + \int_{y \in R^d} c(x(t),y)q(dt,dy) .$$

Here $q(dt,dy) = p(dt,dy) - E[p(dt,dy)]$ with $p(dt,dy)$ being a Poisson random measure with expectation $E[p(dt,dy)] = \frac{dt\,dy}{|y|^{d+1}}$ and $c(x,y)$ is a function from $R^d \times R^d$ to R^d such that $\chi_1(u)\frac{\phi(x,u)}{|u|^{d+\alpha}}du = \int_{\{y:c(x,y)\in du\}}\frac{dy}{|y|^{d+1}}$ [1]. Now, we shall consider the uniqueness of the solution of the equation (0.1). This problem is related to show the existence and the uniqueness of the Markov process of which the weak generator is the operator A.

M. Motoo [3] has shown the existence of such a Markov process under the assumption of continuity of the functions b,ϕ and uniform positivity of ϕ (but, in case $\alpha = 1$, he needs an additional condition on ϕ). Our attempt is to prove the uniqueness of such a process. However, it is solved only under strict conditions (cf. Assumptions (II) and (III)).

§1. Martingale formulation and stochastic equation

Let $\Omega = D([0,\infty)\to R^d)$ be the space of R^d-valued functions on $[0,\infty)$ which are right continuous and have left-hand limits. Put $x(t) = x(t,\omega) = \omega(t)$ for each $\omega\in\Omega$, let $\mathcal{F}_t$ be the σ-field generated by the functions $x(s)$ for $0\le s\le t$ and let $\mathcal{F} = \mathcal{F}_\infty$. Let S^d be the space of symmetric and non-negative definite (d,d)-matrices. For given a, b and ϕ such as a is a function from R^d to S^d and b, ϕ are same as in the operator A, we define

$$X_\theta(t) = \exp\{\theta\cdot(x(t) - x(0) - \int_0^t b(x(s))ds) - \frac{1}{2}\int_0^t a(x(s))\theta\cdot\theta ds$$
$$- \int_0^t\int_{|u|\le 1}(e^{\theta\cdot u} - 1 - \theta\cdot u)\frac{\phi(x(s),u)}{|u|^{d+\alpha}}dsdu\} .$$

---

1) $\chi_n(u) = \begin{cases} 1 & \text{if } |u|\le n \\ 0 & \text{if } |u|>n . \end{cases}$

Let $\mathcal{P}$ be the space of probability measures P on $(\Omega, \mathcal{F})$ such that $(X_\theta(t), \mathcal{F}_t, P)$ is a martingale for each $\theta \in R^d$ and let $\mathcal{P}_x = \mathcal{P} \cap \{P : P(x(0) = x) = 1\}$. When a = 0, we shall say that the uniqueness of the solution of (0.1) holds if the cardinal number of $\mathcal{P}_x$ is at most one for each $x \in R^d$. This formulation is analogous to the one of Stroock and Varadhan (cf.[8]). It is convenient to the proof of uniqueness that there exist regular conditional probabilities on this space Ω.

Definition A d-dimensional stochastic process $\beta(t)$ (resp. $\ell(t)$) defined on $(\Omega, \mathcal{F}, P)$ is called a Brownian motion (resp. symmetirc Cauchy process) on $(\Omega, \mathcal{F}, P; \mathcal{F}_t)$ if it is $(\mathcal{F}_t)$-adapted and the trajectories are continuous (resp. right continuous and have left-hand limits), and moreover

$$E[e^{i\theta\cdot(\beta(t)-\beta(s))} | \mathcal{F}_s] = e^{-(t-s)|\theta|^2} \quad \text{for } t>s \text{ and } \theta \in R^d.$$

(resp. $E[e^{i\theta\cdot(\ell(t)-\ell(s))} | \mathcal{F}_s] = e^{-c(t-s)|\theta|}$, where $c = \pi^{\frac{d+1}{2}}/\Gamma(\frac{d+1}{2})$)

Now, we will make the following assumption on the coefficients.

Assumption (I) a(x) is either bounded measurable and unifomly elliptic (i.e. there exist positive constants C_1 and C_2 such that $C_1|\theta|^2 \le a(x)\theta\cdot\theta \le C_2|\theta|^2$ for all x, $\theta \in R^d$) or $a(x) \equiv 0$. b(x) is bounded measurable and $\phi(x,u)$ is bounded measurable and uniformly positive (i.e. there exist positive constants C_1 and C_2 such as $C_1 \le \phi(x,u) \le C_2$ for all $x \in R^d$ and $u \in S(1)$).

Proposition 1.1 Under the Assumption (I), the following two statements for a probability measure P on $(\Omega, \mathcal{F})$ are equivalent.

(i) $P \in \mathcal{P}$.

(ii) There exist a Brownian motion $\beta(t)$ and a symmetric Cauchy process $\ell(t)$ on $(\Omega, \mathcal{F}, P; \mathcal{F}_t)$ such that the following equation holds with probability one :

$$(1.1) \quad x(t) = x(0) + \int_0^t \sigma(x(s))d\beta(s) + \int_0^t b(x(s))ds + \int_0^t\int c(x(s),y)q(ds,dy)$$

where $\sigma(x)$ is the symmetric positive square root of $a(x)$, and $q(ds,dy) = p(ds,dy) - E[p(ds,dy)]$, $p(ds,dy)$ is the Poisson random measure induced by $\ell(t)$.

<u>Remark</u> Replacing $a(x(s))$, $b(x(s))$ and $\phi(x(s),u)$ by previsible processes $a(s,\omega)$, $b(s,\omega)$ and $\phi(s,\omega,u)$, respectively, we have a similar result.

<u>Proof</u> For the simplicity, we shall explain the case $d = 1$. Assume that $P \in \mathcal{P}$. Under the Assumption (I), we can show that $y(t) = x(t) - x(0) - \int_0^t b(x(s))ds$ is a square integrable right continuous and quasi-left continuous martingale on $(\Omega, \mathcal{F}, P; \mathcal{F}_t)$. Therefore, $y(t)$ is uniquely decomposed into the sum of a continuous martingale $y_1(t)$ and a martingale $y_2(t)$ which is the compensated sum of jumps of $y(t)$. So, $\langle y_1\rangle_t = \frac{1}{2}\int_0^t a(x(s))ds$

$$\langle y_2\rangle_t = \int_0^t\int_{|u|\leq 1} u^2 \frac{\phi(x(s),u)}{|u|^{1+\alpha}} ds du .$$

Let us set $p_y(ds,du) = \sum_{s<t\leq s+ds} \chi(y(t)-y(t-)\in du)$ and

$$q_y(ds,du) = p_y(ds,du) - \chi_1(u)u^2 \frac{\phi(x(s),u)}{|u|^{1+\alpha}} ds du .$$

Then we can show that

$$y_2(t) = \int_0^t\int u q_y(ds,du).$$

Hence, if we set

$\beta(t) = \int_0^t \sigma^{-1}(x(s))dy_1(s)$ and $\ell(t) = \lim_{n\to\infty} \ell_n(t)$, where

$\ell_n(t) = \int_0^t\int_{|u|\leq 1} c^{-1}(x(s),u)\chi_n(c^{-1}(x(s),u))q_y(ds,du)$,[1] then $\beta(t)$ and $\ell(t)$ satisfy the statement (ii). The implication "(ii)→(i)" can be shown by Ito's formula (cf. [2]).

In case $a = 0$, this proposition gives a precise meaning to the

---

1) $c^{-1}(x,\cdot)$ is the inverse function of $c(x,\cdot)$.

equation (0.1).

§2. Some estimates for symmetric stable processes

Let $\xi(t)$ be a temporally homogeneous Lévy process of which the logarithmic characteristic function is

$$-\frac{1}{2}a\theta\cdot\theta + ib\cdot\theta + \int(e^{i\theta\cdot y} - 1 - \frac{i\theta\cdot y}{1+|y|^2})n(dy),$$

where $a\varepsilon S^d$, $b\varepsilon R^d$ and $n(dy)$ is the Lévy measure of $\xi(t)$. Then we shall say that $\xi(t)$ is a Lévy process with components (a,b,n).

Denote by $T_t^{(\alpha)}$ the semigroup of a Lévy process with components $(0,b,\chi_1(u)\frac{\phi(u)}{|u|^{d+\alpha}}du)$ and by $G_\lambda^{(\alpha)}$ the resolvent of a Lévy process with components $(0,0,\chi_1(u)\frac{du}{|u|^{d+\alpha}})$, where $\phi(u)$ is a bounded measurable and uniformly positive function with the same constants C_1 and C_2 as in Assumption (I).

<u>Proposition 2.1</u> (1) For each $\alpha\varepsilon[1,2)$ and $p\varepsilon(1,\infty)$, there exists a constant C, depending only on α,d,p and C_1, such that

$$||T_t^{(\alpha)}f||^{1)} \leq Ct^{-\frac{d}{\alpha p}}||f||_p \quad \text{for any } f\varepsilon L^p(R^d) \text{ and } t>0.$$

(2) (i) For each $\alpha\varepsilon[1,2)$ and $p\varepsilon(1,\infty)$, there exists a constant C, depending only on α,d,C_1,C_2 and p, such that

$$||D_jG_\lambda^{(\alpha)}f||_p \leq C\lambda^{-(1-\frac{1}{\alpha})}||f||_p \qquad \text{for all } f\varepsilon C_K^\infty(R^d) \text{ and } \lambda>0,$$

where $D_j = \frac{\partial}{\partial x_j}$.

(ii) For each $\alpha\varepsilon(1,2)$, $\alpha'<\alpha$ and $p\varepsilon(1,\infty)$, there exists a constant C, depending only on α,α',d,C_1,C_2 and p, such that

$$||D_jG_\lambda^{(\alpha)}f(\cdot+u) - D_jG_\lambda^{(\alpha)}f(\cdot)||_p \leq C|u|^{\alpha'(1-\frac{1}{\alpha})(1-\frac{\alpha'}{\alpha})}||f||_p$$

for all $f\varepsilon C_K^\infty(R^d)$ and $\lambda>0$.

---

1) $||g|| = \sup|g(x)|$.

<u>Proof</u> Denote by $\hat{T}_t^{(\alpha)}(\xi)$ the Fourier transformation of $T_t^{(\alpha)}(dy)$, then there exist positive constants K and L (depending only on C_1 and d) such that

$$|\hat{T}_t^{(\alpha)}(\xi)| \leq Ke^{-Lt|\xi|^\alpha} \quad \text{for all } t>0 \text{ and } \xi\varepsilon R^d.$$

Hence, $T_t^{(\alpha)}(dy)$ has a continuous density $p_t^{(\alpha)}(y)$ with respect to the Lebesgue measure dy such as

$$||p_t^{(\alpha)}||_q \leq Mt^{-\frac{d}{\alpha p}} \quad \text{for any } t>0,$$

where $q = \frac{p}{p-1}$ and M is a constant depending only on K,L,d and p. So, we obtain the statement (1) by making use of Young's inequality.

It suffices to show the assertion (2) for the resolvent of the symmetric stable process with index α. In case $\alpha>1$, we use subordination of Brownian motion after Motoo [3] in the proof of the inequalities in (2). We need the following estimate for the subordinator, which is obtained by Motoo [3] (cf. [6]). Let $\tau(t)$ be a one-sided stable process with index γ $(0<\gamma<1)$, $\tau(0) = 0$ and $\delta>0$, then we have

$$\text{(2.1)} \quad E[\tau(t)^{-\delta}] \leq Dt^{-\frac{\delta}{\gamma}} \quad \text{for any } t>0,$$

where D is a constant depending only on γ and δ. Using estimates analogous to (2) for Brownian motion and the inequality (2.1) for the subordinator with index $\gamma = \frac{\alpha}{2}$, we have the assertions (2)(i),(ii) for $\alpha\varepsilon(1,2)$. In case $\alpha = 1$, the above method cannot be applied, but the arguments of singular integrals yield the assertion (2)(i) (cf. [5]).

§3. Main result

Before we shall state our result, we further make the following assumptions on the coefficients.

<u>Assumption (II)</u> $\phi(x,u) = 1 + \psi(x,u)|u|^{\alpha-\beta}$, $\alpha>\beta$ and $\psi(x,u)$ is a non-negative bounded measurable function.

<u>Assumption (III)</u> $||b|| + \frac{2\omega_d}{1-\beta}||\psi|| < \frac{1}{Cd}$

for some $p\varepsilon(d,\infty)$, where C is the constant in Proposition 2.1 (2)(i) for $\alpha = 1$, ω_d is the surface area of the unit sphere in R^d and $||b|| = \max ||b_j||$ for $b(x) = (b_1(x), \cdot\cdot\cdot, b_d(x))$.

Then, by virtue of Propositions 1.1 and 2.1, our result is obtained by the same method as [5].

<u>Theorem</u> <u>In case</u> $\alpha\varepsilon(1,2)$, <u>the uniqueness of the solution of</u> (0.1) <u>holds under the Assumptions</u> (I),(II). <u>In case</u> $\alpha = 1$, <u>the uniqueness of the solution of</u> (0.1) <u>also holds under the Assumptions</u> (I), (II) <u>and</u> (III).

In this symposium, Professor B. Grigelionis gave the author a comment that he has carried out more general treatments for the similar problem as Proposition 1.1 (cf. [1]).

References

[1] B.Grigelionis: On Markov property of stochastic processes. Lietuvos Matematikos Rinkinys VIII(1968), 489-502. (Russian)

[2] H.Kunita and S.Watanabe: On square integrable martingales. Nagoya Math. Jour. 30(1967), 209-245.

[3] M.Motoo: On the existence of generalized stable processes. Lecture at Summer seminar of Probability and Statistic Group. (July, 1970)

[4] H.Tanaka, M.Tsuchiya and S.Watanabe: Perturbation of drift-type for Lévy processes. (in prep.)

[5] M.Tsuchiya: On a small drift of Cauchy process. Jour. Math. Kyoto Univ. 10(1970), 475-492.

[6] ----------: On the perturbation by drift for stable processes. Sûriken Kôkyûroku (Theory of Markov processes) 112(1971), 87-119. (Japanese)

[7] D.W.Stroock and S.R.S.Varadhan: Diffusion processes with continuous coefficients. I, II, Comm. Pure Appl. Math. XXII (1969), 345-400 and 479-530.

[8] S.Watanabe: On discontinuous additive functionals and Lévy measures of a Markov process. Jap. Jour. Math. 36(1964), 53-70.

Kanazawa University

Kanazawa Japan

SOME RECENT RESULTS ON PROCESSES WITH STATIONARY INDEPENDENT INCREMENTS

Takesi Watanabe

Let (μ_t) be a convolution semi-group of probability measures on the real line $\underset{\sim}{R}$. The characteristic function of μ_t is of the Lévy-Khintchine formula:

$$\int_{\underset{\sim}{R}} e^{ix\xi}\, \mu_t(dx) = e^{-t\psi(\xi)},$$

$$\psi(\xi) = -ia\xi + \frac{\sigma^2}{2}\xi^2 + \int_{\underset{\sim}{R}\setminus\{0\}} \left(1 - e^{i\xi y} + \frac{i\xi y}{1+y^2}\right)\nu(dy),$$

where $a \in \underset{\sim}{R}$, $\sigma \geq 0$ and ν is the so-called Levy measure. Let $(P_t)_{t\geq 0}$ be the Markov semi-group of convolution kernels defined by $P_t f(x) = \tilde{\mu}_t * f(x)$, where $\tilde{\mu}_t$ stands for the reflection of μ_t at the origin and the symbol $*$, for the convolution. It has been known that, for every smooth function f with compact support, $\lim_{t\to 0} t^{-1}[P_t f - f]$ equals

$$(0.1)\quad Af(x) := af'(x) + \frac{\sigma^2}{2} f''(x) + \int_{\underset{\sim}{R}\setminus\{0\}} \left[f(x+y) - f(x) - \frac{y}{1+y^2} f'(x)\right]\nu(dy).$$

This operator and the equation

$$(0.2)\qquad (\lambda - A)u = f, \qquad \lambda \geq 0$$

will play fundamental roles in the study of the semi-group (P_t) and its associated Markov process (= Lévy process).

The main purpose of this note is to analyze the equation

(0.2) from a potential-theoretic point of view.

1. Applications of the Schwartz distribution theory

In this section we will summarize those results of the previous paper [4] and give further information. We follow the notation and terminology of [4].

For a function f [similarly, measure or distribution], $\tilde{f}$ denotes the reflection at the origin. $\mathcal{F}(f)$ is the Fourier transform $\int_{R} e^{ix\xi} f(x)dx$. B^{o} stands for the space of bounded measurable functions, C_o for the space of continuous functions with $f(\pm\infty) = 0$ and M_b for the space of bounded signed measures. (D_{L^p}), $1 \leq p \leq \infty$, denotes the space of C^{∞}-functions ϕ such that the n-th derivative $\phi^{(n)}$ is in L^p for every $n \geq 0$. A sequence $(\phi_j)_{j\geq 1}$ converges to ϕ in (D_{L^p}) if $\| \phi_j^{(n)} - \phi^{(n)} \|_p \rightarrow 0$ for every $n \geq 0$. (D_{L^∞}) is also denoted by B. Space $\dot{B}$ is defined by the subset of B consisting of those functions ϕ such that $\phi^{(n)} \in C_o$ for every $n \geq 0$. L. Schwartz [3 ; Chap. VI, § 8] introduced the spaces of distributions (D'_{L^p}) by

$$(1.1) \qquad (D'_{L^p}) := (D_{L^{p'}})' \quad \text{with} \quad \frac{1}{p'} = 1 - \frac{1}{p} \quad \text{for} \quad 1 < p \leq \infty$$

$$:= (\dot{B})' \quad \text{for} \quad p = 1.$$

Consider the formal adjoint $\tilde{A}$ of A :

$$\tilde{A}f(x) := -af'(x) + \frac{\sigma^2}{2} f''(x) + \int_{R\setminus\{0\}} [f(x-y) - f(x) + \frac{y}{1+y^2} f'(x)]\nu(dy).$$

From the basic results of Schwartz, one can extend the operator A to the one on $(\underline{\underline{D}}'_{L^\infty})$. In fact, A is the convolution operator by $\tilde{A}^o \in (\underline{\underline{D}}'_{L^1})$, where $(\tilde{A}^o, \phi) = \tilde{A}\phi(0)$.

Let $U_\lambda = \int_0^\infty e^{-\lambda t} P_t dt$ and $U_\lambda^o = \int_0^\infty e^{-\lambda t} \mu_t dt$ for $\lambda > 0$. Obviously, $U_\lambda f = \tilde{U}_\lambda^o * f$. Similarly as in A, one can extend P_t and U_λ to those on $(\underline{\underline{D}}'_{L^\infty})$. The following theorem shows that space $(\underline{\underline{D}}'_{L^\infty})$ is the most natural space of distributions as the domain of the operators P_t, U_λ and A.

THEOREM 1.1. *Let* $f \in (\underline{\underline{D}}'_{L^\infty})$. *Then the equation*

$$(1.2) \qquad (\lambda - A)u = f, \quad \lambda > 0$$

has the unique solution $u = U_\lambda f$ *in* $(\underline{\underline{D}}'_{L^\infty})$.

One next considers the infinitesimal generator of $(P_t)_{t \geq 0}$ on various Banach spaces.

THEOREM 1.2. *Let* L *be either of the Banach spaces* L^p, $1 \leq p < \infty$, *or* $\underline{\underline{C}}_o$. *Then* $(P_t)_{t \geq 0}$ *defines a strongly continuous, contraction semi-group of operators on each* L. *The infinitesimal generator* $\dot{A}$ *of* (P_t) *over* L *is given by*

$$(1.3) \qquad D[\dot{A}\ ;\ L] = \{f \in L\ ;\ Af \in L\}$$

$$(1.4) \qquad \dot{A}f = Af \quad \text{for} \quad f \in D[\dot{A}\ ;\ L].$$

This theorem is proved in [4]. A similar result is valid for the spaces $(\underline{\underline{D}}'_{L^p})$.

THEOREM 1.3. <u>For every</u> $1 \leq p \leq \infty$, $M = (\underline{\underline{D}}'_{L^p})$ <u>is a complete</u>(*) <u>topological vector space</u>. (P_t) <u>defines an equi-continuous semi-group of class</u> (C_o) <u>on each</u> M(*). <u>The infinitesimal generator</u> $\dot{A}$ <u>is given by</u>

(1.5) $\quad D[\dot{A}\ ;\ M] = M,$

(1.6) $\quad \dot{A}f = Af \quad$ <u>for</u> $\quad f \in M.$

The proof of this theorem is due to S. Sugitani. For fixing the idea, consider the case $p = \infty$. For each bounded set of $(\underline{\underline{D}}_{L^1})$, define the semi-norm q_B on $(\underline{\underline{D}}'_{L^\infty})$ by $q_B(f) = \sup_{\phi \in B} |(f', \phi)|$. The family (q_B) defines the strong topology of $(\underline{\underline{D}}'_{L^\infty})$. Suppose that (f_i) is a Cauchy directed family in $(\underline{\underline{D}}'_{L^\infty})$. Define a linear functional f by $(f, \phi) = \lim_i (f_i, \phi)$ for each $\phi \in (\underline{\underline{D}}_{L^1})$. It is easy to see that $\lim_i q_B(f - f_i) = 0$ and $q_B(f) < \infty$ for every bounded set B of $(\underline{\underline{D}}_{L^1})$, which implies that $f \in (\underline{\underline{D}}'_{L^\infty})$ and f_i converges to f in $(\underline{\underline{D}}'_{L^\infty})$. Hence $(\underline{\underline{D}}'_{L^\infty})$ is complete.

To prove the second assertion, let us first note that the dual semi-group $(\tilde{P}_t)$ of (P_t)(**) is an equi-continuous semi-group of class (C_o) on $(\underline{\underline{D}}_{L^1})$ and its infinitesimal generator is the operator $\tilde{A}$ on $(\underline{\underline{D}}_{L^1})$. This is an easy consequence of Theorem 1.2 (see T.3.5 (a) of [4]). Since the generator is closed, $\tilde{A}$ is a bounded operator on $(\underline{\underline{D}}_{L^1})$. For each $m \geq 0$, define $p_m(\phi) = \sum_{n=0}^{m} \| \phi^{(m)} \|_1$.

(*) See K. Yosida [6 ; p.105 and 234] for definition.

(**) $(\tilde{P}_t)$ is the Markov semi-group associated with $(\tilde{\mu}_t)$: $\tilde{P}_t f = \mu_t * f$.

One has

$$p_m\Big(\frac{1}{t}\int_0^t \widetilde{P}_s\phi \cdot ds\Big) \leq \frac{1}{t}\int_0^t p_m(\widetilde{P}_s\phi)ds$$

$$\leq \frac{1}{t}\int_0^t p_m(\phi)ds = p_m(\phi),$$

so that, for each bounded set B in $(\underline{\underline{D}}_{L^1})$, the set $B' = \{\frac{1}{t}\int_0^t \widetilde{P}_s\phi \cdot ds\ ;\ \phi \in B,\ t > 0\}$ is also a bounded set. Take any bounded set B in $(\underline{\underline{D}}_{L^1})$. One has to show that

(1.7) $\quad q_B(P_t f - f) \longrightarrow 0 \quad$ for every $f \in (\underline{\underline{D}}'_{L^\infty})$,

and there exists a bounded set B' such that

(1.8) $\quad q_B(P_t f) \leq q_{B'}(f) \quad$ for every $f \in (D'_{L^\infty})$ and $t \geq 0$.

Choose c_f and m such that $|(P_t f - f, \phi)| = |(f, \widetilde{P}_t\phi - \phi)|$ $c_f\, p_m(\widetilde{P}_t\phi - \phi)$ for every $\phi \in (\underline{\underline{D}}_{L^1})$. Since $\widetilde{P}_t\phi - \phi = \int_0^t \widetilde{P}_s\widetilde{A}\phi \cdot ds$ for every $\phi \in (\underline{\underline{D}}_{L^1})$ and $\widetilde{A}$ is bounded, the set $B' = \{\frac{1}{t}(\widetilde{P}_t\phi - \phi);\ \phi \in B,\ t > 0\}$ is a bounded set by virtue of the result of the preceding paragraph. Hence $p_m(\frac{1}{t}(\widetilde{P}_t\phi - \phi))$ is bounded (say, by c_B) for $\phi \in B$. One concludes that

$$q_B(P_t f - f) \leq c_f \sup_{\phi\in B} p_m(\widetilde{P}_t\phi - \phi)$$

$$\leq c_f\, c_B\ \ t \rightarrow 0 \quad (t \rightarrow 0).$$

To show (1.8), let $B' = \{\widetilde{P}_t\phi\ ;\ \phi \in B,\ t \geq 0\}$. Since $p_m(\widetilde{P}_t\phi) \leq p_m(\phi)$, B' is bounded. One has

$$q_B(P_t f) = \sup_{\phi \in B} |(P_t f, \phi)| = \sup_{\phi \in B} |(f, \tilde{P}_t \phi)|$$

$$\leq \sup_{\psi \in B'} |(f, \psi)| = q_{B'}(f).$$

The last assertion is proved in the same way as in Theorem 1.2. Observe only that A makes each M invariant by virtue of the result of Schwartz.

Let G be an open set of $\underset{\sim}{R}$. One next will discuss the equation

(1.9) $\qquad (\lambda - A)\, u = 0 \quad$ on $\quad G \quad$ (in the distribution sense).

Consider first the case $G = \underset{\sim}{R}$. If $\lambda > 0$, then $u = 0$ obviously. To discuss the solution of

(1.10) $\qquad A\, u = 0,$

one introduces the following definition. A measure μ is said to be <u>arithmetic</u> if μ is supported in $\{n\delta\ ;\ n = 0, \pm 1, \cdots\}$ for some $\delta > 0$, and the smallest number of such δ is called the <u>period</u> of μ. The operator A $(\neq 0)$ is said to be <u>arithmetic</u> if the Levy measure ν is arithmetic (hence, a finite measure) and

$$\sigma^2 = a - \int_{\underset{\sim}{R}\setminus\{0\}} \frac{y}{1+y^2}\, \nu(dy) = 0.$$

This condition is equivalent to the condition that each μ_t is arithmetic with the same period as ν. By Theorem 1.1 and a theorem of Choquet-Deny, one has the following result.

THEOREM 1.4. <u>Suppose that</u> $A \neq 0$. <u>Then the solution</u> u <u>in</u> $(\underline{\underline{D}}'_{L^\infty})$ <u>of</u> (1.10) <u>must be a constant if</u> A <u>is non-arithmetic, and a periodic distribution with the period of</u> ν <u>if</u> A <u>is arithmetic</u>.

This theorem implies

THEOREM 1.5. <u>Suppose that</u> $A \neq 0$. <u>Let</u> L <u>be either of the spaces</u> L^p, $1 \leq p < \infty$, $\underline{\underline{C}}_o$ <u>or</u> $(\underline{\underline{D}}'_{L^p})$, $1 \leq p \leq \infty$ <u>and</u> $\dot{A}$, <u>the infinitesimal generator of</u> (P_t) <u>over</u> L <u>and</u> $R(\dot{A})$, <u>the range of</u> $\dot{A}$.

(a) <u>If</u> $L = L^1$ <u>or</u> $(\underline{\underline{D}}'_{L^1})$, $\dot{A}$ <u>is one-to-one but</u> $R(\dot{A})$ <u>is not dense in</u> L.

(b) <u>If</u> $L = (\underline{\underline{D}}'_{L^\infty})$, $\dot{A}$ <u>is not one-to-one</u>.

(c) <u>If</u> $L = \underline{\underline{C}}_o$, L^p <u>or</u> $(\underline{\underline{D}}'_{L^p})$ <u>for</u> $1 < p < \infty$, $\dot{A}$ <u>is one-to-one and</u> $R(\dot{A})$ <u>is dense in</u> L.

We prove the case $L = \underline{\underline{C}}_o$ and $(\underline{\underline{D}}'_{L^p})$ of (c). The rest is proved in the same manner. $\dot{A}f = 0$ implies that $Af = 0$, so that f is a constant or periodic distribution. Hence $f = 0$ if $L = \underline{\underline{C}}_o$ or $(\underline{\underline{D}}'_{L^p})$ for $p \neq \infty$. Note that $\underline{\underline{C}}'_o = \underline{\underline{M}}_b$. Assume that $(\mu, \dot{A}f) = 0$ for every $f \in D(\dot{A})$, where $\mu \in \underline{\underline{M}}_b$. Since $D(\dot{A}) \supset (\underline{\underline{D}})$, $(\mu, A\phi) = (\widetilde{A}\mu, \phi) = 0$ for $\phi \in (\underline{\underline{D}})$. Hence μ is a solution of $\widetilde{A}\mu = 0$. By Theorem 1.4, $\mu = 0$, because $\mu \in \underline{\underline{M}}_b$. If $L = (\underline{\underline{D}}'_{L^p})$, $1 < p < \infty$, note that $(\underline{\underline{D}}'_{L^p})' = (\underline{\underline{D}}_{L^q})$, where $\frac{1}{q} = 1 - \frac{1}{p}$. Suppose that $(\dot{A}f, \phi) = 0$ for every $f \in (\underline{\underline{D}}'_{L^p})$, where $\phi \in (\underline{\underline{D}}_{L^q})$. Since $(\dot{A}f, \phi) = (f, \widetilde{A}\phi) = 0$, $\widetilde{A}\phi = 0$. By the same reason as above one gets $\phi = 0$.

For the spaces L in (c), one can define the potential operator $\dot{U}$ in the sense of K. Yosida [7] by

(1.11) $\quad \dot{U} = -\dot{A}^{-1}, \; D(\dot{U}) = R(\dot{A}).$

This fact was proved by K. Sato [2] when $L = \underline{\underline{C}}_o$ by another method. He calculated the kernels defining the potential operators for various special convolution semi-groups. The same kernels act as the potential operators over every L of the preceding theorem. From those general results by Yosida and Sato, the strong limit of $U_\lambda f$ as $\lambda \longrightarrow 0$ exists for (and only for) $f \in D(\dot{U})$ and equals $\dot{U}f$. Moreover, this condition is equivalent to the condition that $\lambda U_\lambda f \xrightarrow[\lambda \to 0]{} 0$ in L for every $f \in L$.

Let us now consider the standard Markov process X having (P_t) as its transition function:

(1.12) $\quad X = (\Omega, \underline{\underline{F}}, \underline{\underline{F}}_t, X(t), \underset{\sim}{P}^x).$

Let $H_B^\lambda(x, \cdot)$ be the λ-hitting measure of the set B with respect to the process $X(t)$ starting at x. All objects (the resolvent, standard process, hitting measure etc.) with respect to the semi-group $(\widetilde{P}_t)$ are denoted with "$\sim$" over the associated letters. The following is immediate from T.4.4 of [4].

LEMMA 1.6. <u>Let</u> $\lambda > 0$. <u>Let</u> u <u>be the difference of two bounded</u> λ-<u>excessive functions</u>. <u>Then</u>

(1.13) $\quad \mu := (\lambda - A)u \in (\underline{\underline{D}}'_{L^\infty})$

<u>is a signed measure and</u> $u \cdot dx = \mu \widetilde{U}_\lambda$. <u>Moreover, for every open set</u> G,

(1.14) $\quad H_G^\lambda \, u \cdot dx = \mu \widetilde{H}_G^\lambda \widetilde{U}_\lambda .$

Note that every function $u \in (\underline{\underline{C}}_b^2)$ is the difference of two bounded λ-excessive functions, since $u = U_\lambda f_1 - U_\lambda f_2$, where $f_1 [f_2]$ denotes the positive [negative] part of $f = (\lambda - A)u$.

Let $\lambda \geq 0$. Let G be an open set of $\underset{\sim}{R}$. A finite function u(*) is said to be λ-harmonic on G if, for every compact set K in G,

$$(1.15) \qquad u = H^\lambda_{K^c} u \qquad (K^c = \underset{\sim}{R} \setminus K).$$

If equality holds dx-almost surely for each K, one says that u is almost λ-harmonic on G. A function u, not taking the value $-\infty$, is said to be λ-superharmonic [resp. almost λ-superharmonic] on G if, for each K as above,

$$(1.16) \qquad u \geq H^\lambda_{K^c} u \text{ [resp. "}\geq\text{" holds dx-almost surely].}$$

By virtue of Lemma 1.6 one can extend T.5.2 of [4] as follows. (Hereafter we always assume that $A \neq 0$.)

THEOREM 1.7. For each $\lambda \geq 0$, define H_λ be the family of all functions u such that, for each $\lambda' > \lambda$, u is written as the difference of two bounded λ'-excessive functions. Then a function $u \in \underline{\underline{H}}_\lambda$ is almost λ-harmonic on G if and only if

$$(1.17) \qquad (\lambda - A)u = 0 \quad \text{on } G.$$

Note that $\underline{\underline{H}}_o$ contains $(\underline{\underline{C}}^2_b)$.

The proof is quite similar to that of T.5.2 of [4].

For a smooth function u the above theorem is extended as follows.

(*) In [4] we assumed that u is λ-excessive. We here delete this assumption.

THEOREM 1.8. *Let* $\lambda \geq 0$ *and* $u \in (\underline{\underline{C}}_b^2)$. *Then* u *is almost* λ-*superharmonic on* G *if and only if*

$$(1.18) \qquad (\lambda - A)u \geq 0 \quad \text{on } G.$$

Remark. The proof shows that, if G is compact, (1.18) implies that

$$(1.19) \qquad u \geq H^{\lambda}_{\bar{G}^c}\, u \qquad dx\text{-almost surely.}$$

One first proves the "only if" part. Take a countable base $\{V_n\}$ of G consisting of compact neighbourhoods. Then, for almost all x, $u(x) \geq H^{\lambda}_{V_n^c}\, u(x)$ holds for every n. Fix any such point $x \in G$. From Dynkin's formula, one gets

$$(\lambda - A)u(x) = \lim_{\substack{V \downarrow \{x\} \\ V \in \{V_n\}}} \frac{u(x) - H^{\lambda}_{V^c}\, u(x)}{\underset{\sim}{E}^x\, [T_{V^c}]} \geq 0,$$

where T_{V^c} is the hitting time of V^c.

To prove the "if" part, consider first the case $\lambda > 0$. Let K be a compact set in G and $f = (\lambda - A)u$. Let $f_1 = f|_{\bar{G}}$ and $f_2 = f|_{\bar{G}^c}$. Then $u = U_\lambda f = U_\lambda f_1 - U_\lambda f_2$. Since $f_1 \geq 0$, $u_1 = U_\lambda f_1$ is λ-excessive, so that $u_1 \geq H^{\lambda}_{K^c}\, u_1$. Since $f_2 = 0$ on $\bar{G}$, the measure $\mu_2 = f_2 \cdot dx$ satisfies $\mu_2 = \mu_2 \widetilde{H}^{\lambda}_{K^c}$. By Lemma 1.6, $u_2 = U_\lambda f_2$ satisfies

$$u_2 \cdot dx = \mu_2 \widetilde{U}_\lambda = \mu_2 \widetilde{H}^{\lambda}_{K^c}\, U_\lambda = H^{\lambda}_{K^c}\, u_2 \cdot dx,$$

so that $u_2 = H^{\lambda}_{K^c}\, u_2$ dx-almost surely. Next consider the case

$\lambda = 0$. Let $f = -Au$. Since $f \geq 0$ on G, $U_\lambda f \geq H^\lambda_{K^c}\, U_\lambda f$ dx-almost surely by the above result. Hence, for $\phi \in \underline{\underline{C}}^+_c$,

$$([u - H^\lambda_{K^c}\, u],\ \phi) \geq (\lambda U_\lambda u - \lambda H^\lambda_{K^c} U_\lambda u,\ \phi) = (u,\ \lambda \widetilde{U}_\lambda \phi - \lambda \widetilde{H}^\lambda_{K^c} \widetilde{U}_\lambda \phi).$$

The right side converges to zero as $\lambda \longrightarrow 0$, for its modulus is dominated by

$$(|u|\ ,\ I_K \cdot \lambda \widetilde{U}_\lambda \phi) = (\ \lambda U_\lambda (|u|\ I_K),\ \phi) \longrightarrow 0.^{(*)}$$

The "if" part of Theorem 1.8 is valid for more general functions.

THEOREM 1.9. <u>In the following two cases, a function</u> u <u>satisfying</u> (1.18) <u>is almost</u> λ-<u>superharmonic</u>.

(a) u <u>is a bounded continuous function</u>.

(b) U^o_λ <u>is absolutely continuous for</u> dx, <u>and</u> $u \in \underline{\underline{H}}_\lambda$.

Let us prove assertion (a). Take a compact set $K \subset G$ and choose an open set G' such that $K \subset G' \subset \bar{G}' \subset G$. Take a molifier $\rho_\varepsilon(x)$. That is, $\rho_1(x)$ is a positive, C^∞-function, supported in $|x| \leq 1$, such that $\int \rho_1(x)\, dx = 1$ and $\rho_\varepsilon(x)$ is defined by $\varepsilon^{-1} \rho_1(\varepsilon^{-1} x)$. If ε is small enough, $u_\varepsilon = u * \rho_\varepsilon \in (\underline{\underline{C}}^2_b)$ satisfies (1.18) on G'. Hence, by Theorem 1.8, $u_\varepsilon \geq H^\lambda_{K^c}\, u_\varepsilon$ dx-almost surely. Letting $\varepsilon \longrightarrow 0$, $u_\varepsilon(x)$ converges to $u(x)$ for every x. Hence, $u \geq H^\lambda_{K^c}\, u$ dx-almost surely.

We omit the proof of (b).

It looks rather difficult to extend the "only if" part of

(*) This follows from the fact that (P_t) on $\underline{\underline{C}}_o$ admits the potential operator (see the paragraph following (1.11)).

Theorem 1.8 to a more general class of functions. One gives a result in this connection.

Let B be a Borel set of $\underset{\sim}{R}$ and ζ^{o}, the penetration time W_{B^c}. Define

$$X^{o}(t) = X(t) \quad \text{for} \quad t < \zeta^{o}$$
$$= \Delta \quad \text{for} \quad t \geq \zeta^{o}.$$

One can define λ-harmonic functions for the process $X^{o}(t)$ in the same way as for $X(t)$.

THEOREM 1.10. <u>Let</u> u <u>be a bounded function which is</u> λ-<u>excessive for the process</u> $X^{o}(t)$ <u>and</u> λ-<u>harmonic for</u> $X^{o}(t)$ <u>on an open set</u> G <u>included in</u> int-B (= <u>the interior of</u> B). <u>Then</u> u <u>satisfies</u> (1.17).

The proof is omitted.

One gives an application of the above theorem. Consider the sets $G = [0, 1]$ and $A = [1, \infty)$. Let T be the hitting time for set G^c. Consider the function $u(x) = E^x[e^{-\lambda T}; X(T) \in A]$. But this is the λ-hitting probability of the set $A^o = (1, \infty)$ with respect to the process $X^{o}(t)$ on the set $B = [0, \infty)$. Hence, u is λ-excessive for $X^{o}(t)$ and λ-harmonic for $X^{o}(t)$ on int-G. Thus u is a solution of

$$(1.20) \qquad (\lambda - A)u = 0 \quad \text{on} \quad (0, 1).$$

2. The renewal theorem and its converse

In this section we will discuss the potential kernel U of a transient semi-group $(P_t)_{t \geq 0}$. As is well known, the complete principle of the maximum and the renewal theorem are valid for this kernel U. The main result here is to prove the converse the precise description of which is given in Theorem 2.4.

The convolution semi-group (μ_t) is said to be <u>transient</u> if $U^o = \lim\uparrow_{\lambda\to 0} U^o_\lambda$ is a Radon measure. The kernel $Uf = \tilde{U}^o * f$ is called the potential kernel of (P_t). U is different from the potential operator $\dot{U}$. But one can prove that $Uf = \dot{U}f$ for every $f \in D(\dot{U}) \cap (\underline{\underline{D}}'_{L^1})$.

LEMMA 2.1. <u>Let</u> G <u>be a bounded open set and let</u> $f \in (\underline{\underline{C}}^2_b)$ <u>satisfy</u>

(2.1) $(\lambda - A)f \leq 0$ <u>on</u> G.

<u>When</u> $\lambda > 0$, <u>one further assumes that</u> $\sup f > 0$. <u>Then there exists some</u> $y \in G^c$ <u>such that</u>

(2.2) $f(y) \geq \sup_{x\in G} f(x)$.

By the remark for Theorem 1.8, one has

(2.3) $f(x) \leq H^\lambda_{\bar{G}^c} f(x)$ dx-almost surely.

If $\sup_G f > f(y)$ for every $y \in G^c$, choose $x_o \in G$ such that $f(x_o) > f(y)$ for every $y \in G^c$ and (2.3) is valid and if $\lambda > 0$, $f(x_o) \geq 0$. This leads to the contradiction:

$$f(x_o) \leq H^\lambda_{\bar{G}^c} f(x_o) < f(x_o) H^\lambda_{\bar{G}^c} 1(x_o) \leq f(x_o),$$

noting that $H_{\bar{G}^c} 1(x_o) = 1$ if $A \neq 0$.

As a consequence it follows that $U_\lambda (\lambda > 0)$ satisfies the following complete principle of the maximum: for each $f \in (\underline{\underline{D}})$,

(2.4) $[U_\lambda f \leq 1]_{S_{f^+}} \Longmapsto U_\lambda f \leq 1$ on $\underset{\sim}{R}$,

where $S_{f^+} = \{x \mid f(x) > 0\}$. In fact, suppose that $[U_\lambda f \leq 1]_{S_f^+}$. Let $g = U_\lambda f$ and $G = \{x \mid g(x) > 1\}$. If $G \neq \phi$, G is bounded by virtue of $g \in \underline{\underline{C}}_o$ and $G \cap S_{f^+} = \phi$. Hence $(\lambda - A)g = f \leq 0$ on G. By the lemma, there exists some $y \in G^c$ such that $g(y) \geq \sup_G g > 1$. This contradicts to the definition of G.

THEOREM 2.2. (a) Suppose that (μ_t) is transient. Then, (i) $U^o \in (\underline{\underline{D}}'_{L^\infty})$, (ii) U satisfies the complete principle of the maximum and (iii) for every $f \in (\underline{\underline{D}}'_{L^1})^+$, Uf is the positive smallest solution of

(2.5) $$-Au = f, \quad u \in (\underline{\underline{D}}'_{L^\infty}).$$

(b) A convolution semi-group (μ_t) is transient if and only if the equation

(2.6) $$-Af = \delta$$

admits a nonnegative solution $f \in (\underline{\underline{D}}'_{L^\infty})$. In this case, $\tilde{U}^o$ gives the smallest such solution.

This fact is well known. The present form of the theorem was given by C. Herz [1].

Herz also gave a neat proof of the renewal theorem.

THEOREM 2.3. Suppose that (μ_t) is transient and non-arithmetic. Then there exists $\ell^\pm \geq 0$ such that, for every $f \in (\underline{\underline{C}}_c^+)$,

$$(2.7)\qquad \lim_{x\to\infty} Uf(x) = \ell^+ \int f(y)dy, \qquad \lim_{x\to-\infty} Uf(x) = \ell^- \int f(y)dy.$$

The numbers ℓ^+, ℓ^- *are given as follows.*

(a) *If* $\int_{|y|>1} |y|\,\nu\,(dy) = \infty$, *then* $\ell^+ = \ell^- = 0$.

(b) *If* $\int_{|y|>1} |y|\,\nu\,(dy) < \infty$, $Ax(0) = m$ *exists and it is not zero. If* $m > 0$, $\ell^+ = 0$ *and* $\ell^- = \frac{1}{m}$. *If* $m < 0$, *then* $\ell^+ = -\frac{1}{m}$ *and* $\ell^- = 0$.

Let $\mathbf{C}_r$ be the collection of functions $f \in \mathbf{C}_b$ such that $\lim_{x\to\infty} f(x) = 0$ and $\lim_{x\to-\infty} f(x)$ exists and let $\mathbf{C}_\ell = \{f \in \mathbf{C}_b ;\ \check{f} \in \mathbf{C}_r\}$. By the renewal theorem, either U maps $\mathbf{C}_c$ into $\mathbf{C}_r$ or it does into $\mathbf{C}_\ell$. Such a kernel is said to be $\mathbf{C}_r$-continuous [resp. $\mathbf{C}_\ell$-continuous].

In conclusion, the potential kernel U of a transient, non-arithmetic convolution semi-group (μ_t) satisfies the following four conditions.

(α) U is a convolution kernel: $Uf = \tilde{U}^o * f$.

(β) $U^o(\mathbb{R}) = \infty$.

(γ) U is either $\mathbf{C}_r$-continuous or $\mathbf{C}_\ell$-continuous.

(δ) U satisfies the complete principle of the maximum.

One now comes to the converse.

THEOREM 2.4. *If a kernel* U *satisfies the above four conditions, then it is the potential kernel of a unique convolution semi-group which is transient and non-arithmetic.*

This theorem follows from its discrete parameter analogue.

THEOREM 2.5. ([5]) Suppose that a kernel G satisfies (α), (β), (γ) and Meyer's muximum principle: for $f \in (\underline{\underline{D}})$,

$$[Gf \leq 1]_{S_f^+} \Longrightarrow Gf + f^- \leq 1 \quad \text{on } \underset{\sim}{R},$$

where $f^- = \max(-f, 0)$. Then there exists a probability measure μ such that

$$G = \sum_{n \geq 0} N^n \quad \text{with} \quad Nf = \tilde{\mu} * f. \tag{2.8}$$

Assuming this theorem we prove Theorem 2.4. For each $\lambda > 0$, set $G_\lambda = I + \lambda U$. It is easy to verify that G_λ satisfies the conditions of Theorem 2.5. Hence there exists a Markov convolution kernel N_λ such that $G_\lambda = \sum_{n \geq 0} N_\lambda^n$. Set $U_\lambda = \lambda^{-1} N_\lambda$. It follows that $(U_\lambda)_{\lambda>0}$ is a Markov resolvent of convolution kernels and $U = \lim_{\lambda \to 0} U_\lambda$. Let $U_\lambda f = \tilde{\nu}_\lambda * f$. Then (ν_λ) satisfies the equation

$$\nu_\lambda - \nu_\alpha + (\lambda - \alpha)\nu_\lambda * \nu_\alpha = 0 \quad (\lambda, \alpha > 0). \tag{2.9}$$

Taking the Fourier transform one gets

$$\phi_\lambda - \phi_\alpha + (\lambda - \alpha)\phi_\lambda \phi_\alpha = 0. \tag{2.10}$$

The set $A = \{\xi \ ; \ \phi_\lambda(\xi) \neq 0\}$ is independent of λ. Fixing α and letting $\lambda \longrightarrow \infty$, one gets $\lim_{\lambda \to \infty} \lambda \phi_\lambda \phi_\alpha = \phi_\alpha$. Hence

$$\lim_{\lambda \to \infty} \lambda \phi_\lambda(\xi) = \phi(\xi) = \begin{cases} 1 & (\xi \in A), \\ 0 & (\xi \in A^c). \end{cases} \tag{2.11}$$

Since convergence is uniform near $\xi = 0$, ϕ must be a characteristic function of probability measure. Hence $\phi(\xi) \equiv 1$, which proves that $\lambda\nu_\lambda$ converges vaguely to δ as $\lambda \to \infty$.

One has proved that $(U_\lambda)_{\lambda>0}$ is a resolvent on $\underline{\underline{C}}_o$ and satisfies $\lambda U_\lambda f \longrightarrow f (\lambda \longrightarrow \infty)$ for every $f \in (\underline{\underline{C}}_o)$. Hence, by the Hille-Yosida theorem, there exists a strongly continuous, Markov semi-group (P_t) on $\underline{\underline{C}}_o$ such that $U_\lambda = \int_0^\infty e^{-\lambda t} P_t dt$. By the uniqueness it follows that each P_t is a convolution kernel by a probability measure μ_t.

References

[1] C.S. Herz: Les théorèmes de renouvellement, Ann. Inst. Fourier, Grenoble, 15 (1963), 169-188.

[2] K. Sato: Potential operators for Markov processes, to appear in Proc. of the Fifth Berkeley Symp. on Prob. and Stat.

[3] L. Schwartz: Théorie des distributions, Hermann, Paris, 1966.

[4] T. Watanabe: Some potential theory of processes with stationary independent increments by means of the Schwartz distribution theory, J. Math. Soc. Japan, 24, (1972), 213-231.

[5] T. Watanabe: On the maximum principle for elementary kernels (in Japanese), Maximum principles in potential theory, Kôkyuroku No.146, R.I.M.S. Kyoto University (1972).

[6] K. Yosida: Functional Analysis, Springer, Berlin, 1965.

[7] K. Yosida: The existence of the potential operator associated with an equi-continuous semi-group of class (C_o), Studia Math., 31 (1968), 531-533.

Department of Mathematics
Faculty of Science
Osaka University
Toyonaka, Osaka, JAPAN

EXTENSIONS OF MEASURES. STOCHASTIC EQUATIONS

M.P. Yershov

Introduction

1. Necessary and sufficient conditions are well-known for a non-negative additive set function on a field to be extendable to a non-negative σ-additive set function (measure) on the σ-field generated by the original field. Namely, for the existence of at least one σ-additive extension, it is necessary and sufficient that the original set function would be σ-additive on the field. Moreover, this extension, if it exists, is unique.

Now, any measure on a σ-field is trivially extendable to the completion of this σ-field with respect to the given measure.

Finally, any measure on an arbitrary σ-field is known (see Section 5, No. 4) to be extendable to a measure on the σ-field generated by the original σ-field and an arbitrary finite system of sets. Moreover, if the finite system of sets is not contained in the completion of the original σ-field with respect to the original measure, the extension is essentially non-unique.

Extension of measures from a σ-field to the σ-field generated by the original σ-field and an infinite system of sets is, principally different and much more complicated. There is no general result in this direction we know of.

One of the most important and interesting "particular" cases is that of an underlying topological space with the σ-field of Borel sets and its sub-σ-field a measure to be extended is given on.

It turns out that, without any additional assumptions about topological properties of the space, the extension problem can be unsolvable if the original σ-field "differs" from the Borel σ-field by an essentially infinite system of sets. The first negative example, as far as we know, was constructed by E. Szpilrajn-Marczewski in 1938 [1].

In this example, the given measure was non-separable x. In 1946 E. Szpilrajn-Marczewski [2] posed the following problem (Problem 6 in [2]):

<u>A separable measure on a sub-σ-field of the Borel σ-field of a topological space being given, whether it is extendable to a measure on the Borel σ-field?</u>

We will show in Section 1 (Theorem 1.1) that, under some assumptions about the underlying space, E. Szpilrajn-Marczewski's problem has a positive solution. We will also consider there the question of uniqueness of an extension.

<u>2.</u> To the problem of extension we were led by the following problem of probability theory.

Let $(X, \mathcal{X})$ and $(Y, \mathcal{Y})$ be measurable spaces, $F : (X, \mathcal{X}) \to (Y, \mathcal{Y})$ a measurable mapping and ν a probability measure on $(Y, \mathcal{Y})$. <u>Does there exist a probability measure</u> μ <u>on</u> $(X, \mathcal{X})$ <u>such that</u>

$$\mu(F^{-1}(B)) = \nu(B) \tag{0.1}$$

<u>for any</u> $B \in \mathcal{Y}$?

---

x A measure μ on a σ-field is called <u>separable</u> if so is the σ-field obtained from the original one by the factorization with respect to the class of μ-null sets and considered as a metric space with the factor-μ-measure of symmetrical differences as the distance.

If the answer is "yes", it is natural to call μ a <u>solution of the stochastic equation</u>

(0.2) $$F \circ \mu = \nu$$

where $F \circ \mu$ denotes the measure in (Y , $\mathcal{Y}$) defined by (0.1).

This problem can also be formulated as follows:

<u>when do there exist a probability space and random elements ξ and η on it with values in measurable spaces (X , $\mathcal{X}$) and (Y, $\mathcal{Y}$) respectively such that the distribution of η is ν and, with probability 1 , F (ξ) = η</u> ?

For the existence of a solution of the stochastic equation (0.2), it is obviously necessary that

(0.3) $$\nu^* (F(X))=1$$

where ν^* is the outer measure on subsets of Y corresponding to ($\mathcal{Y}$, ν). It can be easily shown that, under the condition (0.3), the equality (0.1) correctly defines a measure μ on the σ-field F^{-1} ($\mathcal{Y}$). Hence, in order to solve the stochastic equation (0.2), one only needs to anyhow extend μ from F^{-1} ($\mathcal{Y}$) to a measure on $\mathcal{X}$. Thus the problem of solving the stochastic equation (0.2) is reduced to that of extension of measures, and, in the case of topological spaces X and Y with the Borel σ-fields $\mathcal{X}$ and $\mathcal{Y}$ respectively (it is just the most interesting and important case), we may apply the main theorem on extension (Theorem 1.1).

Results on existence and uniqueness of stochastic equations are obtained in Section 1. Section 2 contains auxiliary theorems.

1. Extension of measures. Stochastic equations

<u>1.</u> Let X be a topological space, $\mathcal{X}$ be the σ-field of Borel sets in X , a σ-field $\mathcal{X}_0 \subseteq \mathcal{X}$ and μ be an arbitrary measure on $\mathcal{X}_0$.

Theorem 1.1. Let X be a Hausdorf regular space with countable base, and there exist a countable system of sets E_1, E_2, ... $\in \mathcal{X}$ such that [x]

(1.1)
$$\mathcal{E} \subseteq \mathcal{X}_0 \subseteq \mathcal{E}^{\mu}$$

where $\mathcal{E}$ is the σ-field generated by E_1, E_2, ... Then μ can be extended from $\mathcal{X}_0$ to $\mathcal{X}$.

Remark 1. Condition (1.1) is, obviously, equivalent to separability of μ (see the footnote in Introduction). For us (1.1) is a little more convenient.

Remark 2. For the sake of simplicity we use in the proof a weakened result due to M. Sion. To make it easier to the reader to see how our statement can be strengthened, is the only reason why we do not simply write that X should be a separable metric space (under the assumptions of the theorem, X is known to be metrizable).

Proof. Note first that it suffices to prove the theorem in case when

(1.1')
$$\mathcal{E} = \mathcal{X}_0$$

is satisfied instead of (1.1). In fact, if we show that the restriction of μ on $\mathcal{E}$ has an extension to $\mathcal{X}$, then the latter, in view of (1.1), is well known necessarily to coincide with μ on $\mathcal{X}_0$, i.e. it is also an extension of μ from $\mathcal{X}_0$ to $\mathcal{X}$.

Thus we will assume (1.1') to be satisfied.

Let c(x) be the characteristic function of the sequence (E_n) :

$$c(x) = 2\sum_{n=1}^{\infty} 1_{E_n}(x)/3^n.$$

By Theorem on the characteristic function (Section 2, No. 1), $\mathcal{X}_0$=

---

[x] For an arbitrary σ-field $\mathcal{E}$ with a measure μ on it, $\mathcal{E}^{\mu}$ stands for the completion of $\mathcal{E}$ relative μ.

$= c^{-1}(\mathcal{C})$, where $\mathcal{C}$ is the σ-field of Borel subsets of Cantor's discontinuum C with countable open base $\mathcal{O}$ induced by the usual metric of the real line.

Let $C_0 = C \cap c(X)$, $\mathcal{O}_0 = (\, O \cap c(X),\ O \in \mathcal{O}\,)$ and $\mathcal{C}_0$ be the σ-field of subsets of C_0 generated by $\mathcal{O}_0$. It is obvious that

(i) C_0 can be considered as a Hausdorf space with countable base $\mathcal{O}_0$,

(ii) c maps X <u>onto</u> C_0

and

(iii) $\mathcal{X}_0 = c^{-1}(\mathcal{C}_0)$.

Define, on the measurable space $(C_0, \mathcal{C}_0)$, the measure ν :

$$\nu(B) = \mu(c^{-1}(B)), \quad B \in \mathcal{C}_0 ,$$

and let $\mathcal{C}_0^{\nu}$ be the completion of $\mathcal{C}_0$ relative ν .

Since $c : (X, \mathcal{X}) \longrightarrow (C_0, \mathcal{C}_0)$ is a Borel function, by Theorem on the graph of a Borel function (Section 2, No. 2) in view of (i),

$$\Gamma = \bigcup_{x \in X} (x, c(x))$$

is a Borel set in the direct procuct of toplological spaces X and C_0

By Corollary of Theorem on uniformization of sets (Section 2, No. 3), there exists a function $f : C_0 \longrightarrow X$ such that

(1.2) $$cf(y) = y \quad \forall y \in C_0$$

and $f^{-1}(\mathcal{X}) \subseteq \mathcal{C}_0^{\nu}$.

Now define

$$\tilde{\mu}(E) = \nu(f^{-1}(E)), \quad E \in \mathcal{X}.$$

It is obvious that $\tilde{\mu}$ is a measure on $\mathcal{X}$. If $E \in \mathcal{X}_0$, there exists $B \in \mathcal{C}_0$ such that $E = c^{-1}(B)$, and therefore, by (1.2),

$$\tilde{\mu}(E) = \tilde{\mu}(c^{-1}(B)) = \nu(f^{-1}c^{-1}(B)) = \nu(B) ;$$

on the other hand, by the definition

$$\nu(B) = \mu(c^{-1}(B)) = \mu(E) .$$

Thus $\tilde{\mu}$ is an extension of μ to $\mathcal{X}$. Q.E.D.

Remark 3. The reader can observe that the extension $\tilde{\mu}$ we have got possesses an additional property: the completion of $\mathcal{X}_0$ with respect to ($\mathcal{X}, \tilde{\mu}$) contains $\mathcal{X}$. In the (not yet stable) terminology of stochastic differential equation theory, this corresponds to the fact that, under the condition (0.3), we always have a strong sense solution.

Theorem 1.2. Let the conditions of Theorem 1.1 be satisfied, and let there exist $E \in \mathcal{X} \setminus \mathcal{X}_0^{\mu}$ where $\mathcal{X}_0^{\mu}$ is the completion of $\mathcal{X}_0$ relative μ . Then there exist at least continuum different extensions of μ from $\mathcal{X}_0$ to $\mathcal{X}$.

Proof. If $E \in \mathcal{X} \setminus \mathcal{X}_0^{\mu}$, then, for any number m between the inner and outer measures of E corresponding to $(\mathcal{X}_0, \mu)$, one can easily (cf. Theorem on extension in Section 2, No. 4) construct an extension μ_m of μ to the σ-field generated by $\mathcal{X}_0$ and E such that $\mu_m(E) = m$. It only remains, using Theorem 1.1, to anyhow extend μ_m to $\mathcal{X}$. Q.E.D.

2. Let X be a topological space, $\mathcal{X}$ the σ-field of Borel sets in X , (Y , $\mathcal{Y}$) an arbitrary measurable space, F a measurable mapping (X , $\mathcal{X}$) $\longrightarrow$ (Y , $\mathcal{Y}$) and ν a probability measure on (Y , $\mathcal{Y}$) . Consider the stochastic equation (0.2):

$$F \circ \mu = \nu .$$

Theorem 1.3. Let X be a Hausdorf regular space with countable base, there exist a countable system of subsets of Y generating

$\mathcal{Y}$ and the condition (0.3):

$$\nu^*(F(X)) = 1$$

be satisfied.

Then the stochastic equation (0.2) has at least one solution.

A solution of (0.2) is unique iff $(F^{-1}(\mathcal{Y}))^{\mu} \supseteq \mathcal{X}$ where the measure μ on $F^{-1}(\mathcal{Y})$ is defined by

$$(1.3) \qquad \mu(F^{-1}(B)) = \nu(B), \quad B \in \mathcal{Y}.$$

Remark. The reader will easily see that, instead of $\mathcal{Y}$ being countably generated, we could suppose ν to be separable.

Proof. Existence. As it has been noted in Introduction, condition (0.3) enables to correctly define a measure μ on $(X, F^{-1}(\mathcal{Y}))$ by the equality (1.3). By the assumption, $F^{-1}(\mathcal{Y}) \subseteq \mathcal{X}$ and, since $\mathcal{Y}$ is countably generated, so is σ-field $F^{-1}(\mathcal{Y})$. By Theorem 1.1 there exists an extension $\tilde{\mu}$ of μ from $F^{-1}(\mathcal{Y})$ to $\mathcal{X}$. By the definition, any extension of μ from $F^{-1}(\mathcal{Y})$ to $\mathcal{X}$ is a solution of the stochastic equation (0.2).

Uniqueness of the solution follows immediately from Theorem 1.2.

Q.E.D.

2. Auxiliary theorems

1. The characteristic function of a sequence of sets. Let X be an arbitrary space, (E_n) a sequence of its subsets, and $\mathcal{E}$ the σ-field generated by (E_n). Let $1_E(x)$ denote the characteristic function of $E \subseteq X$.

The function

$$c(x) = 2\sum_{n=1}^{\infty} 1_{E_n}(x)/\,3^n$$

is called the characteristic function of the sequence (E_n).

Theorem on characteristic function. (E. Szpilrajn-Marczewski[3]) The function $c(x)$ maps X into the Cantor discontinuum C, and $\mathcal{E} = c^{-1}(\mathcal{C})$ where $\mathcal{C}$ is the σ-field of Borel sets in C with topology induced by the usual metric on the real line.

2. The graph of a Borel function. Let X and Y be topological spaces with the Borel σ-fields $\mathcal{X}$ and $\mathcal{Y}$ respectively. The following fact is well-known; we prove it here being unable to indicate a direct reference.

Theorem on the graph of a Borel function. Let g be a measurable mapping of $(X, \mathcal{X})$ into $(Y, \mathcal{Y})$. If Y is a Hausdorf space with countable base, the graph of g

$$\Gamma = \bigcup_{x \in X} (x, g(x))$$

is a Borel set in the product space $X \times Y$.

Proof. Consider the mapping $G : X \times Y \quad Y \times Y$ defined as

$$G(x, y) = (g(x), y).$$

Denote by D the diagonal of $Y \times Y$,

$$D = \bigcup_{y \in Y} (y, y).$$

Since Y is Hausdorf, D is known to be closed in $Y \times Y$ (cf. Theorem 2, No. III, Section 15 of [4]). It follows that

$$(Y \times Y) \setminus D = \bigcup_h (Q_h \times R_h)$$

where Q_h, R_h are sets from the open base of Y. The latter being countable, we can suppose h to run through a countable set.

Obviously $\Gamma = G^{-1}(D)$. We have

$$\Gamma = G^{-1}(D) = G^{-1}((Y \times Y) \setminus (\bigcup_h (Q_h \times R_h)))$$
$$= (G^{-1}(Y \times Y)) \setminus (\bigcup_h G^{-1}(Q_h \times R_h))$$
$$= (X \times Y) \setminus (\bigcup_h (g^{-1}(Q_h) \times R_h)).$$

Now, the union above being countable, g being a Borel function, and the Borel σ-field in $X \times Y$ containing the product $\mathcal{X} \times \mathcal{Y}$, it follows that Γ is Borel in $X \times Y$.

3. Uniformization of sets in topological spaces. Let X and Y be topological spaces, Q be a set in $X \times Y$ and P be its projection onto Y. Let $\mathcal{X}$ and $\mathcal{Y}$ be the Borel σ-fields of X and Y respectively.

A uniformization of Q is any function $f : P \to X$ such that $(f(y), y) \in Q$ for any $y \in P$.

Now we state an important result due to M. Sion (in a convenient for us form) which plays a decisive role in our Theorem 1.1. Note that our statement is weaker than that of M. Sion's paper [5]. With some modifications in the proof using the result of [5] and the choice axiom, one gets a slightly more general theorem than Theorem 1.1.

Theorem on uniformization of sets. (M. Sion [5]). Let X be a Hausdorf regular space with countable base, Y be a Hausdorf space, and Q be a Borel set in $X \times Y$. Then there exists a uniformization f such that

$$f^{-1}(E) \in \bigcap_{\nu \in \mathfrak{N}} \mathcal{Y}_0^{\nu}$$

for any Borel $E \subseteq X$ where $\mathcal{Y}_0^{\nu}$ is the completion relative ν of the σ-field $\mathcal{Y}_0$ of subsets of P induced by $\mathcal{Y}$ and $\mathfrak{N}$ is the set of all probability measures on $\mathcal{Y}_0$.

Let F be a mapping of X into Y and $\Gamma = \bigcup_{x \in X}(x, F(x))$ be its graph.

<u>Corollary</u>. Let X be a Hausdorf regular space with countable base, Y be a Hausdorf space and Γ be a Borel set in $X \times Y$. Then there exists such a function $f : F(X) \to X$ that

(i) $F(f(y)) = y \qquad \forall y \in F(X)$,

(ii) $f^{-1}(E) \in \bigcap_{\nu \in \mathfrak{N}} \mathcal{Y}_0^{\nu} \qquad \forall E \in \mathcal{X}$,

$\mathcal{Y}_0^{\nu}$ and $\mathfrak{N}$ being the same as in the preceeding theorem.

<u>Proof</u>. By Theorem on uniformization of sets there exist a function $f : F(X) \to X$ such that (ii) is satisfied and,

$$\forall y \in F(X), \quad (f(y), y) \in \Gamma,$$

i.e., for any $y \in F(X)$ there exists an $x \in X$ such that

$$f(y) = x, \quad F(x) = y$$

which implies (i).

The function f may be considered as a measurable branch of the multivalued function F^{-1}.

4. <u>Extension of a measure to a system of non-measurable sets</u>.

Let $(X, \mathcal{X}_0)$ be a measurable space with a measure μ, $Y \subset X$ and $\mathcal{X}$ be the σ-field generated by $\mathcal{X}_0$ and Y.

<u>Theorem on extension of measures</u>. (J. Łos, E. Szpilrajn-Marczewski [6]). <u>The measure</u> μ <u>has an extension from</u> $\mathcal{X}_0$ <u>to</u> $\mathcal{X}$.

<u>Proof</u> is simple. In an explicite form an extension may be constructed as follows. For any $E_0 \in \mathcal{X}_0$, put

$$\mu_\theta(E_0 \cap Y) = \mu_*(E_0 \cap Y) + \theta\,(\mu^*(E_0 \cap Y) - \mu_*(E_0 \cap Y))$$

(here μ^* and μ_* are respectively the outer and inner measures relative $(\mathcal{X}_0, \mu)$, θ is any number, $0 \leq \theta \leq 1$; in the case

$\mu^*(Y) = \infty$, we put $\mu_\theta(E_o \cap Y) = \mu_*(E_o \cap Y)$ or $\mu^*(E_o \cap Y))$ and define

$$\mu_\theta(E_o \cap Y) = \mu(E_o) - \mu_\theta(E_o \cap Y) .$$

Now, any set $E \in \mathcal{X}$ may be represented in the form

$$E = (E_1 \cap Y) \cup (E_2 \setminus Y) , \; E_i \in \mathcal{X}_o .$$

Put

$$\mu_\theta(E) = \mu_\theta(E_1 \cap Y) + \mu_\theta(E_2 \setminus Y) .$$

It is easy to check that this definition is consistent and, for any θ , μ_θ is a measure on $\mathcal{X}$ coinciding with μ on $\mathcal{X}_o$. Q.E.D.

Directly from the definition of μ_θ , one sees that, if $Y \notin \mathcal{X}_o$ (i.e. $\mu^*(Y) \neq \mu_*(Y)$) and $\mu^*(Y) < \infty$, then $\mu_\theta(Y)$, $0 \leq \theta \leq 1$, runs through the continuum $[\mu_*(Y) , \mu^*(Y)]$.

References

1. E.Szpilrajn, Ensembles indépendants et mesures non séparables, C.R.Acad.Sci. Paris, 207 (1938), 768 - 770.
2. Э.Шпильрайн, К проблематике теории меры, Успехи матем.наук, I, 2(I2)(I946),I79-I88.
3. E.Szpilrajn, The caracteristic function of a sequence of sets and some of its applications, Fund.Math., 31 (1938), 207 - 223.
4. К.Куратовский, Топология, том I, М., изд-во "Мир", I966.

5. Maurice Sion, On uniformization of sets in topological spaces, Trans.Amer.Math.Soc., 96, 2 (1960), 237 - 245.
6. J.Łoś, E.Marczewski, Extensions of measure, Fund.Math., 36(1949), 267 - 276.

Steklov Mathematical Institute
of the Academy of Sciences of the USSR, Moscow

ON STOCHASTIC EQUATIONS

M.P. Yershov

In the first part of this note, we propose a new set-up of the problem of finding a "diffusion" with given "drift" and "diffusion" co-efficients. In section 2 a simple example is considered.

1. In P.A. Meyer [1], [2], the following fundamental result is contained:

Let $X_t, t \geq 0$, be a continuous square integrable martingale on a probability space $(\Omega, \mathcal{F}, P)$ relative to an increasing family $\{\mathcal{F}_t\}$ of sub-σ-algebras of $\mathcal{F}$. Then there exists a unique continuous increasing $\mathcal{F}_t$-well adapted process $\langle X\rangle_t$ such that $\langle X\rangle_0 \equiv 0$ and $\forall s<t$

$$E\{(X_t - X_s)^2 \mid \mathcal{F}_s\} = E\{\langle X\rangle_t - \langle X\rangle_s \mid \mathcal{F}_s\}.$$

Taking $\mathcal{F}_t = \sigma\{X_s, s \leq t\}$, one can show that there exists such a functional $A(t, x)$ on $[0,\infty[\times C$ that [x]

(i) $A(0, x) \equiv 0$,

(ii) $\forall x \in C$, $A(t, x)$ is non-decreasing and continuous in t,

(iii) $\forall t$, $A(t, x)$ is C_t-measurable,

(iv) $P\{A(t, X) \neq \langle X\rangle_t\} = 0$.

A natural <u>Problem I</u> arises:

<u>How wide is the class $\mathcal{O}\!\mathcal{l}$ of functionals $A(t, x)$ satisfying the conditions (i) - (iii) for which there exist continuous martingales X_t with respect to $\sigma\{X_s, s \leq t\}$ such that (iv) holds true?</u>

<u>Problem II</u>:

---

[x] C is the space of continuous functions $x = \{x_t, 0 \leq t < \infty\}$, $C_t = \sigma\{x_s, s \leq t\}$.

Let $A(\cdot,\cdot) \in \mathfrak{A}$; when do the measures in $(C, \mathcal{C}_\infty)$ corresponding to any martingales $\{X_t, \sigma\{X_s, s \le t\}\}$ with "diffusion" determined by $A(t,X)$ coincide?

It is easily seen that, in case when

$$A(t,x) = \int_0^t \delta^2(s,x)\,ds$$

where $\delta(s,x)$ is a measurable functional on $[0,\infty[\times C$ such that, $\forall s$, $\delta(s,x)$ is $\mathcal{C}_s$-measurable, Problems I and II turn into problems of existence and uniqueness in distribution (cf. [3]) of a solution of the equation

$$dX_t = \delta(t,X)\,dW_t$$

(W_t is a Wiener process).

Problems I and II may be naturally generalized to the case of continuous local quasi-martingales (i.e. processes representable as sums of continuous local martingales and processes with sample paths of bounded variation) with values in $\mathbb{R}^n$. Moreover, one may set analogous problems for processes continuous from the right.

2. In this section, we solve Problems I and II for functionals $A(t,x)$ of the form

$$A(t,x) = \int_0^t \delta^2(s,x_s)\,ds$$

where $\delta(t,y)$ is supposed (in contrast to usual conditions of regularity in y) to be sufficiently smooth in t , while its dependence on y may be arbitrary.

Theorem. Let $\delta(t,y)$ be a measurable function on $[0,\infty[\times \mathbb{R}^1$ such that

(a) $\forall t,y$

$$\frac{1}{K} \le \delta^2(t,y) \le K < \infty,$$

(b) $\forall y \in \mathbb{R}^1$

$$|\delta(t,y) - \delta(s,y)| \leq K(t-s) \qquad \forall s<t.$$

<u>Then there exists (on some probability space) a continuous martingale</u> X_t <u>such that</u>

$$\langle X \rangle_t \equiv \int_0^t \delta^2(s, X_s)\,ds. \tag{1}$$

<u>All the continuous martingales</u> X_t <u>satisfying</u> (1) <u>have the same distribution in</u> $(C, \mathcal{C}_\infty)$.

<u>Proof.</u> Let x be a fixed function from C. Consider the equation (relative to $\tau(t)$)

$$\tau(t) = \int_0^t \delta^{-2}(\tau(s), x_s)\,ds. \tag{2}$$

Making use of the conditions (a) and (b), by the sequential approximation method and iteration, one can easily establish the existence of a unique solution of (2); moreover, from the construction procedure, one easily sees that $\tau(t) = \tau(t,x)$, for each fixed t, is $\mathcal{C}_t$-measurable. In view of (a), this solution is a strictly monotone function of t for each $x \in C$, and let $\tau^{-1}(t,x)$ be the inverse function.

Let W_t, $t \geq 0$, be a standard Wiener process. It is obvious that $\tau^{-1}(t,W)$ is a random time change relative to the family of σ-algebras $\mathcal{F}_t^W = \sigma\{W_s, s \leq t\}$ (i.e., $\forall u$, $\{\tau^{-1}(t,W) > u \in \mathcal{F}_u^W\}$). Therefore the process $X_t = W_{\tau^{-1}(t,W)}$ is a (continuous) martingale with respect to $\mathcal{F}_t^X = \sigma\{X_s, s \leq t\}$. Using the formula for the inverse function derivative, from (2) one gets:

$$\tau^{-1}(t,W) = \int_0^t \delta^2(s, W_{\tau^{-1}(s,W)})\,ds,$$

and hence

$$E\{(X_t - X_s)^2 \mid \mathcal{F}_s^X\} = E\{\int_s^t \delta^2(u, X_u)\,du \mid \mathcal{F}_s^X\},$$

i.e.

$$\langle X \rangle_t \equiv \int_0^t \delta^2(s, X_s)\,ds.$$

Thus X_t is a solution of Problem I for the functional $A(t,x) = \int_0^t \delta^2(s,x_s)\,ds$. The uniqueness of the distribution of X follows from the fact that the inverse function $A^{-1}(t,x)$ satisfies the equation

$$A^{-1}(t,X) = \int_0^t \delta^{-2}\left(A^{-1}(s,X), X_{A^{-1}(s,X)}\right) ds$$

from which it is uniquely determined as a measurable function of $W_s = X_{A^{-1}(s,X)}$, $s \leq t$, and hence X_t may be obtained from the Wiener process W_t by a random time change $A(t,X)$ relative to the family $\{\mathcal{F}_t^W\}$ $(\{A(u,X)>t\} = \{A^{-1}(t,X)<u\} \in \mathcal{F}_t^W)$.

The proof is completed.

The theorem can be generalized to the case when $A(t,x) = \int_0^t \delta^2(s,x)\,ds$ where $\delta(s,x)$ is now a functional on $[0,\infty[\times C$ dependent on the whole "past" $\{x_u, u \leq s\}$ satisfying some "invariance" conditions relative to non-anticipative time change.

References

1. P.A. Meyer, A decomposition theorem for supermartingales, Illinois J. Math., 6,2(1962), 193-205.
2. P.A. Meyer, Decompositions of supermartingales; The uniqueness theorem, Illinois J. Math., 7,1(1963), 1-17.
3. Toshio Yamada, Shinzo Watanabe, On the uniqueness of solutions of stochastic differential equations, J.Math. Kyoto Univ., 11, 1(1971), 155-167.

Steklov Mathematical Institute
of the Academy of Sciences of
the USSR
Moscow

EXACTNESS OF AN APPROXIMATION IN THE CENTRAL LIMIT THEOREM

V.M.Zolotarev

1. Problem and results. Consider a sum of independent and identically distributed random variables

$$\xi_1 + \xi_2 + \dots + \xi_n$$

with $E\xi_j = 0$ and $E\xi_j^2 = 1$. Denote by F the distribution function of summands ξ_j, and put

$$F_n(x) = F^{n*}(x\sqrt{n}) \quad \text{and} \quad \Phi(x) = \frac{1}{\sqrt{2\pi}} \int_{-\infty}^{x} \exp(-t^2/2)\,dt .$$

It is well known that as $n \to \infty$

$$\rho_n = \rho(F_n, \Phi) = \sup_x |F_n(x) - \Phi(x)| \to 0 .$$

An estimation of the real order of distance ρ_n was one of old problems in probability theory. The famous result due to A.Berry and C.Esseen is:

$$\rho_n \leqslant c\beta n^{-1/2} \quad \text{for all } n \geqslant 1, \tag{1}$$

where c is a positive constant (it will have the same meaning in the sequel) and $\beta = E|\xi_j|^3$. There are many works concerning the problem, but almost all of them consider one of two main reasons of closeness of F_n to Φ only. Namely the effect of summing uniformly small random summands. The influence of the second reason firstly was noted by P.Lévy [1]. The nature of the second reason may be shown by the following simple inequality:

$$\rho_n = \sup_x |F^{n*}(x) - \Phi^{n*}(x)| \leqslant n\rho(F, \Phi) .$$

That is the distance ρ_n between F_n and Φ can be small if F is sufficient close to Φ in the same metric. The first result of a new kind was proved by the author [2]. Namely,

$$\rho_n \leqslant c \left(\nu n^{-1/2} \right)^{1/4}, \tag{2}$$

where the variable

$$\nu = \int |x|^3 \left| d\left(F - \Phi\right) \right| \quad \text{; (it is always } \nu \leqslant c\beta \text{),}$$

can express, in many cases, a closeness of function F to Φ. After some time V.Paulauskas proved in [3] so good variant of (2) that (1) became its corollary. His inequality is the following:

$$\rho_n \leqslant c \max \left(\nu, \nu^{1/4} \right) n^{-1/2}. \tag{3}$$

Unfortunately the variable ν as a measure of a closeness is not always convenient, For instance, if the distribution F is singular, then

$$\nu = \beta + 4/\sqrt{2\pi}.$$

That is why the author replaced in [4] the characteristics of the type

$$\nu(r) = \int |x|^r \left| d\left(U - V\right) \right|,$$

where $r \geqslant 0$ and U, V are distribution functions, by

$$\varkappa(r) = r \int |x|^{r-1} \left| U - V \right| dx ,$$

where $r \geqslant 0$ and $\varkappa(0)$ is defined as the limit when $r \to 0+0$. Characteristics $\varkappa(r)$ are free from the defect noted above for characteristics $\nu(r)$ and, in addition for all $r \geqslant 0$, we have

$$\varkappa(r) \leqslant \nu(r). \tag{4}$$

It was said in [5] that in inequality (3) we may replace the variable $\nu = \nu(3)$ by the more convenient variable $\varkappa = \varkappa(3)$, that is the following inequality is true

$$\rho_n \leqslant c \max\left(\varkappa, \varkappa^{1/4}\right) n^{-1/2}$$

Very specifical form of the right hand side of this inequality sets some questions about appearance of the exponent 1/4, about possibilities to use other characteristics instead of ν and $\varkappa$ etc. We shall discuss the general picture of the problem in the end of this paper. Now we formulate the main result. Define before two new characteristics

$$\varkappa_0 = \int \max\left(1, 3x^2\right) |F - \Phi| \, dx ,$$

$$\nu_0 = \int \max\left(1, |x|^3\right) |d(F - \Phi)| .$$

<u>Theorem.</u> <u>For all</u> $n \geqslant 1$ <u>the following upper estimates of the distance</u> ρ_n <u>are true</u>

(5) $$\rho_n \leqslant c \max\left(\varkappa, \varkappa^{\frac{n}{3n+1}}\right) n^{-1/2},$$

(6) $$\rho_n \leqslant c \max\left(\varkappa_0, \varkappa_0^{\frac{n}{n+1}}\right) n^{-1/2};$$

(7) $$\rho_n \leqslant c\, \nu_0\, n^{-1/2}.$$

We shall obtain the inequalities of the theorem as corollaries of the following two lemmas. Denote by f_n, f and g the characteristic functions corresponding to the distribution functions F_n, F

and Φ respectively. Define also for any positive θ and s such that $0 \leq s \leq 3$ the following variable

$$p = \begin{cases} 1 & \text{, when } \theta \geq 1 \\ \min\left(1, \frac{n}{sn+1}\right) & \text{, when } \theta < 1 . \end{cases}$$

Lemma 1. Let, for all real t

$$\omega(t) = |f(t) - g(t)| \leq \theta \min\left(|t|^s, |t|^3\right).$$

Then, for all $n \geq 2$,

$$\rho_n \leq c \max\left(\theta, \theta^p\right) n^{-1/2} .$$

In the case $n = 1$ each of the following inequalities is true

$$\rho_1 \leq c \left(1 + \frac{1}{s}\right) \max\left(\theta, \theta^p\right),$$

$$\rho_1 \leq c \max\left(1, -\log\theta\right) \max\left(\theta, \theta^p\right)$$

Lemma 2. For all real t, the following inequalities are true

$$\omega(t) \leq \varkappa |t|^3,$$

$$\omega(t) \leq \varkappa_0 \min\left(|t|, |t|^3\right),$$

$$\omega(t) \leq \nu_0 \min\left(1, |t|^3\right).$$

2. Proofs. In view of the well known Berry-Esseen's inequality, we have

$$\rho_n \leqslant c \min_{T>0} \max \left\{ \int_0^T |f_n - g| \frac{dt}{t} , \frac{1}{T} \right\}.$$

Transform the right hand side of the inequality putting

$$t = \sqrt{2nx} \qquad \text{and} \qquad X = T^2/2n .$$

We have

$$(8) \qquad \rho_n \leqslant c \min_{X>0} \max \left\{ \mathcal{I}(X) , (nX)^{-1/2} \right\},$$

where

$$\mathcal{I}(X) = \int_0^X \Delta_n(x) \frac{dx}{x} \qquad \text{and} \qquad \Delta_n(x) = |f^n(\sqrt{2x}) - g^n(\sqrt{2x})|.$$

We obtain an estimate of the integral $\mathcal{I}(X)$ by using the following simple inequality which is true for all positive integers n and any pair of complex numbers u, v :

$$|u^n - v^n| \leqslant 2(n-1)|v||u-v|\left(|u| + |u-v|\right)^{n-2} + |u-v|^n.$$

Denote by Δ the function $\Delta_1(t)$. Since $g(\sqrt{2t}) = e^{-t}$, we have from the last inequality

$$(9) \quad \mathcal{I}(X) \leqslant 2(n-1) \int_0^X \Delta e^{-t} \left(e^{-t} + \Delta\right)^{n-2} \frac{dt}{t} + \int_0^X \Delta^n \frac{dt}{t}.$$

By the condition of the lemma 1 we have for all $t \geqslant 0$

$$\Delta \leqslant 3\,\theta \min\left(t^{5/2}, t^{3/2}\right) = 3\,\theta\,\lambda(t).$$

It is not difficult to see that we may use a somewhat more simple condition $\Delta \leqslant \theta\lambda$, for in this case we need only change the constant c in the statement of the lemma 1. Hence we can restrict ourselves by consideration of upper estimates of the following integrals

$$L_1 = 2(n-1)\theta \int_0^X \lambda e^{-t} \left(e^{-t} + \theta\lambda\right)^{n-2} \frac{dt}{t},$$

$$L_2 = \theta^n \int_0^X \lambda^n \frac{dt}{t}.$$

Let

$$c_1 = 1/2, \quad c_2 = (1 - e^{-c_1})/c_1 \approx 0.79, \quad c_3 = c_2 - c_1 \approx 0.29,$$

and put

$$X = c_1^5 \theta^{-2p}, \qquad X_1 = \min(c_1, X).$$

It can be shown that in the interval $0 \leqslant t \leqslant X$

$$\theta\lambda(t) \leqslant c_1 \min(c_1, t). \tag{10}$$

In fact, if $X_1 = X$, that is $\theta \geqslant c_1^{2/p}$, then

$$\theta\lambda(t)t^{-1} \leqslant \theta t^{1/2} \leqslant \theta X^{1/2} = c_1^{5/2}\theta^{1-p} < c_1,$$

since we have always $p \leqslant 1$, and the case $p < 1$ corresponds to values $\theta < 1$. On the other hand, if $X_1 = c_1$, that is $\theta \leqslant c_1^{2/p}$, then in the interval $0 \leqslant t \leqslant X_1$ we have

$$\theta\lambda(t)t^{-1} \leqslant \theta X_1^{1/2} \leqslant c_1^{2/p + 1/2} \leqslant c_1^{5/2} < c_1.$$

In the interval $X_1 \leqslant t \leqslant X$ we find, using the condition $\theta \leqslant c_1^{2/p}$ and the inequality $n(1 - sp) \geqslant p > 0$ which is implied by it, that

$$\theta\lambda(t)c_1^{-1} \leqslant \theta t^{s/2} c_1^{-1} \leqslant \theta X^{s/2} c_1^{-1} = c_1^{\frac{5}{2}s-1}\theta^{1-sp} \leqslant$$

$$\leqslant c_1^{\frac{5}{2}s-1+(1-sp)2/p} \leqslant c_1 .$$

Consequently, inequality (10) is true. Farther, we have from (10) that, for all $t \geqslant 0$, $t \leqslant X$:

$$e^{-t} + \theta\lambda \leqslant \exp\{e^{-t} - 1 + \theta\lambda\} \leqslant \exp\{-c_2 \min(c_1, t) + \theta\lambda\} \leqslant$$

$$\leqslant \exp\{-c_2 \min(c_1,t) + c_1 \min(c_1,t)\} = \exp\{-c_3 \min(c_1,t)\}$$

Now we obtain for any $n \geqslant 2$

$$L_1 \leqslant 2(n-1)\theta \int_0^X \lambda \exp\{-t-(n-2)c_3 \min(c_1,t)\} \frac{dt}{t} \leqslant$$

$$\leqslant 2n\theta \int_0^{X_1} t^{1/2} \exp\{-t[1+(n-2)c_3]\}\, dt +$$

$$+ 2n\theta \int_{X_1}^{X} \lambda t^{-1} \exp\{-t-(n-2)c_1c_3\}\, dt \leqslant \sqrt{\pi}\,[1+(3/e)^{3/2}]\, c_3^{-3/2}\, \theta n^{-1/2}.$$

Since the integral $L_1 = 0$ for $n = 1$, we obtain for all $n \geqslant 1$

(11) $$L_1 \leqslant 25\, \theta\, n^{-1/2}.$$

Estimate now the integral L_2. We have

$$L_2 = \theta^n \int_0^{X_1} \lambda^n \frac{dt}{t} + \theta^n \int_{X_1}^{X} \lambda^n \frac{dt}{t} = V_1 + V_2 .$$

$$V_1 \leqslant \theta^n \int_0^{X_1} t^{\frac{3}{2}n-1} dt = \frac{2}{3n}\left(\theta X_1^{3/2}\right)^n \leqslant$$

$$\leqslant \begin{cases} \frac{1}{n}\left(\theta^{1-3p} c_1^{15/2}\right)^n & \text{, if } X_1 = X \text{, that is } \theta \geqslant c_1^{2/p} \\ \frac{1}{n}\left(\theta\, c_1^{3/2}\right)^n & \text{, if } X_1 = c_1 \text{, that is } \theta \leqslant c_1^{2/p} . \end{cases}$$

Since the number $1-3p$ can be positive for $\theta < 1$ only, we obtain for all $n \geqslant 1$

$$(12) \qquad V_1 \leqslant \frac{1}{n} c_1^{3/2}\, \theta \leqslant \theta n^{-1/2} .$$

Estimating V_2, we can consider the case $X_1 = c_1$ only, because, in the opposite case $X_1 = X$, we have $V_2 = 0$. Thus

$$(13) \qquad V_2 \leqslant \theta^n \int_{c_1}^{X} t^{\frac{sn}{2}-1} dt = \frac{2}{sn}\theta^n \left\{ X^{sn/2} - c_1^{sn/2} \right\} .$$

Since in the considering case $\theta < 1$ and $n(1-sp) \geqslant p$ always, we have using (13), that for $s \geqslant 1/3$ and for all $n \geqslant 1$

$$(14) \qquad V_2 \leqslant \frac{6}{n} c_1^{\frac{5}{2}sn} \theta^{(1-sp)n} \leqslant 6\theta^p n^{-1/2} .$$

If $s < 1/3$, then we have obviously, for $n \geqslant 2$; $p = 1$ and $\frac{1}{2}(n-1-sn) \geqslant 1/6$. Using now the representation $\theta = c_1^2 Y^{-1/2}$ where $Y = X/c_1 \geqslant 1$ and the estimate,

$$V_2 \leqslant \theta^n X^{sn/2} \log (X/c_1) ,$$

which is implied from (13), we obtain for $n \geqslant 2$

$$(15) \qquad V_2 \leqslant \theta\, c_1^{2(n-1)+\frac{5}{2}n}\, Y^{-\frac{1}{2}(n-1-sn)} \log Y \leqslant$$

$$\leqslant \theta c_1^2 Y^{-1/6} \log Y \leqslant \theta n^{-1/2} .$$

Thus we see from (11)-(12), (14)-(15) that, for $n \geq 2$

$$L_1 + L_2 \leq c \max(\theta, \theta^p) n^{-1/2},$$

and hence the integral $J(X)$ has the same upper estimate. Moreover, for all $n \geq 1$

$$(16) \qquad (nX)^{-1/2} = c_1^{-s/2} \theta^p n^{-1/2} \leq 6\theta^p n^{-1/2}.$$

Thus on the account of the inequality (8) we have the first part of the statement of the lemma 1. In the case $n = 1$, obviously, $L_1 = 0$, and we need only to estimate the variable L_2. Moreover, as the estimate (12) is true, for $n = 1$ also, we may restrict ourselves by estimating V_2 only.

In the case $X_1 = X$, we have $V_2 = 0$. In the opposite case $X_1 = c_1$, we have as above

$$V_2 \leq \frac{2}{s}\theta\{X^{s/2} - c_1^{s/2}\} \leq \frac{2}{s}\theta X^{s/2} \leq \frac{2}{s} c_1^{\frac{5}{2}s} \theta^{1-ps} \leq \frac{2}{s}\theta^p$$

and by other way

$$V_2 \leq \theta X^{s/2} \log(X/c_1) \leq \theta^p \log(c_1^4 \theta^{-2p}).$$

These inequalities together with (11), (12) imply the second part of the statement of the lemma 1.

<u>Remark 1.</u> As a corollary of the lemma 1, we have the following statement.

<u>Let the parameter $s < 1$, then, for all</u>

$$n \geq \max\left(2, \frac{1}{1-s}\right),$$

(for instance for $n \geqslant 2$ if $s \leqslant 1/2$)

$$\rho_n \leqslant c\,\theta\, n^{-1/2} .$$

Remark 2. It is interesting to note that the unique condition of the lemma 1 $\omega(t) \leqslant c \min\left(|t|^s, |t|^3\right)$ on the one hand does not guarantee the existance of the third moment of the distribution F but on the other hand it guarantees the order $n^{-1/2}$ of the distance ρ_n as $n \to \infty$.

Prove now the lemma 2. It is not difficult to check that the following equalities and inequalities are true, for all real t :

$$\omega(t) = \left| \int e^{itx} d(F-\Phi) \right| = \left| t \int e^{itx} (F-\Phi)\, dx \right| =$$

$$= \left| \int \left(e^{itx} - 1 - itx - \tfrac{1}{2}(itx)^2\right) d(F-\Phi) \right| = \left| t \int \left(e^{itx} - 1 - itx\right)(F-\Phi)\, dx \right| .$$

$$\omega(t) \leqslant \int |d(F-\Phi)| , \qquad \omega(t) \leqslant |t| \int |F-\Phi|\, dx ;$$

$$\omega(t) \leqslant \frac{|t|^3}{6} \int |x|^3 |d(F-\Phi)| , \quad \omega(t) \leqslant \frac{|t|^3}{2} \int x^2 |F-\Phi|\, dx$$

We obtain as a result the statement of the lemma 2 from these estimates and the following general inequality, which is true for any nonnegative numbers a_1, a_2, b_1, b_2 :

$$\min\left(a_1 a_2,\, b_1 b_2\right) \leqslant \min\left(a_1, b_1\right) \max\left(a_2, b_2\right).$$

Thus the statements of the lemmas 1 and 2 imply the inequalities (5), (6) for $n \geqslant 1$ and (7) for $n \geqslant 2$. In the case $n = 1$ we have separately that

$$\rho_1 = \rho(F,\Phi) \leqslant \mathrm{Var}(F-\Phi) \leqslant \nu_0 .$$

The theorem is proved completely.

3. Discussion. As we have seen, the exponent p for a fixed value of n , increases from $\frac{n}{3n+1}$ to $\frac{n}{n+1}$ when the characteristic $\varkappa$ is replaced by $\varkappa_0$. This fact can be explaned, on the one hand as a result of the loss of information about closeness of F to Φ in the characteristic $\varkappa$ because of wanishing at the point $x = 0$ of the kernel x^2 in $\varkappa$. The characteristic $\varkappa_0$ is free from such a defect. On the other hand, for $n = 1$, we have in fact inequalities (5)-(7) between metrics $\varkappa, \varkappa_0, \nu_0$ and ρ . It is well known that the uniform metric ρ and mean metrics $\varkappa, \varkappa_0$ are not comparable. Thus the exponent $p = 1$ can not be put in the inequalities (5), (6). But we have a different situation of course in the case of the characteristic ν_0, as it was noted above. Namely, for $n = 1$, we have $\rho(F,\Phi) \leqslant \nu_0$, and, for $n > 1$, the effect of summing of independent summands must help to construct the same inequalities. Hence the appearance of the inequality (7) with exponent $p = 1$ in general case is not strange. Moreover, as for $n = 1$ we have (it is not difficult to prove):

$$\rho_0(F,\Phi) = \max_x \left\{ \max\left(1, |x|^3\right) |F(x) - \Phi(x)| \right\} \leqslant 2\nu_0 ,$$

we can be assure that the following inequality will be true, for all $n \geqslant 1$:

$$\rho_0(F_n,\Phi) \leqslant c\,\nu_0\, n^{-1/2}.$$

A noncomplicate example shows farther that the exponents

$p = \frac{n}{3n+1}$ and $p = \frac{n}{n+1}$ are optimal, for $n = 1$, in (5) and (6) respectively.

Example. Define the symmetric distribution function F as follows

$$F(x) = \begin{cases} 1/2 & \text{, if } \quad 0 < x \leq a\varepsilon , \\ \Phi(\varepsilon) & \text{, if } \quad a\varepsilon < x \leq \varepsilon , \\ \Phi(x) & \text{, if } \quad x > \varepsilon , \end{cases}$$

where ε is any positive number and a is the unique positive solution of the equation

$$\int_0^{\varepsilon} x^2 \, d\Phi = (a\varepsilon)^2 \int_0^{\varepsilon} d\Phi .$$

For small values of ε, we have $a \sim 1/\sqrt{3}$ and

$$\rho(F, \Phi) \sim \frac{\varepsilon}{\sqrt{6\pi}}$$

$$\varkappa \sim \frac{1}{\sqrt{2\pi}} \left(\frac{5}{36} - \frac{\sqrt{3}}{27} \right) \varepsilon^4 , \quad \varkappa_0 \sim \frac{1}{\sqrt{2\pi}} \left(\frac{5}{6} - \frac{\sqrt{3}}{3} \right) \varepsilon^2 .$$

It follows from have that the exponents in $\varkappa$ and $\varkappa_0$, provided the latter are less then 1, can not be grater than 1/4 and 1/2 respectively.

REFERENCES

[1] P.Lévy. Théorie de l'addition des variables aléatoires, Paris, 1937

[2] V.M.Zolotarev. On the closeness of the distributions of the two sums of independent random variables, Theor. Prob. and its Appl. 10 (1965), 519-526

[3] V.Paulauskas. One intensification of the Liapunov's theorem, Lituan, Math. J., 9 (1969), 323-328

[4] V.M.Zolotarev. Some new inequalities in probability connected with Lévy's metric, Dokl. Akad. Nauk SSSR, 190 (1970), 1019-1021

[5] V.M.Zolotarev. Estimations of difference of the distributions in Lévy's metric, Trans. Steklov Math. Inst., 112 (1971), 224-231

Steklov Mathematical Institute
of the Academy of Sciences of the USSR
Moscow

RECORD OF MEETINGS

AUGUST 2: Opening Ceremony

Morning Session

A.A.Borovkov, E.A.Pechersky: General theorems of convergence for random processes.

T.Sirao: On a branching Markov process with derivative.

Afternoon Session

(a) Branching processes

Y.Ogura: Spectral representation of continuous state branching processes.

T.Fujimagari: Controlled Galton-Watson process and its asymptotic behavior.

K.Kawazu: Multitype age-dependent branching process with immigration.

(b) Information theory

S.Takano: On some transmissionary invariant properties and information stability.

S.Ihara: The entropy of a message and the capacity of a channel in the Gaussian information transmission.

Y.Itoh: Linnik's information functional and characterization of normal distribution.

Y.Baba: Entropy and Hausdorff dimension for a sequence of coordinate functions in base-r expansion.

(c) Statistics

K.Takeuchi: On location parameter family of distributions with uniformly minimum variance unbiased estimator of location.

M.Huzii: Some contributions to the estimation theory of correlogram.

H.Kudō: On a convergence of σ-algebras and its application to the theory of asymptotic sufficiency.

AUGUST 3: Morning Session

H.Tanaka: On Markov process corresponding to Boltzmann's equation of Maxwellian gas.

D.M.Chibisov: Limit theorems in statistics.

Afternoon Session

(a) Statistical mechanics (non-equilibrium states)

T.Ueno: A stochastic model associated with Boltzmann equation.

Y.Takahashi: Markov processes with simple interaction — reversions of branching processes —.

S.Nishimura: Discrimination and random collision.

(b) Ergodic theory

Y.Ito: Orbit preserving transformations and invariant measures.

M.Kowada: On the equation $h(Tx) - h(x) = f(x)$ and its applications.

T.Hamachi: Construction of the finite center-valued relative dimension function of a W*-algebra, and invariant measures.

M.Osikawa: Local structure of ergodic measure preserving transformations.

I.Kubo: Ergodicity of the dynamical system of a particle on a domain with irregular walls.

Sh.Ito: A construction of transversal flows for maximal Markov automorphisms.

(c) Statistics

S.H.Siraždinov: About the selection of the optimal sampling plan for statistical control by quantity variables.

T.Okuno: Construction and application of fractional factorial designs.

K.Takeuchi, H.Morimoto: Prediction and sufficiency.

AUGUST 4: Morning Session

V.A.Statulevičius: On limit theorems for random processes.

Y.Takahashi: β-transformations and symbolic dynamics.

Afternoon Session

(a) Probability distributions and limit theorems

V.M.Zolotarev: New estimation of the remainder term in the central limit theorem.

V.M.Kruglov: Generalization of the central limit theorem. (read by V.M. Zolotarev)

R.Shimizu: On the remainder term in the central limit theorem for the unimodal distributions.

T.A.Azlarov: Uniform estimates in limit theorems for the busy period of one queueing system.

I.Shiokawa: On some properties of the dyadic Champernowne numbers.

AUGUST 5: Excursion

AUGUST 6: Morning Session

M.Fukushima: On the generation of Markov processes by symmetric forms.

Yu.V.Prokhorov, V.V.Sazonov: On some problems on distributions in R^k.

Afternoon Session

(a) Markov processes and potentials

D.V.Gusak: On continuous reaching the fixed level by homogeneous process with independent increments on finite Markov chain.

T.Watanabe: Some recent results on processes with stationary independent increments.

K.Sato: Remarks on potential operators for processes with stationary independent increments.

M.Kanda: Remarks related to the monotone property of Green functions in certain Markov processes.

R.Kondō, Y.Ōshima: Potential operators satisfying the semicomplete principle of maximum and Markov resolvents.

M.Takano: On a fine capacity related to a symmetric Markov process.

(b) Sample functions of stochastic processes

T.Kawata: On a class of linear processes.

S.Nagaev: On necessary and sufficient conditions for the strong law of large numbers. (read by A.A.Borovkov)

H.Oodaira: On the log log law for certain dependent random sequences.

T.Kawada, N.Kôno: On the variation of Gaussian processes.

AUGUST 7: Morning Session

B.A.Rogozin: Concentration functions of sums of independent random variables.

N.Ikeda, S.Watanabe: The local structure of a class of diffusions and related problems.

Afternoon Session

(a) Stochastic differential equations and stochastic control

M.Nisio: On stochastic differential equations associated with certain quasilinear parabolic eqations.

M.Tsuchiya: On some perturbations of stable processes.

S.Nakao: Remarks on one-dimensional stochastic differential equations.

T.Yamada: On a comparison theorem for solutions of stochastic differential equations and its applications.

N.Kazamaki: On the existence of solutions of martingale integral equations.

M.P.Yershov: Stochastic equations.

(b) Ergodic theory

G.Maruyama: Applications of Ornstein's theory to stationary processes.

E.Kin: The random ergodic theorem.

T.Kamae: Subsequences of a normal sequence.

H.Shirakawa: A proof of strongly mixing property of a horocycle flow.

I.Kubo, H.Murata, H.Totoki: On the isomorphism problem for not invertible measure preserving transformations.

AUGUST 8: Morning Session

Y.Okabe: On a Markovian property of Gaussian processes.

B.I.Grigelionis: On non-linear filtering theory and absolute continuity of measures, corresponding to the stochastic processes.

Afternoon Session

(a) Stochastic differential equations and stochastic control

M.P.Yershov: On extension of measures and stochastic equations.

T.Komatsu: On absolute continuity of probability solutions of certain martingale equations.

Y.Miyahara: Invariant measures of ultimately bounded stochastic processes.

A.Shimizu: On the stability of a discrete time Markov chain.

To.Watanabe: Random perturbation of two-dimensional real autonomous systems.

(b) Gaussian processes

S.Kotani: On a Markov property for stationary Gaussian processes with a multidimensional parameter.

K.Inoue: On equivalence of measures corresponding to Gaussian processes with a multi-dimensional parameter.

AUGUST 9: Morning Session

H.Nomoto: On white noise and infinite dimensional orthogonal group.

Afternoon Session

(a) Stochastic differential equations and stochastic control

H.Kunita: Asymptotic behavior of the non linear filtering errors of Markov processes.

B.I.Grigelionis: A process representable as stochastic integrals.

M.Hitsuda: Formula for Brownian partial derivatives.

S.Watanabe: A remark on the integrability of $\sup_t X_t$ for martingales.

S.Ihara: A remark on Shiryayev's result.

(b) White noise and measures on topological vector spaces

T.Hida: A probabilistic approach to infinite dimensional unitary group.

H.Sato: On Lie group structure of transformation group of the White Noise.

Closing Ceremony

ORGANIZING COMMITTEE

Soviet side: Yu.V.Prokhorov (Moscow, Chairman), A.N.Shiryayev (Moscow, Deputy Chairman), Yu.V.Linnik (Leningrad), A.V.Skorokhod (Kiev), V.V.Sazonov (Moscow), B.A.Sevastyanov (Moscow), V.A.Statulevičius (Vilnius), M.P.Yershov (Moscow, Secretary).

Japanese side: G.Maruyama (Tokyo, Chairman), T.Ueno (Tokyo), I.Shiokawa (Tokyo), R.Kondō (Shizuoka), H.Kunita (Nagoya), Y.Miyahara (Nagoya), S.Watanabe (Kyoto), M.Miyamoto (Kyoto), N.Kôno (Kyoto), H.Totoki (Kyoto, Secretary), T.Shiga (Nara), Y.Okabe (Osaka), H.Murata (Hiroshima), M.Osikawa (Fukuoka).

LIST OF PARTICIPANTS

SOVIET PARTICIPANTS

Institute of Mathematics of the Academy of Sciences of Ukrainian SSR, Kiev:
GUSAK,D.V.

Institute of Mathematics of the Siberian Branch of the Academy of Sciences of USSR, Novosibirsk: BOROVKOV,A.A., NAGAEV,S.V.*, PECHERSKY,E.A.*, ROGOZIN,B.A.

Institute of Physics and Mathematics of the Academy of Sciences of Lithuanian SSR, and University of Vilnius, Vilnius: GRIGELIONIS,B.I., STATULEVIČIUS,V.A.

Leningrad Branch of Steklov Mathematical Institute of the Academy of Sciences of USSR, Leningrad: LINNIK,Yu.V.*, RUKHIN,A.L.*, SUDAKOV,V.N.*

Leningrad State University, Leningrad: KLEBANOV,L.B.*

Moscow State University, Moscow: KRYLOV,N.V.*

Romanovskii Institute of Mathematics of the Academy of Sciences of Uzbek SSR, Tashkent: SIRAŽDINOV,S.H.

Steklov Mathematical Institute of the Academy of Sciences of USSR, Moscow:
CHIBISOV.D.M., HOLEVO,A.S.*, KRUGLOV,V.M.*, NOVIKOV,A.A.*, PROKHOROV,Yu.V.*, SAZONOV,V.V., SEVASTYANOV,B.A.*, SHIRYAYEV,A.N.*, YERSHOV,M.P., ZOLOTAREV,V.M.

Tashkent State University, Tashkent: AZLAROV,T.A.

---

* Asterisks indicate participants only by paper.

JAPANESE PARTICIPANTS

Aichi University of Education, Kariya Aichi: WATANABE Toitsu

Chiba Institute of Technology, Narashino Chiba: OHGUCHI Takeshi

Gakushuin University, Tokyo: YOSIDA Kôsaku

Hiroshima University, Hiroshima: KANDA Mamoru, MURATA Hiroshi, TAKAHASHI Yōichirō TANAKA Hiroshi

IBM Japan LTD, Tokyo: SIBUYA Masaaki

Institute of JUSE Inc., Tokyo: HONDA Masaru, YAJIMA Keiji

Institute of Statistical Mathematics, Tokyo: NAGASAKA Kenji, SHIMIZU Ryoichi, SUZUKI Giitiro

Kanazawa University, Kanazawa: TSUCHIYA Masaaki

Keio University, Yokohama: KAWATA Tatsuo, WASHIO Yasutoshi

Kobe College of Commerce, Kobe: FUJISAKI Masatoshi, KAWADA Takayuki, TAKASU Kiyosumi, YAMAMOTO Yutaka

Kobe University, Kobe: NAKAO Shintaro, NISIO Makiko, OKAMOTO Chiyo

Kumamoto University, Kumamoto: ŌSHIMA Yōichi

Kyoto Sangyo University, Kyoto: MORI Takakazu

Kyoto University, Kyoto: HIGUCHI Yasunari, ISHITANI Hiroshi, KÔNO Norio, KOTANI Shinichi, MIYAMOTO Munemi, TOTOKI Haruo, WATANABE Shinzo, YOSHIOKA Yoshiake

Kyushu University, Fukuoka: HAMACHI Toshihiro, KAWAZU Kiyoshi, KUDÔ Akio, MITOMA Itaru, OSIKAWA Motosige, SATO Hiroshi, TERADA Nobuaki, TOMISAKI Matuyo, WATANABE Hisao, YAMADA Toshio

Nagoya City University, Nagoya: IHARA Shunsuke

Nagoya Institute of Technology, Nagoya: HITSUDA Masuyuki, SHIMIZU Akinobu

Nagoya University, Nagoya: HASEGAWA Yoshihei, HIDA Takeyuki, ISHIHARA Kanji, KUBO Izumi, KUNITA Hiroshi, MIYAHARA Yoshio, NODA Akio, NOMOTO Hisao

Nara Women's University, Nara: SHIGA Tokuzo

National Institute of Agricultual Sciences, Tokyo: OKUNO Tadakazu

Numazu Kogyō Kōtō Senmon Gakō, Numazu Shizuoka: KIKUTI Ituo

Ochanomizu Women's University, Tokyo: TAKEUCHI Junji

Osaka University, Suita Osaka: TASAKA Masao

Osaka University, Toyonaka Osaka: IKEDA Nobuyuki, MANABE Shôjiro, OKABE Yasunori, SUGITANI Sadao, WATANABE Takesi

Osaka City University, Osaka: KAMAE Teturo, KOMATSU Takashi, KUDŌ Hirokichi, SUZUKI Takeru

Otaru University of Commerce, Otaru: SHIRAKAWA Hiroshi

Saga University, Saga: OGURA Yukio

Shiga Daigaku, Hikone Shiga: ORISHIKIDA Yûji

Shimane University, Matsue: ASOO Yasuhiro

Shinshu University, Matsumoto: INOUE Kazuyuki

Shinshu University, Nagano: TAZAWA Koichiro

Shizuoka University, Shizuoka: BABA Yoshikazu, KONDŌ Ryōji, NEGORO Akira, TANAKA Shigeru

Tohoku University, Sendai: KAZAMAKI Norihiko

Tokyo College of Economics, Tokyo: ISHIDA Nozomu

Tokyo Institute of Technology, Tokyo: FUJIMAGARI Tetsuo, HORI Motoo, HUZII Mituaki, ITOH Yoshiaki, KUDO Yukio, NAGASAWA Masao, NISHIMURA Shōichi, NISHIOKA Kunio, SHIMURA Michio, TAKAHASHI Yukio, TAKANO Seiji

Tokyo Metropolitian University, Tokyo: AOKI Nobuo, KIN Ei Jun, SIRAO Tunekiti

Tokyo University of Education, Tokyo: FUKUSHIMA Masatoshi, HIJIGURO Kayoko, ITO Shunji, KOBAYASHI Shigeru, MARUYAMA Gisiro, SATO Ken-iti, SHIOKAWA Ietaka, TAKANO Masaru

Tsuda College, Kodaira Tokyo: KOWADA Masashi, ONOYAMA Takuji

University of Electro-communications, Chofu Tokyo: AKAHIRA Masafumi

University of Osaka Prefecture, Sakai Osaka: YAMAGATA Hideo

University of Tokyo, Tokyo: SHIBATA Yoshisada, SHINOHARA Masahiko, TAKEUCHI Kei, TANAKA Kenichi, UENO Tadashi

Yokohama National University, Yokohama: OODAIRA Hiroshi

Brown University, Providence USA: ITO Yuji

Vol. 212: B. Scarpellini, Proof Theory and Intuitionistic Systems. VII, 291 pages. 1971. DM 24,–

Vol. 213: H. Hogbe-Nlend, Théorie des Bornologies et Applications. V, 168 pages. 1971. DM 18,–

Vol. 214: M. Smorodinsky, Ergodic Theory, Entropy. V, 64 pages. 1971. DM 16,–

Vol. 215: P. Antonelli, D. Burghelea and P. J. Kahn, The Concordance-Homotopy Groups of Geometric Automorphism Groups. X, 140 pages. 1971. DM 16,–

Vol. 216: H. Maaß, Siegel's Modular Forms and Dirichlet Series. VII, 328 pages. 1971. DM 20,–

Vol. 217: T. J. Jech, Lectures in Set Theory with Particular Emphasis on the Method of Forcing. V, 137 pages. 1971. DM 16,–

Vol. 218: C. P. Schnorr, Zufälligkeit und Wahrscheinlichkeit. IV, 212 Seiten 1971. DM 20,–

Vol. 219: N. L. Alling and N. Greenleaf, Foundations of the Theory of Klein Surfaces. IX, 117 pages. 1971. DM 16,–

Vol. 220: W. A. Coppel, Disconjugacy. V, 148 pages. 1971. DM 16,–

Vol. 221: P. Gabriel und F. Ulmer, Lokal präsentierbare Kategorien. V, 200 Seiten. 1971. DM 18,–

Vol. 222: C. Meghea, Compactification des Espaces Harmoniques. III, 108 pages. 1971. DM 16,–

Vol. 223: U. Felgner, Models of ZF-Set Theory. VI, 173 pages. 1971. DM 16,–

Vol. 224: Revêtements Etales et Groupe Fondamental. (SGA 1). Dirigé par A. Grothendieck XXII, 447 pages. 1971. DM 30,–

Vol. 225: Théorie des Intersections et Théorème de Riemann-Roch. (SGA 6). Dirigé par P. Berthelot, A. Grothendieck et L. Illusie. XII, 700 pages. 1971. DM 40,–

Vol. 226: Seminar on Potential Theory, II. Edited by H. Bauer. IV, 170 pages. 1971. DM 18,–

Vol. 227: H. L. Montgomery, Topics in Multiplicative Number Theory. IX, 178 pages. 1971. DM 18,–

Vol. 228: Conference on Applications of Numerical Analysis. Edited by J. Ll. Morris. X, 358 pages. 1971. DM 26,–

Vol. 229: J. Väisälä, Lectures on n-Dimensional Quasiconformal Mappings. XIV, 144 pages. 1971. DM 16,–

Vol. 230: L. Waelbroeck, Topological Vector Spaces and Algebras. VII, 158 pages. 1971. DM 16,–

Vol. 231: H. Reiter, L^1-Algebras and Segal Algebras. XI, 113 pages. 1971. DM 16,–

Vol. 232: T. H. Ganelius, Tauberian Remainder Theorems. VI, 75 pages. 1971. DM 16,–

Vol. 233: C. P. Tsokos and W. J. Padgett. Random Integral Equations with Applications to Stochastic Systems. VII, 174 pages. 1971. DM 18,–

Vol. 234: A. Andreotti and W. Stoll. Analytic and Algebraic Dependence of Meromorphic Functions. III, 390 pages. 1971. DM 26,–

Vol. 235: Global Differentiable Dynamics. Edited by O. Hájek, A. J. Lohwater, and R. McCann. X, 140 pages. 1971. DM 16,–

Vol. 236: M. Barr, P. A. Grillet, and D. H. van Osdol. Exact Categories and Categories of Sheaves. VII, 239 pages. 1971, DM 20,–

Vol. 237: B. Stenström. Rings and Modules of Quotients. VII, 136 pages. 1971. DM 16,–

Vol. 238: Der kanonische Modul eines Cohen-Macaulay-Rings. Herausgegeben von Jürgen Herzog und Ernst Kunz. VI, 103 Seiten. 1971. DM 16,–

Vol. 239: L. Illusie, Complexe Cotangent et Déformations I. XV, 355 pages. 1971. DM 26,–

Vol. 240: A. Kerber, Representations of Permutation Groups I. VII, 192 pages. 1971. DM 18,–

Vol. 241: S. Kaneyuki, Homogeneous Bounded Domains and Siegel Domains. V, 89 pages. 1971. DM 16,–

Vol. 242: R. R. Coifman et G. Weiss, Analyse Harmonique Non-Commutative sur Certains Espaces. V, 160 pages. 1971. DM 16,–

Vol. 243: Japan-United States Seminar on Ordinary Differential and Functional Equations. Edited by M. Urabe. VIII, 332 pages. 1971. DM 26,–

Vol. 244: Séminaire Bourbaki – vol. 1970/71. Exposés 382–399. IV, 356 pages. 1971. DM 26,–

Vol. 245: D. E. Cohen, Groups of Cohomological Dimension One. V, 99 pages. 1972. DM 16,–

Vol. 246: Lectures on Rings and Modules. Tulane University Ring and Operator Theory Year, 1970–1971. Volume I. X, 661 pages. 1972. DM 40,–

Vol. 247: Lectures on Operator Algebras. Tulane University Ring and Operator Theory Year, 1970–1971. Volume II. XI, 786 pages. 1972. DM 40,–

Vol. 248: Lectures on the Applications of Sheaves to Ring Theory. Tulane University Ring and Operator Theory Year, 1970–1971. Volume III. VIII, 315 pages. 1971. DM 26,–

Vol. 249: Symposium on Algebraic Topology. Edited by P. J. Hilton. VII, 111 pages. 1971. DM 16,–

Vol. 250: B. Jónsson, Topics in Universal Algebra. VI, 220 pages. 1972. DM 20,–

Vol. 251: The Theory of Arithmetic Functions. Edited by A. A. Gioia and D. L. Goldsmith VI, 287 pages. 1972. DM 24,–

Vol. 252: D. A. Stone, Stratified Polyhedra. IX, 193 pages. 1972. DM 18,–

Vol. 253: V. Komkov, Optimal Control Theory for the Damping of Vibrations of Simple Elastic Systems. V, 240 pages. 1972. DM 20,–

Vol. 254: C. U. Jensen, Les Foncteurs Dérivés de $\varprojlim$ et leurs Applications en Théorie des Modules. V, 103 pages. 1972. DM 16,–

Vol. 255: Conference in Mathematical Logic – London '70. Edited by W. Hodges. VIII, 351 pages. 1972. DM 26,–

Vol. 256: C. A. Berenstein and M. A. Dostal, Analytically Uniform Spaces and their Applications to Convolution Equations. VII, 130 pages. 1972. DM 16,–

Vol. 257: R. B. Holmes, A Course on Optimization and Best Approximation. VIII, 233 pages. 1972. DM 20,–

Vol. 258: Séminaire de Probabilités VI. Edited by P. A. Meyer. VI, 253 pages. 1972. DM 22,–

Vol. 259: N. Moulis, Structures de Fredholm sur les Variétés Hilbertiennes. V, 123 pages. 1972. DM 16,–

Vol. 260: R. Godement and H. Jacquet, Zeta Functions of Simple Algebras. IX, 188 pages. 1972. DM 18,–

Vol. 261: A. Guichardet, Symmetric Hilbert Spaces and Related Topics. V, 197 pages. 1972. DM 18,–

Vol. 262: H. G. Zimmer, Computational Problems, Methods, and Results in Algebraic Number Theory. V, 103 pages. 1972. DM 16,–

Vol. 263: T. Parthasarathy, Selection Theorems and their Applications. VII, 101 pages. 1972. DM 16,–

Vol. 264: W. Messing, The Crystals Associated to Barsotti-Tate Groups: with Applications to Abelian Schemes. III, 190 pages. 1972. DM 18,–

Vol. 265: N. Saavedra Rivano, Catégories Tannakiennes. II, 418 pages. 1972. DM 26,–

Vol. 266: Conference on Harmonic Analysis. Edited by D. Gulick and R. L. Lipsman. VI, 323 pages. 1972. DM 24,–

Vol. 267: Numerische Lösung nichtlinearer partieller Differential- und Integro-Differentialgleichungen. Herausgegeben von R. Ansorge und W. Törnig, VI, 339 Seiten. 1972. DM 26,–

Vol. 268: C. G. Simader, On Dirichlet's Boundary Value Problem. IV, 238 pages. 1972. DM 20,–

Vol. 269: Théorie des Topos et Cohomologie Etale des Schémas. (SGA 4). Dirigé par M. Artin, A. Grothendieck et J. L. Verdier. XIX, 525 pages. 1972. DM 50,–

Vol. 270: Théorie des Topos et Cohomologie Etle des Schémas. Tome 2. (SGA 4). Dirige par M. Artin, A. Grothendieck et J. L. Verdier. V, 418 pages. 1972. DM 50,–

Vol. 271: J. P. May, The Geometry of Iterated Loop Spaces. IX, 175 pages. 1972. DM 18,–

Vol. 272: K. R. Parthasarathy and K. Schmidt, Positive Definite Kernels, Continuous Tensor Products, and Central Limit Theorems of Probability Theory. VI, 107 pages. 1972. DM 16,–

Vol. 273: U. Seip, Kompakt erzeugte Vektorräume und Analysis. IX, 119 Seiten. 1972. DM 16,–

Vol. 274: Toposes, Algebraic Geometry and Logic. Edited by. F. W. Lawvere. VI, 189 pages. 1972. DM 18,–

Vol. 275: Séminaire Pierre Lelong (Analyse) Année 1970–1971. VI, 181 pages. 1972. DM 18,–

Vol. 276: A. Borel, Représentations de Groupes Localement Compacts. V, 98 pages. 1972. DM 16,–

Vol. 277: Séminaire Banach. Edité par C. Houzel. VII, 229 pages. 1972. DM 20,–